Omics of Climate Resilient Small Millets

Ramesh Namdeo Pudake •
Amolkumar U. Solanke •
Amitha Mithra Sevanthi • P. Rajendrakumar
Editors

Omics of Climate Resilient Small Millets

Springer

Editors
Ramesh Namdeo Pudake
Amity Institute of Nanotechnology, Amity University Uttar Pradesh
Noida, India

Amolkumar U. Solanke
ICAR - National Institute for Plant Biotechnology
New Delhi, India

Amitha Mithra Sevanthi
ICAR - National Institute for Plant Biotechnology
New Delhi, India

P. Rajendrakumar
ICAR-Indian Institute of Millets Research (ICAR-IIMR)
Hyderabad, India

ISBN 978-981-19-3906-8 ISBN 978-981-19-3907-5 (eBook)
https://doi.org/10.1007/978-981-19-3907-5

This Springer imprint is published by the registered company Springer Nature Singapore Pte Ltd.
The registered company address is: 152 Beach Road, #21-01/04 Gateway East, Singapore 189721, Singapore

Preface

Millets are the collection of small-grained cereal grasses that are grown for human carbohydrate needs. They are among the oldest crops, mainly divided into two groups—major and small millets based on seed size. Small millets are composed of six species that includes finger millet (*Eleusine coracana* L.), little millet (*Panicum sumatrense*), kodo millet (*Paspalum scrobiculatum* L.), foxtail millet (*Setaria italica* L.), barnyard millet (*Echinochloa frumentacea* L.), and proso millet (*Panicum miliaceum* L.). These crops are earlier considered as orphan crops in terms of research work and their marginal inclusion in human diet, but in recent times, they are fetching higher market value due to their taste, texture, and nutritional properties. Considering their growing importance in human nutrition, the year 2023 has been declared as the International Year of Millets. Small millets possess better adaptability to harsh environmental conditions and require minimal agricultural inputs for production. The genomics, transcriptomics, proteomics, metabolomics, bioinformatics, and other omics tools are being widely used to get a clear understanding of mechanistic approaches taken by millets' genomes to tolerate stress. Various reports on elucidation of molecular mechanisms of climate-resilient traits are published in these crops. And recently, the genome of some small millets has been decoded, and a few studies were published related to its application in varietal improvements. This book sheds light on omics studies in small millets for achieving abiotic stress tolerance.

In this book, there are a total of 17 chapters, which broadly focus on the recent advances in the genomics, transcriptomics, proteomics, phenomics, comparative mapping, gene discovery, and molecular breeding of small millets. Chapter 1 is mainly focused on the need of integration of proven and climate-smart technologies for the successful harvest of small millets for the fulfilment of the future needs of the ever-growing population. Chapter 2 discusses the application of the available foxtail millet genome sequence for revealing numerous of its unique properties, such as abiotic stress tolerance, which may aid in a better understanding of its evolution, stress physiology, and adaptation. Chapter 3 discusses the role of omics strategies in barnyard millet improvement for abiotic stress tolerance. In Chap. 4, the authors have discussed how gene promoters and transcription factors contribute to abiotic stress tolerance in small millets. Chapter 5 is focused on various classes of ncRNAs in small millets and their putative roles in modulating plant responses against drought, salt, cold, etc. Further, databases and online tools pertaining to the identification and analysis of ncRNAs are presented. In Chap. 6, the authors reviewed the efforts carried out to explore the insights into abiotic stress tolerance on small millets through the transcriptomics approach. In the recent past, proteomics tools have been used for understanding the details of abiotic stress tolerance in crop plants. Chapter 7 is dedicated to the use of proteomics to help researchers better understand the abiotic stress tolerance in minor millets. Chapter 8 describes plant regeneration and transgenic approaches. In this chapter, biotic and abiotic factors that inhibit the growth of small millets, various plant regeneration methods, transformation studies, potential genes for abiotic stress tolerance, and transgenic approaches to produce improved abiotic stress-tolerant millets have also been elaborated. Chapter 9 is dedicated to the details on genome sequences of small millets, how to mine genes from the draft genome sequences of millets with finger millet as example and comparative genomic studies of small millets. This chapter also discusses the current progress of research work related to comparative genomic approaches and mining genes across small millets. Chapter 10 emphasizes on the scope of functional genomics and molecular marker approaches for improving nutrient-use efficiency in millets and gives future direction for millet improvement. In this chapter, the importance of the implication of functional genomics for crop improvement is also highlighted. Authors of Chap. 11 have presented the bioinformatics resources which are available for small millets. Chapter 12 is about the most important and relevant advances in omics for enhancing abiotic stress tolerance in finger millet. Genome-editing approaches are being increasingly used to develop stress-tolerant varieties in different crops, and Chap. 13 has reviewed its potential application in millets. One of the major attributes of small millets is their high nutritional value as compared to other cereal crops. Chapter 14 describes the different approaches used to dissect the climate-resilient traits in small millets, emphasizing the role of genomics and phenomics tools. Also, this chapter describes how to integrate genomics and phenomics for improving the climate resilience of these crops. Chapter 15 is completely dedicated to present state of knowledge about how small millets can be established as the climate-sustainable model system, and how lncRNAs are involved in abiotic stress management. It also discusses the utilization of several bioinformatics pipelines in unravelling the molecular mystery behind this aspect. Chapter 16 is about understanding the effects of climate change on nutritional quality of millets. Chapter 17 emphasizes on the use of different heat shock proteins in small millets as potential candidates for crop improvement under abiotic stress. This book reviewed all those recent studies and is of interest to plant breeding scientists, research students, and teachers working in agriculture and plant biotechnology. This book can also serve as a reference material for undergraduate and graduate students of agriculture and biotechnology. National and international agricultural scientists and policymakers will also find this to be a useful read.

Reviewing the current research and writing it in the chapter form demands a lot of effort and dedication, and for that we thank our authors for their valuable contribution. Also, we extend our sincere thanks to everyone in the Springer Nature team for their constant help and constructive suggestions particularly to Mrs. Aakanksha

Tyagi and Mr. Selvakumar Rajendran and their team. The editors are highly thankful to Science and Engineering Research Board (SERB), Department of Science and Technology, Government of India, New Delhi, for providing "SERB-TARE fellowship" to Dr. Ramesh Pudake under the mentorship of Dr. Amolkumar Solanke. We hope that the book will be useful for all the readers to find the relevant information on the latest research and advances in effective use of omics techniques for improvement of small millets.

Noida, India — Ramesh Namdeo Pudake
New Delhi, India — Amolkumar U. Solanke
New Delhi, India — Amitha Mithra Sevanthi
Hyderabad, India — P. Rajendrakumar

Contents

Editors and Contributors

About the Editors

Ramesh Namdeo Pudake is Assistant Professor at Amity University Uttar Pradesh, Noida, and is a Ph.D. in Crop Genetics and Breeding from China Agricultural University, Beijing, PRC. After his Ph.D., he is engaged in research in a range of organisms but with a focus on crop plant genomics. He has also worked in Iowa State University Ames, IA, USA, on host–pathogen interaction and gene mapping. Currently, he is focusing on the research on different applications of genomics in host–microbe interaction and abiotic stress tolerance. Dr. Pudake has published more than 35 research publications, 3 books, 18 book chapters, and has 1 scholarship award from Chinese Government, and 1 fellowship from Science and Engineering Research Board, Department of Science and Technology, Government of India, to his credit. He is also an expert reviewer for several journals of repute.

Amolkumar U. Solanke is Senior Scientist, at ICAR-National Institute of Plant Biotechnology, New Delhi. Dr. Solanke has done his Ph.D. in Plant Molecular Biology from Delhi University South Campus, New Delhi, and later joined Indian Council of Agricultural Research, India, as Scientist. His research interest is functional genomics for biotic and abiotic stress management in crop plants, especially rice and finger millet. He was also a part of tomato and cluster bean genome-sequencing projects. He has published more than 70 papers in peer-reviewed journals, besides 15 book chapters. He is executing a few funded research projects on finger millet and guiding M.Sc., Ph.D., and post-doctoral students. He is an expert reviewer in the area of genomics and plant molecular biology for several journals of repute.

Amitha Mithra Sevanthi is Senior Scientist at ICAR-National Institute of Plant Biotechnology, New Delhi. She has done her doctorate in Genetics and Plant Breeding from Indian Agricultural Research Institute, New Delhi. She is working in the area of genomics for developing molecular markers, whole genome sequencing of major crop plants, and identification/mapping of genes for abiotic stress tolerance and their characterization and use of EMS-induced mutants for identification of novel genes and alleles with focus on rice as well as pigeon pea and cluster bean. She has published more than 50 research papers in peer-reviewed journals, contributed to 7 book chapters, and is a supervisor for M.Sc. and Ph.D. students. She is an expert reviewer in the area of molecular mapping, DNA marker development, and applications and genomics in several international journals of repute.

P. Rajendrakumar is Principal Scientist at ICAR-Indian Institute of Millets Research, Hyderabad. He has a Doctorate degree in Plant Breeding and Genetics from Tamil Nadu Agricultural University, Coimbatore. He has developed SSR markers in coffee, rice, and safflower. His research focus is on development of DNA markers in millets, whole genome sequencing, marker-trait associations for grain quality, and molecular breeding for genotypes suitable for lignocellulosic biofuel production in sorghum. He has published more than 30 research papers in peer-reviewed journals, a book, and 10 book chapters. He is also an expert reviewer for journals like Frontiers in Genetics, PLoS One and Crop Science.

Contributors

Mahendra Laxman Ahire Department of Botany, Yashavantrao Chavan Institute of Science, Satara, Maharashtra, India

Muthukrishnan Arun Department of Biotechnology, Bharathiar University, Coimbatore, Tamil Nadu, India

Sharmistha Barthakur ICAR-National Institute for Plant Biotechnology, New Delhi, India

Vitthal Tanaji Barvkar Department of Botany, Savitribai Phule Pune University, Pune, Maharashtra, India

Nandita Bharadwaj Department of Environmental Science, Tezpur University, Tezpur, Assam, India

Mahender Singh Bhinda ICAR-Vivekananda Parvatiya Krishi Anusandhan Sansthan, Almora, Uttarakhand, India

Stanislaus Antony Ceasar Department of Biosciences, Rajagiri College of Social Sciences, Kalamassery, Cochin, Kerala, India

Pooja Choudhary Department of Plant Sciences, School of Life Sciences, University of Hyderabad, Hyderabad, Telangana, India

Pronomita Das Department of Plant Sciences, School of Life Sciences, University of Hyderabad, Hyderabad, Telangana, India

Ashim Debnath Department of Genetics and Plant Breeding, College of Agriculture, Acharya Narendra Deva University of Agriculture and Technology, Ayodhya, Uttar Pradesh, India

Indrajeet Ghodke Department of Microbiology, KTHM College, Nashik, Maharashtra, India

Harun I. Gitari School of Agriculture and Enterprise Development, Kenyatta University, Nairobi, Kenya

Nitin Tanaji Gore Department of Botany, Yashavantrao Chavan Institute of Science, Satara, Maharashtra, India

Jayachandran Halka Department of Biotechnology, Bharathiar University, Coimbatore, Tamil Nadu, India

Akbar Hossain Bangladesh Wheat and Maize Research Institute, Dinajpur, Bangladesh

Savarimuthu Ignacimuthu Xavier Research Foundation, St. Xavier's College, Palayamkottai, Tirunelveli, Tamil Nadu, India

Priyanka Jain ICAR - National Institute for Plant Biotechnology, New Delhi, India

Gitanjali Jiwani ICAR, National Institute for Plant Biotechnology, New Delhi, India

Laxmi Kant ICAR-Vivekananda Parvatiya Krishi Anusandhan Sansthan, Almora, Uttarakhand, India

Vasundhara Karthi Department of Plant Sciences, School of Life Sciences, University of Hyderabad, Hyderabad, Telangana, India

A. Karthikeyan Subtropical Horticulture Research Institute, Jeju National University, Jeju, South Korea
Department of Biotechnology, Centre of Excellence in Innovation, Agricultural College and Research Institute, Tamil Nadu Agricultural University, Madurai, Tamil Nadu, India

Rishi Kesh Laboratory of Molecular Genetics and Plant Immunity, Department of Life Science, National Institute of Technology Rourkela, Rourkela, Odisha, India

Kumaresan Kowsalya Department of Biotechnology, Bharathiar University, Coimbatore, Tamil Nadu, India

Thumadath Palayullaparambil Ajeesh Krishna Department of Biosciences, Rajagiri College of Social Sciences, Kalamassery, Cochin, Kerala, India

Vishesh Kumar ICAR - National Institute for Plant Biotechnology, New Delhi, India

Asha Kumari ICAR-Vivekananda Parvatiya Krishi Anusandhan Sansthan, Almora, Uttarakhand, India

Mayuri D. Mahalle ICAR, National Institute for Plant Biotechnology, New Delhi, India

Theivanayagam Maharajan Department of Biosciences, Rajagiri College of Social Sciences, Kalamassery, Cochin, Kerala, India

H. B. Mahesh Department of Genetics and Plant Breeding, College of Agriculture, Mandya, University of Agricultural Sciences, Bangalore, Karnataka, India

Sagar Maitra Department of Agronomy and Agroforestry, Centurion University of Technology and Management, Paralakhemundi, Odisha, India

Chittabrata Mal Maulana Abul Kalam Azad University of Technology (Formerly Known as West Bengal University of Technology), Nadia, West Bengal, India

Navonil Mal Department of Botany, University of Calcutta, Kolkata, Kolkata, West Bengal, India

Abhinav Arvind Mali Department of Botany, Yashavantrao Chavan Institute of Science, Satara, Maharashtra, India
Department of Botany, Savitribai Phule Pune University, Pune, Maharashtra, India

Vivekananda Mohanta Laboratory of Molecular Genetics and Plant Immunity, Department of Life Science, National Institute of Technology Rourkela, Rourkela, Odisha, India

Mareyam Mukhtar ICAR, National Institute for Plant Biotechnology, New Delhi, India

Pankaj Shivnarayan Mundada Department of Biotechnology, Yashavantrao Chavan Institute of Science, Satara, Maharashtra, India

Mehanthan Muthamilarasan Department of Plant Sciences, School of Life Sciences, University of Hyderabad, Hyderabad, Telangana, India

Raveendran Muthurajan Department of Plant Biotechnology, Centre for Plant Molecular Biology and Biotechnology, Tamil Nadu Agricultural University, Coimbatore, Tamil Nadu, India

Archana Ashokrao Naik Department of Botany, Savitribai Phule Pune University, Pune, Maharashtra, India

Saranya Nallusamy Department of Plant Molecular Biology and Bioinformatics, Centre for Plant Molecular Biology and Biotechnology, Tamil Nadu Agricultural University, Coimbatore, Tamil Nadu, India

Tukaram Dayaram Nikam Department of Botany, Savitribai Phule Pune University, Pune, Maharashtra, India

Jnana Bharati Palai School of Agriculture and Enterprise Development, Kenyatta University, Nairobi, Kenya

Debasish Pati Laboratory of Molecular Genetics and Plant Immunity, Department of Life Science, National Institute of Technology Rourkela, Rourkela, Odisha, India

T. S. S. K. Patro Agricultural Research Station, Acharya N. G. Ranga Agricultural University, Vizianagaram, India

Shubhasisha Praharaj Department of Agronomy and Agroforestry, Centurion University of Technology and Management, Paralakhemundi, Odisha, India

Biswajit Pramanick Department of Agronomy, Dr. Rajendra Prasad Central Agricultural University, Samastipur, India

Lydia Pramitha School of Agriculture and Biosciences, Karunya Institute of Technology and Sciences, Coimbatore, Tamil Nadu, India

Jaganathan Sakthi Yazhini Preetha Department of Biotechnology, Bharathiar University, Coimbatore, Tamil Nadu, India

Ramesh Namdeo Pudake Amity Institute of Nanotechnology, Amity University Uttar Pradesh, Noida, India

N. R. Raghavendra Department of Genetics and Plant Breeding, College of Agriculture, Mandya, University of Agricultural Sciences, Bangalore, Karnataka, India

P. Rajendrakumar ICAR-Indian Institute of Millets Research (IIMR), Hyderabad, India

Raman Renuka Department of Biotechnology, Centre of Innovation, Agricultural College and Research Institute, Tamil Nadu Agricultural University, Madurai, Tamil Nadu, India

V. G. Renganathan Department of Biotechnology, Centre of Excellence in Innovation, Agricultural College and Research Institute, Tamil Nadu Agricultural University, Madurai, Tamil Nadu, India

Vellaichamy Gandhimeyyan Renganathan Department of Plant Breeding and Genetics, Agricultural College and Research Institute, Tamil Nadu Agricultural University, Madurai, Tamil Nadu, India
Department of Biotechnology, Centre of Innovation, Agricultural College and Research Institute, Tamil Nadu Agricultural University, Madurai, Tamil Nadu, India

Lalichetti Sagar Department of Agronomy and Agroforestry, Centurion University of Technology and Management, Paralakhemundi, Odisha, India

Upasana Sahoo Department of Agronomy and Agroforestry, Centurion University of Technology and Management, Paralakhemundi, Odisha, India

Binod Bihari Sahu Laboratory of Molecular Genetics and Plant Immunity, Department of Life Science, National Institute of Technology Rourkela, Rourkela, Odisha, India

Masina Sairam Department of Agronomy and Agroforestry, Centurion University of Technology and Management, Paralakhemundi, Odisha, India

Sanjay Kumar Sanadya Department of Genetics and Plant Breeding, College of Agriculture, Chaudhary Sarwan Kumar Himachal Pradesh Krishi Vishwavidyalaya, Palampur, HP, India

Krishnagowdu Saravanan Department of Biotechnology, Bharathiar University, Coimbatore, Tamil Nadu, India

Selva babu Selvamani Department of Plant Molecular Biology and Bioinformatics, Centre for Plant Molecular Biology and Biotechnology, Tamil Nadu Agricultural University, Coimbatore, Tamil Nadu, India

N. Senthil Department of Plant Molecular Biology and Bioinformatics, Center for Plant Molecular Biology and Biotechnology, Tamil Nadu Agricultural University, Coimbatore, Tamil Nadu, India

Amitha Mithra Sevanthi ICAR - National Institute for Plant Biotechnology, New Delhi, India

Sumaiya Shahabuddin Shaikh Department of Botany, Yashavantrao Chavan Institute of Science, Satara, Maharashtra, India

Tanmoy Shankar Department of Agronomy and Agroforestry, Centurion University of Technology and Management, Paralakhemundi, Odisha, India

Shriya Sharma Department of Plant Sciences, School of Life Sciences, University of Hyderabad, Hyderabad, Telangana, India

Meghana Deepak Shirke Department of Biotechnology, KTHM College, Nashik, Maharashtra, India

Sonam Singh ICAR, National Institute for Plant Biotechnology, New Delhi, India

Amolkumar U. Solanke ICAR - National Institute for Plant Biotechnology, New Delhi, India

Thiyagarajan Thulasinathan ICAR - National Institute for Plant Biotechnology, New Delhi, India

Suraj Dhanyakumar Umdale Department of Botany, Jaysingpur College (Affiliated to Shivaji University, Kolhapur), Jaysingpur, Maharashtra, India

Chockalingam Vanniarajan Department of Plant Breeding and Genetics, Agricultural College and Research Institute, Tamil Nadu Agricultural University, Madurai, Tamil Nadu, India

Hindu Vemuri Department of Plant Sciences, School of Life Sciences, University of Hyderabad, Hyderabad, Telangana, India

Kannan Veni Department of Plant Breeding and Genetics, Agricultural College and Research Institute, Tamil Nadu Agricultural University, Madurai, Tamil Nadu, India

Mani Vetriventhan International Crops Research Institute for the Semi-Arid Tropics, Hyderabad, Telangana, India

Nandakumar Vidya Department of Biotechnology, Bharathiar University, Coimbatore, Tamil Nadu, India

Arvind Kumar Yadav ICAR - National Institute for Plant Biotechnology, New Delhi, India

Small Millets: The Next-Generation Smart Crops in the Modern Era of Climate Change

1

Sagar Maitra, Shubhasisha Praharaj, Akbar Hossain, T. S. S. K. Patro, Biswajit Pramanick, Tanmoy Shankar, Ramesh Namdeo Pudake, Harun I. Gitari, Jnana Bharati Palai, Masina Sairam, Lalichetti Sagar, and Upasana Sahoo

Abstract

Millets are coarse cereals belonging to the family Poaceae, which is cultivated since the ancient period of civilization. Among different millets, small or minor millets are treated as neglected crops due to their low-yield potential compared to major millets (sorghum and pearl millet) and fine cereals (rice, wheat and maize). In spite of their versatile qualities, small millets remained underutilized due to institutional promotion in favour of fine cereals. Recently, these coarse cereals are re-evaluated as 'nutri-cereals' considering their composition and nutritional value. In the present consequences of adverse impacts of climate change, the small millets also attracted the attention of growers and policy-makers as they are less demanding to external inputs, drought-tolerant and register a comparatively lower carbon footprint than other cereals. These beneficial impacts ensured the

S. Maitra (✉) · S. Praharaj · T. Shankar · J. B. Palai · M. Sairam · L. Sagar · U. Sahoo
Department of Agronomy and Agroforestry, Centurion University of Technology and Management, Paralakhemundi, Odisha, India
e-mail: sagar.maitra@cutm.ac.in

A. Hossain
Bangladesh Wheat and Maize Research Institute, Dinajpur, Bangladesh

T. S. S. K. Patro
Agricultural Research Station, Acharya N. G. Ranga Agricultural University, Vizianagaram, India

B. Pramanick
Department of Agronomy, Dr. Rajendra Prasad Central Agricultural University, Pusa, Samastipur, India

R. N. Pudake
Amity Institute of Nanotechnology, Amity University Uttar Pradesh, Noida, India

H. I. Gitari
School of Agriculture and Enterprise Development, Kenyatta University, Nairobi, Kenya

R. N. Pudake et al. (eds.), *Omics of Climate Resilient Small Millets*,
https://doi.org/10.1007/978-981-19-3907-5_1

comeback of small millets after the institutional neglect for a few decades in the developing countries. Considering the food and nutritional security of the common people, small millets can be considered as suitable staples. The emerging health consciousness and food demand for the future pushed small millets to the forefront because of their ecological soundness and mitigating ability to climate change. However, the successful harvest of small millets warrants an integration of proven and climate-smart technologies for the fulfilment of the future needs of the ever-growing population. The chapter focused on all these aspects. Moreover, the research scope mentioned in the chapter implies future directions for enhancing small millet-based agriculture viable in diversifying food baskets and achieving food and nutritional security in a hunger-free society.

Keywords

Small millets · Climate-smart agriculture · Food security · Nutritional security · Cultivation technologies

1.1 Introduction

The agricultural production system has evolved over time and many significant changes have happened in the field of agriculture. The growing population and low resource base necessitate resource efficient yet high-yielding agricultural production system (FAO 2017; Struik and Kuyper 2017). Though productivity has been given utmost focus in the past, nutritional security holds equal importance (Swaminathan and Bhavani 2013; Garcia et al. 2020; FAO, IFAD, UNICEF, WFP and WHO 2021). Providing safe, sufficient and nutritious food have to be achieved sustainably without damaging the resource base (El Bilali et al. 2019; FAO and WHO 2019). With the advent of green revolution in India, the production of cereals like rice and wheat increased significantly (Maitra et al. 2018; Maitra 2020a, b). Moreover, assured market price for these cereals, government procurement system and institutional support, high productivity etc. prompted the farmers to focus largely on fine cereals-based cropping system. This has led to the further simplification of crop diversity and the genetic diversity has become even narrower.

Climate change is one of the major issues challenging present-day farming. Rising temperature, uncertainty in rainfall events, increasing carbon dioxide levels and frequent weather anomalies have made agriculture highly risk prone (Hossain et al. 2021; Bhadra et al. 2021). Since agriculture is the primary occupation of a large section of the population in India, the threat of climate change puts agriculture as well as the livelihood of millions in vulnerable conditions (Mbow et al. 2019). As agriculture in developing or low-income countries are subsistence type, with the food consumption pattern of farm families closely matching the food they grow, the food and nutritional security becomes a function of crop productivity and nutritional quality of their own farm produce (Brahmachari et al. 2018; FAO 2019; Pradhan et al. 2021). In this context, growing nutrient-rich crops as a part of crop production

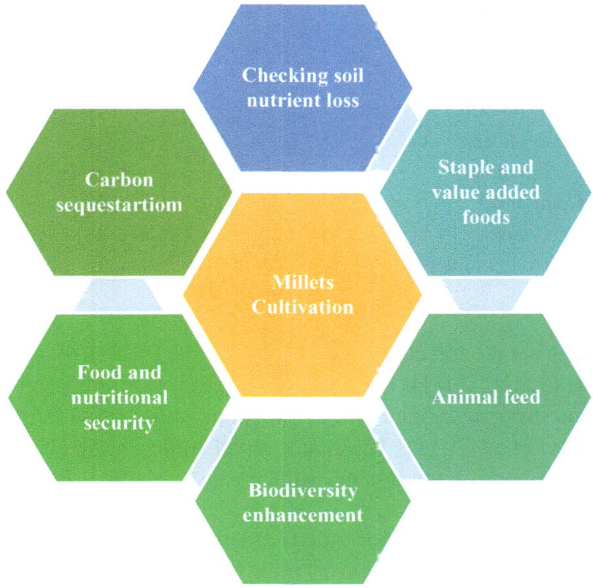

Fig. 1.1 Multiple benefits of millet cultivation

holds a greater significance in improving the health and well-being of farm families (Pradhan et al. 2019).

Millets, the term derived from 'mil' or 'a thousand' denotes the number of grains generated from a single seed (Maitra 2020a, b). They are small-grained coarse cereals and consumed for a long time. Due to the multiple health benefits millets provide and their resilience to unfavourable climatic conditions, they are considered as 'miracle crops' (Banerjee and Maitra 2020). Millets produce nutrient-dense and gluten-free grains with high dietary fibre, while, the stover can be used as nutrient-rich fodder (Maitra and Shankar 2019). These low water-consuming millets offer an excellent option to utilize water-scarce dryland region for its cultivation and also add organic matter to the soil, thus enhancing carbon sequestration. In addition to being low water requiring, the production of millets also emits very less amount of GHG (greenhouse gas), use very less chemical or industrial inputs and hence has a very low carbon footprint (Saxena et al. 2018). Millets can also reduce the erosion problem in sloppy lands. Incorporation of the stover into soil can also replenish the nutrients to a certain extent, add organic matter and improve the infiltration capacity of the soil, thus sustaining the soil health in long run (Fig. 1.1). Among a rich diversity of millets, sorghum (*Sorghum bicolor* L. Moench) and pearl millet (*Pennisetum glaucum* L.) are considered as major millets; while small or minor millets are foxtail or Italian millet (*Setaria italica* L.), finger millet or ragi (*Eleusine coracana* L. Gaertn), kodo millet (*Paspalum scrobiculatum* L.), barnyard millet (*Echinochloa frumentacea* L.), proso millet (*Panicum miliaceum* L.), little millet (*Panicum sumatrense* L.) and brown-top millet (*Brachiaria ramosa* L. Stapf; *Panicum ramosum* L.) (Table 1.1). In the global small millets production map, India

Table 1.1 Common small millets and their botanical name

Small millets	Botanical name
Finger millet	*Eleusine coracana* L. Gaertn.
Barnyard millet	*Echinochloa frumentacea* L.
Proso millet	*Panicum miliaceum* L.
Foxtail millet	*Setaria italica* L.
Kodo millet	*Paspalum scrobiculatum* L.
Little millet	*Panicum sumatrense* L.
Brown-top millet	*Brachiaria ramosa* L. Stapf; *Panicum ramosum* L.

ranks at the top covering an area of 7.0 lakh ha with about 80% of Asia's production (Rao et al. 2011). Due to the presence of different antioxidants, detoxifying agents and immune modulators in the grains, these small millets are known as nutri-cereals (Rao et al. 2011). The major small millets growing states in India are Karnataka, Tamilnadu, Uttarakhand, Maharashtra, Madhya Pradesh, Andhra Pradesh, Odisha and Bihar.

Even though minor millets are highly nutritious and possess high-stress tolerance ability as discussed above, their low productivity and lack of assured market price have made them relatively less popular among farmers. Developing agro-industries that can use millets, developing value-added and market-friendly nutritious products etc. can help in improving the market demand for the crop further and hence the farmers may get a relatively higher price for their produce. Improved agricultural technologies inclusive of smart agriculture or the concept of 'Agriculture 5.0' further create a potential for enhancement of small millet production and productivity (Zambon et al. 2019; Saiz-Rubio and Rovira-Más 2020).

Considering the above points, it can be assumed that millets can be an excellent option for climate-smart agriculture that can address the issues of food and nutritional security to a great extent. Moreover, it helps in diversifying the agroecosystem. Being less input-intensive, climate-resilient and nutritionally super-rich, it can be the answer to climate change, malnutrition and unsustainable resource use.

1.2 Salient Features of Small Millets

The salient features of the small millets have been briefly discussed below.

1.2.1 Finger Millet (*Eleusine coracana* L. Gaertn)

East African highlands is considered as the origin of finger millet. In India, it is also commonly known as *ragi*, *mandua*, *marua*, *nagli* and *kapai*. It is only during the bronze era that finger millet found its entry into India (Fuller et al. 2011). Ten different species (including annuals and perennials) come under the genus *Eleusine*.

Fig. 1.2 Finger millet standing crop (**a**) and (**b**) seeds

The most widely cultivated species of *Eleusine*, i.e. *E. coracana* is tetraploid ($2n = 4x = 36$) and self-pollinated in nature. Finger millet has an erect plant type with a height ranging from 60 to 130 cm. The crop has a shallow and fibrous root system. It has profuse tillering habit and the stem is compressed. The ear heads have spikes, in which the spikelets are arranged. The seeds of this crop are small (Fig. 1.2) and its colour varies from whitish, red-yellow to pale brown. Being a drought-tolerant crop, it can be grown in water-scarce environments and produce a sizable yield (Harika et al. 2019).

1.2.2 Foxtail Millet (*Setaria Italica* L.)

The historical evidence suggested that foxtail millet was domesticated in central China for harvesting grain and fodder yields (Miller et al. 2016). It is also known as Italian and German millet. The millet has several vernacular names in India such as *kakun*, *kangni*, *navane*, *thinai*, *kang* and *rala*. Foxtail millet is a member of the family *Poaceae* and *Panicoideae* subfamily, and is a diploid ($2n = 18$) plant. Like other small millets, it can also be grown in the dryland region. The stem length varies from 80 to 150 cm. The stem is slender and erect. Leaves are narrow, flat and length varies from 30 to 45 cm. The inflorescence is cylindrical. Each spikelet comprises a single or maximum of four bristles, looking like foxtail. The seed is small, convex and enclosed in a glume. Colour variation is observed in the seeds, but creamy white and orange red-coloured seeds are more common (Fig. 1.3). Some varieties also produce purple-coloured seeds. As foxtail millet is tolerant to drought, it can be cultivated in moisture scarce situations as a rainfed crop considering its ability to withstand soil moisture stress.

Fig. 1.3 Foxtail millet standing crop (**a**) and (**b**) seeds

Fig. 1.4 Proso millet standing crop (**a**) and (**b**) seeds

1.2.3 Proso Millet (*Panicum miliaceum* L.)

Proso millet is reported to be first domesticated in North China (Hunt et al. 2008). Recently, it is being grown in some Asian countries such as India, Afghanistan and China and European countries, namely Romania and Turkey. It has some other common names such as broomcorn millet, common millet and hog millet. In India, proso millet is commonly known as *cheena*, *panivaragu*, *variga* and *baragu* in different states. The crop is usually grown in low fertile soils with minimum use of external inputs. It belongs *Poaceae* family ($2n = 36$). Basically, proso millet is a self-pollinated crop, but cross-pollination may occur to an extent of 10% or a little higher. The plant is erect with profuse tillering. The plant is 45–100 cm tall with a fibrous root system. The stem is slender and the leaves are linear. The inflorescence is branched; spikelets are located at the tip of the inflorescence branch. Proso millet produce grains of different colours, namely, yellow, white, yellow, black and reddish (Fig. 1.4). Being a less water-requiring crop, it can be grown in warm regions of the world, where rainfall is low or scanty.

1.2.4 Barnyard Millet (*Echinochloa frumentacea* L.)

The origin of barnyard millet is considered as Japan and historical evidence indicated that the cultivation of barnyard millet was there in China around 10 thousand years ago (Sood et al. 2015). Currently, the area under barnyard millet is largely confined to the Indian subcontinent and China. In Indian vernacular languages, it has some other names such as *madira*, *sawa*, *sawan*, *kudraivali* and *oodalu*. Barnyard millet comes under *Panicoideae* subfamily (family *Poaceae*) and it is hexaploid (with $2n = 6x = 54$) (Clayton and Renvoize 2006). Barnyard millet has wider adaptability and can be grown in higher altitudes also (2000 m above MSL) (Gupta et al. 2009). The crop has the quality of drought tolerance (Maitra et al. 2020). The variation in colour and shape is observed in the panicles of barnyard millet (Kuraloviya et al. 2019) with raceme numbers of 22–64 in every inflorescence (Renganathan et al. 2020). Each spikelet has two florets. It is self-pollinated, however, there is the possibility of cross-pollination. The seed colour is whitish to grey and the seeds are soft (Fig. 1.5) (Maitra et al. 2020).

1.2.5 Little Millet (*Panicum sumatrense* Roth. ex Roem. and Schult)

The probable origin of little millet is India (Maitra and Shankar 2019). Archaeological evidence showed that it was grown in western India during 2000 BC (Venkatesh Bhat et al. 2018). Presently, the crop is cultivated in India, the Philippines, China and Malaysia. In India, it is also known as *kutki*, *sawai*, *samulu* and *same*. It belongs to the family *Poaceae* and is grown in tropical and subtropical climates. Being a drought-tolerant and short-duration crop, it has wider adaptability even at high altitudes. The leaves are 30–100 cm long. The crop has awned panicles and round-shaped brown grains (Fig. 1.6).

Fig. 1.5 Barnyard millet standing crop (**a**) and (**b**) seeds

Fig. 1.6 Little millet standing crop (**a**) and (**b**) seeds

Fig. 1.7 Kodo millet standing crop (**a**) and (**b**) seeds

1.2.6 Kodo Millet

The domestication of kodo millet (*Paspalum scrobiculatum* L.) began in India about 3000 years back. In the Indian language, kodo millet is also known as *varagu*, *haraka* and *arikalu*. Like other millets, it also comes under *Poaceae* family with the subfamily of *Panicoideae*, and it is a tetraploid crop ($2n = 4x = 40$) (Jarret et al. 1995). Kodo millet plants are erect, 45–90 cm tall with purple-coloured leaf pigmentation. It is self-pollinated with single-flowered spikelets. Brown-coloured grain is covered by lemma and palea (Fig. 1.7). Under severe drought conditions, it exhibits a high tolerance to abiotic stresses, namely, scanty soil moisture and heat; and thus, it is considered a suitable crop for drylands.

1.2.7 Brown-Top Millet

The probable region for the domestication of brown-top millet is the Deccan of India (Kingwell-Banham and Fuller 2014) and it has the heritage of an important cereal

Fig. 1.8 Brown-top millet standing crop (**a**) and (**b**) seeds

since 3000 BCE (Fuller et al. 2010). In the ancient period, in agro-pastoral practices, it was a common crop with legumes (Maitra 2020a). At present, it is grown in India, China, Arabia, Australia and a few African countries (Clayton and Renvoize 2006). Presently, in Karnataka–Andhra Pradesh adjacent dry areas this millet is cultivated as a food crop and it is commonly known as *karole* in the Deccan. The *Rayalseema* region of Andhra Pradesh (especially in *Ananthpur* district) and *Chitradurga*, *Gulbarga*, *Tumkuru*, *Dharwar* and *Chikkaballapura* districts of Karnataka are known for its cultivation. The Bundelkhand region of Central India is also recognized for growing this crop (Niyogi 2018) in marginal lands. Unlike other small millets, it is suitable for production in partially shaded areas, ensuring suitability to grow in fruit orchards under limited sunlight conditions. Moreover, in a little millet field, its presence is observed as a weed (Sakamoto 1985).

Brown-top millet is an annual and perennial coarse cereal that belongs to the *Poaceae* family (Fig. 1.8). The crop has an erect stem (culm) or prostrate type growing laterally on the ground. The plant height may be up to 90 cm (Maitra 2020b) and experimental results showed that the plant height of this crop was 68–74 cm in Chhattisgarh, India (Thakur et al. 2019). In another experiment, Saikishore et al. (2020) recorded the plant height of 134–153 cm. The crop has fibrous roots with a maximum of 60 cm depth. The nodes have minute hairs, but leaf blades do not contain hairs (Maitra 2020b). In general, the length and width of leaf blades are 2–25 cm and 4–14 mm, respectively (Clayton and Renvoize 2006). Brown-top millet bears indeterminate white flowers, which are stalked. The inflorescences are 3–15 with a length of 1–8 cm long, originating from a central axis, open and spreading. Panicles are looser and non-bristly. The seeds are ellipsoid and tan coloured. Grains of brown-top millet are ovate to round with a long embryo. The husk is fine-beaded and rough (Kingwell-Banham and Fuller 2014). The average duration of the crop is 75–90 days.

It is clear from the above paragraphs that all the small millets have abiotic stress tolerance ability to a large extent. This feature can be exploited to make it a suitable crop under climate-resilient agricultural practice.

1.3 Small Millets as Functional Foods

As discussed in the introduction, small millets are nutritionally rich and hence can alleviate the issue of malnutrition and nutrient deficiency to a great extent. Their nutritional superiority is not only due to the presence of a high amount of macro- and micronutrients, but also the presence of other compounds of nutraceutical significance that act as a protectant against different diseases. The nutritional values of small millets are presented in Table 1.2.

Millets are an excellent source of nutrition. As millets are generally gluten-free, hence it can be a food alternative for people with gluten allergies. Millets also contain some important phytochemicals like polyphenols, phytosterols and lignans, which have potential health benefits. Millets are also rich in dietary fibres and vitamin B complexes. The antioxidants and immune-modulator activities of millets act as a protective barrier against diseases like Parkinson's disease, cardiovascular disease and respiratory diseases (Rao et al. 2011; Chandrasekara et al. 2012). Several studies have reported on the beneficial role of low glycaemic index (GI—a measure of carbohydrate quality) foods and diets in the nutritional management of diabetes and several other chronic diseases. The rate of glucose absorption is usually decreased by low GI foods causing reduced insulin demand (Shobana et al. 2013). The small millets have low GI than rice and wheat and hence, can be considered a good dietary choice for diabetic people (Patil et al. 2015; Rao et al. 2017). The fibre richness of finger millet helps in the reduction of blood cholesterol and slow release of glucose to the bloodstream during digestion and prevents constipation (Rao et al. 2017).

Considering all the benefits millets can also be considered as nutraceuticals (Banerjee and Ray 2019). Millets are also easy to digest and non-allergenic. Phytochemicals such as polyphenols act as antioxidants and give protection against oxidative damage. Seed coats of most of the millets are found to have phenolics with antioxidant properties (Chandrasekara and Shahidi 2010).

Table 1.2 Nutritional value of small millets (per 100 g of edible portion)

Crop	Carbohydrate (g)	Crude fibre (g)	Protein (g)	Fat (g)	Mineral (g)
Finger millet	72.0	3.6	7.3	1.3	2.7
Barnyard millet	74.3	14.7	11.6	5.8	4.7
Proso millet	70.4	2.2	12.5	1.1	1.9
Foxtail millet	60.0	8.0	12.3	4.3	3.3
Kodo millet	65.9	9.0	8.3	1.4	2.6
Little millet	75.7	8.6	8.7	5.3	1.7
Brown-top millet	71.3	16.1	11.5	1.9	3.9

Sources: Baptist and Perera (1956); Kering and Broderick (2018); Maitra (2020a); Banerjee and Maitra (2020); Hemamalini et al. (2020), IIMR (2021)

1.4 Small Millets as Climate-Smart Crops

In the present context of climate change and associated aberrations, there is an urgent need for the consideration of climate-smart crop production technologies for uninterrupted and sustainable farm output. In this regard, climate-smart crops are to be chosen that can withstand various ill effects of climatic abnormalities and ensure satisfactory farm outputs. Small millets are of automatic choice as they are climate resilient because they can tolerate high temperature and soil moisture deficit to a large extent, which is very common in tropical and subtropical conditions in developing countries. Being less nutrient-demanding crops, they are grown under minimal management ensuring less carbon footprint in agriculture. The small millets possess a C_4 type of photosynthesis and sequester carbon, thereby adding CO_2 abatement opportunities under the elevated CO_2 levels in the future (Bacastow and Keeling 1973; Prentice et al. 2001; Balbinot et al. 2021). Further, small millets enrich the agro-biodiversity (which has been lost due to the adoption of industrialized agriculture-dominated monocropping) and are suitable for intercropping with other important crops inclusive of legumes (Maitra 2020a) millets-based intercropping.

In general, climate-smart agriculture has three basic objectives (Fig. 1.9) and these are as follows:

1. Productivity: Sustainably improving productivity and income.
2. Adaptation: Strengthening the resilience of food systems to climate changes and variability.

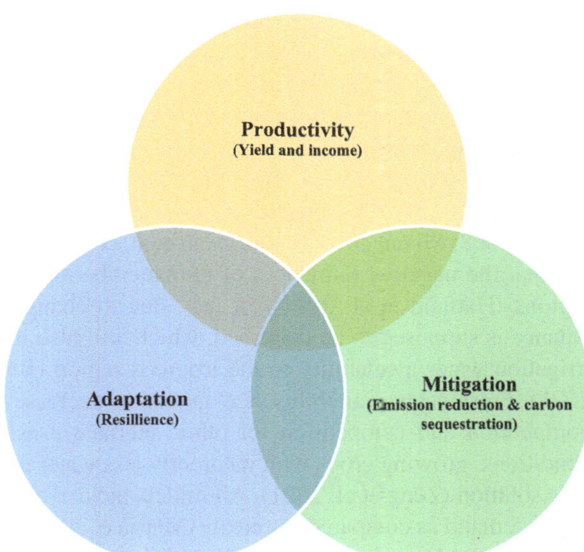

Fig. 1.9 Objectives of climate-smart agriculture

3. Mitigation: Reduce GHG emissions from agricultural activity and sequester carbon in the farmland.

1.4.1 Millets as a Driver of Climate-Smart Agriculture

As discussed in the previous section, climate-smart agriculture aims at attaining sustainable crop production and income generation in the face of changing climatic scenarios by adapting to the changing climate as well as mitigating climate change (FAO 2013; IPCC 2018). These factors are discussed here in detail.

1.4.1.1 Productivity of Small Millets

The small millets have low productivity which can be attributed to poor agronomic management. They are mostly grown in dryland regions where other crops cannot be grown successfully and with very low or minimum use of external inputs (Prasanna Kumar et al. 2019). Proper irrigation and nutrient management can help in improving millet productivity further in these regions. The stress resistance attribute of millets makes it a perfect choice for dryland regions. Considering climate change scenarios, where abiotic stresses such as drought and heat stress are expected to be more frequent, millets can provide production stability and food and nutritional security (Maitra et al. 2022). In addition to agronomic management, genetic interventions can also be extremely useful in improving small millet productivity. The small millet varieties can be bred for high productivity, especially under stressed environments.

1.4.1.2 Adapting to Changing Climate

Climate change is expected to negatively affect freshwater availability for agricultural activity (Singh et al. 2014; Zaman et al. 2017; Boretti and Rosa 2019). Dryland agriculture or rainfed agriculture is expected to be even more affected by the vagaries of climate change events (Ashalatha et al. 2012; Hossain et al. 2021; Tui et al. 2021). Recurring drought, uncertain rainfall events such as the late onset of monsoon, early withdrawal of monsoon and the prolonged dry period within the crop season are going to affect the rainfed agriculture more as compared to irrigated agriculture where there is assured irrigation to counteract the shortage of water at any stage (Turral 2008; Miyan 2015; Sanjeevaiah et al. 2021). However, it is not appropriate to say that the negative influences of climatic aberration will be confined to dryland regions (Hatfield et al. 2011). In fact, due to changing climate, the hydrological balance is supposed to be disturbed, which will also have a negative impact on the irrigation water availability in the irrigated region (Bhave et al. 2018; Zeng et al. 2021). Freshwater availability is expected to decrease further due to inter-sectoral competition and competition for other alternate uses (OECD 2009). Under such conditions, growing crops with inherently less water demand can provide a sustainable solution (Zeng et al. 2021). All millets inclusive of small millets have very less water demand as compared to cereals (Maitra et al. 1997; Ramya et al. 2020; Saxena et al. 2018). Hence, growing millets may result in higher productivity per unit amount of freshwater consumed and hence, they enhance water productivity (Mekonnen and Hoekstra 2014). The performance of millets under the water-scarce condition in the dryland region can also be due to their adaptation at the physiological level because of their C_4 photosynthetic mechanism (Wang et al. 2018; Hatfield and Dold 2019; Ghatak et al. 2021). Small millets, because of their C_4 mechanism, can fix sufficient carbon dioxide for photosynthesis even when the stomata are partially closed in response to moisture stress (Hao et al. 2017). This allows for better photosynthetic activity under moisture stress.

Change in soil fertility is another outcome of climate change. As the temperature is expected to rise due to climate change, there will be rapid oxidation of soil organic matter (Eric et al. 2013; Karmakar et al. 2016). Moreover, due to heavy and intense rainfall events, soil fertility may be lost because of leaching of nutrients and erosion of fertile topsoils. Millets can counteract this negative impact of climate change in two ways. First, being capable of growing well under less fertile conditions, millets can give some yield. Secondly, millets can protect against erosion and save fertile topsoil (Saxena et al. 2018). Small millets when grown in those soil can also add organic matter through degraded root biomass after crop harvest. Moreover, retaining millet residues on the soil surface can also enhance organic matter and nutrients in the soil. Further, it can improve water infiltration, thus enhancing in situ soil conservation. With the addition of organic matter, better soil moisture promotes better microbial activity, which further improves soil fertility through nutrient addition or solubilization. Further, intercropping small millets with legumes helps in the improvement of soil quality because of the combination of cereal legumes in polyculture (Maitra 2020a, b).

As small millets are better adapted to harsh climatic conditions, they provide better stability in production over years/seasons as compared to fine cereals. Also, millets are getting proper attention in the recent past due to their high nutritional quality and health benefits (Saleh et al. 2013; Kumar et al. 2020). This has led to a decent market price for these produces and their associated processed and value-added products. Considering the market demand, yield stability and low cost of cultivation, millets can be a safe bet for resource-poor farmers under resource-starved growing conditions. Production of value-added products can further improve the economic prospect of millet production. To a great extent, it can help in breaking the vicious circle of dearth. It is significant to record that, while calculating the economics, the benefits of good health and well-being are often ignored. As good health brought about by food and nutritional security improves the human resource potential further, it further improves wealth creation. Moreover, good health and well-being also reduce expenditure associated with health-related issues.

Considering the above positive facts associated with small millets, it is clear that they can be an excellent option for adapting to conditions under changing climates. In addition to food security and nutritional security as well as economic profitability, small millets also provide much-needed agroecosystem diversity. Diversity in agroecosystem provides resilience, improves productivity, minimizes risk and provides multiple sources of income as well as profit (Parmentier 2014; Altieri et al. 2015).

1.4.1.3 Mitigating Climate Change

The mitigation of climate change may be achieved by reducing GHG emissions and enhancing soil carbon sequestration (Amelung et al. 2020; Fawzy et al. 2020; Navarro-Pedreño et al. 2021). The rice production system is one of the major contributors to GHG emission when GHG emission from croplands is considered (Boateng et al. 2017; Vetter et al. 2017; Arenas-Calle et al. 2019). The anaerobic environment in the paddy ecosystem is favourable for the emission of GHG such as methane and nitrous oxide (Oertel et al. 2016; Wang et al. 2019). Additionally, frequent tillage also oxidizes the soil organic carbon and hence carbon dioxide emission from cropland increases (Haddaway et al. 2017; Krauss et al. 2017). Unlike cereals, small millets emit lesser GHGs. Also, millets have a comparatively lesser carbon footprint as compared to fine cereals (Kane-Potaka et al. 2021). Moreover, fine cereals use a huge amount of chemical fertilizer, which in its due course of industrial production produces a high amount of GHG (Maitra et al. 2022). An estimate mentions that for fulfilling the global chemical N fertilizers, annually 300 teragrams (Tg) of CO_2 is released into the atmosphere (Jensen et al. 2012). Unlike cereals, millets being less nutrient demanding have a lesser carbon footprint. Millets can also help in sequestering carbon through their shoot and root biomass production and addition to the soil post-harvest. The technological developments facilitated all aspects of farm sector to align towards the direction of smart or precision agriculture. The smart technologies have enough potential to maximize input use efficiency and the productivity of crops can be enhanced with an efficient management. There is enough scope for inclusion of the concept of Agriculture 5.0 with the advent of technological supports such as Internet of Things (IOT) in smart irrigation, robotics and drones and some forms of artificial intelligence and machine learning (AI and ML) in crop management targeting a higher productivity of climate-smart small millets (Zambon et al. 2019; Saiz-Rubio and Rovira-Más 2020; Maitra et al. 2022).

1.5 Climate-Smart Small Millets Production Practices

Climate-smart agronomic practices largely rely on good agriculture practices (GAP) that improves input use efficiency, reduce emission from the system, saves resource, avoid agroecosystem pollution and promotes ecosystem services. The targets are achieved through many practices such as integrated nutrient management (INM), integrated pest management (IPM) and conservation agriculture. In addition to these, available precision agriculture tools and decision management systems can further improve in making decisions regarding climate-smart crop production practices.

1.5.1 Integrated Nutrient Management (INM)

INM integrates all the available nutrient sources judiciously and compatibly to supply essential plant nutrients. It reduces the overdependence on chemical fertilizer and utilizes all the available low-cost local resources to meet the crop nutrient requirement. INM improves soil structure, improves soil fertility, improves soil biological activity, reduces cost of nutrient management, promotes ecosystem services and makes better utilization of inherent soil fertility (Kumara et al. 2007).

Considering the fact that, production of chemical fertilizer emits huge amount of greenhouse gas, INM, through substitution of chemical fertilizers, can help in the reduction of total carbon footprint of the production system. It can also help in soil carbon sequestration through the addition of organic matter. Unlike other crops, where nutrient demand is very high, the application of organic manure can practically replace a large portion of crop nutrient demand in millets. Research evidence suggests the beneficial effects of INM (Table 1.3).

Table 1.3 Beneficial effects of INM in small millets

Crop	Salient findings	References
Finger millet	Manures applied previously yielded more grains of finger millet than only chemical fertilizers	Pilbeam et al. (2002)
	Application of farmyard manure (FYM) enriched with P and recommended dose of N produced more grain yield than the recommended dose of fertilizer (RDF) in clay loamy soil	Jagathjothi et al. (2011)
	A combined application of FYM (10 t ha^{-1}) +100% NPK and maize residue incorporation (5 t ha^{-1}) + 100% NPK produced more grain yield in semi-arid tropical Alfisol	Sankar et al. (2011)
	FYM (10 t ha^{-1}) + bioinoculant consortia (60 g kg^{-1} seed each) + $ZnSO_4$ (12.5 kg ha^{-1}) + borax (kg ha^{-1}) + 100% RDF (50:30:25) yielded more grain and straw of kharif finger millet than recommended dose of nutrients	Roy et al. (2018)
Foxtail millet	50% RDF + 25% N as neem cake + biofertilizer (*Azophos*) recorded more growth and productivity than RDF	Monisha et al. (2019)
	Adoption of INM with FYM + RDF + 3% *Panchagavya* resulted in maximum grain yield	Kumaran and Parasuraman (2019)
	Application of 75% recommended dose of nitrogen (RDN) through chemicals +25% N through poultry manure + biofertilizer (*Azospirillum* seed inoculation) enhanced the grain yield	Selectstar Marwein et al. (2019)
Little millet	INM with FYM @ 7.5 t ha^{-1}, N 40:P 20:K10 kg ha^{-1}, calcium carbonate, zinc sulphate and borax resulted in more grain yield	Parihar et al. (2010)
	Combined application of 100% RDF + neem cake @ 1 t ha^{-1} resulted in maximum productivity	Sandhya Rani et al. (2017)
	Application of 75% RDN (chemical fertilizer) and 25% RDN (vermicompost) recorded higher growth and grain yield than RDF	Thesiya et al. (2019)
Kodo millet	A combination of 125% RDF + soil application of *Azospirillum* @ 2 kg ha^{-1} + vermicompost @ 2 t ha^{-1} + foliar spray of 1% nutrient supplement increased grain productivity	Prabudoss et al. (2014)

1.5.2 Soil Test Crop Response (STCR)-Based Nutrient Management

Soil test crop response (STCR)-based nutrient management supplies nutrients to the crop based on soil test value and response conditions for a target yield. STCR aims at balanced fertilization by considering the contribution of soil and applied nutrients. STCR approach to nutrient management improves yield and it is environment friendly as well as economical (Das et al. 2015). STCR also improves nutrient-use efficiency (Gangwar et al. 2016; Jemila et al. 2017). As per the STCR equation, for achieving a target yield of 4 t ha^{-1} in finger millet, a combination of fertilizers applied against the recommended dose and a nutrient combination of 200% nitrogen, 100% phosphorus, 100% potassium, 25% zinc, 25% sulphur and 25% boron integrated with 5 t ha^{-1} farmyard manure (FYM) produced satisfactory yield output in the soil with low available N, high P and medium K (Sandhya Rani et al. 2017).

1.5.3 Resource Conserving Technologies

Resource conservation technologies (RCTs) aim at saving resources and improving resource-use efficiency. Conservation agriculture (CA) is a RCT and it has three basic principles, i.e. maintaining crop residues on soil surface, zero tillage and diversified crop rotation. Conservation agriculture improves soil organic matter content, nutrient-use efficiency and soil moisture storage. It also reduces energy use in agriculture. Greenhouse gas (GHG) emission from CA has been found to be lesser as compared to conventional agriculture. As more organic matter is maintained in the field, the microbial activity and overall soil health are also improved. Residue retained on the soil surface improves opportune time for water infiltration and hence helps in situ moisture storage. Soil temperature is also optimized under conservation agriculture because of surface residue retention.

In contrast to conventional tillage where land is ploughed causing a global loss of soil organic carbon (SOC) as much as 60–90 Pg (Lal 1999), in CA, the stored organic matter is not rapidly oxidized as the soil is not disturbed frequently. This helps in long-term carbon storage in the soil. Further, CA improves soil physico-chemical and biological properties (Lal 2004). Finger millet yielded more under substitution of 50% of the recommended N with organic source in Alfisol of Karnataka, India, however, reduced tillage enhanced SOC as recorded by Prasad et al. (2016). Malviya et al. (2019) recommended conservation tillage and crop residue mulching to raise the sole crop of kodo millet in the Rewa region of Madhya Pradesh India. RCTs like conservation agriculture have a large scope in climate-smart agriculture as it reduces the emission of GHG, and improves the soil environment for better resilience to climate change-induced alterations.

In addition to conservation agriculture, other RCTs such as bed planting and laser land levelling may also be applied for improving yield and enhancing climate resilience.

1.5.4 Breeding of Suitable Varieties

Improved varieties of small millets must be bred for their better adaptation to climate change and other biotic and abiotic stresses. Breeding may take a longer time and involve a high initial cost. However, in long run, it can be a very cost-effective strategy for climate change adaptation. Many varieties tolerant against abiotic stresses and biotic stresses have been developed. Developing varieties with high yielding ability, better nutritional bioavailability, tolerance to multiple stresses (both biotic and abiotic), high nutrient efficiency, higher photosynthetic efficiency etc. can help to improve crop productivity and quality. Both conventional breeding and biotechnology approaches may be used for breeding small millets. In this regard, germplasms are the keys to crop improvement as they provide the desired variability. Worldwide, 133,849 cultivated germplasms of small millets are conserved in addition to 30,627 accessions, of which most of them are collected from Asia and Africa (Vetriventhan et al. 2020). There are several small millets germplasms containing promising traits such as nutritional quality and tolerance to biotic and abiotic stresses. The conventional breeding programmes through selection and hybridization have already developed different small millet varieties (Nandini et al. 2019). As a leading producer of small millets, India has developed about 248 varieties of six small millets, namely, finger millet (121), foxtail millet (32), proso millet (24), kodo millet (33), barnyard millet (18) and little millet (20) (AICSMIP 2014). The genome sequence and gene mapping are two advanced methods of crop improvement considered for crop improvement of small millets. The genomes of some small millets have been already sequenced with prior mapping of desired quality traits by following germplasm characterization and marker trait association inclusive of quantitative trait nucleotides (QTNs). Recently, biotechnological tools as well as omics approaches are also included in breeding of small millets. Initiatives have already been taken through transcriptome-based gene expression profiling, proteomics and metabolomics and *Agrobacterium*-mediated system for transformation of small millets for qualitative improvement of these climate-smart crops (Vetriventhan et al. 2020). However, inadequate number of germplasms and insufficient information on genetic diversity are major limitations for crop improvement. Biotechnological processes are involved in high value and major crops and there is a need for future intervention of omics approaches in the improvement of small millets.

1.5.5 Agronomic Practice Adjustment

Sowing time, plant spacing, nutrient application etc. may be manipulated to make the plant more suitable to face different stresses. For example, manipulation of sowing time can help to avoid terminal heat stress to some extent. Selecting a suitable crop can also be an approach to avoid or minimize the effect of stress. In finger millet yield loss due to terminal drought stress is maximum, however, considerable yield loss was noticed in proso millet (34.6%), little millet (80.1%)

and pearl millet (60.1%) (Bidinger et al. 1987; Goron and Raizada 2015; Tadele 2016). Priming of seeds can also be an effective strategy for adapting to moisture stress, especially during the early period of crop growth (Maitra et al. 1997). Further, cropping systems can play a great role in this regard. Intercropping systems with small millets must be considered in this regard (Maitra 2020b). Intercropping small millets with legumes are one of the suitable options for smallholders in the drylands. The multifaceted benefits of intercropping small millets with legumes have been evidenced by researchers in terms of a higher productivity, resource-use efficiency, natural insurance against crop failure under extreme climatic conditions, and food and nutritional security (Maitra et al. 2020).

1.5.6 System of Millet Intensification

Principles of the system of crop intensification (SCI) can also be applied to millet cultivation for improving productivity and resource-use efficiency. Research on the system of finger millet intensification showed that transplanting 10 days old seedlings with square planting of 25 cm yielded more than conventional planting (Bhatta et al. 2017). In some areas of Karnataka such as Dhadwad, Haveri, Kolar and Shimoga, farmers are familiar to raise finger millet with square planting by adopting a traditional method and it is colloquially known as '*guni*' which is nothing but a form of System of Finger Millet Intensification. In the *guni* method, 3-weeks-old seedlings are planted with two seedlings per hill. In between the third and sixth weeks of transplanting, the crop is planked by animal-drawn implement to enhance tillering and growth of adventitious roots. Researchers recorded that guni method yielded more grains of finger millet (Sukanya et al. 2021). The SCI practices may be standardized for more millets and can be an effective strategy for improving productivity.

1.6 Conclusion

Small millets are an excellent source of macro and micronutrients as well as dietary fibre. As millets are low resource consuming crops, it can be grown under resource-scarce conditions, where resources for most other crops seem sub-optimal. As the climate is changing and resource availability is under pressure, agriculture needs to be climate-resilient and resource-efficient for sustainable development. Under such conditions, small millets can be grown to counteract the negative impacts of climate change to a great extent due to their inherent capacity to survive under low moisture, low nutrient demand, C_4 photosynthetic pathway etc. Developing suitable agronomic practices for millets, developing varieties with better stress tolerance and high nutrient bioavailability, identifying suitable microbial strains for improving nutrient cycle or growth-promoting ability etc. need immediate research attention to further improve the productivity and quality of millets. More importantly, millets-centric policies for better storage facilities, good processing platforms and assured market price can further promote millet cultivation.

Further, the present chapter offers the following future scope of research which can also be considered. An integration of multidisciplinary approaches can truly offer possible scope and opportunity for small millets to exploit their real potential as climate-smart crops.

1. Agronomic practices suitable for resource-limited environments need to be standardized. As resource scarcity is one of the prime negative outcomes of climate change; agronomic practices to counteract such environment and resource-scarce conditions need to be standardized.
2. The response of different millets to elevated atmospheric carbon levels may be studied. Alteration in phenological, physiological and biochemical parameters can be monitored. Such knowledge can be utilized for predicting crop response to climate change and developing a more accurate crop model.
3. System of crop intensification knowledge can be applied to small millets and the practice needs to be standardized for different environments.
4. Nutraceutical benefits of millet need to be studied further.
5. Bioavailability of nutrients after processing and value addition needs to be evaluated. Care must be taken to improve the bioavailability of nutrients.
6. Awareness regarding nutraceutical and health benefits of millets and millet products need to be created. It will not only improve millet consumption but also create a good market for millets and its products.

References

AICSMIP (2014) Report on compendium of released varieties in small millets [Internet]. Banglore, India; 2014. http://www.dhan.org/smallmillets/docs/report/Compendium_of_Released_Varieties_in_Small_millets.pdf. Accessed 18 Mar 2020

Altieri MA, Nicholls CI, Henao A, Lana MA (2015) Agroecology and the design of climate changeresilient farming systems. Agron Sustain Dev 35(3):869–890. https://doi.org/10.1007/s13593-015-0285-2

Amelung W, Bossio D, de Vries W, Kögel-Knabner I, Lehmann J, Amundson R, Bol R, Collins C, Lal R, Leifeld J, Minasny B (2020) Towards a global-scale soil climate mitigation strategy. Nat Commun 11(1):1–10

Arenas-Calle LN, Whitfield S, Challinor AJ (2019) A climate smartness index (csi) based on greenhouse gas intensity and water productivity: application to irrigated rice. Front Sustain Food Syst 3:105. https://doi.org/10.3389/fsufs.2019.00105

Ashalatha KV, Munisamy G, Bhat AR (2012) Impact of climate change on rainfed agriculture in India: a case study of Dharwad. Int J Environ Sci Dev 3(4):368–371

Bacastow RO, Keeling CD (1973) Atmospheric carbon dioxide and radiocarbon in the natural carbon cycle: II. Changes from AD 1700 to 2070 as deduced from a geochemical model. In: Brookhaven Symposia in Biology 24:86–135

Balbinot A, da Rosa FA, Fipke MV, Rockenbach D, Massey JH, Camargo ER, Mesko MF, Scaglioni PT, de Avila LA (2021) Effects of elevated atmospheric CO_2 concentration and water regime on rice yield, water use efficiency, and arsenic and cadmium accumulation in grain. Agriculture 11:705. https://doi.org/10.3390/agriculture11080705

Banerjee P, Maitra S (2020) The role of small millets as functional food to combat malnutrition in developing countries. Ind J Nat Sci 10(60):20412–20417
Banerjee P, Ray DP (2019) Functional food: a brief overview. Int J Biores Sci 6:57–60. https://doi.org/10.30954/2347-9655.02.2019.2
Baptist NG, Perera BPM (1956) Essential amino-acids of some tropical cereal millets. www.cambridge.org/core/terms. https://doi.org/10.1079/BJN19560050, (Accessed 12 Aug 2021)
Bhadra P, Maitra S, Shankar T, Hossain A, Praharaj S, Aftab T (2021) Climate change impact on plants: plant responses and adaptations. In: Aftab T, Roychoudhury A (eds) Plant perspectives to global climate changes. Elsevier Inc., Academic Press, pp 1–24. https://doi.org/10.1016/B978-0-323-85665-2.00004-2
Bhatta LR, Subedi R, Joshi P, Gurung SB (2017) Effect of crop establishment methods and varieties on tillering habit, growth rate and yield of finger-millet. Agric Res Tech J 11(5):555826. https://doi.org/10.19080/ARTOAJ.2017.11.555826
Bhave AG, Conway D, Dessai S, Stainforth DA (2018) Water resource planning under future climate and socioeconomic uncertainty in the Cauvery River basin in Karnataka, India. Water Resour Res 54(2):708–728
Bidinger FR, Mahalakshmi V, Rao GDP (1987) Assessment of drought resistance in pearl millet (*Pennisetum americanum*). 2. Estimation of genotype response to stress. Aust J Agric Res 38(1):49–59
Boateng GKK, Obeng GY, Mensah E (2017) Rice cultivation and greenhouse gas emissions: a review and conceptual framework with reference to Ghana. Agriculture 7:7. https://doi.org/10.3390/agriculture7010007
Boretti A, Rosa L (2019) Reassessing the projections of the world water development report. NPJ Clean Water 2:15. https://doi.org/10.1038/s41545-019-0039-9
Brahmachari K, Sarkar S, Santra DK, Maitra S (2018) Millet for food and nutritional security in drought prone and red laterite region of eastern India. Int J Plant Soil Sci 26(6):1–7
Chandrasekara A, Shahidi F (2010) Content of insoluble bound phenolics in millets and their contribution to antioxidant capacity. J Agric Food Chem 58(11):6706–6714. https://doi.org/10.1021/jf100868b
Chandrasekara A, Naczk M, Shahidi F (2012) Effect of processing on the antioxidant activity of millet grains. Food Chem 133(1):1–9
Clayton WD, Renvoize SA (2006) Genera *Graminum*: grasses of the world. Kew Bulletin Additional Series XIII, Royal Botanical Gardens Kew, Her Majesty Stationery Office, London
Das D, Dwivedi B, Meena M, Singh VK, Tiwari KN (2015) Integrated nutrient management for improving soil health and crop productivity. Ind J Fert 11:64–83
El Bilali H, Callenius C, Strassner C, Probst L (2019) Food and nutrition security and sustainability transitions in food systems. Food Energy Secur 8:e00154. https://doi.org/10.1002/fes3.154
Eric GO, Lagat JK, Ithinji GK, Mutai BK, Kenneth SW, Joseph MK (2013) Maize farmers perceptions towards organic soil management practices in Bungoma County, Kenya. Res J Environ Earth Sci 5(2):41–48
FAO (2013) Climate-Smart Agriculture Sourcebook, Food and Agriculture Organization, Rome, Italy, https://www.fao.org/3/i3325e/i3325e.pdf (accessed 15 November 2021)
FAO (2017). The future of food and agriculture—Trends and challenges, Rome, Italy, pp.163
FAO (2019) The state of food and agriculture 2019. Moving forward on food loss and waste reduction. Rome. License: CC BY-NC-SA 3.0 IGO
FAO, IFAD, UNICEF, WFP, WHO (2021) The state of food security and nutrition in the world 2021. Transforming food systems for food security, improved nutrition and affordable healthy diets for all. Rome, FAO. https://doi.org/10.4060/cb4474en
FAO, WHO (2019). Sustainable healthy diets—Guiding principles. Rome, pp. 37
Fawzy S, Osman AI, Doran J, Roony D (2020) Strategies for mitigation of climate change: a review. Environ Chem Lett 18:2069–2094. https://doi.org/10.1007/s10311-020-01059-w
Fuller DQ, Boivin N, Hoogervorst T, Allaby R (2011) Across the Indian Ocean: the prehistoric movement of plants and animals. Antiquity 85:544–558

Fuller DQ, Sato Y-I, Castillo C, Qin L, Weisskopf AR, KingwellBanham EJ, Song J, Ahn S-M, van Etten J (2010) Consilience of genetics and archaeobotany in the entangled history of rice. Archaeol Anthropol Sci 2(2):115–131

Gangwar S, Naik KR, Jha A, Bajpai A (2016) Soil properties as influenced by organic nutrient management practices under rice based cropping systems. Res Crops 17(1):8–12

Garcia SN, Osburn BI, Jay-Russell MT (2020) One health for food safety, food security, and sustainable food production. Front Sustain Food Syst 4:1. https://doi.org/10.3389/fsufs.2020.00001

Ghatak A, Chaturvedi P, Bachmann G, Valledor L, Ramšak Ž, Bazargani MM, Bajaj P, Jegadeesan S, Li W, Sun X, Gruden K, Varshney RK, Weckwerth W (2021) Physiological and proteomic signatures reveal mechanisms of superior drought resilience in pearl millet compared to wheat. Front Plant Sci 11:600278. https://doi.org/10.3389/fpls.2020.600278

Goron TL, Raizada MN (2015) Genetic diversity and genomic resources available for the small millet crops to accelerate a new green revolution. Front Plant Sci 6:157. https://doi.org/10.3389/fpls.2015.00157

Gupta A, Mahajan V, Kumar M, Gupta HS (2009) Biodiversity in the barnyard millet (*Echinochloa frumentacea* Link, Poaceae) germplasm in India. Genet Resour Crop 56:883–889

Haddaway NR, Hedlund K, Jackson LE, Kätterer T, Lugato E, Thomsen IK, Jørgensen HB, Isberg PE (2017) How does tillage intensity affect soil organic carbon? A systematic review. Environ Evidence 6:1–48. https://doi.org/10.1186/s13750-017-0108-9

Hao XY, Li P, Li HY, Zong YZ, Zhang B, Zhao JZ, Han YH (2017) Elevated CO_2 increased photosynthesis and yield without decreasing stomatal conductance in broomcorn millet. Photosynthetica 55:176–183

Harika JV, Maitra S, Shankar T, Bera M, Manasa P (2019) Effect of integrated nutrient management on productivity, nutrient uptake and economics of finger millet (*Eleusine coracana* L. Gaertn). Int J Agric Environ Biotechnol 12(3):273–279

Hatfield JL, Dold C (2019) Water-use efficiency: advances and challenges in a changing climate. Front Plant Sci 10:103. https://doi.org/10.3389/fpls.2019.00103

Hatfield JL, Boote KJ, Kimball BA, Ziska LH, Izaurralde RC, Ort D, Thomson AM, Wolfe D (2011) Climate impacts on agriculture: implications for crop production. Agron J 103(2):351–370. https://doi.org/10.2134/agronj2010.0303

Hemamalini C, Patro TSSK, Anuradha N, Triveni U, Jogarao P, Sandhya Rani Y (2020) Estimation of nutritive composition of seven small millets. J Pharmagcog Phytochem 9(3):1871–1875

Hossain A, Skalicky M, Brestic M, Maitra S, Ashraful Alam M, Syed MA, Hossain J, Sarkar S, Saha S, Bhadra P, Shankar T (2021) Consequences and mitigation strategies of abiotic stresses in wheat (*Triticum aestivum* L.) under the changing climate. Agronomy 11:241. https://doi.org/10.3390/agronomy11020241

Hunt HV, Vander Linden M, Liu X, Motuzaite-Matuzeviciute G, Colledge S, Jones MK (2008) Millets across Eurasia: chronology and context of early records of the genera *Panicum* and *Setaria* from archaeological sites in the old world. Veg Hist Archaeobot 17:5–18

IIMR (2021) Indian Institute of Millet Research. Nutritional benefits of millets (for 100g of each millet). https://www.millets.res.in/millets_info.php (Accessed 01 August, 2021)

IPCC, 2018: Global Warming of 1.5°C. An IPCC Special Report on the impacts of global warming of 1.5°C above pre-industrial levels and related global greenhouse gas emission pathways, in the context of strengthening the global response to the threat of climate change, sustainable development, and efforts to eradicate poverty, Masson-Delmotte V, Zhai P, Pörtner HO, Roberts D, Skea J, Shukla PR, Pirani A, Moufouma-Okia W, Péan C, Pidcock R, Connors S, Matthews JBR, Chen Y, Zhou X, Gomis MI, Lonnoy E, Maycock T, Tignor M, Waterfield T. (eds.), https://www.ipcc.ch/site/assets/uploads/sites/2/2019/06/SR15_Full_Report_High_Res.pdf (accessed 15 November, 2021)

Jagathjothi N, Ramamoorthy K, Kuttimani R (2011) Integrated nutrient management on growth and yield of rainfed direct sown finger millet. Res Crop 12:79–81

Jarret RL, Ozias-Akins P, Phatak S, Nadimpalli R, Duncan R, Hiliard S (1995) DNA contents in *Paspalum* spp. determined by flow cytometry. Genet Res Crop 42:237–242

Jemila C, Saliha BB, Udayakumar S (2017) Evaluating the effect of phosphatic fertilizers on soil and plant P availability and maximising rice crop yield. Oryza 54:305–313

Jensen ES, Peoples MB, Boddey RM, Gresshoff PM, Hauggaard-Nielsen H, Alves BJ, Morrison MJ (2012) Legumes for mitigation of climate change and the provision of feedstock for biofuels and biorefineries—a review. Agron Sustain Dev 32:329–364

Kane-Potaka J, Anitha S, Tsusaka T, Botha R, Budumuru M, Upadhyay S, Kumar P, Mallesh K, Hunasgi R, Jalagam AK (2021) Assessing millets and sorghum consumption behavior in urban India: a large-scale survey. Front Sustain Food Syst 5:260

Karmakar R, Das I, Dutta D, Rakshit A (2016) Potential effects of climate change on soil properties: a review. Sci Int 4:51–73. https://doi.org/10.17311/sciintl.2016.51.73

Kering MK, Broderick C (2018) Potassium and manganese fertilization and the effects on millet seed yield seed quality and forage potential of residual stalks. Agric Sci 09(07):888–900. https://doi.org/10.4236/as.2018.97061

Kingwell-Banham E, Fuller DQ (2014) Brown top millet: origins and development. Encyclopaedia of Global Archaeology. Springer, New York, pp 1021–1024

Krauss M, Ruser R, Müller T, Hansen S, Mäder P, Gattinger A (2017) Impact of reduced tillage on greenhouse gas emissions and soil carbon stocks in an organic grass-clover ley-winter wheat cropping sequence. Agric Ecosyst Environ 239:324–333

Kumar HV, Gattupalli N, Babu SC, Bhatia A (2020) Climate-smart small millets (CSSM): a way to ensure sustainable nutritional security. In: Venkatramanan V et al (eds) Global climate change: resilient and smart agriculture. Springer Nature Singapore Pte Ltd., pp 137–154. https://doi.org/10.1007/978-981-32-9856-9_7

Kumara O, Naik TB, Palaiah P (2007) Effect of weed management practices and fertility levels on growth and yield parameters in finger millet. Karnataka J Agric Sci 20:230–233

Kumaran G, Parasuraman P (2019) Effect of enriched FYM and *Panchagavya* spray on foxtail millet (*Setaria italica*) under rainfed conditions. Int J Chem Stud 7(2):2121–2123

Kuraloviya M, Vanniarajan C, Vetriventhan M, Babu C, Kanchana S, Sudhagar R (2019) Qualitative characterization and clustering of early maturing barnyard millet (*Echinochloa* spp.) germplasm. Elec J Plant Breeding 10:535. https://doi.org/10.5958/0975-928x.2019.00067.x

Lal R (1999) Soil management and restoration for C sequestration to mitigate the accelerated greenhouse effect. Prog Environ Sci 1:307–326

Lal R (2004) Soil carbon sequestration impacts on global climate change and food security. Science 304:1623–1627

Maitra S (2020a) Potential horizon of brown-top millet cultivation in drylands: a review. Crop Res 55(1–2):57–63. https://doi.org/10.31830/2454-1761.2020.012

Maitra S (2020b) Intercropping of small millets for agricultural sustainability in drylands : a review. Crop Res 55(3–4):162–171

Maitra S, Panda P, Panda SK, Behera D, Shankar T, Nanda SP (2020) Relevance of barnyard millet (*Echinochloa frumentacea* L) cultivation and agronomic management for production sustainability. Int J Bioinform Biol Sci 8:27–32

Maitra S, Pine S, Banerjee P, Shankar T (2022) Millets: robust entrants to functional food sector. In: Pirzadah TB, Malik B, Bhat A, Hakeem KR (eds) Bioresource technology: concept, tools and experiences. Wiley Online Library. https://doi.org/10.1002/9781119789444.ch1

Maitra S, Shankar T (2019) Agronomic management in little millet (*Panicum sumatrense* L.) for enhancement of productivity and sustainability. Int J Bioresour Sci 6:91–96

Maitra S, Sounda S, Ghosh DC, Jana PK (1997) Effect of seed treatment on finger millet (*Eleusine coracana*) varieties in rainfed upland. Ind J Agric Sci 67(10):478–480

Maitra S, Zaman A, Mandal TK, Palai JB (2018) Green manures in agriculture: a review. J Pharma Phytochem 7(5):1319–1327

Malviya KS, Bakoriya L, Kumar S, Aske S, Mahajan G, Malviya KD (2019) Effect of tillage and cultural practices on growth, yield and economics of *kodo* millet. Int J Curr Microbiol App Sci 8(06):890–895. https://doi.org/10.20546/ijcmas.2019.806.107

Mbow C, Rosenzweig C, Barioni LG, Benton TG, Herrero M, Krishnapillai M, Liwenga E, Pradhan P, Rivera-Ferre MG, Sapkota T, Tubiello FN (2019) Food security. In: Shukla PR, Skea J, Buendia EC, Masson-Delmotte V, Pörtner H-O, Roberts DC, Zhai P, Slade R, Connors S, van Diemen R, Ferrat M, Haughey E, Luz S, Neogi S, Pathak M, Petzold J, Pereira JP, Vyas P, Huntley E, Kissick K, Belkacemi M, Malley J (eds) Climate change and land: an IPCC special report on climate change, desertification, land degradation, sustainable land management, food security, and greenhouse gas fluxes in terrestrial ecosystems. https://www.ipcc.ch/site/assets/uploads/sites/4/2021/02/08_Chapter-5_3.pdf, accessed 15 November, 2021

Mekonnen MM, Hoekstra AY (2014) Water footprint benchmarks for crop production: a first global assessment. Ecol Indic 46:214–223. https://doi.org/10.1016/j.ecolind.2014.06.013

Miller NF, Spengler RN, Frachetti M (2016) Millet cultivation across Eurasia: origins, spread, and the influence of seasonal climate. The Holocene:1–10. https://doi.org/10.1177/0959683616641742

Miyan MA (2015) Droughts in Asian least developed countries: Vulnerability and sustainability, weather and climate extremes, 7:8–23. https://doi.org/10.1016/j.wace.2014.06.003

Monisha V, Rathinaswamy A, Mahendran PP, Kumutha K (2019) Influence of integrated nutrient management on growth attributes and yield of foxtail millet in red soil. Int J Chem Stud 7(3): 3536–3539

Nandini C, Bhat S, Reddy S, Jayramegowda P (2019) Modified crossing (SMUASB) method for artificial hybridization in proso millet (*Panicum miliaceum* L.) and little millet (*Panicum sumatrense*). Electron J Plant Breed 10(3):1161–1170

Navarro-Pedreño J, Almendro-Candel MB, Zorpas AA (2021) The increase of soil organic matter reduces global warming, myth or reality? Science 3:18. https://doi.org/10.3390/sci3010018

Niyogi D (2018) Millets back in our fields and plates, The Millennium Post. http://www.millenniumpost.in/opinion/millets-back-in-our-fields-and-plates-317237, (Accessed 19 December, 2021)

OECD (2009) Integrating climate change adaptation into development co-operation, policy guidance. OECD Publishing, ISBN 978–92–64-05476-9, p.193, Paris, France, https://www.oecd.org/env/cc/44887764.pdf (accessed 15 November 2021)

Oertel C, Matschullat J, Zurba K, Zimmermann F, Erasmi S (2016) Greenhouse gas emissions from soils—a review. Geochemistry 76(3):327–352

Parihar SK, Dwivedi BS, Khan IM, Tiwari RK (2010) Effect of integrated nutrient management on yield and economics of little millet. J Soils Crops 20(2):211–215

Parmentier S (2014) Scaling-up agroecological approaches: what, why and how. Oxfam-Solidarity, Brussels, pp 472–480

Patil KB, Chimmad BV, Itagi S (2015) Glycemic index and quality evaluation of little millet (*Panicum miliare*) flakes with enhanced shelf life. J Food Sci Technol 52(9):6078–6082. https://doi.org/10.1007/s13197-014-1663-5

Pilbeam CJ, Gregory PJ, Tripathi BP, Munankarmy RC (2002) Fate of nitrogen-15-labelled fertilizer applied to maize-millet cropping systems in the mid-hills of Nepal. Biol Fertil Soils 35:27–34

Prabudoss V, Jawahar S, Shanmugaraja P, Dhanam K (2014) Effect of integrated nutrient management on yield and nutrient uptake of transplanted Kodo millet. Eur J Biotechnol Biosci 1(5): 30–32

Pradhan A, Panda AK, Bhavani RV (2019) Finger millet in tribal farming systems contributes to increased availability of nutritious food at household level: insights from India. Agric Res 8: 540–547. https://doi.org/10.1007/s40003-018-0395-6

Pradhan ADJN, Panda AK, Wagh RD, Maske MRRVB (2021) Farming system for nutrition—a pathway to dietary diversity: evidence from India. PLoS One 16(3):e0248698. https://doi.org/10.1371/journal.pone.0248698

Prasad JVNS, Srinivasa RC, Srinivasa K, Naga Jyothia C, Venkateswarlub B, Ramachandrappa BK, Dhanapal GN, Ravichandra K, Mishra PK (2016) Effect of ten years of reduced tillage and recycling of organic matter on crop yields, soil organic carbon and its fractions in Alfisols of semi-arid tropics of southern India. Soil Till Res 156:131–139. https://doi.org/10.1016/j.still.2015.10.013

Prasanna Kumar D, Maitra S, Shankar T, Ganesh P (2019) Effect of crop geometry and age of seedlings on productivity and nutrient uptake of finger millet (*Eleusine coracana* L. Gaertn.). Int J Agric Environ Biotechnol 12(3):267–272

Prentice IC, Farquhar GD, Fasham MJ, Goulden ML, Heimann M, Jaramillo VJ, Kheshgi HS, Le Quéré C, Scholes RJ, Wallace DW, Archer D (2001) The carbon cycle and atmospheric carbon dioxide, 183–237, https://www.ipcc.ch/site/assets/uploads/2018/02/TAR-03.pdf (Accessed 12 November, 2021)

Ramya P, Maitra S, Shankar T, Adhikary R, Palai JB (2020) Growth and productivity of finger millet (*Eleusine coracana* L. Gaertn) as influenced by integrated nutrient management. Agron Econ 7:19–24

Rao BR, Nagasampige MH, Ravikiran M (2011) Evaluation of nutraceutical properties of selected small millets. J Pharm Biol Sci 3(2):277–279

Rao DB, Bhaskarachary K, Arlene Christina GD, Sudha Devi G, Tonapi VA (2017) Nutritional and health benefits of millets. ICAR_Indian Institute of Millets Research (IIMR), Rajendranagar, Hyderabad, p 112

Renganathan VG, Vanniarajan C, Karthikeyan A, Ramalingam J (2020) Barnyard millet for food and nutritional security: current status and future research direction. Front Genet 11. https://doi.org/10.3389/fgene.2020.00500

Roy AK, Ali N, Lakra RK, Alam P, Mahapatra P, Narayan R (2018) Effect of integrated nutrient management practices on nutrient uptake, yield of finger millet (*Eleusine coracana* L. Gaertn.) and post-harvest nutrient availability under rainfed condition of Jharkhand. Int J Curr Microbiol App Sci 7(08):339–347. https://doi.org/10.20546/ijcmas.2018.708.038

Saikishore A, Bhanu Rekha K, Hussain SA, Madhavi A (2020) Growth and yield of browntop millet as influenced by dates of sowing and nitrogen levels. Int J Chem Stud 8(5):1812–1815. https://doi.org/10.22271/chemi.2020.v8.i5y.10564

Saiz-Rubio V, Rovira-Más F (2020) From smart farming towards agriculture 5.0: a review on crop data management. Agronomy 10:207. https://doi.org/10.3390/agronomy10020207

Sakamoto S (1985) A preliminary repost of the studies on millet cultivation and its agro-pastoral culture complex in the Indian subcontinent. Studies on millet cultivation and its agro-pastoral culture complex in the Indian subcontinent, Kyoto University Research Team, Japan. pp.139

Saleh AS, Zhang Q, Chen J, Shen Q (2013) Millet grains: nutritional quality, processing, and potential health benefits. Compr Rev Food Sci Food Saf 12(3):281–295

Sandhya Rani Y, Triveni U, Patro TSSK, Divya M, Anuradha N (2017) Revisiting of fertilizer doses in finger millet (*Eleusine coracana* (L.) Garten.) through targeted yield and soil test crop response (STCR) approach. Int J Curr Microbiol App Sci 6(7):2211–2221

Sanjeevaiah SH, Rudrappa KS, Lakshminarasappa MT, Huggi L, Hanumanthaiah MM, Venkatappa SD, Lingegowda N, Sreeman SM (2021) Understanding the temporal variability of rainfall for estimating agro-climatic onset of cropping season over south interior Karnataka. India Agron 11:1135. https://doi.org/10.3390/agronomy11061135

Sankar GRM, Sharma KL, Dhanapal GN, Shankar MA, Mishra PK, Venkateswarlu B, Grace JK (2011) Influence of soil and fertilizer nutrients on sustainability of rainfed finger millet yieldand soil fertility in semi-arid Alfisols. Commun Soil Sci Plant Ann 42:1462–1483

Saxena R, Vanga SK, Wang J, Orsat V, Raghavan V (2018) Millets for food security in the context of climate change: a review. Sustainability 10:2228. https://doi.org/10.3390/su10072228

Selectstar Marwein B, Singh R, Chhetri P (2019) Effect of integrated nitrogen management on yield and economics of foxtail millet genotypes. Int J Curr Microbiol App Sci 8(08):2543–2546

Shobana S, Krishnaswamy K, Sudha V, Malleshi NG, Anjana RM, Palaniappan L, Mohan V (2013) Finger millet (Ragi, *Eleusine coracana* L.): a review of its nutritional properties, processing, and plausible health benefits. Adv Food Nutr Res 69:1–39

Singh VP, Mishra AK, Chowdhary H, Khedun CP (2014) Climate change and its impact on water resources. In: Wang L, Yang C (eds) Modern water resources engineering. Handbook of Environmental Engineering, vol 15. Humana Press, Totowa, NJ. https://doi.org/10.1007/978-1-62703-595-8_11

Sood S, Khulbe RK, Gupta AK, Agrawal PK, Upadhyaya HD, Bhatt JC (2015) Barnyard millet—a potential food and feed crop of future. Plant Breed 134:135–147

Struik PC, Kuyper TW (2017) Sustainable intensification in agriculture: the richer shade of green. A review. Agron Sustain Dev 37:39. https://doi.org/10.1007/s13593-017-0445-7

Sukanya TS, Chaithra C, Pratima NM (2021) Guni cultivation of finger millet: an indigenous practice for sustained productivity and scientific evaluation. Front Crop Improv 9:1000–1004

Swaminathan MS, Bhavani RV (2013) Food production & availability—essential prerequisites for sustainable food security. Ind J Med Res 138(3):383–391

Tadele Z (2016) Drought adaptation in millets. In: Shanker AK, Shanker C (eds) Abiotic and biotic stress in plants—recent advances and future perspectives. IntechOpen, London, pp 639–662

Thakur AK, Kumar P, Netam PS (2019) Effect of different nitrogen levels and plant geometry, in relation to growth characters and yield of browntop millet [*Brachiaria ramosa* (L.)] at Bastar Plateau Zone of Chhattisgarh. Int J Curr Microbiol App Sci 8(02):2789–2794. https://doi.org/10.20546/ijcmas.2019.802.327

Thesiya NM, Dobariya JB, Patel JG (2019) Effect of integrated nutrient management on growth and yield parameters of kharif little millet under little millet-green gram cropping sequence. Int J Pure App Biosci 7(3):294–298. https://doi.org/10.18782/2320-7051.7392

Tui SHK, Descheemaeker K, Valdivia RO, Masikati P, Sisito G, Moyo EN, Crespo O, Ruane AC, Rosenzweig C (2021) Climate change impacts and adaptation for dryland farming systems in Zimbabwe: a stakeholder-driven integrated multi-model assessment. Clim Chang 168:10. https://doi.org/10.1007/s10584-021-03151-8

Turral H (2008) Climate change, water and food security. Food and Agriculture Organization, Water Reports 36, Rome, Italy, p. 175

Venkatesh Bhat B, Dayakar Rao B, Tonapi VA (2018) The story of millets. (Ed). Karnataka State Department of Agriculture, Bengaluru and ICAR-Indian Institute of Millets Research, Hyderabad, India, pp. 110

Vetriventhan M, Azevedo VCR, Upadhyaya HD et al (2020) Genetic and genomic resources, and breeding for accelerating improvement of small millets: current status and future interventions. Nucleus 63:217–239. https://doi.org/10.1007/s13237-020-00322-3

Vetter SH, Sapkota TB, Hillier J, Stirling CM, Macdiarmid JI, Aleksandrowicz L, Green R, Joy EJ, Dangour AD, Smith P (2017) Greenhouse gas emissions from agricultural food production to supply Indian diets: implications for climate change mitigation. Agric Ecosyst Environ 16:234–241. https://doi.org/10.1016/j.agee.2016.12.024. PMID: 28148994; PMCID: PMC5268357

Wang A, Ma X, Xu J, Lu W (2019) Methane and nitrous oxide emissions in rice-crab culture systems of Northeast China. Aquacult Fish 4(4):134–141. https://doi.org/10.1016/j.aaf.2018.12.006

Wang J, Vanga SK, Saxena R, Orsat V, Raghavan V (2018) Effect of climate change on the yield of cereal crops: a review. Climate 6:41. https://doi.org/10.3390/cli6020041

Zaman K, Abdullah I, Ali M (2017) Decomposing the linkages between energy consumption air pollution climate change and natural resource depletion in Pakistan. Environ Prog Sustain Energy 36(2):638–648. https://doi.org/10.1002/ep.12519

Zambon I, Cecchini M, Egidi G, Saporito MG, Colantoni A (2019) Revolution 4.0: industry vs. agriculture in a future development for SMEs. PRO 7:36. https://doi.org/10.3390/pr7010036

Zeng Y, Liu D, Guo S, Xiong L, Liu P, Yin J, Tian J, Deng L, Zhang J (2021) Impacts of water resources allocation on water environmental capacity under climate change. Water 13:1187

Omics for Abiotic Stress Tolerance in Foxtail Millet

2

Mahender Singh Bhinda, Sanjay Kumar Sanadya, Asha Kumari, Laxmi Kant, and Ashim Debnath

Abstract

Panicoid grasses and crops include finger millet, switch grass, foxtail millet, napier grass, and pearl millet. Foxtail millet is an important C_4 panicoid crop with a small genome size (~490 Mb), a short life cycle, inbreeding characteristics, and great abiotic stress tolerance. It is an important food and fodder crop in arid and semiarid parts of Asia and Africa, including North China and India. *Setaria italica* (cultivated) and *Setaria viridis* (wild) are two well-known *Setaria* species that have served as ideal model systems for evolutionary, architectural, and physiological studies in related potential bioenergy. Foxtail millet has high genetic diversity, with numerous core and mini-core collections holding a wide range of genotypes. Significant phenotypic variations enable the discovery and mining of new abiotic stress tolerance variants for crop improvement. Several foxtail millet accessions may be drought and salinity tolerant, or rather exploiting their agronomic and stress-tolerant traits may be especially important for marker-assisted selection and genetic engineering. In addition, the release and availability of the foxtail millet genome sequence have revealed numerous of its unique properties, such as abiotic stress tolerance, which may aid in a better understanding of its evolution, stress physiology, and adaptation. The foxtail millet genome

M. S. Bhinda · A. Kumari (✉) · L. Kant
ICAR-Vivekananda Parvatiya Krishi Anusandhan Sansthan, Almora, Uttarakhand, India
e-mail: asha.kumari@icar.gov.in

S. K. Sanadya
Department of Genetics and Plant Breeding, College of Agriculture, Chaudhary Sarwan Kumar Himachal Pradesh Krishi Vishwavidyalaya, Palampur, HP, India

A. Debnath
Department of Genetics and Plant Breeding, College of Agriculture, Acharya Narendra Deva University of Agriculture and Technology, Ayodhya, Uttar Pradesh, India

R. N. Pudake et al. (eds.), *Omics of Climate Resilient Small Millets*,
https://doi.org/10.1007/978-981-19-3907-5_2

sequence thus aids not only in the identification and introgression of agronomically important traits but also in the deciphering of this exceptionally stress-tolerant crop's abiotic stress tolerance mechanisms, in addition to the evolution of climate-resilient crops, which is critical in this epoch of global climate change.

Keywords

Foxtail millet · Genetic diversity · Abiotic stress tolerance · Climate change

2.1 Introduction

Agriculture's advancement in some crop species was the most pivotal point in human history, and plant breeding plays a key role in supporting agricultural production by altering the genetic constituents of crop species in relation to their economic use. Systematic and planned studies on plant breeding began in the nineteenth century, following the rediscovery of Mendel's laws of inheritance. The most significant limitations of traditional breeding methods are that selection is based on phenotype, which produces misleading results because it is heavily influenced by the environment. However, environmental effects have been lessened through the discovery of molecular markers that aid in the selection of desirable traits in an indirect manner. After the 1990s, the emergence of recombinant DNA technology, DNA sequencing methodologies, and other advanced breeding techniques marked turning points in several omics fields (Fig. 2.1). The genomics toolbox contains linkage and physical maps, genome sequencing, bioinformatics software and databases, and new generations of omics tools like transcriptomics, proteomics, metabolomics, genomics, and high-throughput phenomics. Indeed, over the last few decades, it had made spectacular radiances in genomics research. For example, at the dawn of the genomics era, *Arabidopsis* was chosen as the first model for whole-genome sequencing, followed quickly by the sequencing of other model genomes (rice for monocots, *Medicago* and Lotus for legume crops, and poplar for tree species) and crop species (soybean, cassava, sorghum, maize, pigeon pea, wheat, finger millet, foxtail millet, etc.). Generated sequence data is being analyzed in parallel for characterizing genes and validating their functions using comparative and functional genomics approaches. This information can be applied to evolutionary significance and crop improvement programs. There are currently a plethora of high-throughput genome sequencing technologies and tools, omics platforms, and strategies available for using genomics in plant breeding. However, the high costs of omics research, both invested in and associated with it, currently limit the implementation of omics-based crop breeding, particularly for underutilized, potential, minor, and orphan crops.

Foxtail millet (*Setaria italica*) is a small diploid C_4 crop species whose genome has been sequenced collaboratively by the Joint Genome Institute (JGI), Berkeley, California, USA, and the Beijing Genome Initiative (BGI), China (Singh et al.

Fig. 2.1 Historical timeline of plant breeding with respect to Omics

2017). It serves as a model for other biofuel grasses such as napier grass, switch grass, and pearl millet. The foxtail millet's final drought genome size is (490 Mb), fastened onto nine chromosomes, and functional 38,801 genes are annotated. Because of its small genome size, short life cycle, and inbreeding nature, foxtail millet is an excellent model crop for studies in plant nutritional biology, stress response studies, and C_4 metabolism research.

2.2 Characterization of Abiotic Stress Tolerance Mechanism in Foxtail Millet

Drought, salt, water logging, high temperatures, and mineral toxicity all limit growth and development of plants, resulting in potential yield losses of more than half of major agricultural crop production. Positive feedback and tolerance to survive and reproduce under such stressful conditions is a complicated performance that is largely coordinated and maintained by correcting crops' physiological, cellular, and molecular activities (Chinnusamy et al. 2004; Ahuja et al. 2010). As a result, addressing intricate regulatory pathways and the biological theories governing abiotic stress tolerance in plants necessitates a thorough understanding of genome-scale metabolic responses at critical growth stages of development regarding external triggers. Advanced technologies, particularly omics approaches, are of top priority and promising approaches for improving foxtail millet abiotic stress tolerance (Fig. 2.2).

Omics refers to all a molecular biology applications that require a molecular understanding of stress response, viz., genomics, functional genomics, genetic engineering, gene expression, proteomics or metabolomics(s), and to understand their ultimate phenotypic consequences in response to environmental changes, which are usually followed by appropriate modifications in the crop transcriptome, proteome, or metabolome (Ahuja et al. 2010). Because of its dwarf genome size and diploidy, foxtail millet has been studied and characterized more thoroughly than other millet crops under abiotic stress conditions. Genome-wide studies and expression profiling studies on foxtail millet are underway. Furthermore, proteomic and metabolomic studies on millets in response to abiotic stress are still far behind. Gene

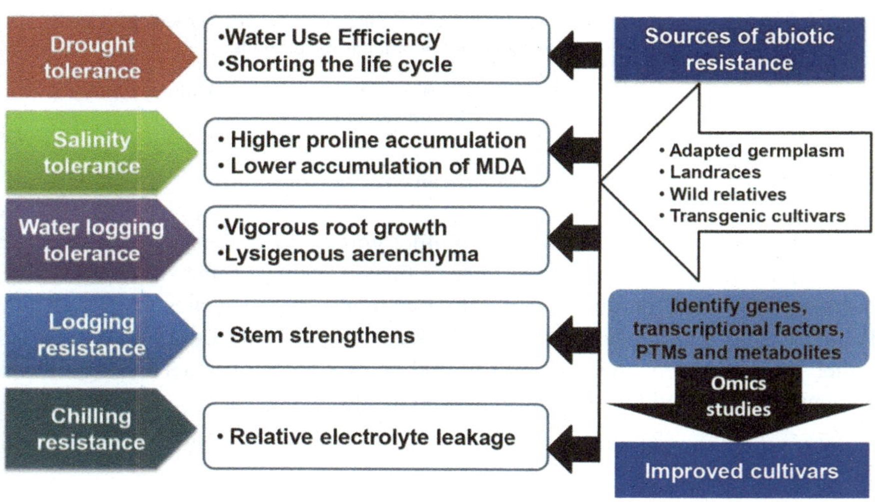

Fig. 2.2 Abiotic stress resistance responses to develop improved foxtail millet cultivars through different omics research

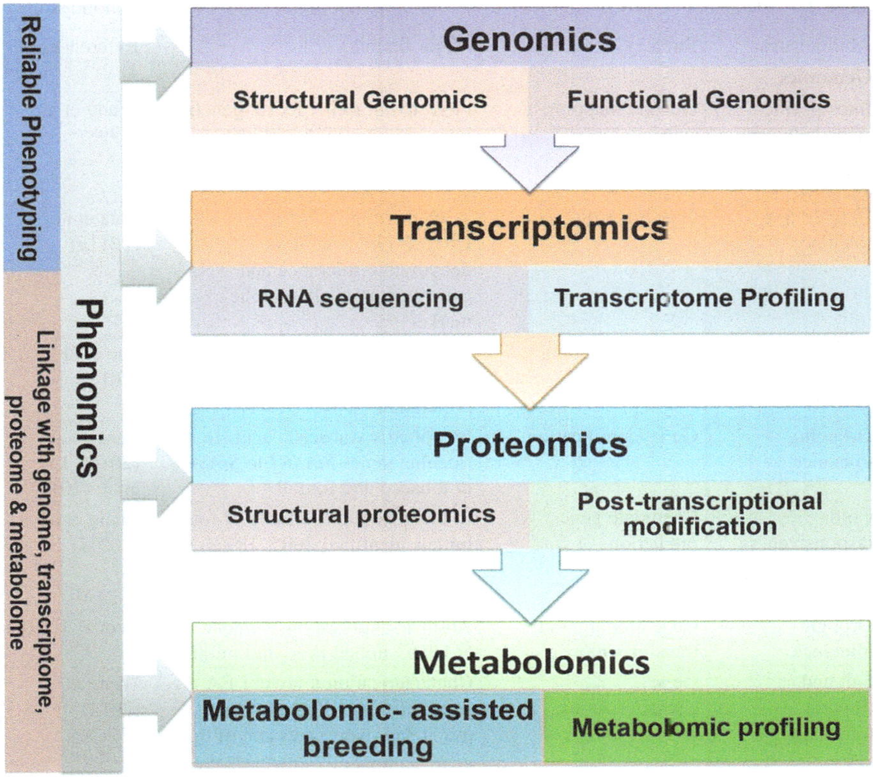

Fig. 2.3 Flow diagram showing different omics strategies

modification, including regulatory genes, has proven to be more responsive than using mono or polygenes for improving abiotic stress tolerance in millets. As a result, recent advances in "omics" technology related to genome sequences have created a plethora of opportunities for increasing foxtail millet crop stress tolerance against abiotic stressors (Fig. 2.3) (Table 2.1).

2.2.1 Genomics Studies in Foxtail Millets to Understand Abiotic Stress

The study of the structure and function of a living organism's genome is known as genomics. The foxtail millet is an excellent model plant for a wide range of plant nutritional biology, stress-responsive investigations, and C_4 metabolism research (Zhang et al. 2012). Because of its small genome size (490 Mbp), small plant size, and quick generation time, the foxtail millet genome sequencing provides a platform for organizing knowledge on various millets and comprehending the shared and independent features of C_4 plant evolution. Characterization of plant genomes and

Table 2.1 Details of "OMICS" studies carried out in foxtail millet under different abiotic stresses

Abiotic stress	Strategy	Major findings	References
Genomics			
Interspecific cross between *S. italica* x *S. virdis*	Genetic mapping	First foxtail millet RFLP genetic map and identified 160 loci	Wang et al. (1998)
Dehydration tolerance	Gene identification	A synonymous SNP in a novel DREB2-like gene linked to dehydration tolerance and generated an allele-specific marker	Lata et al. (2011a)
Drought tolerance	Linkage mapping	Identified 18 QTLs for a number of drought-tolerant traits in RILs population	Qie et al. (2014)
Drought tolerance	Gene identification	Identified six abscisic acid stress ripening genes *SiASR1* to *SiASR6* in tobacco and foxtail	Feng et al. (2016); Li et al. (2017)
Cold responsiveness	Candidate gene prediction	In the genomic assembly of four millets identified 2605, 1872, 4888, and 3148 cold-responsive genes	Meng et al. (2021)
Drought tolerance	Gene cloning and transformation	Abscisic acid-responsive gene *SiARDP* cloned in foxtail millet	Li et al. (2014)
Salt and drought tolerances	Gene characterization	Characterization a novel LEA gene (*SiLEA14*) from foxtail millet and overexpression of this gene significantly enhances the salt and drought tolerances of foxtail millet	Wang et al. (2014)
Numerous environmental stresses	Gene characterization	Characterization two genes (*SiNRT1.11* and *SiNRT1.12*) for NRT and one gene for ammonium transporters (*SiAMT1.1*) that accelerates N acquisition by upregulating these genes expression	Nadeem et al. (2018)
Transcriptomics			
Drought tolerance	Illumina HiSeq 2000 platform	Transcriptome analysis of 14-day old var. Yugu1 seedlings exposed to 20% polyethylene glycol (PEG) 6000 stress at different time points using shoot tissues	Qi et al. (2013)
Salinity tolerance	Suppression subtractive hybridization (SSH) and dot blot array	Comparative transcriptome analysis of contrasting foxtail millet cultivars under short-term salinity stress (250 mM NaCl; 6 h) using whole seedling	Puranik et al. (2011)

(continued)

Table 2.1 (continued)

Abiotic stress	Strategy	Major findings	References
Drought tolerance	SSH and dot blot array	Comparative transcriptome analysis of a tolerant foxtail millet cv. Prasad at early (0.5 h) and late (6 h) time points of20% PEG-induced drought stress using whole seedling	Lata et al. (2010)
Salinity tolerance	cDNA-AFLP	Comparative cDNA-AFLP analysis between contrasting foxtail millet cultivars differing for salinity tolerance using leaves	Jayaraman et al. (2008)
Drought tolerance	SSH and microarray	Comprehensive transcriptome analysis of a tolerant foxtail millet cv. Mar1 under drought stress (20% PEG)using root and shoot tissues	Zhang et al. (2007)
Salinity tolerance	cDNA macroarray	A barley cDNA macroarray used for monitoring transcript abundance in contrasting foxtail millet seedlings exposed to long-term high salinity stress using whole seedling	Sreenivasulu et al. (2004)
Proteomics			
Drought tolerance	Expression analysis	Identified many differentially expressed genes (DEGs) in foxtail millet using SSH and microarray	Zhang et al. (2007)
Salinity tolerance	Transcript profiling	Identified salt-responsive differentially expressed transcripts (DETs) using the cDNA-AFLP approach in two foxtail millet germplasm	Jayaraman et al. (2008)
Dehydration tolerance	Gene identification	Revealed 86 upregulated genes in the resistant foxtail millet cultivar 'Prasad'	Lata et al. (2010)
Salinity tolerance	Comparative transcriptomics	Discovered 159 differentially expressed transcripts (DETs)	Puranik et al. (2011)
Drought tolerance	Transcriptome profiling	Discovered 2824 drought-responsive genes with variable expression patterns.	Qi et al. (2013)
Drought tolerance	Transcript profiling	F-box protein was discovered to be implicated in drought and ABA response	Yin et al. (2014)
Drought tolerance	In silico study	Identified 2297 putative TFs and categorized them in 55 families in foxtail millet.	Bonthala et al. (2014)
Drought tolerance	Transcriptome profiling	Discovered 2 TFs namely SiAREB1 and SiAREB2 were able to bind *SiARDP* gene	Li et al. (2017)

(continued)

Table 2.1 (continued)

Abiotic stress	Strategy	Major findings	References
Drought tolerance	Gene expression profiling and cluster analysis	20 potential candidate genes were identified and modulate the genotype-specific drought stress responses of foxtail millet	Tang et al. (2017)
Drought tolerance	Transcriptomic analysis	Revealed 17 GO enrichments and 14 KEGG pathways and RNA sequencing analysis identified numerous DEGs	Shi et al. (2018); Pan et al. (2020)
Abiotic stresses	Transcriptomic analysis	Explained overexpression of glutathione-S-transferase (GST)/ glutathione peroxidase (GPX) from different species to enhance stress tolerance toward abiotic stresses through assisting peroxide scavenging and oxidative stress metabolism	Kumar and Trivedi (2018); Pan et al. (2020)
Salinity tolerance	Transcriptomic analysis	Identified CAMTA6 that is a key negative regulator of most salt-responsive genes and ion homeostasis during early germination	Shkolnik et al. (2019); Pan et al. (2020)
Salinity tolerance	Transcript profiling	Identified five R2R3 MYB TFs which increased in DEGs that responsible for anthocyanin biosynthetic genes, and further contribute to flavonoids elevating and salt resistance	Pan et al. (2020)
Salinity tolerance	Expression analysis	Identified five PIN genes and six IAA genes which can reduce auxin accumulation and growth inhibition in the root	Pan et al. (2020)
Drought tolerance	Transcript profiling	Evaluation of different physiological parameters under drought condition and expression of genes *SiP5CS1* and *SiP5CS2* in foxtail millet and results revealed upregulation of *SiP5CS1* and *SiP5CS2* under drought stress conditions	Qin et al. (2020)
Proteomics			
Salinity tolerance	2D electrophoresis	7-day old salt-treated tolerant cultivar "Prasad" seedlings to perform a proteome study, which led to the identification of 29 distinct expressed proteins (DEPs)	Canovas et al. (2004)

(continued)

Table 2.1 (continued)

Abiotic stress	Strategy	Major findings	References
Drought tolerance	qProteomic analysis	Identified numerous DEPsin foxtail millet seedlings and some of proteins exhibited significant expression changes. The qRT-PCR was also carried out to analyze the transcription levels of 21 DEPs	Pan et al. (2018)
Metabolomics			
Salinity tolerance	Metabolomic profiling analysis	Foxtail roots showed significantly upregulation of 17 flavonoids synthesis genes and also obtained 27 flavonoids metabolites	Pan et al. (2020)

the genes contained within them will aid geneticists and molecular biologists in their quest to understand C_4 metabolism physiology and will help plant breeders in their goal of developing better products. It serves as a reference genome for molecular biological studies in the cereals, for the following reasons:

(i) Availability of complete genome sequences of both genotypes, Yugu1 and Zhang gu (Bennetzen et al. 2012; Zhang et al. 2012), which enabled investigations of transcriptome activity in a range of tissues and developmental; stages.
(ii) Large genetic resources for Millet Genome Database (https://db.cngb.org/datamart/plant/DATApla6/).
(iii) Identification of candidate stress-responsive genes (Sood and Prasad 2017).
(iv) Well-established genetic transformation techniques (Sood et al. 2020).
(v) Strong genome databases facilitating access and depositing of information (Bonthala et al. 2013).
(vi) The high-resolution mapping of epigenetic modifications for chromosomes (Ni et al. 2017; Wang et al. 2017).
(vii) The genome-wide identification of genetic variation in gene expression between different *Setaria* species (Jaiswal et al. 2019).

Among the most important genomic resources for genetic studies and molecular breeding approaches are molecular markers, genetic maps, and sequence information. As a result, genomic research is needed to develop molecular markers and identify quantitative trait loci (QTLs), as well as to use marker-assisted breeding to develop improved cultivars. Furthermore, these genetic resources aid crop development efforts by reducing the time and labor-intensive direct screening of germplasm cultivated in field and greenhouse conditions. Furthermore, genetic maps and molecular markers are used to help with comparative genomic research in agricultural plants. The foxtail millet genome sequences, as well as genetic and physical maps, are valuable tools for the discovery and characterization of genes and QTLs for a wide range of agronomic traits, which can help with marker-aided breeding (MAB) and crop improvement. Wang et al. (1998) identified 160 loci in the first foxtail millet genetic map based on restriction fragment length polymorphism (RFLP) markers in an inter-varietal cross. Since then, numerous researchers and scientific communities have been involved in the development of linkage maps, as well as the identification of numerous genes in foxtail millet using genomic or EST-based molecular markers for further use in genotyping, phylogenetic relationship, and transferability studies (Jia et al. 2007; Jia et al. 2009; Heng et al. 2011; Gupta et al. 2012; Gupta et al. 2013). Using referenced-based foxtail millet genome assembly and de novo assembly, Xu et al. (2013) created new 9576 and 7056 microsatellites, respectively. Several other researchers used Setaria genome sequence data to identify microsatellite repeat motifs (Pandey et al. 2013; Zhang et al. 2014), eSSRs (Kumari et al. 2013), and TE-based markers (Kumari et al. 2013; Yadav et al. 2014a). Gupta et al. (2011) and Muthamilarasan et al. (2014a) generated 98 and 5123 intron length polymorphic (ILP) markers in foxtail millet, respectively, and demonstrated the utility of ILP markers in cross-species transferability, germplasm characterization, comparative mapping, and genomic studies among millets or non-millet crop plants, respectively. Another noteworthy study produced 176 microRNA (miRNA)-based molecular markers from 335 mature foxtail millet miRNAs (Khan et al. 2014, Yadav et al. 2014b). Lata et al. (2011a) identified a synonymous SNP in a novel DREB2-like gene associated with dehydration tolerance and developed an allele-specific marker (ASM). The ASM was also validated in a core set of over 100 foxtail millet accessions, with regressions of lipid peroxidation (LP) and relative water content (RWC) on the ASM revealing that the SiDREB2-associated trait accounted for roughly 27% and 20% of total variation, respectively. As a result, the significance of this QTL for foxtail millet dehydration tolerance is demonstrated (Lata and Prasad 2011; Lata and Prasad 2013).

Furthermore, Jia et al. (2013) developed a haplotype map using 916 diverse foxtail millet accessions that yielded a total of 2,584,083 SNPs, of which 845,787 SNPs (minor allele frequency > 0.05) were responsible for haplotype map. In genomics-assisted crop breeding, these marker resources are used efficiently from time to time. Other than these markers, amplified fragment length polymorphism (AFLP) inter simple sequence repeat (ISSR), and orthologous microsatellite markers have been employed to analyze genetic relationships, investigate domestication-related geographical structure, and phylogenetic relationships between foxtail millet and wild relatives (Le Thierry d'Ennequin et al. 2000; Li and Brutnell 2011; Zhang et al. 2012). In addition to these, these are also utilized to evaluate genetic diversity, population structure, and linkage disequilibrium (LD) in various foxtail accessions (Li and Brutnell 2011; Vetriventhan et al. 2014) as well as for examining important agronomic traits (Mauro-Herrera et al. 2013; Gupta et al. 2014). These markers and markers associated studies if once validated, might be used to identify new genes/QTLs linked to traits of interest and, consequently, in the marker-assisted breeding in foxtail millet and related crops (Muthamilarasan and Prasad 2015). Qie et al. (2014) identified QTLs for a number of traits namely promptness index, radical root length, coleoptile length, lateral root number at germinating stage, and seedling survival rate in 190 F_7 RILs population derived from foxtail millet cultivar (Yugu1) x wild *Setaria viridis* genotype. A total of eighteen QTLs were detected in nine linkage groups and among them, nine QTLs explained more than 10% of phenotypic variance for a given trait. Feng et al. (2016) and Li et al. (2017) identified six abscisic acid stress ripening genes *SiASR1* to *SiASR6* in tobacco and foxtail millet which enhanced drought and oxidative tolerance in plants through decreased H_2O_2 accumulation and improved antioxidant enzyme activities. Meng et al. (2021) conducted research to predict the cold responsiveness in different cold-tolerant grass species namely foxtail, proso, pearl millet, and switchgrass. In genomic assembly of four species identified 2605, 1872, 4888, and 3148 cold-responsive genes, among them 1154, 992, 1761, and 1609 cold-responsive genes showed comparative syntenic relationships. Non-syntenic genes, on the other hand, responded to cold stress in a more varied way across species than syntenic genes. Chen et al. (2021) identified ARF gene family which participated in stress tolerance in foxtail millet. Twenty-four *SiARF* genes were identified using GWAS and their expression not only revealed their potential functions in stress response but also indicates their functional redundancy (Chen et al. 2021). Stomatal density and canopy temperature are the most important traits that could be reduced to improve water use efficiency (WUE) in C_4 crops. Therefore, quantitative trait loci (QTL) mapping was performed in RILs population derived from a *Setaria italica* × *Setaria viridis* cross under control and drought conditions and identified multiple QTLs that highlight the potential of Setaria as a model to understand the stress biology in C_4 species (Prakash et al. 2021).

2.2.2 Functional Genomics

Extensive research using functional genomics techniques like overexpression, gene silencing, insertional mutagenesis, and targeted induced local lesion in genome (TILLING) has been critical in refining our understanding of complex regulatory networks involved in stress response, adaptability, and tolerance in plants throughout the post-genomic era. Various omics technologies are also generating massive data sets in millets, which could be used to identify important candidate genes for crop development projects utilizing MAS or transgenic technology. Hundreds of discovered genes or proteins still need to be functionally verified, despite the advent of genome sequencing programs and proteomics. Various reverse genetics methods will be essential in these instances. Overexpression of a gene in a homologous or heterologous system via a constitutive or stress-inducible promoter is a useful method for determining gene function. Knocking down or reducing the expression of a gene is another important method for determining its function. Transgenomics, often known as transgenic technology, is a gene-based functional genomics method that can provide crucial information on the regulatory mechanisms that support plant stress tolerance. Transgenomics in millets is still neglected, despite the economic and nutritional importance.

The majority of candidate genes functionally validated in *Arabidopsis* or tobacco in the case of millet crops are transcription factors (TFs), which can influence the expression of downstream stress-sensitive genes and thus play an important role in stress tolerance. Because of its close similarity to possibly bioenergy grasses, the foxtail millet is an ideal candidate for genetic change among millet crops. A novel gene from foxtail millet namely *SiPf40* that is the member of *ZIP* gene family was characterized (Liu et al. 2009) and revealed that its overexpression causes vessel enlarging and xylary fibers enhancing with significantly decreased IAA content that could be enhanced cold tolerance in foxtail millet in lesser extent. Despite the fact that *Agrobacterium*-mediated transformation is becoming more common, there has only been one report of improved abiotic stress tolerance in foxtail millet. The SiARDP overexpression in foxtail millet improved drought tolerance, whereas constitutive expression of SiARDP in *Arabidopsis thaliana* improved drought tolerance throughout seed germination and seedling development. The expression levels of target genes of SiARDP were upregulated in transgenic Arabidopsis and foxtail millet. These finding revealed that SiARDP, one of the genes targeted by SiAREB, is involved in ABA-dependent signalling pathways, and plays a critical role in the abiotic stress response in plants (Li et al. 2014). Gene silencing is a well-known reverse genetics approach for discovering the function of a known gene in an organism that produces morphological changes. As a result, virus-induced gene silencing (VIGS) with dsRNA has emerged as a valuable and simple functional genomics method (Baulcombe 2004). VIGS works by utilizing viral vectors that contain a target region of the host gene to decrease the expression of target plant genes. The foxtail millet has been chosen as a high-throughput platform for virus-induced gene silencing (VIGS) to anticipate gene inhibition findings due to its proximity to a number of bioenergy grasses (Doust et al. 2009). The VIGS technique for functional assessment of cell wall associated genes in foxtail millet was developed by Ramanna et al. (2013) using the Brome mosaic virus as a vector. Late embryogenesis abundant (LEA) proteins played important roles in protecting higher plants from by environmental stresses. Wang et al. (2014) characterized a novel LEA gene (*SiLEA14*) from foxtail millet which was expressed in roots, stems, leaves, inflorescences, and seeds at different levels under normal growth conditions. Overexpression of this gene significantly enhances the salt and drought tolerances of foxtail millet due to responsiveness to ABA, PEG, and NaCl. Furthermore, *SiLEA14* could be used in crops genetic engineering with the aim of improving stress tolerance. Nitrogen (N) is the macronutrient and is one of the most limiting factors in agricultural plant production which deficiency or excess tends to slow down plant growth or to produce toxicity that indirectly invites drought vulnerability in crop. Nitrogen is obtained in the form of nitrate and ammonium, which are regulated by nitrate redistribution transporters (NRT) and ammonium transporters (AMT) (AMT) (Loqué and von Wirén 2004; Nadeem et al. 2020). Recently research was conducted to identify the functional genes that respond to NRT and AMT in foxtail millet genotypes. Nadeem et al. (2018) characterized two genes (*SiNRT1.11* and *SiNRT1.12*) for NRT and one gene for ammonium transporters (*SiAMT1.1*) and reported that N acquisition was accelerated by upregulating these genes expression.

Muthamilarasan et al. (2020) performed expression profiling in the Setaria model species namely *Setaria italica* and *Setaria viridis* for distinct stress specific genes which were subjected to abiotic stresses showed upregulation of *SiαCaH1*, *SiβCaH5*, *SiPEPC2*, *SiPPDK2*, *SiMDH8*, and *SiNADP-ME5* in the tolerant cultivar. Overexpression of *SiNADP-ME5* in yeast in dehydration stress conditions highlights the putative role in conferring dehydration tolerance. Methylation on nitrogenous bases of DNA may also exhibit stress tolerance. Therefore, research had conducted by Hao et al. (2020) to analyze the differences of millet seedlings under osmotic stress using qRT-PCR, bisulfate sequencing PCR, DNA methyltransferase (DNMT). Findings revealed that osmotic stress induced the transcriptional expression and activity of key enzymes in H_2S biosynthesis which may improve osmotic stress tolerance of foxtail millet by mediating DNA methylation. Phosphorus (P), absorbed from soil solutions, is a limiting nutrient for plant growth and yield. PHT1 family genes seem to play a key role for low Pi stress tolerance in foxtail millet. Therefore, twenty genotypes of foxtail millet with contrasting degree of growth and Pi uptake responses under low Pi (LP) and high Pi (HP) supply were screened and identified *SiPHT1* genes in shoot tissues of the LP best-performing genotypes (LPBG) (Roch et al. 2020). Gene families namely LEA, DREB, PIP, HSP, and P5CS played crucial roles in salt tolerance and dehydration tolerance. Therefore, da Cunha Valença et al. (2020) exploited qPCR analysis to evaluate the expression of selected genes in roots of *Setaria virdis* genotypes on three key sampling days (3, 6, and 10) under the stress conditions and stated that all the genes were significantly upregulated on day 3 (da Cunha Valença et al. 2020). Gene family containing CCT motif played a key role in photoperiod-regulation flowering pathway that can enhance multiple stress resistance. Studied phylogenetic relationship among foxtail millet, sorghum, and maize for CCT motif genes. Real-time PCR analysis showed that the expression level of *SiCCT* gene was higher in leaf than in young panicle and leaf sheath that can exhibit stress tolerance in foxtail millet. PTI1 (Pto-interacting 1) protein kinase proteins participate in a variety of abiotic stress and defense responses against pest or pathogens in several plant species. Huangfu et al. (2021) identified 12 *SiPTI1* genes from the foxtail genome using qPCR analysis after induction of multiple stresses. Heterologous expression of *SiPTI1–5* in yeast and *E. coli* enhanced tolerance to salt stress (Huangfu et al. 2021). Plants regulate salt uptake via calcineurin B-like proteins (CBLs). Yan et al. (2021) had done expression analysis and revealed that overexpression of *SiCBL5* increased its salt tolerance in foxtail millet through lower accumulations of Na^+ and stronger Na^+ efflux but its knockdown increased salt hypersensitivity.

2.2.3 Transcriptomics

The term "transcriptome" refers to an organism's whole set of transcripts (mRNA), whereas "transcriptomics" refers to the study of these mRNAs. The transcriptome approach, in and of itself, provides for a comprehensive analysis and quantification of abiotic stress-induced alterations at the whole-organism level The use of genome-wide expression profiling to discover candidate genes involved in a variety of biological processes and stress regulation networks has been proven to be beneficial. Transcriptome research is crucial for the development of molecular markers such as microsatellites and single nucleotide polymorphisms (Lata 2015). When whole-genome sequencing is not possible, transcriptome sequencing, or RNA-Seq, becomes more important. It provides cost-effective, functionally relevant information and the production of a large number of genetic resources in a reasonable amount of time as compared to traditional approaches. Differential gene expression, isoforms or allele-specific expression, regulatory element characterization, alternative splicing, RNA editing, population genetics, system biology, etc. are areas that may be benefited from the information generated using transcriptomics analysis (Arya et al. 2015). Sreenivasulu et al. (2004) used a microarray filter with 711 cDNA spots suggesting 620 unigenes from a barley EST collection to identify many differentially expressed transcripts in salt-treated tolerant and sensitive foxtail millet seedlings. Following that, Zhang et al. (2007) used suppression subtractive hybridization and cDNA microarray methods to identify many drought-induced differentially expressed genes in foxtail millet. The expression analysis of these genes revealed that in response to drought stress, different sets of genes are activated in the roots and shoots of foxtail millet, with most of them contributing to the protein degradation pathway. Jayaraman et al. (2008), on the other hand, conducted the first comprehensive study to identify salt-responsive differentially expressed transcripts using the cDNA-AFLP approach in two foxtail millet cultivars, tolerant and sensitive to salinity tolerance, and discovered 90 transcripts derived fragments. Lata et al. (2010) discovered 86 upregulated genes in the resistant foxtail millet cultivar "Prasad" under polyethylene-induced dehydration stress. Puranik et al. (2011) discovered 159 differentially expressed transcripts in a comparative transcriptome study of salinity tolerant and sensitive cultivars in order to better understand the salt tolerance pathway. Deep sequencing was used to analyze the whole transcriptome of foxtail millet by Qi et al. (2013), who discovered 2824 drought-responsive genes with variable expression patterns. Up to 48.23% and 51.77% of the genes were up- and downregulated, respectively. The most frequently upregulated proteins were LEA proteins, dehydrins, HSPs, aquaporins, and phosphatase 2C, indicating a possible role in foxtail millet drought tolerance. Yin et al. (2014) performed transcript profiling on drought-stressed foxtail millet as well. In this study, an F-box protein was discovered to be involved in drought and ABA response. Apart from the aforementioned, thorough genome-wide analyses and expression profiling examinations of numerous TF families such as AP2/ERFs, NAC, C_2H_2 zinc finger, and MYB (Lata et al. 2014; Puranik et al. 2013; Muthamilarasan et al. 2014b, c) and stress-responsive genes such as WD40, DCL, AGO, and RDR polymerases and ALDH (Mishra et al. 2014; Yadav et al. 2015; Zhu et al. 2014) have also been found in foxtail millet. Studies These were conducted to characterize gene regulatory networks and molecular cross linking for abiotic stress responses in foxtail millet, as well as to analyze gene families involved in stress tolerance, recognize relevant candidate genes, and identify relevant candidate genes.

Bonthala et al. (2014) undertaken in silico study and identified 2297 putative TFs and categorized them in 55 families in foxtail millet genome. Lata et al. (2011b) identified a DREB2-like TF (SiDREB2) that is associated with drought tolerance in foxtail millet accessions that can also be useful for other millet species for the purposes of functional complementation of orthologous genes. Similarly, another abscisic acid (ABA)-responsive DREB-binding protein gene (*SiARDP*), cloned from foxtail millet that increases tolerance to drought and high salinity stress (Li et al. 2014). Furthermore, also found that two TFs namely SiAREB1 and SiAREB2 were able to bind *SiARDP* (Li et al. 2017). Tang et al. (2017) studied phenotypic and transcriptomic differences between two foxtail millet drought-sensitive varieties (An04 and Yugu1). Gene expression profiling and a cluster analysis suggested that transcriptomic regulatory mechanisms controlling stress responses in these cultivars were significantly affected by genotype x environment interactions. After combining the RNA sequencing data with previously mapped QTLs (Qie et al. 2014), 20 potential candidate genes were identified. A functional enrichment analysis indicated that genes involved in phytohormone metabolism and signalling, TFs, cellular detoxification, and osmotic adjustment may play important roles in modulating the genotype-specific drought stress responses of foxtail millet. Drought-resistant foxtail millet F_1 hybrid (M79) and its parental lines E1 and H1 were subjected to transcriptomic analysis by Shi et al. (2018) and revealed 17 Gene Ontology (GO) enrichments and 14 Kyoto Encyclopedia of Genes and Genomes (KEGG) pathways in hybrid, including photosystem II (PSII) complex, peroxidase activity, plant hormone signal transduction, and chlorophyll biosynthesis. RNA sequencing analysis identified 5258 DEGs between M79 and E1, and 3594 between M79 and H1, indicating that the drought-tolerant cultivar M79 and its parental lines had different transcriptional profiles. These results indicate that various molecular pathways, including photosynthesis, respond to drought stress in M79, and provide abundant molecular information for further analysis of the underlying mechanism responding to this stress. Similarly, recently Pan et al. (2020) employed transcriptomics in foxtail millet cultivars Yugu2 and An04, respectively, and results revealed that 8887 and 12,249 DEGs identified in both cultivars in response to salinity, respectively, and 3149 overlapped between two varieties. Shkolnik et al. (2019) reported that calmodulin-binding transcription activator 6 (CAMTA6) is a key negative regulator of most salt-responsive genes and ion homeostasis during early germination. Recently, in foxtail millet, Pan et al. (2020) identified one CAMTA gene (*Si000282m.g*), and its decrease also confirmed that Yugu2 is less sensitive to salt. Kumar and Trivedi (2018) explained that overexpression of glutathione-S-transferase (GST)/ glutathione peroxidase (GPX) from different species enhances tolerance toward abiotic stress through assisting peroxide scavenging and oxidative stress metabolism. Therefore, a study was conducted by Pan et al. (2020) to check the expression of GST/GPX in foxtail cultivar and reported higher expression levels of GST, GPX, and peroxidases (POD) possessing stronger antioxidant capacity. Flavonoids produced in foxtail germplasm act as the main anti-salinity effectors which are regulated by a series of transcription factors (TFs). Pan et al. (2020) identified five R2R3 MYB TFs which increased in differentially expressed genes (DEGs) YG2TvsAn04T and YG2TvsYG2C. The identified DEGs of TFs might be regulators for activating the transcription of the anthocyanin biosynthetic genes, and further contribute to flavonoids elevating and salt resistance improvement of foxtail millet. Salt stress inhibits root meristem development by stabilizing IAA17 and reducing PINFORMED (PIN) gene expression, lowering auxin levels and suppressing auxin signalling, respectively ((Liu et al. 2015). Pan et al. (2020) identified five PIN genes and six IAA genes were also evidently repressed in both foxtail millet cultivars An04 and Yugu2, which can result in reduction of auxin accumulation and growth inhibition in root. Stress-induced DELLA accumulation could hold back growth and enhance the tolerance by upregulation of ROS detoxification enzymes under stress. The higher expression level of DELLA genes (Si009972m.g and Si000959m.g) with higher expression level of ROS scavenging related genes conferred that Yugu2 possessing higher salt tolerance than An04. Qin et al. (2020) researched to evaluate different physiological parameters under drought conditions and expression of Δ1-pyrroline-5-carboxylate synthetase genes (*Si*P5CS1 and *Si*P5CS2) in foxtail millet cultivar, Jigu 16. RNA sequencing analysis results revealed upregulation of *Si*P5CS1 and *Si*P5CS2 under drought stress conditions and described that these genes involve in various physiological functions such as photosynthesis, signal transduction, phenylpropanoid biosynthesis, starch and sucrose metabolism, and osmotic adjustment. Micro RNAs (miRNAs) were playing crucial role in developmental gene expression regulation in plants and act as a important regulator in abiotic stress adaptation. Therefore, a small RNA sequencing approach was employed to profile the salt stress-responsive miRNA landscapes of the shoot and root tissues of two *Setaria viridis* accessions. RT-PCR analysis identified two regulatory modules (miR160 and miR167) that were accession- and tissue-specific responses in accessions shoot and root tissues and provide tolerance against salt stress (Pegler et al. 2020).

2.2.4 Proteomics

The term "proteome" refers to the entire set of proteins produced or modified by an organism, whereas "proteomics" refers to a large-scale analysis of proteins expressed in a cell, tissue, or organism at a specific stage or condition in order to better understand biology or causal regulatory mechanisms. Every organism's current state is determined by the proteome, which is essentially a link between the transcriptome and the metabolome. Because mRNA levels are rarely associated with protein accumulation, and proteins have direct effects on biochemical processes, a proteomics approach to evaluating plant stress responses is becoming more important (Gygi et al. 1999). Despite advancements in proteomics research, very few reports on the application of proteomics to crops have been published in comparison to other organisms (Reddy et al. 2012). Furthermore, compared to transcriptome analysis, proteome analysis in plants in response to various stimuli is still relatively limited (Canovas et al. 2004), with only one study reported in foxtail millet among all millets. Veeranagamallaiah et al. (2008) performed proteomic study in 7-day old salt-treated tolerant foxtail millet cultivar "Prasad" using 2D electrophoresis that identified 29 distinct expressed proteins among them some of involved in stress response. Pan et al. (2018) identified a total of 2474 differentially expressed proteins (DEPs) by quantitative proteomic analysis in foxtail millet seedlings for drought tolerance. There were 321 proteins with significant expression changes, including 252 proteins that were upregulated and 69 proteins that were downregulated. Western blot analysis was used to confirm the expression patterns of five drought-responsive proteins. The transcription levels of 21 differentially expressed proteins were also investigated using qRT-PCR. The findings revealed significant inconsistency in the variation between proteins and their corresponding mRNAs, demonstrating once again that posttranscriptional modification plays a critical role in gene expression regulation.

Grain protein contents (GPCs) of seeds is a main indicator of nutritional values of foxtail millet also play roles in drought tolerance. The GPC were also played important roles in different metabolic pathways such as starch and sucrose metabolism, glycolysis/gluconeogenesis, amino acids biosynthesis, detoxification and defense, protein degradation, tricarboxylic acid (TCA) cycle, protein synthesis, energy metabolism, transporter, pentose phosphate pathway, and signal transduction pathway. Li et al. (2019) had done 2D electrophoresis coupled with matrix-assisted laser desorption/ionization-tandem time-of-flight (MALDI-TOF) analysis to characterize the response of foxtail millet under drought stress at a proteomic level. A total of 104 differentially abundant protein spots (DAPs) were identified; among them, 57 were upregulated and 47 were downregulated under drought treatment. This study provides new information on proteomic changes in foxtail millet. Xu et al. (2020) studied proteomic profiles of mature seeds under three drought stress levels, conducted on two varieties, Longgu6 and Huangjinmiao (HJM) of foxtail millet by 2D electrophoresis. Using proteomics study identified levels of DEPs moderate level of drought system. Protein biosynthesis was upregulated, and drought stress-related pathways were enhanced in HJM under moderate level of drought system, thereby resulting in the increase of GPC in HJM under this condition (Xu et al. 2020). Selenium response in foxtail could be useful for analyzing the Se response pathway and producing Se-enriched foxtail millet in the future. Liang et al. (2020) investigated Selenium (Se) effect on foxtail millet proteomic profiling, identified 123 differentially expressed proteins and results showed that foliar sodium selenite (Na_2SeO_3) applied at critical growth stages significantly improved the Se content in foxtail millet. C-terminally encoded peptides (CEPs) are small peptides, typically posttranslationally modified, and highly conserved in many species. CEPs are known to inhibit plant growth and development. Zhang et al. (2021) identified 14 CEPs in foxtail millet. Moreover, expression of all *SiCEPs* was induced by abiotic stresses and phytohormones. In the presence of abscisic acid (ABA), growth inhibition and ABA content in seedlings increased with the concentration of *SiCEP3* gene.

2.2.5 Metabolomics

The term "metabolome" refers to the totality of low molecular weight biochemicals found in a sample that are either substrates or byproducts of enzymatic processes and have a direct influence on the phenotype of a cell, tissue, or organism. Metabolomics, on the other hand, attempts to assess a sample's metabolite profile at a specific time, stage, or environment, and thus provides a functional account of an organism's physiological state. Metabolomics concepts such as target analysis, metabolite profiling, and fingerprinting are examples of metabolomics concepts that can be applied to large-scale applications such as phenotyping transgenics, gene function analysis, significant equivalence testing, and stress response tracking. As a result, metabolomics may be regarded as bridging the genotype–phenotype divide. A complete transcriptome, proteome, and metabolome study is necessary to have a detailed knowledge of both gene function and molecular mechanisms mediating complicated biological processes. Only one research used gas chromatography-time-of-flight mass spectrometry (GC-TOFMS) to evaluate the variety of primary metabolites and phenolic acids in three proso millet types (Kim et al. 2013). A variety of metabolites, including suitable solutes, nitrogen-containing chemicals, antioxidants, and signal macromolecules, have been discovered to be involved in salt tolerance (Kim et al. 2007). Li et al. (2018) recognized and quantified more than 300 secondary metabolites using LC-MS, and inheritance patterns of metabolome in foxtail hybrids. Among them secondary metabolites, flavonoids and its derivatives are also identified that play a key role in drought resistance. The integrative omics analysis suggested that phenyl propanoid, flavonoid, and lignin biosynthesis pathways, as well as lysophospholipids, were important in determining foxtail millet salinity tolerance. Importantly, Yugu2 tolerance was linked to improved ion channel and antioxidant system efficiency. The results of an integrated analysis of metabolomic and transcriptomic data from foxtail millet roots revealed that 17 flavonoids biosynthesis-related genes were significantly upregulated in the Yugu2 cultivar when exposed to salinity. The overabundance of 27 flavonoids metabolites was obtained in accordance with gene expression (Pan et al. 2020).

2.2.6 Phenomics

Phenomics is the high-throughput study of phenotype through the evaluation of an organism's morphological, physiological, and biological features, as well as the linkage of genetic, epigenetic, and environmental factors (Furbank and Tester 2011; Deshmukh et al. 2014). Phenotyping technology advances have been widely used to analyze germplasm collections in order to predict phenotype based on genetic markers. This method is referred to as "genetic symptoms" (Deshmukh et al. 2014). When combined with other omics tools, phenomics can reveal a great deal about how cell biochemical or biophysical processes result in the final phenotype. Plant phenotypes are inherently complex because they result from genotypes interacting with a wide range of environmental factors (Bilder et al. 2009). Automatic and high-throughput phenotyping platforms have been developed to investigate background variation, trait relationships, plant growth behaviors, and multi-parametric phenotypic data. Automated imaging tools are used to generate nondestructive phenotypic data for quantitative studies such as growth, tolerance, resistance, physiology, and yield that are required to link to molecular studies such as germplasm collections, population mapping, breeding materials, and the design of "omics" and quantitative trait locus (QTLs) studies (Rahaman et al. 2015). The field of plant-based phenotyping is being transformed by automated high-throughput hardware platforms and associated software packages (Rahaman et al. 2015). Fahlgren et al. (2015), for example, studied drought responses in Setaria using a conveyer belt-based platform and scan 3D images. The authors discovered that *Setaria virdis* grows faster than foxtail millet using image analysis. Specialized imaging systems, such as near infrared (NIR) imaging, spectroscopy imaging, and fluorescence imaging, can also be used to measure physiological traits. For example, Fahlgren et al. (2015) discovered significant differences in water content in Setaria species using NIR imaging. Furthermore, fluorescence imaging accurately measures photosynthesis rate in 2D leaves (Cruz et al. 2016). Spectroscopy imaging can also be used to investigate stress responses (Rahaman et al. 2015), but this technology has yet to be used in millet research. It is critical to have a method for obtaining root images. Rhizotrons are root visualization systems that consist of a thin layer of soil or nutrient substrates sandwiched between two plastic sheets. This system has been used to assess the rate of root growth in pearl millet (Passot et al. 2016). Growth and Luminescence Observatory for Roots (GLO-Roots), an imaging system with a cleaner background, aids in the detection of luciferase reporter expression in *S. viridis* transgenic lines (Sebastian et al. 2016). As phenotyping systems rapidly develop, it is important to have software packages that can efficiently extract biologically meaningful information from images. Software such as ImageJ, a new generation of high-throughput, customizable and open-source software is available (Singh et al. 2016). PlantCV is another Image Analysis package that has been developed at the Danforth Plant Science Center, Missouri, USA, which detected different timely related changes to water availability between the wild and domesticated *Setaria* species (Fahlgren et al. 2015) combined with the Bellwether Phenotyping Platform helps to collect data. *Setaria* dataset of 79,000 images available on the iPlant Data Store and through the Bio-Image Semantic Query User Environment

(Bis Que) platform (Kvilekval et al. 2010; http://plantcv.danforthcenter.org/pages/data.html; publicly; Goff et al. 2011). Proximal sensing techniques, which can be used to obtain more accurate plant-level measurements, can be a useful tool in studying plant stress biology. As a result, a study was conducted to assess water use efficiency (WUE) in foxtail millet using thermal imaging and visual-near infrared spectroscopy. The findings indicated that visible-near-infrared reflectance data and thermal images could be used as a potential rapid technique for assessing plant physiological responses such as WUE (Wang et al. 2016).

2.3 Conclusion

In conclusion, foxtail millet can be used as a genetic model to study the domestication process, population structure, comparative mapping, evolution, and physiology of C4 photosynthesis, and biomass production in panicoid grasses. By comparing the genome sequences of foxtail millet and panicoid grasses, potential genes and molecular markers linked with useful agronomic traits can be identified for utilization in crop improvement programs. Foxtail millet can be used to study the regulation of C4 photosynthesis. So, extensive research must be conducted to collect and identify potential germplasm resources of foxtail millet for useful agronomic traits and abiotic stress tolerance. Further studies have to be done on the nutritional and medicinal properties of foxtail millet in order to popularize and increase production globally.

References

Ahuja I, de Vos RCH, Bones AM, Hall RD (2010) Plant molecular stress responses face climate change. Trends Plant Sci 15:1360–1385

Arya L, Verma M, Ram C (2015) Small millets transcriptomics. Transcriptomics 3:e114. https://doi.org/10.4172/2329-8936.1000e114

Baulcombe D (2004) RNA silencing in plants. Nature 431:356–363. https://doi.org/10.1038/nature02874

Bennetzen JL, Schmutz J, Wang H, Percifield R, Hawkins J, Pontaroli AC, Estep M, Feng L, Vaughn JN, Grimwood J, Jenkins J (2012) Reference genome sequence of the model plant *Setaria*. Nat Biotechnol 30(6):555–561

Bilder RM, Sabb FW, Cannon TD, London ED, Jentsch JD, Parker DS, Poldrack RA, Evans C, Freimer NB (2009) Phenomics: the systematic study of phenotypes on a genome-wide scale. Neuroscience 164(1):30–42

Bonthala VS, Muthamilarasan M, Misra G, Prasad M (2013) FmMDb: a versatile database of foxtail millet markers for millets and bioenergy grasses research. PLoS One 8(8):1–7. https://doi.org/10.1371/journal.pone.0071418

Bonthala VS, Muthamilarasan M, Roy R, Prasad M (2014) FmTFDb: a foxtail millet transcription factors database for expediting functional genomics in millets. Mol Biol Rep 41(10):6343–6348

Canovas F, Dumas-Gaudot E, Recorbet G, Jorrin J, Mock H-P, Rossignol M (2004) Plant proteome analysis. Proteomics 4:285–298

Chen Y, Liu B, Zhao Y, Yu W, Si W (2021) Whole-genome duplication and purifying selection contributes to the functional redundancy of auxin response factor (ARF) genes in foxtail millet (Setaria italica L.). Int J Genomics 2021

Chinnusamy V, Schumaker K, Zhu JK (2004) Molecular genetics perspectives on cross-talk and specificity in abiotic stress signalling in plants. J Exp Bot 55:225–236

Cruz JA, Savage LJ, Zegarac R, Hall CC, Satoh-Cruz M, Davis GA, Kovac WK, Chen J, Kramer DM (2016) Dynamic environmental photosynthetic imaging reveals emergent phenotypes article dynamic environmental photosynthetic imaging reveals emergent phenotypes. Cell Systems 2(6):365–377. https://doi.org/10.1016/j.cels.2016.06.001

da Cunha Valença D, de Moura SM, Travassos-Lins J, Alves-Ferreira M, Medici LO, Ortiz-Silva B, Macrae A, Reinert F (2020) Physiological and molecular responses of Setaria viridis to osmotic stress. Plant Physiol Biochem 155:114–125

Deshmukh R, Sonah H, Patil G, Chen W, Prince S, Mutava R, Vuong T, Valliyodan B, Nguyen HT (2014) Integrating omic approaches for abiotic stress tolerance in soybean. Front Plant Sci 5:1–12

Doust AN, Kellogg EA, Devos KM, Bennetzen JL (2009) Foxtail millet a sequence-driven grass model system. Plant Physiol 149(1):137–141. https://doi.org/10.1104/pp.108.129627

Fahlgren N, Feldman M, Gehan MA, Wilson MS, Shyu C, Bryant DW, Hill ST, McEntee CJ, Warnasooriya SN, Kumar I, Ficor T (2015) A versatile phenotyping system and analytics platform reveals diverse temporal responses to water availability in *Setaria*. Mol Plant 8(10): 1520–1535

Feng ZJ, Xu ZS, Sun J, Li LC, Chen M, Yang GX, He GY, Ma YZ (2016) Investigation of the asr family in foxtail millet and the role of ASR1 in drought/oxidative stress tolerance. Plant Cell Rep 35(1):115–128. https://doi.org/10.1007/s00299-015-1873-y

Furbank RT, Tester M (2011) Phenomics—technologies to relieve the phenotyping bottleneck. Trends Plant Sci 16:635–644

Goff SA, Vaughn M, McKay S, Lyons E, Stapleton AE, Gessler D, Matasci N, Wang L, Hanlon M, Lenards A, Muir A (2011) The iPlant collaborative : cyber infrastructure for plant biology. Front Plant Sci 2:1–16

Gupta S, Kumari K, Das J, Lata C, Puranik S et al (2011) Development and utilization of novel intron length polymorphic markers in foxtail millet (*Setaria italica* (L.) P. Beauv.). Genome 54: 586–602

Gupta S, Kumari K, Muthamilarasan M, Parida SK, Prasad M (2014) Population structure and association mapping of yield contributing agronomic traits in foxtail millet. Plant Cell Rep 33(6):881–893

Gupta S, Kumari K, Sahu PP, Vidapu S, Prasad M (2012) Sequence-based novel genomic microsatellite markers for robust genotyping purposes in foxtail millet [*Setaria italica* (L.) P. Beauv]. Plant Cell Rep 31:323–337

Gupta S, Kumari K, Sahu PP, Vidapu S, Prasad M (2013) Sequence-based novel genomic microsatellite markers for robust genotyping purposes in foxtail millet [Setaria italica (L.) P. Beauv]. Plant Cell Rep 31:323–337

Gygi SP, Rochon Y, Franza BR, Aebersold R (1999) Correlation between protein and mRNA abundance in yeast. Mol Cell Biol 19:1720–1730

Hao X, Jin Z, Wang Z, Qin W, Pei Y (2020) Hydrogen sulfide mediates DNA methylation to enhance osmotic stress tolerance in Setaria italica L. Plant Soil 453(1):355–370

Heng L, Chih C, Song C, Chang K (2011) Development of simple sequence repeats (SSR) markers in *Setaria italica* (Poaceae) and cross-amplification in related species. Int J Mol Sci 12:7835–7845

Huangfu Y, Pan J, Li Z, Wang Q, Mastouri F, Li Y, Yang S, Liu M, Dai S, Liu W (2021) Genome-wide identification of PTI1 family in Setaria Italica and salinity-responsive functional analysis of SiPTI1-5. BMC Plant Biol 21(1):1–17

Jaiswal V, Gupta S, Gahlaut V, Muthamilarasan M, Bandyopadhyay T, Ramchiary N, Prasad M (2019) Genome-wide association study of major agronomic traits in foxtail millet (*Setaria italica* L.) using DdRAD sequencing. Sci Rep 9(1):1–11

Jayaraman A, Puranik S, Rai NK, Vidapu S, Sahu PP, Lata C, Prasad M (2008) cDNA-AFLP analysis reveals differential gene expression in response to salt stress in foxtail millet (*Setariaitalica* L.). Mol Biotechnol 40:241–251

Jia G, Huang X, Zhi H, Zhao Y, Zhao Q, Li W, Chai Y, Yang L, Liu K, Lu H, Zhu C (2013) A haplotype map of genomic variations and genome-wide association studies of agronomic traits in foxtail millet (Setaria italica). Nat Genet 45(8):957–961

Jia X, Zhang Z, Liu Y, Zhang C, Shi Y, Song Y, Wang T, Li Y (2009) Development and genetic mapping of SSR markers in foxtail millet [*Setaria italica* (L.) P. Beauv.]. Theor Appl Genet 118:821–829

Jia XP, Shi YS, Song YC, Wang GY, Wang TY, Li Y (2007) Development of EST-SSR in foxtail millet (Setaria italica). Genet Resour Crop Evol 54:233–236. https://doi.org/10.1007/s10722-006-9139-8

Khan Y, Yadav A, Suresh VB, Muthamilarasan M, Yadav CB, Prasad M (2014) Comprehensive genome-wide identification and expression profiling of foxtail millet [Setaria italica (L.)] miRNAs in response to abiotic stress and development of miRNA database. Plant Cell Tiss Organ. https://doi.org/10.1007/s11240-014-0480-x

Kim JK, Bamba T, Harada K, Fukusaki E, Kobayashi A (2007) Time-course metabolic profiling in *Arabidopsis thaliana* cell cultures after salt stress treatment. J Exp Bot 58:415–424

Kim JK, Park S-Y, Yeo Y, Cho HS, Kim YB, Bae H, Park CH, LeeJ-H and Park SU (2013) Metabolic profiling of millet(*Panicum miliaceum*) using gas chromatography–time-offlightmass spectrometry (GC-TOFMS) for qualityassessment Plant Omics J 6: 73–78

Kumar S, Trivedi PK (2018) Glutathione S-transferases: role in combating abiotic stresses including arsenic detoxification in plants. Front Plant Sci 9:751

Kumari K, Muthamilarasan M, Misra G, Gupta S, Subramanian A et al (2013) Development of eSSR-markers in Setaria italica and their applicability in studying genetic diversity, cross-transferability and comparative mapping in millet and non-millet species. PLoS One 8:e67742

Kvilekval K, Fedorov D, Obara B, Singh A, Manjunath BS (2010) Bisque: a platform for bioimage analysis and management. Bioinformatics 26:544–552

Lata C (2015) Advances in omics for enhancing abiotic stress tolerance in millets. Proc Indian Nat Sci Acad 81(2):397–417

Lata C, Bhutty S, Bahadur RP, Majee M, Prasad M (2011a) Association of an SNP in a novel DREB2-like gene *SiDREB2* with stress tolerance in foxtail millet [*Setaria italica* (L.)]. J Exp Bot 62(10):3387–3401

Lata C, Bhutty S, Bahadur RP, Majee M, Prasad M (2011b) Association of an SNP in a novel DREB2-like gene SiDREB2 with stress tolerance in foxtail millet [*Setaria italica* (L.)]. J Exp Bot 62:3387–3401

Lata C, Mishra AK, Muthamilarasan M, Bonthala VS, Khan Y, Prasad M (2014) Genome-wide investigation and expression profiling of AP2/ERF transcription factor superfamily in foxtail millet (*Setaria italica* L.). PLoS One 9:e113092

Lata C, Prasad M (2011) Role of DREBs in regulation of abiotic stress responses in plants. J Exp Bot 62:4731–4748

Lata C, Prasad M (2013) Setaria genome sequencing: an overview. J Plant Biochem Biotechnol 22: 257–260. https://doi.org/10.1007/s13562-013-0216-8

Lata C, Sahu PP, Prasad M (2010) Comparative transcriptome analysis of differentially expressed genes in foxtail millet (*Setaria italica* L.) during dehydration stress. Biochem Biophys Res Commun 393:720–727

Le Thierry d'Ennequin M, Panaud O, Toupance B, Sarr A (2000) Assessment of genetic relationships between Setaria italica and its wild relative S. viridis using AFLP markers. Theor Appl Genet 100:1061–1066

Li C, Yue J, Wu X, Xu C, Yu J (2014) An ABA-responsive DRE-binding protein gene from *Setaria italica*, *SiARDP*, the target gene of *SiAREB*, plays a critical role under drought stress. J Exp Bot 65:5415–5427

Li J, Dong Y, Li C, Pan Y, Yu J (2017) *SiASR4*, the target gene of *SiARDP* from *Setaria italica*, improves abiotic stress adaption in plants. Front Plant Sci 7:2053

Li J, Li X, Yang Q, Luo Y, Gong X, Zhang W, Hu Y, Yang T, Dong K, Feng B (2019) Proteomic changes in the grains of foxtail millet (Setaria italica (L.) Beau) under drought stress. Span J Agric Res 17(2):802

Li P, Brutnell TP (2011) *Setaria viridis* and *Setaria italica*, model genetic systems for panicoid grasses. J Exp Bot 62:3031–3037. https://doi.org/10.1093/jxb/err096. PubMed: 21459768

Li S, Dong X, Fan G, Yang Q, Shi J, Wei W, Zhao F, Li N, Wang X, Wang F, Feng X (2018) Comprehensive profiling and inheritance patterns of metabolites in foxtail millet. Front Plant Sci 9:1716

Liang K, Liang S, Zhu H (2020) Comparative proteomics analysis of the effect of selenium treatment on the quality of foxtail millet. Lwt 131:109691

Liu W, Li RJ, Han TT, Cai W, Fu ZW, Lu YT (2015) Salt stress reduces root meristem size by nitric oxide mediated modulation of auxin accumulation and signaling in *Arabidopsis*. Plant Physiol 168(1):343–356

Liu Y, Feng X, Xu Y, Yu J, Ao G, Peng Z et al (2009) Overexpression of millet ZIP-like gene (SiPf40) affects lateral bud outgrowth in tobacco and millet. Plant Physiol Biochem 47:1051–1060

Loqué D, von Wirén N (2004) Regulatory levels for the transport of ammonium in plant roots. J Exp Bot 55:1293–1305

Mauro-Herrera M, Wang X, Barbier H, Brutnell TP, Devos KM, Doust AN (2013) Genetic control and comparative genomic analysis of flowering time in Setaria (Poaceae). G3 (3):283–295

Meng X, Liang Z, Dai X, Zhang Y, Mahboub S, Ngu DW, Roston RL, Schnable JC (2021) Predicting transcriptional responses to cold stress across plant species. Proc Natl Acad Sci U S A 118(10):1–9

Mishra AK, Muthamilarasan M, Khan Y, Parida SK, Prasad M (2014) Genome-wide investigation and expression analyses of WD40 protein family in the model plant foxtail millet (Setaria italica L.). PLoS One. https://doi.org/10.1371/journal.pone.0086852

Muthamilarasan M, Bonthala VS, Mishra AK, Khandelwal R, Khan Y, Roy R et al (2014b) C_2H_2type of zinc finger transcription factors in foxtail millet define response to abiotic stresses. Funct Integr Genomics 14:531–543

Muthamilarasan M, Khandelwal R, Yadav CB, Bonthala VS, Khan Y, Prasad M (2014c) Identification and molecular characterization of MYB transcription factor superfamily in C_4 model plant foxtail millet (*Setaria italica* L.). PLoS One 9:e109920

Muthamilarasan M, Prasad M (2015) Advances in Setaria genomics for genetic improvement of cereals and bioenergy grasses. Theor Appl Genet 128(1):1–14

Muthamilarasan M, Singh RK, Suresh BV, Rana S, Dulani P, Prasad M (2020) Genomic dissection and expression analysis of stress-responsive genes in C4 panicoid models, Setaria italica and *Setaria viridis*. J Biotechnol 318:57–67

Muthamilarasan ME, Venkata Suresh B, Pandey GA, Kumari KA, Parida SK, Prasad MA (2014a) Development of 5123 intron-length polymorphic markers for large-scale genotyping applications in foxtail millet. DNA Res 21(1):41–52

Nadeem F, Ahmad Z, Ul Hassan M, Wang R, Diao X, Li X (2020) Adaptation of foxtail millet (Setaria italica L.) to abiotic stresses: a special perspective of responses to nitrogen and phosphate limitations. Front Plant Sci 11:187

Nadeem F, Ahmad Z, Wang R, Han J, Shen Q, Chang F, Diao X, Zhang F, Li X (2018) Foxtail millet [*Setaria italica* (L.) Beauv.] grown under low nitrogen shows a smaller root system, enhanced biomass accumulation, and nitrate transporter expression. Front Plant Sci 9 (February):1–12

Ni X, Xia Q, Zhang H, Cheng S, Li H, Fan G, Guo T, Huang P, Xiang H, Chen Q, Li N (2017) Updated foxtail millet genome assembly and gene mapping of nine key agronomic traits by resequencing a RIL population. Giga Sci 6(2):1–8

Pan J, Li Z, Dai S, Ding H, Wang Q, Li X, Ding G, Wang P, Guan Y, Liu W (2020) Integrative analyses of transcriptomics and metabolomics upon seed germination of foxtail millet in response to salinity. Sci Rep 10(1):1–16

Pan J, Li Z, Wang Q, Garrell AK, Liu M, Guan Y, Zhou W, Liu W (2018) Comparative proteomic investigation of drought responses in foxtail millet. BMC Plant Biol 18(1):1–19

Pandey G, Misra G, Kumari K, Gupta S, Parida SK et al (2013) Genome-wide development and use of microsatellite markers for large-scale genotyping applications in foxtail millet [*Setaria italica* (L.)]. DNA Res 20:197–207

Passot S, Gnacko F, Moukouanga D, Lucas M, Guyomarc'h S, Ortega BM, Atkinson JA, Belko MN, Bennett MJ, Gantet P, Wells DM (2016) Characterization of pearl millet root architecture and anatomy reveals three types of lateral roots. Front Plant Sci 7:1–11

Pegler JL, Nguyen DQ, Grof CP, Eamens AL (2020) Profiling of the salt stress responsive microRNA landscape of C4 genetic model species Setaria viridis (L.) Beauv. Agronomy 10(6):837

Prakash PT, Banan D, Paul RE, Feldman MJ, Xie D, Freyfogle L, Baxter I, Leakey AD (2021) Correlation and co-localization of QTL for stomatal density, canopy temperature, and productivity with and without drought stress in Setaria. J Exp Bot 72(13):5024–5037

Puranik S, Jha S, Srivastava PS, Sreenivasulu N, Prasad M (2011) Comparative transcriptome analysis of contrasting foxtail millet cultivars in response to short-term salinity stress. J Plant Physiol 168:280–287

Puranik S, Sahu PP, Mandal SN, Parida SK, Prasad M (2013) Comprehensive genome-wide survey, genomic constitution, and expression profiling of the NAC transcription factor family in foxtail millet (*Setaria italica* L.). PLoS One 8:e64594

Qi X, Xie S, Liu Y, Yi F, Yu J (2013) Genome-wide annotation of genes and noncoding RNAs of foxtail millet in response to simulated drought stress by deep sequencing. Plant Mol Biol 83: 459–473

Qie L, Jia G, Zhang W, Schnable J, Shang Z, Li W, Liu B, Li M, Chai Y, Zhi H, Diao X (2014) Mapping of quantitative trait locus (QTLs) that contribute to germination and early seedling drought tolerance in the interspecific cross *Setaria italica* x *Setaria viridis*. PLoS One 9(7):3–10. https://doi.org/10.1371/journal.pone.0101868

Qin L, Chen E, Li F, Yu X, Liu Z, Yang Y, Wang R, Zhang H, Wang H, Liu B, Guan YA (2020) Genome-wide gene expression profiles analysis reveal novel insights into drought stress in foxtail millet (*Setaria italica* L.). Int J Mol Sci 21(22):1–21

Rahaman M, Chen D, Gillani Z, Klukas C, Chen M (2015) Advanced phenotyping and phenotype data analysis for the study of plant growth and development. Front Plant Sci 6:1–15. https://doi.org/10.3389/fpls.2015.00619

Ramanna H, Ding XS, Nelson RS (2013) Rationale for developing new virus vectors to analyze gene function in grasses through virus-induced gene silencing. Methods Mol Biol 975:15–32

Reddy DS, Bhatnagar-Mathur P, Vadez V, Sharma KK (2012) Grain legumes (soybean, chickpea, and Peanut): omics approaches to enhance abiotic stress tolerance. In: Tuteja N, Gill SS, Tiburcio AF, Tuteja R (eds) Improving crop resistance to abiotic stress. Wiley-VCH Verlag GmbH & Co.KGaA. https://doi.org/10.1002/9783527632930.ch39

Roch GV, Maharajan T, Krishna TA, Ignacimuthu S, Ceasar SA (2020) Expression of PHT1 family transporter genes contributes for low phosphate stress tolerance in foxtail millet (Setaria italica) genotypes. Planta 252(6):1–9

Sebastian J, Yee MC, Viana WG, Rellán-Álvarez R, Feldman M, Priest HD, Trontin C, Lee T, Jiang H, Baxter I, Mockler TC (2016) Grasses suppress shoot-borne roots to conserve water during drought. PNAS USA 113(31):8861–8866. https://doi.org/10.1073/pnas.1604021113

Shi W, Cheng J, Wen X, Wang J, Shi G, Yao J, Hou L, Sun Q, Xiang P, Yuan X, Dong S (2018) Transcriptomic studies reveal a key metabolic pathway contributing to a well-maintained photosynthetic system under drought stress in foxtail millet (*Setaria italica* L.). PeerJ 6:e4752

Shkolnik D, Finkler A, Pasmanik-Chor M, Fromm H (2019) Calmodulin-binding transcription activator 6: a key regulator of Na^+ homeostasis during germination. Plant Physiol 180:1101–1118

Singh A, Ganapathysubramanian B, Singh AK, Sarkar S (2016) Machine learning for high-throughput stress phenotyping in plants. Trends Plant Sci 21:110–124

Singh RK, Muthamilarasan M, Prasad M (2017) Foxtail millet: An introduction. In: Prasad M (ed) The foxtail millet genome. Compendium of plant genomes. Springer, Cham, pp 1–9. https://doi.org/10.1007/978-3-319-65617-5_1

Sood P, Prasad M (2017) Genetic transformation of *Setaria*: a new perspective. In: Prasad M (ed) The foxtail millet genome, compendium of plant genomes. Springer, Berlin, pp 105–121

Sood P, Singh RK, Prasad M (2020) An efficient *agrobacterium*-mediated genetic transformation method for foxtail millet (*Setaria italica* L.). Plant Cell Rep 39(4):511–525

Sreenivasulu N, Miranda M, Prakash HS, Wobus U, Weschke W (2004) Transcriptome changes in foxtail millet genotypes at high salinity: identification and characterization of aPHGPX gene specifically upregulated by NaCl in a salt tolerant line. J Plant Physiol 161:467–477

Tang S, Li L, Wang Y, Chen Q, Zhang W, Jia G, Zhi H, Zhao B, Diao X (2017) Genotype-specific physiological and transcriptomic responses to drought stress in *Setaria italica* (an emerging model for panicoideae grasses). Sci Rep 7(1):1–15. https://doi.org/10.1038/s41598-017-08854-6

Veeranagamallaiah G, Jyothsnakumari G, Thippeswamy M, Reddy PC, Surabhi GK, Sriranganayakulu G, Mahesh Y, Rajasekhar B, Madhurarekha C, Sudhakar C (2008) Proteomic analysis of salt stress responses in foxtail millet (*Setaria italica* L. cv. Prasad) seedlings. Plant Sci 175(5):631–641

Vetriventhan M, Upadhyaya HD, Anandakumar CR, Senthilvel S, Varshney RK, Parzies HK (2014) Population structure and linkage disequilibrium of ICRISAT foxtail millet (*Setaria italica* (L.) P. Beauv.) core collection. Euphytica 196:423–435. https://doi.org/10.1007/s10681-013-1044-6

Wang J, Wang Z, Du X, Yang H, Han F, Han Y, Yuan F, Zhang L, Peng S, Guo E (2017) A high-density genetic map and qtl analysis of agronomic traits in foxtail millet [*Setaria italica* (L.) P. Beauv.] using RAD-Seq. PLoS One 12(6):1–15

Wang M, Ellsworth PZ, Zhou J, Cousins AB, Sankaran S (2016) Evaluation of water-use efficiency in foxtail millet (Setaria italica) using visible-near infrared and thermal spectral sensing techniques. Talanta 152:531–539

Wang M, Li P, Li C, Pan Y, Jiang X, Zhu D, Zhao Q, Yu J (2014) *SiLEA14*, a novel atypical LEA protein, confers abiotic stress resistance in foxtail millet. BMC Plant Biol 14(1):1–16

Wang ZM, Devos KM, Liu CJ, Wang RQ, Gale MD (1998) Construction of RFLP-based maps of foxtail millet, *Setaria italica* (L.) P. Beauv. Theor Appl Genet 96:31–36

Xu B, Gao X, Dong K, Li X, Yang P, Yang T, Feng B (2020) Grain protein content comparison and proteomic analysis of foxtail millet (*Setaria italica* L.) seed response to different drought stress levels. Acta Physiol Plant 42(2):1–1

Xu J, Li Y, Ma X, Ding J, Wang K, Wang S, Tian Y, Zhang H, Zhu XG (2013) Whole transcriptome analysis using next-generation sequencing of model species *Setaria viridis* to support C4 photosynthesis research. Plant Mol Biol 83:77–87

Yadav CB, Bonthala VS, Muthamilarasan M, Pandey G, Khan Y, Prasad M (2014a) Genome-wide development of transposable elements-based markers in foxtail millet and construction of an integrated database. DNA Res. https://doi.org/10.1093/dnares/dsu039

Yadav CB, Muthamilarasan M, Pandey G, Khan Y, Prasad M (2014b) Development of novel microRNA-based genetic markers in foxtail millet for genotyping applications in related grass species. Mol Breed 34:2219–2224

Yadav CB, Muthamilarasan M, Pandey G, Prasad M (2015) Identification, characterization and expression profiling of dicer-like, argonaute and RNA-dependent RNA polymerase gene families in foxtail millet. Plant Mol Biol Rep 33:43–55

Yan J, Yang L, Liu Y, Zhao Y, Han T, Miao X, Zhang A (2021) Calcineurin B-like protein 5 (SiCBL5) in Setaria italica enhances salt tolerance by regulating Na+ homeostasis. Crop J 15. https://doi.org/10.1016/j.cj.2021.06.006

Yin H, Yu Q, Dai Y, An L and Li W (2014). Transcript profiling of foxtail millet seedlings under drought condition and discovery of a new drought-tolerant gene. In: 1[st]International *Setaria* Genetics Conference, pp 24

Zhang G, Liu X, Quan Z, Cheng S, Xu X et al (2012) Genome sequence of foxtail millet (*Setaria italica*) provides insights into grass evolution and biofuel potential. Nat Biotechnol 30:549–554

Zhang J, Liu T, Fu J, Zhu Y, Jia J, Zheng J, Zhao Y, Zhang Y, Wang G (2007) Construction and application of EST library from *Setaria italica* in response to dehydration stress. Genomics 90: 121–131

Zhang L, Ren Y, Wan Y, Xu Q, Yang G, Zhang S, Huang J, Yan K, Zheng C, Wu C (2021) SiCEP3, a C-terminally encoded peptide from *Setaria italica*, promotes ABA import and signaling. J Exp Bot 72(18):6260–6273

Zhang S, Tang C, Zhao Q, Li J, Yang L, Qie L, Fan X, Li L, Zhang N, Zhao M, Liu X (2014) Development of highly polymorphic simple sequence repeat markers using genome wide microsatellite variant analysis in foxtail millet [*Setaria italica* (L.) P. Beauv]. BMC Genomics 15:78

Zhu C, Ming C, Zhao-shi X, Li L-C, Xue-ping C, Ma Y-z (2014) Characteristics and expression patterns of the aldehyde dehydrogenase (ALDH) gene superfamily of foxtail millet (*Setaria italica* L.). PLoS One 9:e101136

Current Status and Future Prospects of Omics Strategies in Barnyard Millet

3

Vellaichamy Gandhimeyyan Renganathan, Chockalingam Vanniarajan, Raman Renuka, Kannan Veni, and Mani Vetriventhan

Abstract

Barnyard millet (*Echinochloa* spp) is a wide-adaptive crop that produces larger seeds compared to other minor millets. It is one of the most important minor millet crops that is predominantly cultivated for human consumption in Asia and livestock feed in the United States. Barnyard millet has incredible resilience to biotic and abiotic stresses. Grains are highly nutritious with significant amount of protein, carbohydrate, fiber, and, most notably, micronutrients (iron and zinc) than other staple cereals. Despite these benefits, barnyard millet has received little attention in genetic and genomic studies for many years. Progress has been made in developing genetic and genomic resources over last few years, thanks to the reducing cost of sequencing technologies in polyploid and orphan millets species as well. Recently, the genome of wild and transcriptome sequences of cultivated *Echinochloa* species has facilitated the understanding of the genetic architecture of important agronomic and micronutrient traits. In this chapter, we discuss the

V. G. Renganathan (✉)
Department of Plant Breeding and Genetics, Agricultural College and Research Institute, Tamil Nadu Agricultural University, Madurai, Tamil Nadu, India

Department of Biotechnology, Centre of Innovation, Agricultural College and Research Institute, Tamil Nadu Agricultural University, Madurai, Tamil Nadu, India

C. Vanniarajan · K. Veni
Department of Plant Breeding and Genetics, Agricultural College and Research Institute, Tamil Nadu Agricultural University, Madurai, Tamil Nadu, India

R. Renuka
Department of Biotechnology, Centre of Innovation, Agricultural College and Research Institute, Tamil Nadu Agricultural University, Madurai, Tamil Nadu, India

M. Vetriventhan
International Crops Research Institute for the Semi-Arid Tropics, Hyderabad, Telangana, India

R. N. Pudake et al. (eds.), *Omics of Climate Resilient Small Millets*,
https://doi.org/10.1007/978-981-19-3907-5_3

importance of barnyard millet in the current climatic scenario and highlight the up-to-date status of genetic and genomics research and the research gaps that need to be addressed in this crop. This chapter also provides an overview of deploying omics approaches to study the barnyard millet for better understanding of crops' nutritional richness and tolerance to environmental stresses.

Keywords

Barnyard millet · *Echinochloa* · Omics · Nutritional security · Small millets

3.1 Introduction

The modern agriculture production system has had a dramatic impact on environment and its contribution to sustainable agriculture and healthy human diet is minimal. Over the century, the system increased the global dependence on selective cereal species which resulted not only in the loss of biodiversity but also loss in the traditional knowledge of orphan crops that adequately supplies a daily healthy diet. Climate change, on the other hand warns to shift over to the next-generation crop plants that possess inherent water and nutrient use efficiency and sustainable productivity in wide array of environmental conditions (Pareek et al. 2020). Small millets are central to food security of many livelihoods, especially in Asia and Africa, among which, barnyard millet is an ancient millet crop cultivated in Asia, mainly India, China, Japan, and Korea. Barnyard millet belongs to the genus *Echinochloa*, which consists of almost 250 annual and perennial species that are widely distributed in the warmer and temperate parts of the world (Bajwa et al. 2015). *Echinochloa* species have very few cultivatable forms and among these are the two most popular species namely, *Echinochloa frumentacea* (Indian barnyard millet) and *Echinochloa esculenta* (Japanese barnyard millet) (Sood et al. 2015). These cultivated barnyard millet species are annual, robust, and tall that grow up to a height of 220 cm with a short-growing cycle (Padulosi et al. 2009). Barnyard millet is the fourth most produced minor millet, providing food and nutritional security to many resources poor people of developing countries. India is the biggest producer of barnyard millet, both in terms of area (0.146 m ha^{-1}) and production (0.147 mt) with average productivity of 1034 kg ha^{-1} in recent past (IIMR 2018).

Barnyard millet can grow with almost no inputs and can withstand various biotic and abiotic stresses. Besides, the grains are highly valued for their nutritional value and health benefits compared to major cereals like rice, wheat, and maize (Renganathan et al. 2020). Most notably, grains are rich in dietary fiber and micronutrients, and this substantiates barnyard millet as ideal food for people with lifestyle diseases and anemic patients in developing countries. The carbohydrate-to-crude fiber ratio ensures the slower release of sugars in the blood [low Glycemic Index (GI)], and thus aids in maintaining blood sugar level and prevents type II diabetics (Ugare et al. 2014). A significant amount of resistant starch lowers the blood glucose, serum cholesterol, and triglycerides in rats (Kumari and

Thayumanavan 1998). The low level of phytate that is often removed during the dehulling process, further favors the bioavailability of minerals on consumption. The significant amount of polyphenols and carotenoids in barnyard millet have various therapeutic properties like antioxidant, anticarcinogenic, anti-inflammatory, antimicrobial, wound healing, biliousness, and alleviating constipation-associated diseases (reviewed by Renganathan et al. 2020). All these features favor barnyard millet as ideal supplementary and alternative crop for vagaries of climate. Despite these, the lack of awareness among scientific and farming communities has led this crop to be considered as a neglected and underutilized crop. Over the decades, only few attempts have been made on exploring the diversity within the germplasm through morphological and molecular markers. Very few studies explored the nutritional profile particularly micronutrients in the grains.

The genome research in barnyard millet is still in the early stage and far behind other minor millets. Recently, second and third-generation sequencing technologies facilitated the identification of the genome and transcriptome sequences of wild species (Li et al. 2013a, b; Yang et al. 2013; Nah et al. 2015; Guo et al. 2017) and transcriptome sequences of cultivated species (Murukarthick et al. 2019). However, comprehensive research is needed to identify trait-specific donors, develop appropriate mapping population, and discover the quantitative trait locus (QTLs)/gene governing the major traits. Here, we discuss the current status of genomic resources available in barnyard millet and outline the scope of "omics" strategy in understanding the molecular basis of growth and development, stress response, and micronutrient content to speed up the breeding processes such as marker-assisted selection and gene pyramiding.

3.2 Omics in Barnyard Millet

Advances in sequencing technologies, especially in next-generation sequencing (NGS), have significantly improved our understanding of many complex millet species like pearl millet, finger millet, and tef over the past one decade. Today, several "omics" (genomics, transcriptomics, proteomics, metabolomics, and phenomics) approaches were developed and applied to identify the candidate gene (s)/QTL(s) associated with traits of agricultural significance in various crop plants. However, such studies are scarce in barnyard millet due to its genome complexity and lack of research funding on this orphan crop. The recent availability of the draft genome of wild and transcriptome of cultivated species provides the opportunity for genetic diversity, trait mapping, and comparative genomics studies (Table 3.1). The evolution of omics strategy in barnyard millet is depicted in Fig. 3.1.

3.2.1 Studies on Nuclear Genome

Knowledge of *Echinochloa* genomes initially lagged behind other millets but the unique nature of invasiveness and adaptation of weedy wild species (*E. crus-galli*)

Table 3.1 Nuclear, chloroplast genome, and transcriptome sequence resources in *Echinochloa* spp.

Genome composition	Genome or data size	No. of genes predicted/putative protein-coding transcripts
Nuclear genome		
E. crus-galli[a]	1.27 (Gb)	108,771
Chloroplast genome		
E. crus-galli[d]	139,851 (bp)	131
E. colona[d]	139,592 (bp)	131
E. esculenta[d]	139,851 (bp)	111
E. frumentacea[d]	139,593 (bp)	112
Transcriptome		
E. frumentacea[b]	48.6 (Gb)	65,276
E. esculenta[d]	11.6 (Gb)	–
E. crus-galli[c]	32.8 (Gb)	68,480
E. colona[d]	11.6 (Gb)	–
E. stagnina[d]	02.4 (Gb)	–
E. glabrescens[d]	05.0 (Gb)	–

[a] Guo et al. (2017)
[b] Murukarthick et al. (2019)
[c] Nah et al. (2015)
[d] https://www.ncbi.nlm.nih.gov/search/all/?term=echinochloa

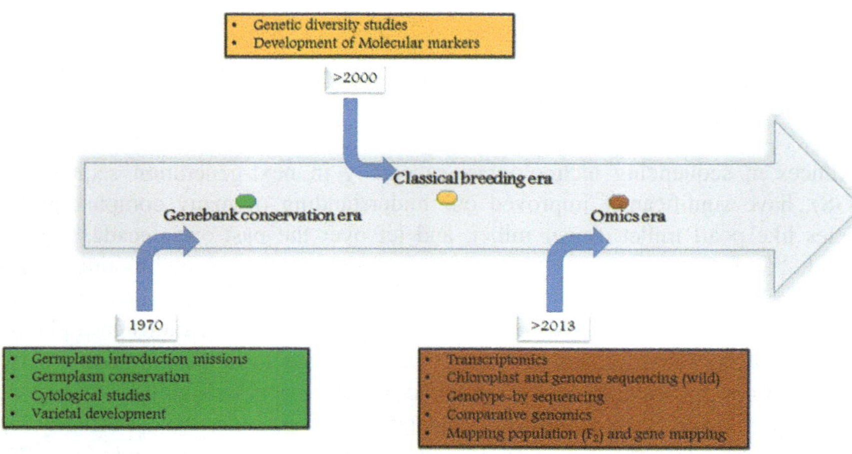

Fig. 3.1 Evolution of omics in barnyard millet

has facilitated the genomic research (whole genome sequencing) in 2017. A group of scientists from China sequenced the weedy *E. crus-galli* genome of 1.27 Gb at 171x sequencing depth representing approximately 90.7% of the predicted genome size (Guo et al. 2017). The average length of genomic libraries ranged between 160 bp and 20 Kb and the total number of contigs were 4534, which ranged between 1 kb and 11.7 Mb. Further gene annotation revealed 514 Mb of repetitive elements, 785 miRNAs, 108,771 protein-coding genes, and many noncoding RNAs. To date, 54% of genomic resources are available for *E. crus-galli*, while cultivated barnyard millet species like *E. frumentacea* and *E. esculenta* have only 4% (Renganathan et al. 2020). As of 2020, the NCBI domain includes 1246 nucleotide sequences, 822 gene sequences, 2468 protein sequences, 105 short read archive sequences (SRA), 74 expressed sequence tags (ESTs), and one gene expression omnibus (GEO) dataset related to *Echinochloa* species which includes the species like *E. crus-galli*, *E. oryzicola*, *E. phyllopogon*, *E. colona*, *E. pyramidalis*, *E. esculenta*, *E. frumentacea*, and *E. oryzoides* (last accessed: September 2021; https://www.ncbi.nlm.nih.gov/search/all/?term=echinochloa). Nevertheless, this is not at par with other millets, in terms of genomic resources. This calls for enrichment of genomic research in cultivated barnyard millet as it is necessary for speeding up crop improvement using biotechnological tools.

3.2.2 Chloroplast Genome Studies

As of 2020, chloroplast genome of seven *Echinochloa* species viz., *E. crus-galli*, *E. ugandensis*, *E. stagnina*, *E. colona*, *E. esculenta*, *E. frumentacea*, *E. haploclada*, and *E. oryzicola* have been sequenced and annotated (Ye et al. 2014; Hereward et al. 2016; Nah et al. 2016; Perumal et al. 2016; Lee et al. 2017; Sebastin et al. 2019; Jiang et al. 2021). The chloroplast genome structure of all the seven *Echinochloa* species revealed that they are highly conserved in their genome structure, organization, and gene order. The chloroplast genome of *Echinochloa* species comprises 38.6% GC regions and 61% AT regions (Sebastin et al. 2019). The genome identity among cultivated and wild was around 99.5% and the genome size was between 139,592 and 139,851 bp. However, the number of genes between cultivated and wild were 111 and 131, respectively, which suggests the reorganization of gene copy number and structure during the course of evolution or speciation (Renganathan et al. 2020). As in other plants, the chloroplast genome of *Echinochloa* is quadripartite structure, which comprises a pair of inverted repeats (IR) separated by a small single-copy region (SSC) and a large-single copy region (LSC). The size of IR, LSC, and SSC regions varied from 22,618 to 22,748 bp, 81,837 to 82,053 bp, and 12,518 to 12,519 bp, respectively. Further, comparative and molecular divergence clock analysis of *Echinochloa* chloroplast genomes with other grass species revealed that *Echinochloa* species are closer to the *Panicum virgatum* and it had older diversification history, i.e., 21.6 Mya. The wild and cultivated *Echinochloa* species have high sequence identity with *Panicum virgatum* and *Sorghum bicolor* followed by *Triticum aestivum* and *Oryza sativa* (Ye et al. 2014). The wild species *E. crus-galli,* diverged from *E. oryzicola* around 3.3 Mya while, *E. colona,* diverged from *E. oryzicola* and *E. crus-galli* between 2.65 and 3.18 Mya, respectively (Lee et al. 2017). In contrast, the cultivated species (*E. frumentacea*) diverged from *E. oryzicola* and *E. crus-galli* was 1.9–2.7 Mya (Perumal et al. 2016).

3.2.3 Transcriptomics Studies

Transcriptomics is the study of RNA present in plant or organ or tissue or cell at any given point of time. Early transcriptomic platforms like microarray are built to capture the differential mRNA content of a sample (control/treated) based on nucleotide sequence complementarity. Rapid developments in the NGS technologies viz., RNA-sequencing (amplification) has now superseded the microarray technologies (hybridization) so as to provide an absolute frequency of each transcript in an accurate, cost, and time-effective manner (Weber 2015). In principle, the total RNA is converted into a library of cDNA molecules with adopters attached to one end (for single-end sequencing) or both ends (for pair-end sequencing). These molecules are then sequenced through RNA-seq high-throughput sequencing platform to generate short sequence reads (typically 300–400 bp) from one end or both ends (Wang et al. 2009). The sequenced reads are then analyzed to identify the specific genes involved in the expression of particular phenotypes in particular environments either through reference or de novo assembly-based approaches. Today, RNA-seq platform is a fundamental tool in generating the molecular markers like EST-SSRs (expressed sequence tags-simple sequence repeats) and SNPs (single nucleotide polymorphism) in various complex millet species. To date, *Echinochloa* has enormous transcript profiles particularly, in weedy species for various traits associated with herbicide resistance, photosynthesis, flooding response, and other housekeeping genes (Li et al. 2013a, 2013b; Yang et al. 2013; Nah et al. 2015; Guo et al. 2017; Gao et al., 2018). Very recently, transcriptomics of *E. frumentacea* (cv. CO (Kv) 2), identified some key genes involved in the regulation of Fe and Zn accumulation and drought tolerance. Besides, the study yielded a wealth of transcripts (97,065) data for future gene discovery in cultivated species (Murukarthick et al. 2019). The study also generated 300 EST-SSR primer pairs from 10,881 SSR loci targeting major repeats of trinucleotide (122) followed by dinucleotide (121), tetra-nucleotide (35), penta-nucleotide (20), and hexa-nucleotide (2).

To date, NCBI database holds 952 gene sequences of *Echinochloa* species generated through transcriptomics approach. The maximum number of genes are available in *E. crus-galli* (170) followed by *E. oryzicola* (132), *E. frumentacea* (130), *E. esculenta* (130), *E. colona* (130), *E. ugandensis* (131), and *E. stagnina* (129). The genes were found to be associated with various adaptive traits including photosynthesis (PS I, PS II, NADH-plastoquinone *oxidoreductase*, and ATP *synthase)*, C4 pathways (phosphoenolpyruvate *carboxylase, aldolase, maturase* K, and *kinase*), micronutrient transportation (Fe^{2+} transport protein 2-like protein (*IRT2*) gene, *nicotianamine synthase* 1 (*NAS1*) and *nicotianamine synthase* 2 (*NAS2*)), herbicide resistance (*1-aminocyclopropane-1-carboxylic acid synthase* 3, *acetolactate synthase*, calcineurin, cyclophilin 2, cytochrome P450, GH31 and *glutathione S-transferase)*, flooding tolerances (*enolase* and *alcohol dehydrogenase)*, waxy grains (*granule-bound starch synthase)*, non-shattering (*sh_4*), ribosomal RNA, RNA and DNA *polymerases* and transfer RNAs, etc. (last accessed: September, 2021; https://www.ncbi.nlm.nih.gov/search/all/?term=echinochloa).

3.2.4 Proteomics Studies

Proteomics is the branch of omics that deals with overall proteome/proteins expressed in cell, tissue, or an organism at any given point of time (Aslam et al. 2017). Although proteome is directly proportional to the transcriptome, the levels and diversity of proteome are not directly proportional to the transcriptome due to complex regulatory systems and other posttranslational modification mechanisms (Aslam et al. 2017). Proteomics supplements the genomic and transcriptomics to large-scale structural and functional analysis of the expressed proteins. Initially, two-dimensional electrophoresis (2-DE), a gel-based technique was the only method for separation and quantification in proteomic studies. Later, advanced tools like Liquid Chromatography-Tandem Mass Spectrometry (LC–MS-MS), Matrix-assisted laser desorption ionization-time-of-flight MALDI-TOF/TOF, X-ray crystallography and NMR spectroscopy are being extensively used in major and minor cereals including rice, wheat, maize, sorghum, pearl millet, finger millet, and foxtail millet. However, in barnyard millet the reports on proteomics are very scanty due to lack of research investment. Nevertheless, to date total of 540 proteins were annotated in *Echinochloa* species for tRNA, ribosomal, and photosystem associated proteins (last accessed: September 2021; https://www.ncbi.nlm.nih.gov/search/all/?term=echinochloa). The comparative analysis further revealed that *Echinochloa* proteins are orthologues to commonly expressed proteins involved in C_4 pathways, calcium-binding protein, photosynthesis, *bZIP* transcription factor 1, translational initiation factors, and transporters of major and minor cereals (Yang et al. 2013). Recently, unique proteins like Cu/Zn superoxide *dismutase*, defensin, quinclorac-resistance, cadmium tolerance, viral nucleoprotein, and antimicrobial peptides in *E. crus-galli* (Odintsova et al. 2008; Wu et al. 2019), multiple-herbicide-resistant protein in *E. phyllopogan* (Iwakami et al. 2014), and *granule-bound starch synthase* in *E. esculenta* (Ishikawa et al. 2013) were also identified. However, more research aimed at understanding the species-specific expressions and other structural analyses in barnyard millet is warranted.

3.3 Application of Omics in Barnyard Millet

3.3.1 Genetic Diversity Studies

Development of molecular marker greatly facilitates genetic diversity, quantitative trait loci (QTL) mapping/analyses, and marker-assisted breeding in crop plants. The early molecular marker studies in *Echinochloa* species were largely based on Random Amplified Polymorphic DNA (RAPD) markers to access the genetic diversity and phylogenetic relationship. Although RAPD markers are relatively difficult to score and reproduce, Hilu (1994) used RAPD to differentiate the cultivated and weedy species of *E. frumentacea* and *E. utilis* populations and found that the cultivated *E. frumentacea* was more diverse than *E. utilis*. Similarly, Rutledge et al. (2000) used RAPD markers to study the genetic diversity among weedy *Echinochloa* populations for propanil resistance and susceptibility. Later, Ruiz-Santaella et al. (2006) used RAPD markers for discriminating the *Echinochloa* species against herbicide tolerance and clearly separated the highly susceptible species *E. oryzicola* from low-susceptible species *E. colona*. Isozyme marker analysis by Prabha et al. (2010) between two cultivated species of *Echinochloa* revealed that *esterase* enzyme produced more polymorphism between and within the species and this indicated the presence of diversity in allelic composition. Later, other studies using Amplified Fragment Length Polymorphism (AFLP) markers documented more alleles per primer compared to RAPD markers. For instance, Danquah et al. (2002) reported 166 polymorphic bands using four primer pairs with an average of 41.5 per primer pair in 28 genotypes. Similarly, Tabacchi et al. (2009) observed 156 polymorphic bands using seven primer pairs with an average of 22.3 alleles per primer in 80 genotypes. Besides, the PCR-based RFLP (polymerase chain reaction–restriction fragment length polymorphism) was also applied for species identification among *Echinochloa* accessions (Yamaguchi et al. 2005; Yasuda et al. 2006). InDel markers developed in *E. colona* have the potential to serve as a cost-effective approach to DNA barcoding of cultivated species (Lee et al. 2017).

With the advancement of next-generation sequencing (NGS) techniques, molecular markers for genetic diversity studies used sequence-based markers such as SSRs and EST-SSRs that offered both high information content per assay besides being codominant and reproducible in many crop plants (Lin et al. 2011). However, the information available on sequenced-based markers in barnyard millet is very scanty compared to other minor millets. For instance, Nozawa et al. (2006) used five SSR markers to successfully differentiate 155 accessions of *Echinochloa* species into three clusters. The ESTs markers on the other hand mined from SSR have also been proven to be a very informative and effective tool for the analysis of genetic diversity in crop plants. Extensive transcriptomics and annotation reports are available in weedy *E. crus-galli* species against herbicide tolerance ((Li et al. 2013b; Yang et al. 2013). However, these resources were used limitedly in the marker development and diversity studies since they are weedy ancestors. To date, 74 ESTs have been discovered and available public in *Echinochloa* species). Besides, Babu and Chauhan (2017) mined 22 pairs of EST-SSR primers from wild for comparative genomics studies. The study revealed that two ESTs were homologous to the chromosomal regions of 2, 5, 6, 8, 9, and 12 of rice and waxy genes of maize. Recently, a restriction-site associated DNA (RAD) approach has also been performed for the rapid and mass discovery of SSR and SNP markers in *E. phyllopogon* (Chen et al. 2017). This study yielded 4710 SSRs and 49,179 SNPs. Out of the 4710 SSR markers annotated, 78 showed polymorphism in four *E. phyllopogon* population and produced 66 alleles with an average of 3.1–4.8 alleles per locus. Similarly, SNP mining from the ESTs revealed one potential SNP and one reliable SNP and two haplotypes. Besides the sequence analysis revealed, barnyard millet has the most frequent motif as AT (>82%) followed by tri, tetra, penta, and hexa. The percentage of GC was high (48.9%) as in other monocots. The increased GC content was mostly documented in the species that grow seasonally cold and/or in dry climates (Smarda et al. 2014). This proved the barnyard millet possesses the characteristic feature of freezing and desiccation tolerance.

In cultivated species, Manimekalai et al. (2018) and Murukarthick et al. (2019) developed a few EST-SSR markers from the transcriptome sequence of *E. frumentacea* for genetic diversity studies. Among 51 EST-SSR markers developed 14 were polymorphic with average PIC values between 0.276 and 0.652 (Manimekalai et al. 2018). Similarly, Murukarthick et al. (2019) identified 10 polymorphic markers in 30 EST-SSRs and showed polymorphism in 30 cultivated barnyard millet accessions. Recently, Wallace et al. (2015) developed 21,000 SNPs through genotyping-by-sequencing (GBS) method in core germplasm comprising 95 barnyard millet accession. Among 21,000 SNPs, 10,816 were spread in *E. colona* and 8217 in *E. crus-galli* biotypes. The discriminating SNPs among *E. colona* and *E. crus-galli* population were 1299 and 1444, respectively. The population structure analysis further separated the *E. colona* biotypes into four different clusters and *E. crus-galli* into three biotypes (Wallace et al. 2015).

3.3.2 Gene/QTL Mapping

Identification of markers linked to the genes/QTLs governing important traits will help in development of genotypes having good agronomic performance and nutritional quality. For any QTL/gene mapping, two strategies have been widely applied in crop plants: (1) linkage/family-based mapping that requires construction of mapping populations such as F_2, RIL (recombinant-inbred lines), NIL (near-isogenic lines), or DH (doubled-haploid) developed from the contrast parents; (2) association mapping/natural population-based mapping that requires natural populations and takes advantage of historic recombination events that were accumulated over generations. Several, attempts have been made in various millets in deciphering the key genes governing important traits using both strategies. However, in barnyard millet only two preliminary mapping studies have been published using linkage-based mapping approach due to complexity of the genome (hexaploid) and minute flower in nature, which often limits the development of mapping populations (Renganathan et al. 2021). In a study, Ishikawa et al. (2013) identified the molecular marker associated with waxy genes in barnyard millet populations. Here, the functional alleles in all the three loci (*EeWx1, EeWx1,* and *EeWx3*) exhibited normal amylose content (wild), one of the mutated alleles produced low amylose, and all the loci with completely homozygous mutant alleles exhibited amylose-free genotype (waxy). The marker identified in this study may be used in marker-assisted breeding for waxy traits in barnyard millet. In another study, bulked segregants analysis (BSA) of F_2 population that varied in anthocyanin pigmentation revealed that EST-SSR marker viz., BMESSR 39, was associated with pigmentation traits (Renganathan et al. 2021). Moreover, this marker segregates according to the pigmentation trait (1:2:1) suggesting that the markers and trait are closely linked. Therefore, the gene controlling pigmentation was reported to be monogenic with dominant expression in barnyard. This information could pave way for marker/genetic studies in barnyard millet.

3.3.3 Comparative Genomics Studies

Due to nonavailability of whole-genome sequencing in cultivated type, the comparative genomics approach has provided significant opportunities for analysis of genetic diversity in barnyard millet (Renganathan et al. 2020). For instance, the genomes of rice, maize, and foxtail millet have served as essential models to study the marker-based syntenic relationships in various millet species including pearl millet and finger millet. Similarly, in barnyard millet, the in silico mining of genomic SSRs (159) developed from the foxtail millet sequence exhibited high degree of cross-transferability (91.3%) in barnyard millet and other minor millet species (Pandey et al. 2013). Further, EST-derived SSR markers from transcriptome sequence of foxtail millet showed consistent amplification and high cross-transferability in barnyard millet (90.6%) and other non-millet species (Kumari et al. 2013). In another study, intron length polymorphic markers (ILP) mined from the foxtail millet genome showed 94% cross-transferability amplification in barnyard millet (Muthamilarasan and Prasad 2014). The genic SSRs derived for calcium transporters and calcium kinases from rice genome showed 100% and 72.2% cross-transferability, respectively, in barnyard millet (Yadav et al. 2014). The cross-species amplification of rice, maize, and finger millet genomic SSRs in barnyard millet genetic diversity revealed that compared to rice, maize, and finger millet SSRs showed higher cross-species amplification and PIC values (Babu et al. 2018a, 2018b). Moreover, the EST-derived SSR markers had higher cross-genome amplification than genomic SSR markers, indicating EST-SSRs are highly conserved than the gSSR in grass family. The summary of cross-transferability markers in barnyard millet is presented in Table 3.2.

Through a comparative genomics approach, the divergence time of wild *E. crus-galli* from rice and sorghum was found between 48.5 and 28.5 Mya, respectively. During 7.8 Mya the *Echinochloa* species undergoes polyploidization and diversification into the cultivated types (Guo et al. 2017). The synteny analysis of the genes of *E. crus-galli*, related to DIMBOA (2, 4-dihydroxy-7-methoxy-1,4-benzoxazin-3-one) revealed that the segments of BX1–5 and BX8 of the maize genome have perfect synteny with *Echinochloa*. Similarly, momilactones of rice (Guo et al. 2017), waxy gene of maize, *granule-bound starch synthase* I (GBSSI-S) gene of foxtail millet, and little millet were orthologous to *E. crus-galli* genes (Babu and Chauhan 2017). Besides, several syntenic regions for the traits viz., seed dormancy, plant height, panicle length, spikelet characters, leaf senescence, seed weight/yield-related traits, shattering character, root traits, blast resistance, brown plant hopper resistance, and amylose content were orthologous to rice (Babu et al. 2018b). Therefore, the conserved sequences for both gene and SSR flanking regions across the cereal species, suggested the possibility of using cross-species markers for various applications in genetics and genomics research of barnyard millet.

Table 3.2 Cross-species markers for barnyard millet genetic diversity studies

Species	Marker	No. of markers studied	No. of amplified markers	Cross-transferability percentage	References
Finger millet	eSSR	104	15	14.2	Arya et al. (2014)
	gSSR	132	20	15.5	Arya et al. (2014)
	eSSR, gSSR	101	33	55.4	Krishna et al. (2018)
	SSR	18	07	85.7	Babu et al. (2018a)
	SSR	32	32	90.6	Babu et al. (2018b)
Pearl millet	SSR	32	10	31.2	Arya et al. (2014)
Foxtail	SSR	58	53	91.3	Pandey et al. (2013)
	eSSR	106	106	90.6	Kumari et al. (2013)
	ILP	100	94	94.1	Muthamilarasan and Prasad (2014)
	SSR	64	46	73.4	Gupta et al. (2013)
	SSR	26	07	65.38	Krishna et al. (2018)
Sorghum	eSSR	42	42	80.9	Yadav et al. (2014)
Rice	eSSR	102	102	72.1	Yadav et al. (2014)
	SSR	120	85	48.2	Babu et al. (2018b)
Maize	SSR	46	32	70.0	Babu et al. (2018a)

SSR simple sequence repeat, *eSSR* EST-derived simple sequence repeat, *gSSR* genomic simple sequence repeat, *ILP* intron length polymorphism, *PIC* polymorphic information content, *NA* not available

3.4 Summary and Future Perspective

To date, genomic research in barnyard millet has relied almost exclusively on comparative genomics, even though the transcriptomic sequences have been quite widely used in other orphan polyploid species including teff. As a wide range of "omics" technologies are available to enhance the genomic research in major and some minor cereals, it is imperative that barnyard millet researchers should focus on these tools to fast-track the development of newer varieties for future needs. The availability of high-throughput phenotyping such as infrared imaging, digital imaging, and drone-based platforms in major millets like sorghum are needed to be tested and validated for accurate mapping of genes/QTLs/alleles responsible for the trait-of-interest in barnyard millet. These techniques reduce the laborious nature of the phenotyping work, and labor cost, besides giving precise phenotyping data at the field level. Reduction in sequencing costs and improvements in bioinformatical softwares for polyploid sequence assembly, as in the case of sugarcane, provide positive examples for sequencing the complex cultivated barnyard millet species.

The availability of draft genome sequence of wild aids in understanding the various adaptive traits of *Echinochloa* species. Several developmental, adaptive, and stress tolerance trait genes have been identified in barnyard millet, but only a few of them are known to be characterized through proteomics approaches. Therefore, proteomic techniques must be applied to catalog and characterize those special proteomes in response to posttranslational modification systems. Since, the barnyard millet grains are rich in micronutrients, particularly, Fe and Zn, high-throughput elemental profiling systems like inductively coupled plasma-mass spectrometry (ICP-MS), inductively coupled plasma-optical emission spectrometry (ICP-OES), X-ray fluorescence (XRF), laser ablation inductively coupled plasma-mass spectrometry

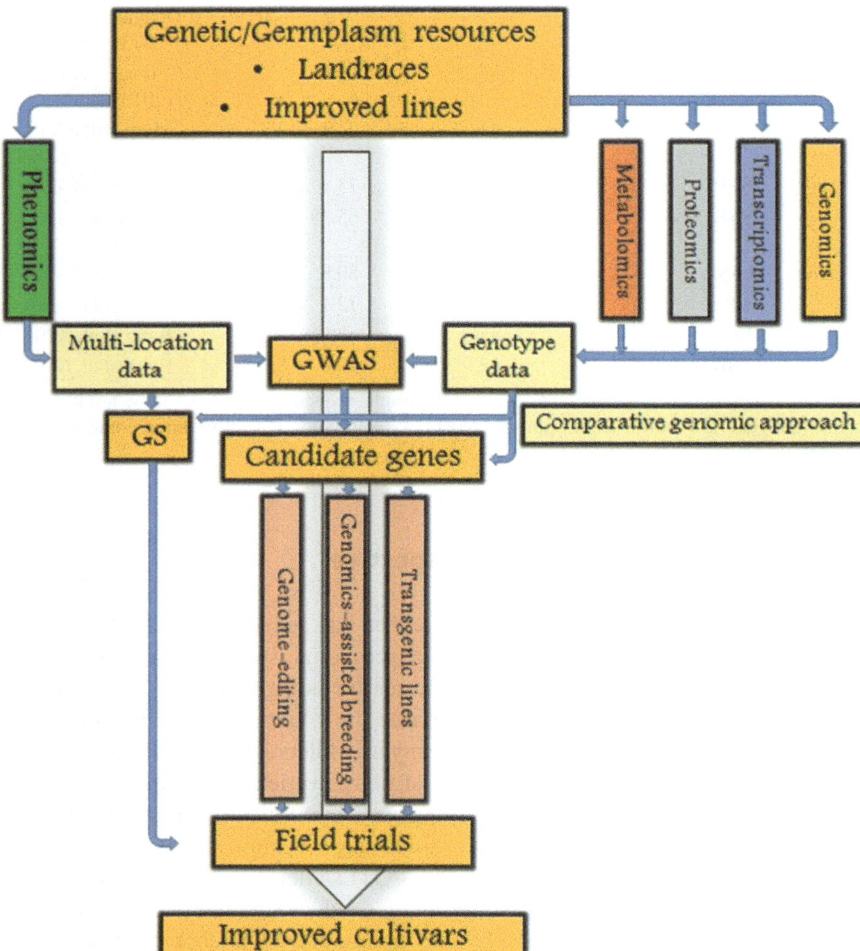

Fig. 3.2 Schematic representation of the proposed omics strategy for genetic improvement of barnyard millet. GWAS, Genome-wide association study; GS, Genomic selection

(LA-ICP-MS), neutron activation analysis (NAA) have to be adopted for studying ionomes. The ionomics approach not only profiles the sum of ionomes present in the cell but also deciphers the dynamics of particular ionomes in affecting plant growth stages under various stresses.

To date, no reports are available on genetic transformation in barnyard millet. This system provides new avenues for the functional genomics studies related to micronutrient accumulation and tolerance mechanisms against biotic and abiotic stresses. The schematic representation of omics strategy to improve the barnyard millet cultivar is presented in Fig. 3.2. Finally, once all these omics data are developed, a web-based resources could be established for retrieving information and other functional analysis frameworks as in foxtail millet. Recently, UN-FAO has declared the year 2023 as "International Year of Millets," intended to increase the consumer and farmers' awareness about millet consumption for healthy life and their persistent productivity under climate change. Omics strategy along with a better investment plan would help to overcome the above said challenges in nutritionally rich and climate-resilient barnyard millet for future food security and sustainable production under different climate change scenarios.

Acknowledgments All the authors sincerely acknowledge Tamil Nadu Agricultural University, Coimbatore, Tamil Nadu, India. The first author extends his thanks to AAIC, the Centre of Excellence in Innovation-Postdoctoral program.

References

Arya L, Chauhan D, Yadav Y, Verma M (2014) Transferability of simple sequence repeat (SSR) markers developed in Finger Millet, and Pearl Millet to Kodo Millet and barnyard Millet. *Innovative Approach in Stem Cell Research, Cancer Biology and Applied Biotechnology*, 60–64

Aslam B, Basit M, Nisar MA, Khurshid M, Rasool MH (2017) Proteomics: technologies and their applications. J Chromatogr Sci 55:182–196. https://doi.org/10.1093/chromsci/bmw167

Babu B, Chauhan R (2017) *In-silico* identification of EST based microsatellite markers and SNPs, and comparative genomic analysis of ESTs in barnyard millet for their omics applications. Curr Agric Res J 5:279–287. https://doi.org/10.12944/CARJ.5.3.03

Babu B, Rashmi C, Sood S (2018a) Cross transferability of finger millet and maize genomic SSR markers for genetic diversity and population structure analysis of barnyard millet. Indian J Genet Plant Breed 78:364–372. https://doi.org/10.31742/IJGPB.78.3.5

Babu B, Sood S, Kumar D, Joshi A, Pattanayak A, Kant L et al (2018b) Cross-genera transferability of rice and finger millet genomic SSRs to barnyard millet (*Echinochloa spp.*). 3 Biotech 8. https://doi.org/10.1007/s13205-018-1118-1

Bajwa A, Jabran K, Shahid M, Ali HH, Chauhan B, Ehsanullah (2015). Eco-biology and management of *Echinochloa crus-galli*. Crop Prot 75, 151–162. doi:https://doi.org/10.1016/j.cropro.2015.06.001

Chen G, Zhang W, Fang J, Dong L (2017) Identification of massive molecular markers in *Echinochloa phyllopogon* using a restriction-site associated DNA approach. Plant Divers 39. https://doi.org/10.1016/j.pld.2017.08.004

Danquah EY, Hanley SJ, Brookes RC, Aldam C, Karp A (2002) Isolation and characterization of microsatellites in *Echinochloa* (L.) Beauv. spp. Mol Ecol Notes 2:54–56

Gao Y, Li J, Pan X, Liu D, Napier R, Dong L (2018) Quinclorac resistance induced by the suppression of the expression of 1-aminocyclopropane-1-carboxylic acid (ACC) synthase and

ACC oxidase genes in Echinochloa crusgall i var. zelayensis. Pestic Biochem Physiol 146:25–32. https://doi.org/10.1016/j.pestbp.2018.02.005

Guo L, Qiu J, Ye CY, Jin G, Lingfeng M, Zhang H et al (2017) *Echinochloa crus-galli* genome analysis provides insight into its adaptation and invasiveness as a weed. Nat Commun 8. https://doi.org/10.1038/s41467-017-01067-5

Gupta S, Kumari K, Muthamilarasan M, Alagesan S, Prasad M (2013) Development and utilization of novel SSRs in foxtail millet [*Setaria italica* (L.) P. Beauv.]. Plant Breed 132. https://doi.org/10.1111/pbr.12070

Hereward JP, Werth JA, Thornby DF, Keenan M, Chauhan BS, Walter GH (2016) Complete chloroplast genome sequences of six lines of *Echinochloa* colona (L.) link. Mitochondrial DNA Part B 1:945–946

Hilu K (1994) Evidence from RAPD markers in the evolution of *Echinochloa* millets (Poaceae). Plant Syst Evol 189:247–257. https://doi.org/10.1007/BF00939730

IIMR (2018) Annual Report 2017–18. Indian Institute of Millets Research, Hyderabad

Ishikawa G, Seimiya Y, Saito M, Nakamura T, Hoshino T (2013) Molecular characterization of spontaneous and induced mutations in the three homoeologous *waxy* genes of Japanese barnyard millet [*Echinochloa esculenta* (A. Braun) H. Scholz]. Mol Breed 31:69–78. https://doi.org/10.1007/s11032-012-9769-9

Iwakami S, Uchino A, Kataoka Y, Shibaike H, Watanabe H, Inamura T (2014) Cytochrome P450 genes induced by bispyribac-sodium treatment in a multiple-herbicide-resistant biotype of *Echinochloa phyllopogon*. Pest Manag Sci 70:549–558. https://doi.org/10.1002/ps.3572

Jiang B, Lao S, Wu D, Fan L, Ye CY (2021) The complete chloroplast genome of Echinochloa haploclada. Mitochondrial DNA Part B 6:3105–3106

Krishna TPA, Maharajan T, David RHA, Ramakrishnan MS, Ceasar A, Duraipandiyan V, Roch GV, Ignacimuthu S (2018) Microsatellite markers of finger millet (*Eleusine coracana* (L.) Gaertn) and foxtail millet (*Setaria italica* (L.) Beauv) provide resources for cross-genome transferability and genetic diversity analyses in other millets. Biocatal Agric Biotechnol 16: 493–450

Kumari K, Muthamilarasan M, Misra G, Gupta S, Alagesan S, Parida S et al (2013) Development of eSSR-markers in *Setaria italica* and their applicability in studying genetic diversity, cross-transferability and comparative mapping in millet and non-millet species. PLoS One 8:e67742. https://doi.org/10.1371/journal.pone.0067742

Kumari KS, Thayumanavan B (1998) Characterization of starches of proso, foxtail, barnyard, kodo, and little millets. Plant Foods Hum Nutr 53:47. https://doi.org/10.1023/A:1008083020810

Lee J, Kim C-S, Lee I-Y (2017) Discrimination of *Echinochloa colona* (L.) Link from other *Echinochloa* species using DNA barcode. Weed Turfgrass Sci 4:225–229. https://doi.org/10.5660/WTS.2015.4.3.225

Li G, Wu S, Cai L, Wang Q, Zhao X, Wu C (2013b) Identification and mRNA expression profile of glutamate receptor-like gene in quinclorac-resistant and susceptible *Echinochloa crus-galli*. Gene 531(2):489–495. https://doi.org/10.1016/j.gene.2013.09.013

Li G, Wu SG, Yu RX, Cang T, Chen LP, Zhao XP, Cai LM, Wu CX (2013a) Identification and expression pattern of a glutathione S-*transferase* in *Echinochloa crus-galli*. Weed Res 53. https://doi.org/10.1111/wre.12031

Lin H-S, Chiang CY, Chang S-B, Kuoh C-S (2011) Development of simple sequence repeats (SSR) markers in *Setaria italica* (Poaceae) and cross-amplification in related species. Int J Mol Sci 12: 7835–7845. https://doi.org/10.3390/ijms12117835

Manimekalai M, Dhasarathan M, Karthikeyan A, Murukarthick J, Renganathan VG, Thangaraj K et al (2018) Genetic diversity in the barnyard millet (*Echinochola frumentacea*) germplasms revealed by morphological traits and simple sequence repeat markers. Curr Plant Biol 14:71–78. https://doi.org/10.1016/j.cpb.2018.09.006

Murukarthick J, Manimekalai M, Karthikeyan A, Perumal S, Dhasarathan M, Kandasamy T et al (2019) Transcriptomes of Indian barnyard millet and barnyard grass reveal putative genes involved in drought adaptation and micronutrient accumulation. Acta Physiol Plant 41:66. https://doi.org/10.1007/s11738-019-2855-4

Muthamilarasan M, Prasad M (2014) Advances in *Setaria* genomics for genetic improvement of cereals and bioenergy grasses. Theor Appl Genet. https://doi.org/10.1007/s00122-014-2399-3

Nah G, Im JH, Kim JW, Kim K, Lim J, Choi AY, Choi IY, Yang TJ, Park TS, Lee D, Kim DS (2016) The complete chloroplast genomes of three Korean *Echinochloa crus-galli* accessions. Mitochondrial DNA Part A 27:4357–4358

Nah G, Im J-H, Kim J-W, Park H-R, Yook M-J, Yang T-J et al (2015) Uncovering the differential molecular basis of adaptive diversity in three *Echinochloa* leaf transcriptomes. PLoS One 10: e0134419. https://doi.org/10.1371/journal.pone.0134419

Nozawa S, Takahashi M, Nakai H, Sato Y-I (2006) Difference in SSR variations between Japanese Barnyard Millet (*Echinochloa esculenta*) and its wild relative *E. crus-galli*. Breed Sci 56:335–340. https://doi.org/10.1270/jsbbs.56.335

Odintsova TI, Rogozhin EA, Baranov Y, Musolyamov AK, Yalpani N, Egorov TA et al (2008) Seed defensins of barnyard grass *Echinochloa crusgalli* (L.) Beauv. Biochimie 90:1667–1673. https://doi.org/10.1016/j.biochi.2008.06.007

Padulosi S, Mal B, Ravi SB, Gowda J, Gowda KTK et al (2009) Food security and climate change: role of plant genetic resources of minor millets. Indian J Plant Genet Resour 22:1–16

Pandey G, Misra G, Kumari K, Gupta S, Kumar Parida S, Chattopadhyay D et al (2013) Genome-wide development and use of microsatellite markers for large-scale genotyping applications in Foxtail Millet [*Setaria italica* (L.)]. DNA Res 20. https://doi.org/10.1093/dnares/dst002

Pareek A, Dhankher OP, Foyer C (2020) Mitigating the impact of climate change on plant productivity and ecosystem sustainability. J Exp Bot 71:451–456. https://doi.org/10.1093/jxb/erz518

Perumal S, Jayakodi M, Kim D-S, Yang T-J, Natesan S (2016) The complete chloroplast genome sequence of Indian barnyard millet, *Echinochloa frumentacea* (Poaceae). Mitochondrial DNA B 1(1):79–80. https://doi.org/10.1080/23802359.2015.1137832

Prabha D, Negi YK, Khanna VK (2010) Morphological and isozyme diversity in the accessions of two cultivated species of barnyard millet. Nat Sci 8:71–76

Renganathan VG, Vanniarajan C, Karthikeyan A, Ramalingam J (2020) Barnyard millet for food and nutritional security: current status and future research direction. Front Genet 11:500. https://doi.org/10.3389/fgene.2020.00500

Renganathan VG, Vanniarajan C, Senthil N, Nirmalakumari A, Karthikeyan A, Veni K, Ramalingam J (2021) Genetics and molecular markers for anthocyanin pigmentation in barnyard millet (Echinochloa frumentacea (Roxb.) link). Plant Breed 140:246–253. https://doi.org/10.1111/pbr.12892

Ruiz-Santaella PJ, Bastida F, Franco A, Prado R (2006) Morphological and molecular characterization of different *Echinochloa* spp. and *Oryza sativa* populations. J Agric Food Chem 54: 1166–1172. https://doi.org/10.1021/jf0520746

Rutledge J, Talbert RE, Sneller CH (2000) RAPD analysis of genetic variation among propanil-resistant and -susceptible *Echinochloa crus-galli* populations in Arkansas. Weed Sci 48:669–674

Sebastin R, Lee KJ, Cho GT, Lee JR, Shin MJ, Kim SH et al (2019) The complete chloroplast genome sequence of Japanese millet *Echinochloa esculenta* (A. braun) H. scholz (*Poaceae*). Mitochondrial DNA Part B Resour 4:1392–1393. https://doi.org/10.1080/23802359.2019.1598787

Smarda P, Bures P, Horova L, Leitch IJ, Mucina L et al (2014) Ecological and evolutionary significance of genomic GC content diversity in monocots. Proc Natl Acad Sci U S A 111: E4096-102. https://doi.org/10.1073/pnas.1321152111

Sood S, Khulbe R, Kumar RA, Agrawal PK, Upadhyaya H (2015) Barnyard millet global core collection evaluation in the sub mountain Himalayan region of India using multivariate analysis. Crop J 3(6):517–525. https://doi.org/10.1016/j.cj.2015.07.005

Tabacchi M, Mantegazza R, Spada A, Ferrero A (2009) Morphological traits and molecular markers for classification of *Echinochloa* species from Italian rice fields. Weed Sci 54:1086–1093. https://doi.org/10.1614/WS-06-018R1.1

Ugare R, Chimmad B, Naik R, Bharati P, Itagi S (2014) Glycemic index and significance of barnyard millet (*Echinochloa frumentacae*) in type II diabetics. J Food Sci Technol 51(2): 392–395. https://doi.org/10.1007/s13197-011-0516-8

Wallace GJ, Upadhyaya H, Vetriventhan M, Buckler E, Hash Charles J, Ramu P (2015) The genetic makeup of a global barnyard millet germplasm collection. Plant Genome 08(01):01–07. https://doi.org/10.3835/plantgenome2014.10.0067

Wang Z, Gerstein M, Snyder M (2009) RNA-Seq: a revolutionary tool for transcriptomics. Nat Rev Genet 10:57–63. https://doi.org/10.1038/nrg2484

Weber A (2015) Discovering new biology through RNA-Seq. Plant Physiol 169:01081.2015. https://doi.org/10.1104/pp.15.01081

Wu LM, Fang Y, Yang HN, Bai LY (2019) Effects of drought-stress on seed germination and growth physiology of quinclorac-resistant *Echinochloa crusgalli*. PLoS One 14(4):e0214480

Yadav S, Gaur V, Jaiswal JP, Kumar A (2014) Simple sequence repeat (SSR) analysis in relation to calcium transport and signaling genes reveals transferability among grasses and a conserved behavior within finger millet genotypes. Plant Syst Evol 300:1–8. https://doi.org/10.1007/s00606-014-0982-3

Yamaguchi H, Utano AYA, Yasuda K, Yano A, Soejima A (2005) A molecular phylogeny of wild and cultivated *Echinochloa* in East Asia inferred from non-coding region sequences of trnT-L-F. Weed Biol Manage 5:210–218. https://doi.org/10.1111/j.1445-6664.2005.00185.x

Yang X, Yu X-Y, Li YF (2013) De novo assembly and characterization of the barnyard grass (*Echinochloa crus-galli*) transcriptome using next-generation pyrosequencing. PLoS One 8: e69168. https://doi.org/10.1371/journal.pone.0069168

Yasuda K, Yano A, Nakayama Y, Yamaguchi H (2006) Molecular identification of *Echinochloa oryzicola* Vasing. And *E. crus-galli* (L.) Beauv. Using a polymerase chain reaction–restriction fragment length polymorphism technique. Weed Biol Manage 2:11–17. https://doi.org/10.1046/j.1445-6664.2002.00041.x

Ye C-Y, Lin Z, Li G, Wang Y, Qiu J, Fu F et al (2014) *Echinochloa* chloroplast genomes: insights into the evolution and taxonomic identification of two weedy species. PLoS One 9(11):e113657. https://doi.org/10.1371/journal.pone.0113657

Role of Inducible Promoters and Transcription Factors in Conferring Abiotic Stress-Tolerance in Small Millets

4

H. B. Mahesh, Meghana Deepak Shirke, Indrajeet Ghodke, and N. R. Raghavendra

Abstract

Millets are a diverse group of small-seeded annual C_4 Panicoid grasses cultivated as food and fodder crops mainly grown in poor and marginal subtropical and tropical regions globally. Generally, the millets are subjected to various abiotic stresses, like drought, heat, waterlogging, and lodging which affect the overall crop yield significantly. Plants exhibit several complex mechanisms to combat such adverse stress conditions. During this time, the gene expressions of stress-responsive genes are either up or downregulated, and several signal cascades are modulated causing an altered accumulation of osmoprotectants. During this signal transduction, transcription factors (TFs) directly regulate the expression of stress-associated genes via serving as molecular switches. In this chapter, we identified TFs across the genomes of millets for which the whole genome and gene repertoire are available. Also, a brief overview of a few well-characterized TFs in some small millets has been also reviewed. Overall, our in silico analysis revealed that *bHLH*, *ERF*, *NAC*, *MYB*, *WRKY*, *bZIP*, *MYB_related* TFs were represented in higher fractions across all the small millets. These TFs are known to impart tolerance against many abiotic stresses and might be the possible reason for small millets being hardy crops that can thrive under adverse climatic conditions.

H. B. Mahesh (✉) · N. R. Raghavendra
Department of Genetics and Plant Breeding, College of Agriculture, Mandya, University of Agricultural Sciences, Bangalore, Karnataka, India

M. D. Shirke
Department of Biotechnology, KTHM College, Nashik, Maharashtra, India

I. Ghodke
Department of Microbiology, KTHM College, Nashik, Maharashtra, India

R. N. Pudake et al. (eds.), *Omics of Climate Resilient Small Millets*,
https://doi.org/10.1007/978-981-19-3907-5_4

Keywords

Small millets · Transcription factors · Inducible promoters · Abiotic stress · Gene regulation · *cis*-regulatory element

4.1 Introduction

Millets are the major cereal crops in the developing countries of semiarid tropical regions of Africa and Asia. Widely grown millets are grouped into two subfamilies, namely Panicoideae (pearl millet, foxtail millets, barnyard millet, kodo millet, little millet, proso millet, and fonio) and Chloridoideae (finger millet and tef) based on their substantial genetic variability coupled with variability for seed size, seed color, seed weight, plant architecture, and shape of their panicles. These are small grain crops (except pearl millet) mainly cultivated in extreme environmental conditions, especially under low moisture and poor soil fertility. Millets are inherently tolerant to many abiotic stresses like drought, salinity, waterlogging, and high temperature. Small millets are very important crops in rainfed areas of semiarid regions being cultivated on limited land. These millets are of local importance that provides reliable yield and contributes significantly to food and nutritional security due to their versatile nature of incorporating into the diets of major food crops. In general, millets are rich in vitamins, calcium, iron, potassium, magnesium, and zinc. In addition, finger millet is rich in methionine, lysine, threonine, and valine while proso millet possesses high leucine, isoleucine, and methionine. The quality of grain protein of proso millet is better than wheat protein. Similarly, fonio grains are also rich in leucine, methionine, and valine. Most of the millet-based diets are considered healthy food due to their low glycemic index, high fiber content, gluten-free, antiproliferative, anticancer, and antioxidant properties.

4.2 Plant Responses to Abiotic Stresses

Abiotic stresses such as salinity, drought, and extreme temperatures severely impact plant growth and metabolism resulting in the loss of a certain amount of grain yield up to 50%. These stresses modulate biochemical, physiological, and molecular activities of a plant cell from seedling to maturity stages. However, plants are evolved with some intricate mechanisms to cope with such stresses. The moment the plants are exposed to abiotic stresses, gene expression patterns are up- or downregulated leading to increased or decreased levels of mRNA, proteins, and metabolites at the cellular level imparting a certain level of tolerance. During this time, multifarious signalling pathways triggered by several stress-responsive genes will be activated to mitigate the stress. During this signal transduction, the expression of stress-associated genes is directly regulated by transcription factors (TFs) by acting as molecular switches. These TFs are temporarily and spatially turned on or off the transcription process by interacting directly with cis-elements located in the promoter region of the target genes they regulate. Usually, these promoter regions are located at upstream region of the transcription start site (TSS) and which is further divided into core promoter, proximal promoter, and distal promoter regions. TFs are proteins coded elsewhere in the genome and will bind to *cis*-regulatory element (CRE) in the promoter region and initiates the transcription process. Promoters are classified into three groups (constitutive, tissue-specific, and inducible promoters) depending on the pattern of gene expression that they produce. Inducible promoters initiate transcription of a particular gene when the plant experiences a specific stimulus, which can be endogenous like plant hormones or external stimuli like abiotic or biotic stresses. In plants, ~10–12% of gene repertoire that are predicted based on whole-genome sequence data potentially encode for TFs, which are classified into different gene families (Franco-Zorrilla et al. 2014; Golldack et al. 2011; Jin et al. 2016; Jin et al. 2014; Hittalmani et al. 2017). The role of TFs in plants during abiotic stresses in plants, in general, is reviewed earlier (Lata et al. 2011) and readers are directed to refer to this review chapter.

4.3 Mining of Transcription Factors (TFs) and Inducible Promoters and their Role in Combating Abiotic Stresses

The whole-genome sequence/assembly is available for most of the small millets in the public databases. Retrieving those assemblies and identification of TFs at whole-genome level will give a global view of TFs content and their sequence diversity within and between species. We retrieved the protein sequences of genes predicted from whole-genome assemblies of finger millet, foxtail millet, proso millet, barnyard millet, teff, fonio, and adlay millet (Kang et al. 2020; Zou et al. 2019; Cannarozzi et al. 2014; Guo et al. 2017; Hittalmani et al. 2017; Wang et al. 2021; Zhang et al. 2012). The TFs were predicted in the PlantTFDB v5.0 prediction tool (http://planttfdb.gao-lab.org/prediction.php) and classified into different families (Table 4.1). The importance of inducible promoters in crop improvement is enormous, hence understanding their precise role in the expression of genes under stress is very critical.

4.3.1 Finger Millet

Out of 85,243 genes in finger millet genome, 11,125 genes were predicted to code for TFs distributed across 56 families (Hittalmani et al. 2017). The bHLH, MYB, FAR1, WRKY, NAC, MYB-related, B3, ERF, bZIP, and HD-ZIP were the top ten TFs represented in higher proportion. Interestingly, M-type and MIKC TFs were present only in finger millet gene repertoire as compared to other millets. Several researchers have attempted to characterize TFs in finger millet, and such studies have been highlighted hereunder.

The NAC (NAM, ATAF1-2, and CUC2) proteins constitute one of the largest families of plant-specific TFs involved in plant growth, defense, and abiotic stress

Table 4.1 Transcription factors in genomes of millets

Sl no.	TF Family	Finger millet	Foxtail millet	Proso millet	Barnyard millet	Teff	Fonio	Adlay millet
1	bHLH	877	233	298	506	165	299	128
2	ERF	505	183	262	405	109	195	114
3	NAC	668	165	214	431	172	249	111
4	MYB	809	155	194	379	128	206	108
5	WRKY	722	126	157	316	121	139	93
7	bZIP	459	141	171	272	93	140	80
8	MYB_related	581	114	119	201	143	138	75
6	C2H2	421	130	200	305	101	172	91
11	G2-like	285	65	99	145	55	97	48
9	FAR1	806	86	76	166	33	23	58
10	GRAS	263	62	117	156	72	127	56
12	B3	505	72	94	153	120	96	45
14	C3H	315	59	85	126	81	73	39
13	HD-ZIP	428	57	83	140	60	94	44
15	M-type_MADS	0	32	34	102	31	18	36
16	Dof	93	37	50	86	33	38	32
17	Trihelix	182	56	54	87	17	44	30
18	GATA	173	37	57	82	23	51	28
19	AP2	65	30	50	79	46	46	28
20	LBD	158	34	58	93	48	55	27
21	ARF	243	38	41	78	53	27	24
22	HSF	150	36	10	77	34	4	24
23	MIKC_MADS	0	51	30	107	52	2	21
24	SBP	60	27	34	59	47	40	20
25	TCP	98	26	37	51	4	16	20

26	TALE	282	43	40	68	38	36	19
27	NF-YB	96	20	28	38	22	15	14
28	ZF-HD	48	21	25	31	2	15	13
29	GeBP	90	16	21	35	2	21	12
30	CO-like	50	16	22	37	4	18	11
31	NF-YC	78	20	27	33	21	15	11
32	Nin-like	78	25	25	48	15	19	10
33	HB-other	129	15	15	32	24	23	9
34	DBB	44	12	15	28	9	20	9
35	GRF	28	13	19	32	3	20	9
36	NF-YA	54	24	21	28	18	18	9
37	YABBY	106	15	17	24	12	18	9
38	ARR-B	33	14	14	30	18	5	9
39	WOX	161	13	22	41	14	21	8
40	CPP	122	13	20	29	16	12	8
41	E2F/DP	72	13	18	26	9	21	7
42	CAMTA	69	9	14	21	25	14	7
43	BES1	58	10	13	25	10	13	7
44	SRS	17	5	13	22	2	11	6
45	EIL	21	6	15	17	7	3	6
46	LSD	11	8	7	13	9	6	4
47	RAV	23	6	6	14	0	7	3
48	NF-X1	31	2	3	3	3	4	3
49	HB-PHD	8	2	4	9	3	1	3
50	Whirly	6	2	3	7	4	4	2
51	VOZ	17	2	4	6	3	4	2
52	LFY	1	1	2	3	2	2	1

(continued)

Table 4.1 (continued)

Sl no.	TF Family	Finger millet	Foxtail millet	Proso millet	Barnyard millet	Teff	Fonio	Adlay millet
53	S1Fa-like	136	1	2	3	2	0	1
54	HRT-like	6	2	2	2	2	0	1
55	STAT	29	2	2	3	1	0	1
56	BBR-BPC	25	7	6	10	4	6	0
57	M-type	188	0	0	0	0	0	0
58	MIKC	142	0	0	0	0	0	0
	Total	**11,125**	**2410**	**3069**	**5320**	**2145**	**2761**	**1594**

tolerance. Expression analysis and characterization of *EcNAC1* transcription factor in finger millet showed increased abiotic stress tolerance in transgenic tobacco plants (Ramegowda et al. 2012). The transgenic tobacco plants expressing *EcNAC1* exhibited tolerance to various abiotic stresses like simulated osmotic stress, by polyethylene glycol (PEG) and mannitol, salinity stress and methyl-viologen (MV)-induced oxidative stress. They also showed reduced levels of reactive oxygen species (ROS) and ROS-induced damage in pot assays. The ROS are the main factors causing cytotoxic damage when plants are exposed to high-intensity light. The root growth under stress and recovery growth after stress was also more in transgenic plants. Many stress-responsive genes like WUSCHEL-related homeobox 13 (*HB-13*), MYB-CC-type TF (*MYB*), protein phosphatase 2C (*PP2C*), serine-threonine protein kinase (*STPK*), NADPH-cytP450 oxidoreductase (*CytOR*), Rop subfamily GTPase (*ROP*), and ATP-binding subunit, *ERD1* were found to be upregulated in transgenic lines expressing *EcNAC1*.

A study of putative NAC TF from a salinity-tolerant finger millet was shown to confer tolerance to drought and salinity in transgenic rice (Rahman et al. 2016). The transgenic rice lines overexpressing *EcNAC67* were generated by *Agrobacterium*-mediated transformation. Transgenic rice plants showed higher root and shoot biomass during stress and showed better revival ability after alleviation of salinity stress. In comparison to non-transgenic cultivar, the *EcNAC67* transgenic lines showed higher relative water content and lesser reduction in grain yield during drought stress. The transgenic lines exhibited much lower drought-induced spikelet sterility than the non-transgenic plants.

The basic leucine zipper (bZIPs) transcription factors have also been studied in finger millet. They have a highly conserved bZIP domain with 60–80 amino acids, with a stretch of 16 basic amino acids, responsible for DNA binding followed by a leucine zipper consisting of heptad repeats of leucine. The leucine zipper residues confer the dimerization specificity. These TFs are known to regulate many cellular responses in plants under stress conditions and play diverse roles in plant development. Researchers have cloned *EcbZIP60* stress-responsive TF from bZIP family and functionally validated in tobacco (Babitha et al. 2015). The expression of *EcbZIP60* was highly upregulated under drought, osmotic, salt, and methyl viologen-induced stress in finger millet. The transgenic plants showed improved tolerance to drought, salinity stress, oxidative stress, and endoplasmic reticulum stress inducers dithiothreitol and tunicamycin. There was a significant upregulation of unfolded protein-responsive pathway genes such as *BiP1*, *CRT1*, and *PDIL* under stress. Thus, improved tolerance to various stresses could be through the unfolded protein-responsive signalling pathway.

Another interesting study has reported the discovery of a novel bZIP transcription factor, *EcbZIP17* tethered to the endoplasmic reticulum (ER) membrane in finger millet (Ramakrishna et al. 2018). The transgenic plants exhibited improved stress tolerance to salt, heat, drought, and osmotic stress. The RT-PCR analysis showed higher expression of the ER stress-responsive genes *BiP*, *PDIL*, and *CRT1*. Interestingly, greater leaf diameter and plant height were observed in transgenic plants as compared to wild type. The overall results indicated that *EcbZIP17* improves plant growth through brassinosteroid signalling under optimal growth conditions and imparts tolerance to various stresses through ER signalling pathways. Overexpression of *EcDREB2A* in transgenic plants showed enhanced heat tolerance through the regulation of ROS homeostasis and upregulation of stress-responsive genes (Ramakrishna et al. 2018; Singh et al. 2021).

4.3.2 Foxtail Millet

Foxtail millet is a climate-resilient model cereal crop to study C_4 photosynthesis, stress tolerance, and bioenergy traits due to its small genome size and short life cycle. The C2H2-type zinc finger TFs were characterized in foxtail millet (*SiC2H2*) and physically mapped on chromosomes (Muthamilarasan et al. 2014a). Zinc fingers are the largest family of TFs in plants, which can be classified into C2H2, C2HC, C2HC5, C3HC4, CCCH, C4, C4HC3, C6, and C8 groups on the basis of the number and location of characteristic residues of zinc finger proteins (Kiełbowicz-Matuk 2012). The expression profiling of nine candidate *SiC2H2* genes in response to salinity, dehydration, and cold stress showed differential expression patterns at different time points. Another study involving genome-wide identification of genes encoding MYB (myeloblastosis) TF was reported in foxtail millet (Muthamilarasan et al. 2014b). The MYB TFs were first identified in *Zea mays* and it was involved in the regulation of anthocyanin biosynthesis (Paz-Ares et al. 1987). MYB proteins in eukaryotes possess a highly conserved MYB DNA-binding domain at N-terminus and 1-4 imperfect repeats (R0, R1, R2, and R3). These repeats contain 50–53 amino acids encoding 3 alpha helices, out of which second and third helices form a helix-turn-helix (HTH) structure. When MYB binds to DNA, the HTH motif intercalates in the major groove of DNA (Lipsick 1996). The activation domain at C-terminus varies significantly between MYB proteins, enabling different regulatory roles of MYB TFs. Muthamilarasan and colleagues have identified 209 *SiMYB* (*S. italica MYB*) protein-encoding genes mapped to nine chromosomes in the foxtail genome. Based on protein domain analysis and the presence of two or three MYB repeats, SiMYB was classified into three structural classes "MYB-related," "MYB-R2R3," and "MYB-R1R2R3." The gene expression pattern of SiMYBs was studied using publicly available Illumina RNA-Seq data for four tissues of foxtail millet, namely root, leaf, stem, and spica. The predominant MYBs like *SiMYB003*, *SiMYB004*, *SiMYB005*, *SiMYB017*, *SiMYB025*, *SiMYB051*, *SiMYB064*, *SiMYB080*, *SiMYB089*, *SiMYB090*, and *SiMYB0100* were expressed in all the tissues while few MYBs like *SiMYB007*, *SiMYB087*, and *SiMYB192* showed tissue-specific expression. Phylogenetic analysis divided the SiMYBs into ten clades, where most of the SiMYB proteins clustered together in the tree with respect to their type (Muthamilarasan et al. 2014b).

A genome-wide investigation of AP2/ERF TF superfamily showed presence of 171 AP2/ERF genes in foxtail millet genome (Lata et al. 2014). APETALA2/ethylene-responsive element-binding factor (AP2/ERF) superfamily is categorized into AP2, ERF, and RAV families based on a number of repetitions and sequence of

AP2/ERF DNA-binding domain. This study has also identified three potential candidate genes viz. *SiAP2/ERF-069*, *SiAP2/ERF-103*, and *SiAP2/ERF-120* for functional validation and improvement of stress resistance. The expression of these genes was upregulated under drought and salinity stresses in ABA-dependent manner.

Late embryogenesis abundant (LEA) proteins which were first identified in cotton plants are very important factors in protecting higher plants from damage due to environmental stresses. With the development of deep sequencing technology, many LEA proteins have been identified. A novel atypical LEA gene named *SiLEA14* was characterized from foxtail millet (Wang et al. 2014). The transgenic Arabidopsis seedlings and foxtail millet showed improved growth under salt and drought stresses as compared to the wild type plants indicating the involvement of *SiLEA14* protein in imparting resistance to abiotic stresses in plants. Nuclear Factor Y (NF-Y), also called heme-activated protein (HAP) or CCAAT binding factor (CBF) is involved in abiotic stress in plants. NF-Y TFs are composed of three distinct subunits NF-YA (HAP2 or CBF-B), NF-YB (HAP3 or CBF-A), and NF-YC (HAP5 or CBF-C). A total of 97 NF-Y genes were identified in foxtail millet. De novo transcriptome analysis showed maximum upregulation of *SiNF-YA1* and *SiNF-YB8* in leaves and/or roots by drought and salt stresses (Feng et al. 2015). Abscisic acid (ABA) and H_2O_2 also played positive roles in the induction of *SiNF-YA1* and *SiNF-YB8*. So, these TFs can be used for the improvement of physiological traits and stress tolerance. Abscisic acid stress ripening (ASR) proteins with ABA/WDS domains are a class of plant-specific TFs, playing an important role in abiotic stress responses and plant development. Six foxtail millet ASR genes were identified (Feng et al. 2016). Overexpression of *SiASR1* showed increased tolerance to drought and oxidative stresses in tobacco, as indicated by developmental and physiological analyses. Expression of *SiASR1* also modulated the transcript levels of oxidation-related genes, including *NtSOD, NtAPX, NtCAT, NtRbohA*, and *NtRbohB*, under drought and oxidative stress conditions.

The homeodomain leucine zipper (HD-Zip) gene family is a major group of TFs in higher plants. They have a DNA-binding domain (HD) at the C-terminal and a leucine-zipper domain (LZ) which can interact with other proteins. The DNA-binding specificity is conferred by homo or heterodimers formed by leucine-zipper domain. Chai and colleagues have reported 47 HD-Zip genes in foxtail millet (Chai et al. 2018). Based on the phylogenetic analysis performed by comparing protein sequences in maize, rice, sorghum, and green foxtail, the HD-Zip genes were classified into four subfamilies (I, II, III, and IV). Conserved motif analysis also divided the HD-Zip gene family into four subfamilies consistent with the classification obtained through the phylogenetic analysis. Segmental as well as tandem duplication events were observed in HD-Zip gene family expansion. Expression profiling carried out to gain insights into the role of HD-Zip TFs during abiotic and biotic stress factors revealed that, all the HD-Zip I genes were drought- and salinity-responsive, with different expression patterns.

GRAS is an important family of plant-specific TFs named after the first three functionally characterized genes, GAI, RGA, and SCR (Zhang et al. 2018). GRAS proteins range from 400 to 700 amino acid residues and can be divided into at least 13 subfamilies including SCR, DLT, OS19, LAS, SCL4/7, OS4, OS43, DELLA, PAT1, SHR, HAM, SCL3, and LISCL (Xu et al. 2016). These TFs play a critical role in the regulation of developmental processes, stress responses, and phytohormonal signalling. A recent genome-wide investigation in foxtail millet identified 57 *GRAS* genes (*SiGRASs*) (Fan et al. 2021b), which were unevenly distributed on nine chromosomes, with the largest number on Chr3 and Chr9 (12 genes, ~21.05%). While there were no *SiGRAS* genes on Chr6. SiGRASs were divided into 13 subfamilies based on the number of conserved domains and gene structure, via phylogenetic tree analysis. Members of the same subfamily had similar gene structures and motif compositions. Genetic structure analysis showed that most *SiGRAS* genes lacked introns. Synteny analysis of *SiGRAS* genes demonstrated the presence of collinear relationships for 47 *SiGRAS* genes with those in Arabidopsis, *F. tataricum*, tomato, *B. distachyon*, rice, and maize. Gene duplication analysis stipulates that segmental duplication events might have contributed more to *GRAS* gene family expansion as compared to tandem duplications. Gene expression analysis showed significant differences in the expression of *SiGRAS* genes in different tissues and stages of fruit development. This study has provided the potential gene targets for genetic improvement and functional genomics research in foxtail millet.

The basic helix-loop-helix (bHLH) TFs family has also been studied in foxtail millet. A total of 187 bHLH genes (SibHLHs) were identified across nine chromosomes in foxtail millet (Fan et al. 2021a). The authors also constructed a phylogenetic tree using Neighbour Joining method by comparing the conserved domains and gene structures of 187 *SibHLHs* and 56 *AtbHLHs* from *Arabidopsis thaliana*. The SibHLH genes were divided into 21 subfamilies and two unclassified orphan SibHLH genes (*SibHLH153* and *SibHLH176*). This analysis also indicated an orthologous relationship and functional conservation with *Arabidopsis thaliana* bHLHs (*AtbHLHs*). The gene duplication analysis revealed a total of 16 tandem repeat events and 38 pairs of segmental duplications which might have contributed to bHLH family expansion in foxtail millet. Gene expression response analysis for 15 typical SibHLH genes, showed a considerable variation in different tissues and during fruit development stages. Analysis of gene expression pattern under eight different abiotic stresses like acid (0.1 M), alkali (0.2 M), PEG (10%), NaCl (5%), heat (40 °C), cold (4 °C), flooding, and darkness showed some SibHLH genes were significantly upregulated or inhibited under different stresses, indicating their involvement in different physiological functions.

Among all millets, there has been abundant research on stress-responsive TFs in foxtail millet through genome-wide studies as well as individual gene characterization. The whole-genome assembly followed by gene prediction by Zhang et al. (2012) reported 38,801 genes in the foxtail millet. As per our analysis, 2410 genes are coding for TFs and distributed across 56 families. The bHLH, ERF, NAC, MYB, bZIP, C2H2, WRKY, MYB_related, FAR1, and B3 TFs were found to be abundant.

4.3.3 Other Small Millets

Proso millet/Broomcorn millet is a highly drought-tolerant C_4 cereal/millet crop widely cultivated in semiarid regions of Asia, Europe, and other continents. It is more efficient in carbon fixation, water, and nitrogen use (Zou et al. 2019). Out of 55,930 genes, 3069 genes were encoding for TFs and bHLH, ERF, NAC, C2H2, MYB, bZIP, WRKY, MYB_related, GRAS, and G2-like TFs were significantly distributed across the proso millet genome. Several researchers have attempted to characterize TFs like WRKY (Yue et al. 2016) NAM, ATAF1/2, and CUC2 (NAC) (Shan et al. 2020), and MYB (Hu et al. 2008) in proso millet under various abiotic stresses but exploring other TFs responsible for imparting abiotic tolerance in this highly valuable crop is the need of the hour.

Barnyard millet is a pernicious weed in agricultural fields worldwide. This is more invasive in rice fields and this tendency is attributed to the presence of cytochrome P450 monooxygenase (CYP450), glutathione S-transferase (GST), and 2,4-dihydroxy-7-methoxy-1,4-benzoxazin-3-one (DIMBOA) genes responsive during allelopathic interactions with rice (Guo et al. 2017). The TFs prediction revealed 5320 genes coding for TFs belong to 56 families and bHLH, NAC, ERF, MYB, WRKY, C2H2, bZIP, MYB_related, FAR1, and GRAS were the major TFs found in barnyard millet. So far, no studies have been attempted to characterize individual TFs in this millet.

Teff is the most important crop in Ethiopia and gaining popularity owing to its nutritional and health benefits. Grain is rich in protein, vitamins, minerals (calcium, iron, and zinc), and gluten-free. Identification of TFs across 41,136 protein-coding genes of teff resulted in 2145 genes encoding for various TFs distributed across 55 families. NAC, bHLH, MYB_related, MYB, WRKY, B3, ERF, C2H2, bZIP, and C3H TFs were in highest fraction compared to other families. Viviparous1 (VP1) TF has a central role in seed development as well as plant response to abiotic stresses during cold and drought was identified in teff (Espelund et al. 2000). Recently, 180 CREs (Cis-acting regulatory elements) associated with WRKY TFs were identified in teff crops which are involved in crosstalks during biotic and abiotic stress-related hormonal responses (Mulat and Sinha 2021b). In silico analysis with other millets resulted in identification of growth regulating factors (GRF), DNA binding with one finger (Dof), vascular plant one zinc finger (VOZ), NF-X1-Like and Hordeum repressor transcription (HRT-like) TFs in teff which are involved in various biological processes and abiotic stress responses (Mulat and Sinha 2020, 2021a, c, 2022). All these TFs were also reported from our in silico prediction analysis (Table 4.1).

Fonio is an annual herbaceous C_4 crop being cultivated under a large range of environments from tropical monsoon climate in western Guinea to arid desert climate in the Sahel zone. It is a drought-tolerant crop which can even thrive under nutrient-poor soils. Exploring its TFs content in the genome of fonio is a feasible option to characterize TFs involved in imparting abiotic stress tolerance. A similar analysis in 59,985 genes of fonio helped us to identify 53 families of TFs distributed across 2761 genes and bHLH, NAC, MYB, ERF, C2H2, bZIP, WRKY,

MYB_related, GRAS, and G2-like were major TFs found. To our knowledge, no TFs are experimentally characterized so far in this crop.

Adlay millet is also known as *Coix lacryma*-jobi or Job's tears cultivated in Asia. Seed is rich in prolamin seed storage protein (Coixin – ~20%) and starch (~67%). In addition, seed extract is reported to have anticancer, antioxidant, anti-inflammatory, anti-allergic, anti-diabetes, and gastroprotective properties (Zhu 2017). Out of 39,574 genes predicted from adlay millet's genome, 1594 genes were encoding for various TFs distributed across 55 families wherein bHLH, ERF, NAC, MYB, WRKY, C2H2, bZIP, MYB_related, FAR1, and GRAS were in higher proportion. Recently, transcriptome sequencing of drought-induced tissue of adlay revealed upregulation of MYB_related, NAC, AP2, ERF WRKY62, and FAR1 and downregulation of WRKY, bHLH, HSF, and ARF TFs (Miao et al. 2021).

Kodo millet is climate-resilient nutri-cereal cultivated in arid and semiarid regions of India, Thailand, Indonesia, Philippines, Vietnam, and West Africa (Upadhyaya et al. 2016). Though whole-genome sequence data is not available for this millet, we surveyed research work conducted on abiotic stresses. Low moisture stress imposed during early seedling stage was assessed for differential expression of genes by transcriptome sequencing. This enabled the identification of several TFs like MYB_related, WRKY, NAC, bHLH, MYB, ERF, C2H2, and bZIP indicating their crucial role in mediating dehydration stress response in kodo millet (Suresh et al. 2022). However, whole-genome sequencing, annotation, and mining of TFs across gene repertoire of this millet is worth attempting in the near future.

Little millet is generally regarded as an orphan crop, small-seeded mostly grown in semiarid tropical and subtropical regions of Asia and Africa. It is native to India and popularly known as "Indian Millet" rich in proteins, fibers, minerals, and good lipids. A recent study on transcriptome profiling of leaf and root tissues under abiotic stress response (drought and salinity) revealed an abundance of WRKY, MYB/MYB_related, C2H2, and bHLH TFs in little millet. In response to stress, more TFs were expressed in leaf than in root (Das et al. 2020). Studies on TFs and whole-genome sequencing should be the future attempt to develop genomic resources for millet breeding community to tap the genetic potential of this highly valued crop.

Overall, bHLH, ERF, NAC, MYB, WRKY, bZIP, MYB_related TFs were represented in higher fractions across all the millets. These TFs are known to impart tolerance against many abiotic stresses and might be the possible reason for millets being hardy crops that can thrive under adverse climatic conditions (Guo et al. 2021; Lata and Prasad 2011; Nakashima et al. 2012; Li et al. 2015; Li et al. 2020; Banerjee and Roychoudhury 2017; Sornaraj et al. 2016). To our knowledge, no studies on TFs involved in abiotic stress tolerance have been conducted in guinea millet and browntop millet.

4.4 Comparative Analysis of TFs across Millets

The identification of orthologous groups is very helpful for cross-species functional annotation, gene evolution, comparative genomics, and identification of species-specific sequences. We used protein sequences of all the TFs predicted from seven millets (28,424 genes) for which the genome information is available (Table 4.2) for orthologous clustering. The protein sequences were uploaded to OrthoVenn2 tool and TFs were clustered as per default parameters (Mahesh et al. 2021). 491 core orthologous groups (COGs) were formed consisting of 5929 genes from seven millets wherein 873 finger millet, 671 foxtail millet, 827 proso millet, 1385 barnyard millet, 744 teff, 902 fonio, and 527 adlay millet genes included in the cluster. Similarly, remaining clusters formed from genes of 6, 5, 4, 3, 2, and 1 species are depicted in Fig. 4.1. In addition, 1231, 125, 443, 687, 449, 292, and 122 singletons were formed in finger millet, foxtail millet, proso millet, barnyard millet, teff, fonio, and adlay millet, respectively. Among COGs, MYB, NAC, bHLH, WRKY, HD-ZIP, ERF, bZIP, G2-like, C2H2, AP2, C3H, B3, GRAS, SBP, MYB_related, and LBD were in major fraction which were conserved across millets.

4.5 Future Prospects

Considering the ever-growing human population and climate change, the development of stress-tolerant varieties is critical for increasing the productivity to meet projected food and energy requirement. The TFs are good candidates for genetic engineering to improve crop tolerance because of their role as master regulators of clusters of stress-tolerant genes. In the last two decades, significant research on identification and characterization of WRKY, NAC, MYB, DREB, and bZIP TFs have been studied in detail in other crops like wheat, rice, maize, barley, and sorghum (Baillo et al. 2019). Similar efforts have been attempted in millets like finger millet and foxtail millet, however, the genetic potential of other millets is still untapped. The millets are a great genomic reservoir of various TFs for regulating abiotic stress tolerance. As we all know, the response of TFs under stress is very complicated and needs an accurate characterization in the lab. Individual characterization of several TFs at molecular level might be a challenging and daunting task but future studies must be undertaken to study multiple TFs and multiple stresses to decipher the crosstalk among various TFs. Abiotic stress-responsive *cis*-regulatory elements are the critical regulators of transcription of stress-related genes. So, well-characterized abiotic stress-inducible promoter have tremendous potential for genetic engineering of millets to improve the tolerance level. Genetic engineering of TFs and inducible promoters in a particular genotype is very important for its expression and effect on other genes. Recently, discovered and characterized TFs in combination with DREB/CBF can be more effective in stress tolerance without compromising yield (Yang et al. 2020). Many TFs can be genetically engineered to produce transgenics with higher tolerance to drought, salinity, heat, and cold using different promoters to improve tolerance to abiotic stresses and productivity of

Table 4.2 Genome size, gene repertoire, and transcription factors of millets

Sl no.	Crop	Botanical name	Ploidy level	Genome size (Mbp)	No. of genes predicted	TFs encoding genes	TFs Families	Comments
1	Finger millet	*Eleusine coracana*	$2n = 4x = 38$	1196	85,243	11,125	56	
2	Foxtail millet	*Setaria italica*	$2n = 2x = 18$	423	38,801	2410	56	
3	Proso millet	*Panicum miliaceum*	$2n = 4x = 36$	855	55,930	3069	56	
4	Barnyard millet	*Echinochloa crus-galli*	$2n = 6x = 54$	1270	1,08,771	5320	56	
5	Kodo millet	*Paspalum scrobiculatum*	$2n = 4x = 40$	–	–	–	–	Genome data unavailable
6	Little millet	*Panicum sumatrense*	$2n = 4x = 36$	–	–	–	–	Genome data unavailable
7	Teff	*Eragrostis tef*	$2n = 4x = 40$	672	41,136	2145	55	
8	Fonio	*Digitaria exilis*	$2n = 4x = 36$	716	59,985	2761	53	
9	Adlay millet	*Coix lacryma-jobi*	$2n = 4x = 20$	1280	39,574	1594	55	
10	Guinea millet	*Brachiaria deflexa*	$2n = 18,36$	–	–	–	–	Genome data unavailable
11	Browntop millet	*Urochloa ramosa*	$2n = 2x = 18$, $2n = 4x = 36, 72$	732	42,232	–	–	Protein sequences unavailable

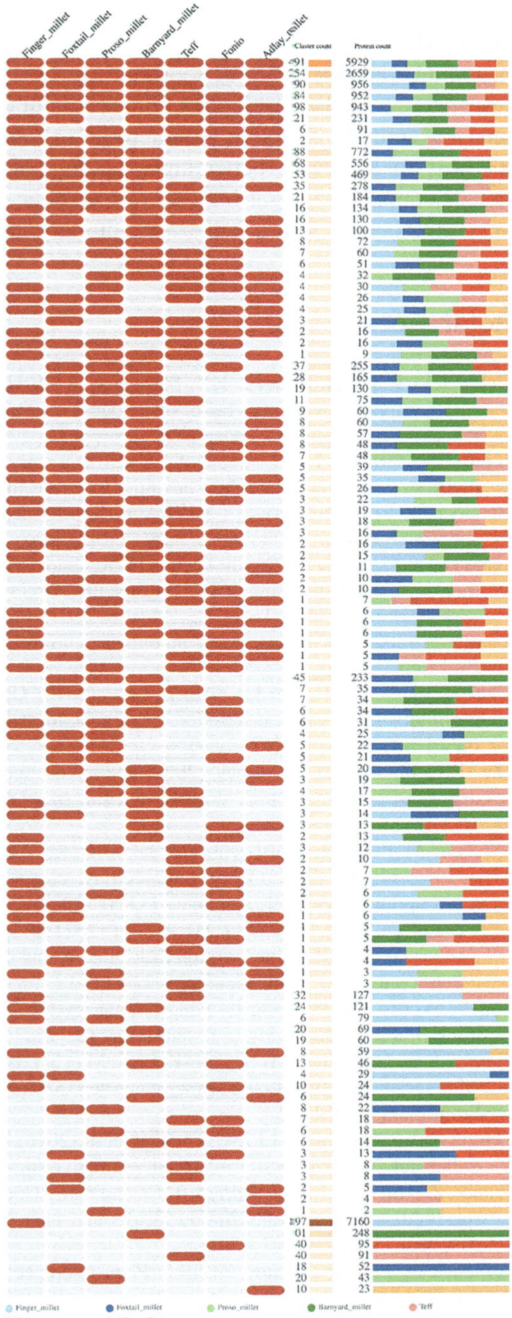

Fig. 4.1 Orthologous clustering of transcription factors across millets

millets. Recent tools like CRISPR/Cas can be deployed to edit/modulate the expression of TFs in millets in near future (Ceasar 2021). As a matter of fact, TFs are becoming valuable genomic tools and are likely to play a prominent role in future research toward the development of successful tailor-made crops. Integrating global analyses from transcription to proteins to metabolites during abiotic stress will pave the way to achieve a better understanding of major metabolic pathways, signal transduction, and insight into future metabolic engineering for abiotic stress tolerance in millets as well as other agriculturally important crops.

References

Babitha K, Ramu S, Nataraja KN, Sheshshayee M, Udayakumar M (2015) EcbZIP60, a basic leucine zipper transcription factor from Eleusine coracana L. improves abiotic stress tolerance in tobacco by activating unfolded protein response pathway. Mol Breed 35(9):1–17

Baillo EH, Kimotho RN, Zhang Z, Xu P (2019) Transcription factors associated with abiotic and biotic stress tolerance and their potential for crops improvement. Genes 10(10):771

Banerjee A, Roychoudhury A (2017) Abscisic-acid-dependent basic leucine zipper (bZIP) transcription factors in plant abiotic stress. Protoplasma 254(1):3–16

Cannarozzi G, Plaza-Wüthrich S, Esfeld K, Larti S, Wilson YS et al (2014) Genome and transcriptome sequencing identifies breeding targets in the orphan crop tef (Eragrostis tef). BMC Genomics 15(1):1–21

Ceasar A (2021) Genome-editing in millets: current knowledge and future perspectives. Mol Biol Rep:1–9

Chai W, Si W, Ji W, Qin Q, Zhao M et al (2018) Genome-wide investigation and expression profiling of HD-zip transcription factors in foxtail millet (Setaria italica L.). Biomed Res Int 2018

Das RR, Pradhan S, Parida A (2020) De-novo transcriptome analysis unveils differentially expressed genes regulating drought and salt stress response in Panicum sumatrense. Sci Rep 10(1):1–14

Espelund M, Bekele E, Holst-Jensen A, Jakobsen KS, Nordal I (2000) A molecular genetic analysis of Eragrostis tef (Zucc.) trotter: non-coding regions of chloroplast DNA, 18S rDNA and the transcription factor VP1. Hereditas 132(3):193–202

Fan Y, Lai D, Yang H, Xue G, He A et al (2021a) Genome-wide identification and expression analysis of the bHLH transcription factor family and its response to abiotic stress in foxtail millet (Setaria italica L.). BMC Genomics 22(1):1–18

Fan Y, Wei X, Lai D, Yang H, Feng L et al (2021b) Genome-wide investigation of the GRAS transcription factor family in foxtail millet (Setaria italica L.). BMC Plant Biol 21(1):1–19

Feng Z-J, He G-H, Zheng W-J, Lu P-P, Chen M et al (2015) Foxtail millet NF-Y families: genome-wide survey and evolution analyses identified two functional genes important in abiotic stresses. Front Plant Sci 6:1142

Feng Z-J, Xu Z-S, Sun J, Li L-C, Chen M et al (2016) Investigation of the ASR family in foxtail millet and the role of ASR1 in drought/oxidative stress tolerance. Plant Cell Rep 35(1):115–128

Franco-Zorrilla JM, López-Vidriero I, Carrasco JL, Godoy M, Vera P et al (2014) DNA-binding specificities of plant transcription factors and their potential to define target genes. Proc Natl Acad Sci 111(6):2367–2372

Golldack D, Lüking I, Yang O (2011) Plant tolerance to drought and salinity: stress regulating transcription factors and their functional significance in the cellular transcriptional network. Plant Cell Rep 30(8):1383–1391

Guo J, Sun B, He H, Zhang Y, Tian H et al (2021) Current understanding of bHLH transcription factors in plant abiotic stress tolerance. Int J Mol Sci 22(9):4921

Guo L, Qiu J, Ye C, Jin G, Mao L et al (2017) Echinochloa crus-galli genome analysis provides insight into its adaptation and invasiveness as a weed. Nat Commun 8(1):1–10
Hittalmani S, Mahesh H, Shirke MD, Biradar H, Uday G et al (2017) Genome and transcriptome sequence of finger millet (Eleusine coracana (L.) Gaertn.) provides insights into drought tolerance and nutraceutical properties. BMC Genomics 18(1):1–16
Hu Y-G, Lin F-Y, Wang S-Q, He B-R (2008) Cloning and expression analysis of drought-tolerant and water-saving related gene PmMYB in broomcorn millet. Yi Chuan=Hereditas 30(3): 373–379
Jin J, Tian F, Yang D-C, Meng Y-Q, Kong L et al (2016) PlantTFDB 4.0: toward a central hub for transcription factors and regulatory interactions in plants. Nucleic Acids Res:gkw982
Jin J, Zhang H, Kong L, Gao G, Luo J (2014) PlantTFDB 3.0: a portal for the functional and evolutionary study of plant transcription factors. Nucleic Acids Res 42(D1):D1182–D1187
Kang S-H, Kim B, Choi B-S, Lee HO, Kim N-H et al (2020) Genome assembly and annotation of soft-shelled adlay (Coix lacryma-jobi variety ma-yuen), a cereal and medicinal crop in the poaceae family. Front Plant Sci 11:630
Kiełbowicz-Matuk A (2012) Involvement of plant C2H2-type zinc finger transcription factors in stress responses. Plant Sci 185:78–85
Lata C, Mishra AK, Muthamilarasan M, Bonthala VS, Khan Y et al (2014) Genome-wide investigation and expression profiling of AP2/ERF transcription factor superfamily in foxtail millet (Setaria italica L.). PLoS One 9(11):e113092
Lata C, Prasad M (2011) Role of DREBs in regulation of abiotic stress responses in plants. J Exp Bot 62(14):4731–4748
Lata C, Yadav A, Prasad M (2011) Role of plant transcription factors in abiotic stress tolerance. Abiotic Stress Response in Plants, INTECH Open Access Publishers 10:269–296
Li C, Ng CK-Y, Fan L-M (2015) MYB transcription factors, active players in abiotic stress signaling. Environ Exp Bot 114:80–91
Li W, Pang S, Lu Z, Jin B (2020) Function and mechanism of WRKY transcription factors in abiotic stress responses of plants. Plan Theory 9(11):1515
Lipsick JS (1996) One billion years of Myb. Oncogene 13(2):223–235
Mahesh H, Prasannakumar M, Manasa K, Perumal S, Khedikar Y et al. (2021) Genome, transcriptome, and germplasm sequencing uncovers functional variation in the warm-season grain legume Horsegram Macrotyloma uniflorum (Lam.) Verdc. Frontiers in plant science 12
Miao G, Qin Y, Guo J, Zhang Q, Bao Y (2021) Transcriptome characterization and expression profile of Coix lacryma-jobi L. in response to drought. PLoS One 16(9):e0256875
Mulat MW, Sinha VB (2020) Identification and characterization of Dof in tef [Eragrostis tef (Zucc.) trotter] using in silico approaches. Gene Rep 19:100590
Mulat MW, Sinha VB (2021a) Comparative in silico analysis of Eragrostis tef (Zucc.) trotter with other species for elucidating presence of growth regulating factors (GRFs). Genet Resour Crop Evol 68(2):499–512
Mulat MW, Sinha VB (2021b) Distribution and abundance of CREs in the promoters depicts crosstalk by WRKYs in Tef [Eragrostis tef (Zucc.) Troetter]. Gene Rep 23:101043
Mulat MW, Sinha VB (2021c) In silico approach for unraveling the structural and functional roles of NF-X1-like proteins in underutilized cereal Eragrostis tef. Biol Bull 48(3):251–262
Mulat MW, Sinha VB (2022) VOZS identification from TEF [Eragrostis tef (Zucc.) trotter] using in silico tools decipher their involvement in abiotic stress. Mater Today Proc 49:3357–3364
Muthamilarasan M, Bonthala VS, Mishra AK, Khandelwal R, Khan Y et al (2014a) C2H2 type of zinc finger transcription factors in foxtail millet define response to abiotic stresses. Funct Integr Genomics 14(3):531–543
Muthamilarasan M, Khandelwal R, Yadav CB, Bonthala VS, Khan Y et al (2014b) Identification and molecular characterization of MYB transcription factor superfamily in C4 model plant foxtail millet (Setaria italica L.). PLoS One 9(10):e109920

Nakashima K, Takasaki H, Mizoi J, Shinozaki K, Yamaguchi-Shinozaki K (2012) NAC transcription factors in plant abiotic stress responses. Biochimica et Biophysica Acta (BBA)-Gene Regulatory Mechanisms 1819(2):97–103

Paz-Ares J, Ghosal D, Wienand U, Peterson P, Saedler H (1987) The regulatory c1 locus of Zea mays encodes a protein with homology to myb proto-oncogene products and with structural similarities to transcriptional activators. EMBO J 6(12):3553–3558

Rahman H, Ramanathan V, Nallathambi J, Duraialagaraja S, Muthurajan R (2016) Over-expression of a NAC 67 transcription factor from finger millet (Eleusine coracana L.) confers tolerance against salinity and drought stress in rice. BMC Biotechnol 16(1):7–20

Ramakrishna C, Singh S, Raghavendrarao S, Padaria JC, Mohanty S et al (2018) The membrane tethered transcription factor EcbZIP17 from finger millet promotes plant growth and enhances tolerance to abiotic stresses. Sci Rep 8(1):1–14

Ramegowda V, Senthil-Kumar M, Nataraja KN, Reddy MK, Mysore KS et al (2012) Expression of a finger millet transcription factor, EcNAC1, in tobacco confers abiotic stress-tolerance. PLoS One 7(7):e40397

Shan Z, Jiang Y, Li H, Guo J, Dong M et al (2020) Genome-wide analysis of the NAC transcription factor family in broomcorn millet (Panicum miliaceum L.) and expression analysis under drought stress. BMC Genomics 21(1):1–13

Singh S, Chopperla R, Shingote P, Chhapekar SS, Deshmukh R et al (2021) Overexpression of EcDREB2A transcription factor from finger millet in tobacco enhances tolerance to heat stress through ROS scavenging. J Biotechnol 336:10–24

Sornaraj P, Luang S, Lopato S, Hrmova M (2016) Basic leucine zipper (bZIP) transcription factors involved in abiotic stresses: a molecular model of a wheat bZIP factor and implications of its structure in function. Biochimica et Biophysica Acta (BBA)-General Subjects 1860(1):46–56

Suresh BV, Choudhary P, Aggarwal PR, Rana S, Singh RK et al (2022) De novo transcriptome analysis identifies key genes involved in dehydration stress response in kodo millet (Paspalum scrobiculatum L.). Genomics:110347

Upadhyaya HD, Vetriventhan M, Dwivedi SL, Pattanashetti SK, Singh SK (2016) Proso, barnyard, little, and kodo millets. In: Genetic and genomic resources for grain cereals improvement. Elsevier, pp 321–343

Wang M, Li P, Li C, Pan Y, Jiang X et al (2014) SiLEA14, a novel atypical LEA protein, confers abiotic stress resistance in foxtail millet. BMC Plant Biol 14(1):1–16

Wang X, Chen S, Ma X, Yssel AE, Chaluvadi SR et al (2021) Genome sequence and genetic diversity analysis of an under-domesticated orphan crop, white fonio (Digitaria exilis). GigaScience 10(3):giab013

Xu W, Chen Z, Ahmed N, Han B, Cui Q et al (2016) Genome-wide identification, evolutionary analysis, and stress responses of the GRAS gene family in castor beans. Int J Mol Sci 17(7):1004

Yang Y, Al-Baidhani HHJ, Harris J, Riboni M, Li Y et al (2020) DREB/CBF expression in wheat and barley using the stress-inducible promoters of HD-zip I genes: impact on plant development, stress tolerance and yield. Plant Biotechnol J 18(3):829–844

Yue H, Wang M, Liu S, Du X, Song W et al (2016) Transcriptome-wide identification and expression profiles of the WRKY transcription factor family in broomcorn millet (Panicum miliaceum L.). BMC Genomics 17(1):1–11

Zhang B, Liu J, Yang ZE, Chen EY, Zhang CJ et al (2018) Genome-wide analysis of GRAS transcription factor gene family in Gossypium hirsutum L. BMC Genomics 19(1):1–12

Zhang G, Liu X, Quan Z, Cheng S, Xu X et al (2012) Genome sequence of foxtail millet (Setaria italica) provides insights into grass evolution and biofuel potential. Nat Biotechnol 30(6): 549–554

Zhu F (2017) Coix: chemical composition and health effects. Trends Food Sci Technol 61:160–175

Zou C, Li L, Miki D, Li D, Tang Q et al (2019) The genome of broomcorn millet. Nat Commun 10(1):1–11

Genome-Wide Identification and Expression Profiling of Noncoding RNAs in Response to Abiotic Stresses in Small Millets

5

Saranya Nallusamy, Selva Babu Selvamani, and Raveendran Muthurajan

Abstract

Small millets are known for their nutritional values, nutrient use efficiency, and extremely high levels of tolerance against abiotic stresses. Small millets exhibit a high level of molecular complexity due to their polyploidy nature and hence unraveling molecular networks controlling the above traits was challenging. Recent reports elucidated the role of noncoding RNAs (ncRNAs) in regulating an array of genes involved in stress tolerance and nutrient use efficiency traits. NcRNAs remained unexplored due to the lack of suitable platforms to identify, analyze, and characterize their expression. Advancements in Next Generation Sequencing and bioinformatics platforms have enabled the completion of several whole genome sequencing and transcriptome profiling projects in small millets that led to the discovery of a large number of ncRNAs and understanding of their expression and regulatory mechanism(s). This chapter discusses various classes of ncRNAs in small millets and their putative roles in modulating plant responses against drought, salt, cold, etc. Further, databases and online tools pertaining to the identification and analysis of Noncoding RNAs are presented.

Keywords

Small millets · Noncoding RNAs · Expression profiling · Gene regulation · Bioinformatics tools · Stress-responsive miRNA

S. Nallusamy · S. B. Selvamani
Department of Plant Molecular Biology and Bioinformatics, Centre for Plant Molecular Biology and Biotechnology, Tamil Nadu Agricultural University, Coimbatore, Tamil Nadu, India

R. Muthurajan (✉)
Department of Plant Biotechnology, Centre for Plant Molecular Biology and Biotechnology, Tamil Nadu Agricultural University, Coimbatore, Tamil Nadu, India

R. N. Pudake et al. (eds.), *Omics of Climate Resilient Small Millets*,
https://doi.org/10.1007/978-981-19-3907-5_5

5.1 Introduction

Small millets, also called as minor millets belong to the grass family "Poaceae" brought into cultivation during the early period of crop domestication. Small millets have ability to grow and yield under an diverse ecological conditions, drought conditions and remains tolerant to insects, pests, herbicides and to many plant diseases (Vetriventhan et al. 2020). Millets are mainly consumed as healthy foods due to the presence of high levels of nutrients, dietary fibers, vitamins, calcium, and low glycemic index (Nkhata et al. 2018). Minor millets or small millets include finger millet (*Eleusine coracana*), foxtail millet (*Setaria italic L.*), proso millet (*Panicum miliaceum*), barnyard millet (*Echinochloa crus-galli*), kodo millet (*Paspalum scrobiculatum*), little millet (*Panicum sumatrense*), tef (*Eragrostis tef*), fonio (*Digitaria exilis*), job's tears (*Coix lacryma-jobi*), guinea millet (*Brachiaria defexa*), and browntop millet (*Brachiaria ramosa*). In recent years, small millets are gaining importance due to their resilience to extremely harsh environments and health-benefiting properties. Hence, small millets like foxtail millet and finger millet are used as model systems in C4 photosynthesis and genomics research. Genome sequences of small millets such as foxtail millet, finger millet, proso millet, tef, and Japanese barnyard millet have been reported and among these foxtail millet with smallest genome size of 423–510 Mb has been explored largely compared to other small millets. Genomic resources for small millets are less available compared with that of other cereal crops. This chapter will explain the role of noncoding RNAs such as long non-coding RNA (lncRNA), microRNA (miRNA), and small interfering RNA (siRNA) in small millets. Further, protocols, tools, and databases that are used for the genome-wide identification of noncoding RNAs are discussed.

5.2 Noncoding RNA

Noncoding RNAs (ncRNA) are observed in all the biological systems and a greater portion of the genome is occupied by noncoding parts as in the case of *Arabidopsis thaliana* (Jampala et al. 2021). Noncoding RNAs are transcribed but not translated to protein products but have vital roles in regulating the expression of genes during various biotic and abiotic stress conditions. Noncoding RNAs such as ribosomal RNAs, transfer RNAs, small nuclear RNAs, and small nucleolar RNAs are classified as housekeeping ncRNAs. Regulatory ncRNAs include micro RNAs (miRNA), small interfering RNA (siRNA), and long noncoding RNA (lncRNA) (Cech and Steitz 2014). MiRNA and siRNA are considered as small ncRNAs with less than 50 nucleotides in length and play important roles in posttranscriptional and transcriptional regulation of eukaryotic gene expression (Kim and Sung 2012).

5.2.1 Role of Millets microRNA during Abiotic Stress Conditions

MicroRNAs (miRNAs) are small, single-stranded, and noncoding regulatory RNA sequences of ~21–24 bp sizes. miRNAs play an important role during seed germination, transition from vegetative to reproductive phase, initiation of flowering, fruit/seed setting and root development, biotic and abiotic stresses, and hormone signalling (Frazier et al. 2011; Jagadeeswaran et al. 2009; Yang et al. 2011), signal transduction, protein degradation (Zhang et al. 2008; Zhou et al. 2010), transgene suppression (Allen et al. 2005), development of disease (Johnson et al. 2005), viral resistance (Bennasser et al. 2004), and nutrient deprivation and heavy metals (Chen et al. 2012). Compared to other noncoding RNAs, miRNA was found to be highly conserved among plant species (Lima et al. 2012; Sunkar 2010).

5.2.1.1 Methods to Identify Abiotic Stress-Related miRNA

Genome-wide identification of miRNA involves different steps in crop plants. There is no single software to perform all these steps in the identification of the abiotic stress-responsive miRNA in crops (Friedländer et al. 2012). By finding the abiotic stress-related miRNAs and their target genes, it is possible to understand the biological mechanism of abiotic stress tolerance in crop plants. Deep sequencing of the plant RNA is required to identify the drought-responsive miRNA and related mRNA targets (Goldman and Domschke 2014). Also, deep sequencing helps to identify the novel miRNAs which play a significant role in abiotic stress, biotic stress, and grain yield.

After sequencing, primary step is preprocessing which involves quality check, quality trimming to remove low-quality reads and adapter sequences. The tools Trimmomatic (Bolger et al. 2014) and Cutadapt (Martin 2011) are mostly used for preprocessing step. FastQc (Andrews 2010) is the widely used desktop application for checking the quality of reads in the high-throughput analysis of sequenced reads. In the next step of mapping, locations of the reads are mapped and indexed in the genome. Hisat2 software (hierarchical indexing for spliced alignment of transcripts 2) (Kim et al. 2019) and TopHat2 (Kim et al. 2013) are well-known mappers used for mapping the reads with the reference genome. Redundant reads are removed to retain the unique reads. High-quality reads are searched against Rfam database (http://www.sanger.ac.uk/Software/Rfam) to identify the noncoding RNAs such as rRNAs, tRNAs, small nuclear RNAs (snRNAs), and small nucleolar RNAs (snoRNAs). Unmapped reads of 18–26 nucleotide length are used for the classification of known and novel miRNA identification. Details of known microRNAs and their precursor sequences can be obtained for different plant species by searching miRNA databases like PNRD—a plant noncoding RNA database (Yi et al. 2015) and miRBase (http://www.miRbase.org/). Those miRNAs that do not find similarities in databases are classified as novel miRNA. The differentially expressed (significantly expressed) miRNA in contrasting conditions are selected for the mRNA target prediction. In this procedure, miRNA-regulated targets can be predicted by (1) complementary matching of differentially expressed miRNA and differentially expressed mRNA in stress conditions; (2) or by evaluating the

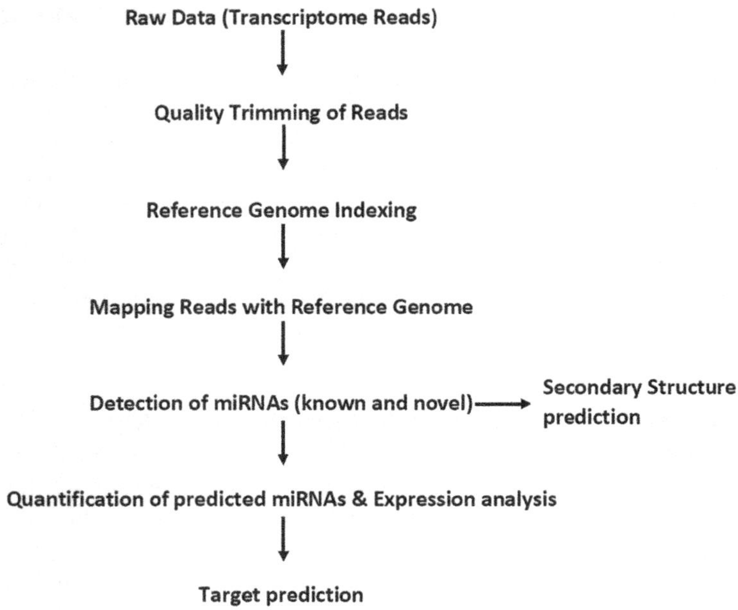

Fig. 5.1 General workflow for the identification of miRNA in small millets

interaction of abiotic stress-related genes with miRNA (Dai and Zhao 2011). The secondary structure prediction of the miRNAs is one of the ways to validate the novel and known miRNAs and to understand abiotic stress-related conserved miRNA gene families. Finally, all the predicted miRNA and related mRNAs can be validated using quantitative real-time PCR (qRT-PCR) analysis. This is a general workflow to predict the miRNA in small millets (Fig. 5.1).

5.2.1.2 Abiotic Stress-Responsive miRNAs in Small Millets

A set of 33 novel and 18 known dehydration-responsive miRNAs have been identified in two foxtail millet cultivars, namely IC403579 (stress tolerant-T) and IC480117 (stress susceptible-S) based on the RNA-seq data (Yadav et al. 2016). After the stress treatment, 32 dehydration-responsive miRNAs were upregulated in tolerant cultivars and 22 miRNAs were downregulated in sensitive cultivars, suggesting that miRNA-mediated molecular regulation might play important role in providing contrasting characteristics to these cultivars. Predicted targets of identified miRNAs were found to encode various transcription factors and functional enzymes, indicating their involvement in broad-spectrum regulatory functions and biological processes. Further, differential expression patterns of seven known miRNAs were validated by northern blot and expression of ten novel dehydration-responsive miRNAs were confirmed by Stem-loop quantitative real-time PCR. Differential expression behavior of five miRNA-target genes was verified under

Table 5.1 Abiotic stress-related miRNA reported in small millets (Chakraborty et al. 2020)

miRNA	Stress-related mRNAs function	Species
miR104, miR105, miR1120, miR123, miR150, miR154, miR158, miR163, miR38, miR41, miR6, miR156, miR171, miR2108, miR164, miR395, miR167, miR170, miR160, miR319, miR169, miR172, miR165	Dehydration stress	Foxtail millet
miR1873, miR10095	Drought stress	Finger millet
miR399	Drought stress	Foxtail millet
miR165, miR395, miR2108, miR1432, miR159, miR160, miR162, miR168, miR169, miR319, miR390, miR393, miR408, miR167, miR156, miR171, miR172, miR399	Drought stress	Proso millet
miR10290	Drought stress, oxidative stress, water-deficit stress	Finger millet
miR2108, miR91	Drought stress, water-deficit stress	Finger millet
miR845, miR8770, miR7748	Water-deficit stress	Finger millet

dehydration stress treatment and two of them were also validated by RNA ligase-mediated rapid amplification of 5' cDNA ends (RLM RACE).

Chakraborty and co-workers (Chakraborty et al. 2020) reported the drought stress-related miRNAs and their target genes by analyzing the genomic data (expressed sequence tags (ESTs), genome survey sequences (GSS), and whole genome sequences (WGS)) retrieved from the NCBI (http //www.ncbi.nlm.nih.gov/) database. A collective study on five millets have been reported in their article which includes three small millet species such as finger millet (accession no. GCA_002180455.1), foxtail millet, (GCA_000263155.2), and proso millet (GCA_002895445.2). They have also reported many of the drought-responsive miRNA families and their interaction with functional enzymes such as protein kinases, peroxidase, amino/carboxy peptidases, stress associated proteins (HSPs, RNA binding proteins), and drought-specific TFs (ARF, NAC family, MAD box, WRKY, bHLH, and ZFs). For example, miRNA gene family miR2108, miR170, and miR171 targeted important genes, such as auxin response factors (ARFs), NTR/PTR, NAC domain, and heat stress factors. Details of abiotic stress-responsive miRNAs reported in small millets are provided in Table 5.1.

Wang et al. (2016) have analyzed extensively the drought responsive miRNAs and their target genes in the foxtail millet (Variety: An04–4783) using Illumina and degradome sequencing approach. Degradome sequencing has been used to identify the accurate cleavage site of miRNA at the global level. On comparing the expression levels of miRNA in the control and drought conditions with log fold change value of 2, 18 miRNAs (14 upregulated and 4 downregulated) belonging to the

16 miRNA families have changed significantly. They have also reported three differentially expressed (DE) novel miRNAs (2 upregulated and 1 downregulated) in response to drought stress. Some of the drought responsive miRNAs such as miR159, miR167, and miR390 were found to be conserved across other plant species under water-deficit condition. Target gene analysis of DE miRNAs showed 56 and 26 genes for known and novel miRNAs, respectively, using the degradome sequencing approach. Among the identified target genes, most of them were found to be transcription factors such as squamosal promoter-binding family, MYB, several auxin response factors (ARF), and no apical meristem (NAC) protein which were regulated by stress responsive conserved miRNAs such as miR156, miR159, miR160, and miR164, respectively. Moreover, authors have insisted on the role of conservation of regulatory events across the plant kingdom under various abiotic stress conditions.

In a similar study of genome wide identification of miRNAs in foxtail millet 355 miRNAs were identified (Khan et al. 2014). Expression profiling of 8 miRNAs (Sit-miR162a, Sit-miR397a, Sit-miR393, Sit-miR167b, Sit-miR156c, Sit-miR171b, Sit-miR160d, and Sit-miR6248) was performed under different abiotic stress conditions which includes drought, salt and cold using northern blot and Stem–loop quantitative real-time PCR experiments for the resistant (IC403579) and sensitive (IC480117) cultivars. Data generated from the above study was developed into a foxtail millet miRNA database (FmMiRNADb: http://59.163.192.91/FmMiRNADb/index.html). Further, this genome-wide study of foxtail millet miRNA provided the insights on regulation of stress-related genes in millets and bioenergy grasses.

5.2.2 Role of lncRNA in Abiotic Stress

LncRNAs are located in the nucleus or cytoplasm and are transcribed by RNA polymerase II or III and polymerase IV/V (Ariel et al. 2014; Herr et al. 2005; Jha et al. 2020; Liu et al. 2015). siRNAs are precursors of lncRNAs transcribed by Pol IV (Herr et al. 2005; Martin 2011). Functions of lncRNAs involves regulation of plant development, disease resistance, nutrient acquisition (Cech and Steitz 2014; Laurent et al. 2015; Mattick and Rinn 2015; Sun et al. 2018), and other biological processes through chromatin remodelling, histone modification, and target mimicry (Corona-Gomez et al. 2020; Franco-Zorrilla et al. 2007; Nejat and Mantri 2018; Shin et al. 2018; Sun et al. 2020). Regulation of function occurs at transcriptional, posttranscriptional, and epigenetic levels. *enod40* (early nodulin 40) was the first lncRNAs identified in plants which has its role in plant growth regulation (Crespi et al. 1994). Based on the position with respect to protein-coding genes, lncRNAs are classified as long intergenic ncRNAs (lincRNAs), intronic ncRNAs (incRNAs), natural antisense transcripts (NATs) (transcribed from complementary DNA strands of sense coding regions), and circular long noncoding RNAs (circRNAs) (transcribed by the back-splicing reaction of internal exons in pre-mRNA). Figure 5.2 describes the position of lncRNA transcribed with respect to genes.

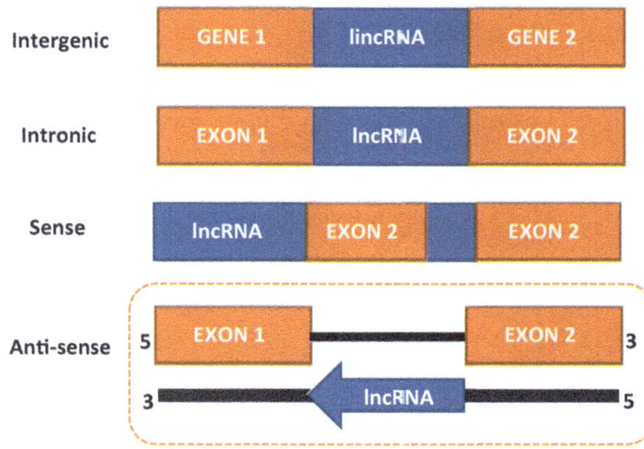

Fig. 5.2 Classification of lncRNA

5.2.3 LncRNA and their Role in Abiotic Stress

Genome-wide lncRNA identification and expression analysis begins with the sequencing of the small millet (foxtail millet, kodo millet, little millet, proso millet, and barnyard millet) (Muthamilarasan and Prasad 2021) or collecting the previously sequenced small millets data from the database like Gene Expression Omnibus (Edgar et al. 2002). LncRNA workflow is similar to miRNA identification in collecting raw data and preprocessing (quality check and trimming to get high-quality reads). In further steps, lncRNA workflow is different from miRNA workflow. Here reads were mapped to the reference genome and the mapped reads were assembled into transcripts using the assembly tools. The assembled transcripts are used for further downstream analysis like identification, expression analysis, and target prediction of lncRNA. Identification of lncRNA: there are some well-defined steps to identify the lncRNA from the assembled transcripts (Fig. 5.3). Steps: I) coding potential calculation, II) removing the transcripts length smaller than 200nts and single exon transcripts, III) finding similarity against the databases: Pfam (El-Gebali et al. 2019), miRbase (Griffiths-Jones et al. 2007), RFam (Kalvari et al. 2018), and plant proteins (Li et al. 2020). The transcripts which have similarities with any of these databases are filtered and the remaining transcripts are taken for further analysis like differential expression analysis (based on FPKM value) and target prediction. LncRNAs with the suitable fold change cutoff value of (mostly) logFC >2 and <-2 and P value $= 0.05$ are considered as differentially expressed and are statistically significant. LncRNAs with logFC value greater than 2 are considered as upregulated and those with logFC less than -2 were considered as downregulated lncRNAs (Zhou et al. 2019).

Based on the location of the lncRNAs it can be classified into intronic, intergenic, generic, and antisense (Kang and Liu 2019). LncRNAs are classified into known and novel lncRNA by similarity searching against the known lncRNA sequences reported in the noncoding databases. Further, identified lncRNAs are explored for their interaction with target genes and miRNA to understand their regulatory role.

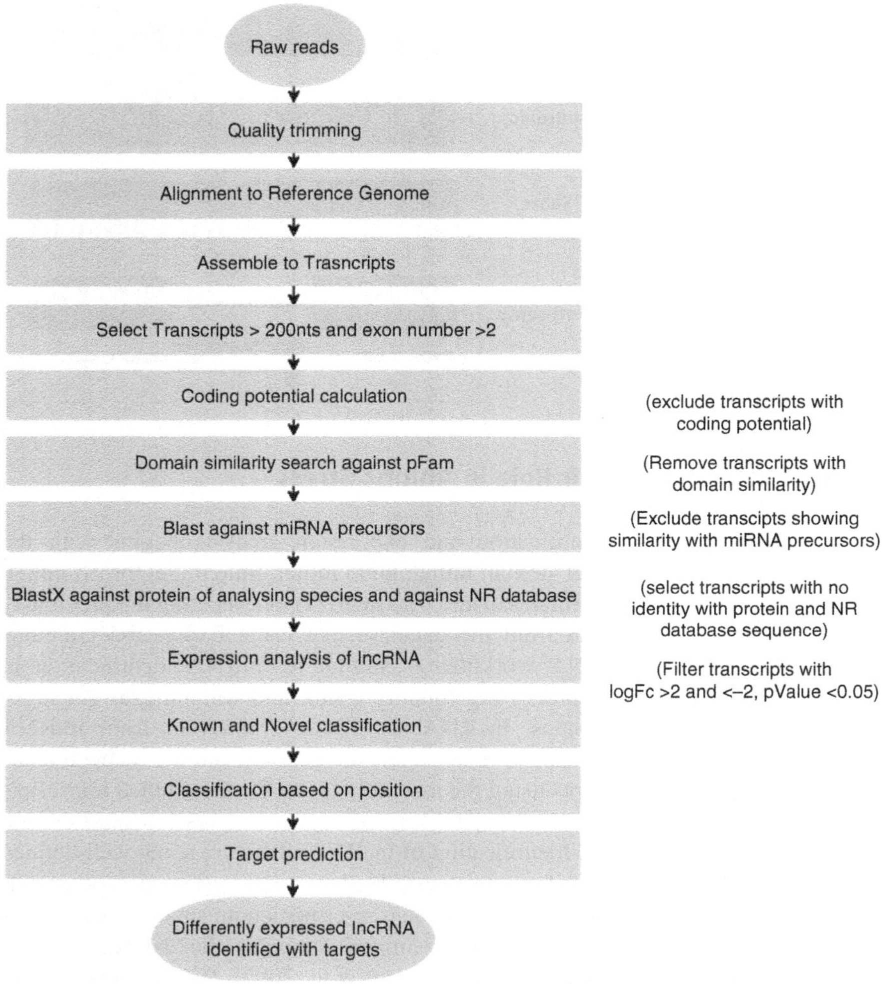

Fig. 5.3 General workflow for the identification and characterization of lncRNA in small millets

Till now only foxtail millet lncRNAs are identified, (Qi et al. 2013) and lncRNAs in other millet varieties are open for exploration which may regulate the stress responsive mRNAs or miRNAs. Differential expression analysis showed that 19 lncRNAs (17 lincRNAs and 2 NATs) in foxtail millet responded to PEG-induced drought stress. They also reported that the target genes of the reported lncRNAs are mostly distantly located in the genome and not in the proximity (upstream and downstream) of the lncRNA being studied.

5.2.4 Drought-Responsive siRNA in Small Millets

miRNA and siRNA belong to the class of small RNA with similar size (20–24 nt) and are known to be important regulators of gene expression, but differ in precursor structures, pathway of biogenesis, and modes of action (Bartel 2004; Hamilton et al. 2002; Hutvágner et al. 2001). Biogenesis of both the small RNAs occurs from long RNA precursors by Dicer-like ribonucleases (Hammond et al. 2000; Lee et al. 1993).

High quality reads after preprocessing are used for the siRNA identification. sRNA reads that matched sequences in either the Rfam (http://rfam.sanger.ac.uk/) (Kalvari et al. 2018) or miRbase (http://www.mirbase.org/) (Kozomara et al. 2019) databases are filtered out and sequences with the size of 18–25 nts are taken for siRNA analysis (Chen et al. 2012). Differentially expressed siRNA candidates are quantified and psRNAtarget server (Dai and Zhao 2011) can be used for predicting the mRNA targets.

One study has identified drought-responsive siRNAs in foxtail millet (Qi et al. 2013). The most abundant siRNAs in the foxtail millet shoots were the 24-nt siRNAs. 24-nt siRNAs are involved in RNA-directed DNA methylation which guides the silencing of transposable elements at the transcriptional level (Cantu et al. 2010; Law and Jacobsen 2010).

5.3 Databases and Tools Used for Identification and Analysis of Noncoding RNA in Small Millets

Many bioinformatics databases and tools are employed to identify the noncoding in small millets. Details of the software and tools for identification and target prediction of miRNA, lncRNA, and siRNA are given in Table 5.2.

There are several databases available to get information about noncoding RNAs in plants. Some of the databases are listed in Table 5.3.

5.4 Conclusion

Noncoding RNA contains wider range of regulatory functions in controlling the expression of genes, miRNA, chromatin remodelling, etc. Compared to the major crops, role of noncoding RNA is less studied in small millets. *Setaria italica is* the only minor millet and the model crop for stress biology, which was explored for the functionality of miRNA, lncRNA, and siRNA under drought conditions. Genomic size of the small millets remains a major challenge in the sequencing, assembly, and annotation process. Genetic level understanding of coding and noncoding functions on small millets that has tolerance to withstand adverse abiotic stress conditions and higher nutrient content, will levitate the genetic engineering research on other major crops with advancing CRISPR-cas technologies. Further, understanding the noncoding RNA role in abiotic stress will provide food security as well plays important role in the eradication of nutrient deficiency that supports ever increasing human population.

Table 5.2 Bioinformatics tools used in the identification and analysis of noncoding RNAs

Tool	Availability	work	References
cufflinks	Standalone	Assemble the reads into a transcriptome	Trapnell et al. (2012)
cuffmerge	Standalone	Merges the cufflinks output .gtf files into master transcriptome file	
cuffdiff	Standalone	Differential expression analysis	
cuffcompare	Standalone	Helps to compare the transcripts assembled from different libraries	
tophat2	Standalone	Mapping reads with the reference genome	Kim et al. (2013)
hisat2	Standalone	Mapping reads with the reference genome	Kim et al. (2019)
bowtie	Standalone	Mapping reads with reference genome and uses to index the reference genome	Langmead and Salzberg (2012)
Bwa – Burrows Wheeler Aligner	Standalone	Software package for mapping the low divergent sequence against the large reference genome	Li (2013)
FastQC	Standalone	Checks the quality of the reads	Andrews (2010)
Trimmomatic	Standalone	Software to trim the low-quality reads	Bolger et al. (2014)
Cutadapt	Standalone	Quality trimming of the reads	Martin (2011)
cummeRbund	Standalone	Cufflinks output downstream analysis	Goff et al. (2012)
DeSeq2	Standalone	Based on the negative binomial distribution it performs the differential gene expression	Love et al. (2014)
edgeR	Standalone	Package to perform the differential gene expression	Robinson et al. (2010)
psRNAtarget	Online server	Predicts the targets mRNA for small RNAs	Dai and Zhao (2011)
Galaxy	Online server	Assembly of transcripts, differential expression analysis	Jalili et al. (2020)
stringtie	Standalone	To assemble the reads into transcripts	Pertea et al. (2015)
mirPRo	Standalone	miRNA identification, quantification, structure prediction	Shi et al. (2015)
miRDeep2	Standalone	miRNA identification, quantification, and structure prediction	Friedländer et al. (2012)
CPC	Online server	Calculates the potential of transcripts	Kong et al. (2007)

Table 5.3 Bioinformatics databases providing information regarding noncoding RNAs

Database	Description	URL	References
PNRD—Plant Noncoding RNA database	Using this database can retrieve the information of different types ncRNAs in plants	http://structuralbiology.cau.edu.cn/PNRD/	Yi et al. (2015)
miRBase—microRNA database	microRNA database contains information about the annotated and published miRNA sequence	https://www.mirbase.org/	Griffiths-Jones et al. (2007)
CANTATAdb—Plant noncoding RNA database	This database will be helpful for purpose of identified lncRNAs annotation	http://cantata.amu.edu.pl/	Szcześniak et al. (2016)
PLncDB—Plant noncoding RNA database	Provides the comprehensive information of plant lncRNAs in genomic view for the plant research community	http://chualab.rockefeller.edu/gbrowse2/homepage.html	Jin et al. (2021)
AlnC	An extensive database for the lncRNA in angiosperms	http://14.139.61.8/AlnC/index.html	Singh et al. (2021)
PeTMbase	Provides information about the miRNA and endogenous target mimics and helps in the functional genomics studies for miRNA regulatory networks	http://tools.ibg.deu.edu.tr/petmbase/	Karakülah et al. (2016)
Plant snoRNA database	For comprehensive information on small nucleolar RNAs in plants	http://bioinf.scri.sari.ac.uk/cgi-bin/plant_snorna/home	Brown et al. (2003)
sRNAanno	Repository for the plant small RNAs and this data will be helpful in the annotation	http://www.plantsrnas.org/	Chen et al. (2021)
Plant Phased Secondary SiRNA Database	Tissue-specific, stress-responsive PhasiRNA information available in the database	http://bioinfo.bti.cornell.edu/cgi-bin/phasiRNA/index.cgi	Fei Bioinformatics Lab, Boyce Thompson Institute and USDA Robert W. Holley Center
TarDB—miRNA target	User friendly server to find the plant miRNA targets and miRNA triggered phased siRNAs	http://www.biosequencing.cn/TarDB	Liu et al. (2021)
PMRD - Plant miRNA database	Functions and target information of miRNAs are available in the	http://bioinformatics.	Zhang et al. (2010)

(continued)

Table 5.3 (continued)

Database	Description	URL	References
	database, helpful in classifying the novel and known miRNAs and also easy to find the target for known miRNAs	cau.edu.cn/PMRD/	
PmiREN (Plant miRNA ENcyclopedia)	Useful in the annotation and in finding the expression of pre published miRNAs	https://www.pmiren.com/	Guo et al. (2020)
EVLncRNAs	Low throughput experimentally validates manually curated lncRNAs database	http://biophy.dzu.edu.cn/EVLncRNAs	Zhou et al. (2018)
CRISPRlnc	Contains manually curated and validated CRISPR/Cas9 sgRNAs for lncRNAs	http://www.crisprlnc.org/	Chen et al. (2019)
GREENC	Database contains plant lncRNAs and used for the annotation of identified lncRNAs	http://greenc.sequentiabiotech.com/wiki/Main_Page	Gallart et al. (2016)
PlaNC-TE	Finds the overlaps between the lncRNA and transposable elements	http://planc-te.cp.utfpr.edu.br/	Pedro et al. (2018)
PLncPRO	Machine learning algorithm tool used to find the lncRNAs and used to find the abiotic stress-responsive lncRNAs in rice and chickpea	http://ccbb.jnu.ac.in/plncpro/	Singh et al. (2017)

References

Allen E, Xie Z, Gustafson AM, Carrington JC (2005) microRNA-directed phasing during trans-acting siRNA biogenesis in plants. Cell 121:207–221

Andrews S (2010) FastQC: a quality control tool for high throughput sequence data. http://www.bioinformatics.babraham.ac.uk/projects/fastqc

Ariel F, Jegu T, Latrasse D, Romero-Barrios N, Christ A, Benhamed M, Crespi M (2014) Noncoding transcription by alternative RNA polymerases dynamically regulates an auxin-driven chromatin loop. Mol Cell 55:383–396

Bartel DP (2004) MicroRNAs: genomics, biogenesis, mechanism, and function. Cell 116:281–297

Bennasser Y, Le S-Y, Yeung ML, Jeang K-T (2004) HIV-1 encoded candidate micro-RNAs and their cellular targets. Retrovirology 1:1–5

Bolger AM, Lohse M, Usadel B (2014) Trimmomatic: a flexible trimmer for Illumina sequence data. Bioinformatics 30:2114–2120

Brown JW et al (2003) Plant snoRNA database. Nucleic Acids Res 31:432–435

Cantu D et al (2010) Small RNAs, DNA methylation and transposable elements in wheat. BMC Genomics 11:1–15

Cech TR, Steitz JA (2014) The noncoding RNA revolution—trashing old rules to forge new ones. Cell 157:77–94

Chakraborty A, Viswanath A, Malipatil R, Rathore A, Thirunavukkarasu N (2020) Structural and functional characteristics of miRNAs in five strategic millet species and their utility in drought tolerance. Front Genet 11:608421. https://doi.org/10.3389/fgene.2020.608421

Chen C et al (2021) sRNAanno—a database repository of uniformly annotated small RNAs in plants. Hortic Res 8:1–8

Chen L, Wang T, Zhao M, Tian Q, Zhang W-H (2012) Identification of aluminum-responsive microRNAs in *Medicago truncatula* by genome-wide high-throughput sequencing. Planta 235: 375–386

Chen W et al (2019) CRISPRlnc: a manually curated database of validated sgRNAs for lncRNAs. Nucleic Acids Res 47:D63–D68

Corona-Gomez JA, Garcia-Lopez IJ, Stadler PF, Fernandez-Valverde SL (2020) Splicing conservation signals in plant long noncoding RNAs. RNA 26:784–793

Crespi M, Jurkevitch E, Poiret M, d'Aubenton-Carafa Y, Petrovics G, Kondorosi E, Kondorosi A (1994) enod40, a gene expressed during nodule organogenesis, codes for a non-translatable RNA involved in plant growth. EMBO J 13:5099–5112

Dai X, Zhao PX (2011) psRNATarget: a plant small RNA target analysis server. Nucleic Acids Res 39:W155–W159

Edgar R, Domrachev M, Lash AE (2002) Gene expression omnibus: NCBI gene expression and hybridization array data repository. Nucleic Acids Res 30:207–210

El-Gebali S, Mistry J, Bateman A, Eddy SR, Luciani A, Potter SC, Qureshi M, Richardson LJ, Salazar GA, Smart A, Sonnhammer EL (2019) The Pfam protein families database in 2019. Nucleic Acids Res 47(D1):D427–D432

Franco-Zorrilla JM et al (2007) Target mimicry provides a new mechanism for regulation of microRNA activity. Nat Genet 39:1033–1037

Frazier TP, Sun G, Burklew CE, Zhang B (2011) Salt and drought stresses induce the aberrant expression of microRNA genes in tobacco. Mol Biotechnol 49:159–165

Friedländer MR, Mackowiak SD, Li N, Chen W, Rajewsky N (2012) miRDeep2 accurately identifies known and hundreds of novel microRNA genes in seven animal clades. Nucleic Acids Res 40:37–52

Gallart AP, Pulido AH, De Lagrán IAM, Sanseverino W, Cigliano RA (2016) GREENC: a wiki-based database of plant lncRNAs. Nucleic Acids Res 44:D1161

Goff LA, Trapnell C, Kelley D (2012) CummeRbund: visualization and exploration of Cufflinks high-throughput sequencing data R package version 2

Goldman D, Domschke K (2014) Making sense of deep sequencing. Int J Neuropsychopharmacol 17:1717–1725. https://doi.org/10.1017/S1461145714000789

Griffiths-Jones S, Saini HK, Van Dongen S, Enright AJ (2007) miRBase: tools for microRNA genomics. Nucleic Acids Res 36:D154–D158

Guo Z et al (2020) PmiREN: a comprehensive encyclopedia of plant miRNAs. Nucleic Acids Res 48:D1114–D1121

Hamilton A, Voinnet O, Chappell L, Baulcombe D (2002) Two classes of short interfering RNA in RNA silencing. EMBO J 21:4671–4679

Hammond SM, Bernstein E, Beach D, Hannon GJ (2000) An RNA-directed nuclease mediates post-transcriptional gene silencing in Drosophila cells. Nature 404:293–296

Herr AJ, Jensen MB, Dalmay T, Baulcombe DC (2005) RNA polymerase IV directs silencing of endogenous DNA. Science 308:118–120

Hutvágner G, McLachlan J, Pasquinelli AE, Bálint É, Tuschl T, Zamore PD (2001) A cellular function for the RNA-interference enzyme dicer in the maturation of the let-7 small temporal RNA. Science 293:834–838

Jagadeeswaran G, Saini A, Sunkar R (2009) Biotic and abiotic stress down-regulate miR398 expression in Arabidopsis. Planta 229:1009–1014

Jalili V et al (2020) The Galaxy platform for accessible, reproducible and collaborative biomedical analyses: 2020 update. Nucleic Acids Res 48:W395–W402

Jampala P, Garhewal A, Lodha M (2021) Functions of long non-coding RNA in Arabidopsis thaliana. Plant Signal Behav 16(9):1925440

Jha UC, Nayyar H, Jha R, Khurshid M, Zhou M, Mantri N, Siddique KH (2020) Long non-coding RNAs: emerging players regulating plant abiotic stress response and adaptation. BMC Plant Biol 20:1–20

Jin J et al (2021) PLncDB V2. 0: a comprehensive encyclopedia of plant long noncoding RNAs. Nucleic Acids Res 49:D1489–D1495

Johnson SM et al (2005) RAS is regulated by the let-7 microRNA family. Cell 120:635–647

Kalvari I et al (2018) Rfam 13.0: shifting to a genome-centric resource for non-coding RNA families. Nucleic Acids Res 46:D335–D342

Kang C, Liu Z (2019) An easy-to-follow pipeline for long noncoding RNA identification: a case study in diploid strawberry Fragaria vesca. In: Plant long non-coding RNAs. Humana Press, New York, NY, pp 223–243

Karakülah G, Yücebilgili Kurtoğlu K, Unver T (2016) PeTMbase: a database of plant endogenous target mimics (eTMs). PLoS One 11:e0167698

Khan Y, Yadav A, Bonthala VS, Muthamilarasan M, Yadav CB, Prasad M (2014) Comprehensive genome-wide identification and expression profiling of foxtail millet [Setaria italica (L.)] miRNAs in response to abiotic stress and development of miRNA database. Plant Cell Tiss Organ Cult (PCTOC) 118:279–292

Kim D, Paggi JM, Park C, Bennett C, Salzberg SL (2019) Graph-based genome alignment and genotyping with HISAT2 and HISAT-genotype. Nat Biotechnol 37:907–915

Kim D, Pertea G, Trapnell C, Pimentel H, Kelley R, Salzberg SL (2013) TopHat2: accurate alignment of transcriptomes in the presence of insertions, deletions and gene fusions. Genome Biol 14:1–13

Kim E-D, Sung S (2012) Long noncoding RNA: unveiling hidden layer of gene regulatory networks. Trends Plant Sci 17:16–21

Kong L, Zhang Y, Ye Z-Q, Liu X-Q, Zhao S-Q, Wei L, Gao G (2007) CPC: assess the protein-coding potential of transcripts using sequence features and support vector machine. Nucleic Acids Res 35:W345–W349

Kozomara A, Birgaoanu M, Griffiths-Jones S (2019) miRBase: from microRNA sequences to function. Nucleic Acids Res 47:D155–D162

Langmead B, Salzberg SL (2012) Fast gapped-read alignment with Bowtie 2. Nat Methods 9:357–359

Laurent GS, Wahlestedt C, Kapranov P (2015) The landscape of long noncoding RNA classification. Trends Genet 31:239–251

Law JA, Jacobsen SE (2010) Establishing, maintaining and modifying DNA methylation patterns in plants and animals. Nat Rev Genet 11:204–220

Lee RC, Feinbaum RL, Ambros V (1993) The *C. elegans* heterochronic gene lin-4 encodes small RNAs with antisense complementarity to lin-14. Cell 75:843–854

Li H (2013) Aligning sequence reads, clone sequences and assembly contigs with BWA-MEM. arXiv:13033997

Li J, Zhang X, Liu C (2020) The computational approaches of lncRNA identification based on coding potential: status quo and challenges. Comput Struct Biotechnol J 18:3666–3677. https://doi.org/10.1016/j.csbj.2020.11.030

Lima JC, Loss-Morais G, Margis R (2012) MicroRNAs play critical roles during plant development and in response to abiotic stresses. Genet Mol Biol 35:1069–1077

Liu J, Liu X, Zhang S, Liang S, Luan W, Ma X (2021) TarDB: an online database for plant miRNA targets and miRNA-triggered phased siRNAs. BMC Genomics 22:1–12

Liu X, Hao L, Li D, Zhu L, Hu S (2015) Long non-coding RNAs and their biological roles in plants. Genomics Proteomics Bioinformatics 13:137–147

Love M, Anders S, Huber W (2014) Differential analysis of count data–the DESeq2 package. Genome Biol 15:10–1186

Martin M (2011) Cutadapt removes adapter sequences from high-throughput sequencing reads. EMBnet J 17:10–12

Mattick JS, Rinn JL (2015) Discovery and annotation of long noncoding RNAs. Nat Struct Mol Biol 22:5–7

Muthamilarasan M, Prasad M (2021) Small millets for enduring food security amidst pandemics. Trends Plant Sci 26:33–40

Nejat N, Mantri N (2018) Emerging roles of long non-coding RNAs in plant response to biotic and abiotic stresses. Crit Rev Biotechnol 38:93–105

Nkhata SG, Ayua E, Kamau EH, Shingiro JB (2018) Fermentation and germination improve nutritional value of cereals and legumes through activation of endogenous enzymes. Food Sci Nutr 6:2446–2458

Pedro DLF, Lorenzetti APR, Domingues DS, Paschoal AR (2018) PlaNC-TE: a comprehensive knowledgebase of non-coding RNAs and transposable elements in plants. Database (Oxford) 2018:1–7. https://doi.org/10.1093/database/bay078

Pertea M, Pertea GM, Antonescu CM, Chang T-C, Mendell JT, Salzberg SL (2015) StringTie enables improved reconstruction of a transcriptome from RNA-seq reads. Nat Biotechnol 33: 290–295

Qi X, Xie S, Liu Y, Yi F, Yu J (2013) Genome-wide annotation of genes and noncoding RNAs of foxtail millet in response to simulated drought stress by deep sequencing. Plant Mol Biol 83: 459–473

Robinson MD, McCarthy DJ, Smyth GK (2010) edgeR: a Bioconductor package for differential expression analysis of digital gene expression data. Bioinformatics 26:139–140

Shi J et al (2015) mirPRo–a novel standalone program for differential expression and variation analysis of miRNAs. Sci Rep 5:1–12

Shin S-Y, Jeong JS, Lim JY, Kim T, Park JH, Kim J-K, Shin C (2018) Transcriptomic analyses of rice (Oryza sativa) genes and non-coding RNAs under nitrogen starvation using multiple omics technologies. BMC Genomics 19:1–20

Singh A, Vivek A, Kumar S (2021) AlnC: an extensive database of long non-coding RNAs in angiosperms. PLoS One 16:e0247215

Singh U, Khemka N, Rajkumar MS, Garg R, Jain M (2017) PLncPRO for prediction of long non-coding RNAs (lncRNAs) in plants and its application for discovery of abiotic stress-responsive lncRNAs in rice and chickpea. Nucleic Acids Res 45:e183–e183

Sun X, Zheng H, Sui N (2018) Regulation mechanism of long non-coding RNA in plant response to stress. Biochem Biophys Res Commun 503:402–407

Sun Z, Huang K, Han Z, Wang P, Fang Y (2020) Genome-wide identification of Arabidopsis long noncoding RNAs in response to the blue light. Sci Rep 10:1–10

Sunkar R (2010) MicroRNAs with macro-effects on plant stress responses. In: Seminars in cell & developmental biology, 2010. Elsevier, pp 805–811

Szcześniak MW, Rosikiewicz W, Makałowska I (2016) CANTATAdb: a collection of plant long non-coding RNAs. Plant Cell Physiol 57:e8

Trapnell C et al (2012) Differential gene and transcript expression analysis of RNA-seq experiments with TopHat and cufflinks. Nat Protoc 7:562–578

Vetriventhan M et al (2020) Genetic and genomic resources, and breeding for accelerating improvement of small millets: current status and future interventions. Nucleus:1–23

Wang Y, Li L, Tang S, Liu J, Zhang H, Zhi H, Jia G, Diao X (2016) Combined small RNA and degradome sequencing to identify miRNAs and their targets in response to drought in foxtail millet. BMC Genet 17(1):1–16

Yadav A, Khan Y, Prasad M (2016) Dehydration-responsive miRNAs in foxtail millet: genome-wide identification, characterization and expression profiling. Planta 243:749–766

Yang L, Conway SR, Poethig RS (2011) Vegetative phase change is mediated by a leaf-derived signal that represses the transcription of miR156. Development 138:245–249

Yi X, Zhang Z, Ling Y, Xu W, Su Z (2015) PNRD: a plant non-coding RNA database. Nucleic Acids Res 43:D982–D989

Zhang B, Pan X, Stellwag EJ (2008) Identification of soybean microRNAs and their targets. Planta 229:161–182

Zhang Z et al (2010) PMRD: plant microRNA database. Nucleic Acids Res 38:D806–D813

Zhou B et al (2018) EVLncRNAs: a manually curated database for long non-coding RNAs validated by low-throughput experiments. Nucleic Acids Res 46:D100–D105

Zhou L, Liu Y, Liu Z, Kong D, Duan M, Luo L (2010) Genome-wide identification and analysis of drought-responsive microRNAs in *Oryza sativa*. J Exp Bot 61:4157–4168

Zhou Y, Cho WK, Byun H-S, Chavan V, Kil E-J, Lee S, Hong S-W (2019) Genome-wide identification of long non-coding RNAs in tomato plants irradiated by neutrons followed by infection with tomato yellow leaf curl virus. PeerJ 7:e6286

Insights into Abiotic Stress Tolerance in Small Millets through Transcriptomics

6

Pankaj Shivnarayan Mundada, Abhinav Arvind Mali, Sumaiya Shahabuddin Shaikh, Nitin Tanaji Gore, Suraj Dhanyakumar Umdale, Archana Ashokrao Naik, Vitthal Tanaji Barvkar, Tukaram Dayaram Nikam, and Mahendra Laxman Ahire

Abstract

Food security is a global challenge and present staple foods are not sufficient to fulfill the increasing demand for food for growing population in terms of yield as well as nutrition. Small millets are nutritionally rich and have the potential to become the new staple. Small millets are a group of crops comprising finger millet, kodo millet, little millet, foxtail millet, barnyard millet, and proso millet. The crops are grown in a variety of agroecological conditions in diverse soils and varying rainfall. The growth and nutrition of the crops are highly influenced by abiotic stresses. Thus, majority of the crop improvement programs focus on the identification and development of stress tolerance mechanisms in crops. However, most of the crop improvement stratagems are carried out in staple cereals such as rice, wheat, and maize. Whereas very less or negligible attention is given toward minor millets. The present chapter summarizes the efforts carried out to

P. S. Mundada
Department of Biotechnology, Yashavantrao Chavan Institute of Science, Satara, Maharashtra, India

A. A. Mali
Department of Botany, Yashavantrao Chavan Institute of Science, Satara, Maharashtra, India

Department of Botany, Savitribai Phule Pune University, Pune, Maharashtra, India

S. S. Shaikh · N. T. Gore · M. L. Ahire (✉)
Department of Botany, Yashavantrao Chavan Institute of Science, Satara, Maharashtra, India

S. D. Umdale
Department of Botany, Jaysingpur College (Affiliated to Shivaji University, Kolhapur), Jaysingpur, Maharashtra, India

A. A. Naik · V. T. Barvkar · T. D. Nikam
Department of Botany, Savitribai Phule Pune University, Pune, Maharashtra, India

R. N. Pudake et al. (eds.), *Omics of Climate Resilient Small Millets*,
https://doi.org/10.1007/978-981-19-3907-5_6

explore the insights into abiotic stress tolerance on small millets through a transcriptomics approach.

Keywords

Small millets · Abiotic stress · Transcriptome · Crop improvement · RNA sequencing · Drought stress · Heat stress

6.1 Introduction

The world is facing agrarian as well as nutritional challenges. Nutritional quality of food is a key component in maintaining overall physical well-being as it forces health, development, and maximization of human genetic potential. Therefore, for solving the problem of deep-rooted food insecurity and malnutrition; dietary quality should be taken into consideration (Singh and Raghuvanshi 2012). Agricultural lands with irrigation facilities have been exploited up to their maximum capacity, and therefore, there is a need to attention toward dry lands for additional grain production. In the present situation, production of sufficient quality grains in dry lands is a big challenge due to low fertility. Millets scores highly over other important cereals such as wheat and rice in terms of nutritional composition and marginal growing conditions (Kumar et al. 2018). Millets are a rich source of several minerals such as iron, calcium and zinc, essential fatty acids, dietary fibers, vitamins, bioactive components such as antioxidants, and phytochemicals like polyphenols, lignins, phytosterols, phytoestrogens, and phytocyanins (Bouis 2000; Rao et al. 2011) that can help to overcome the problems of malnutrition. Cultivation of millets on dry lands can keep marginal lands productive and ensures food and nutritional security (Kumar et al. 2018).

Millet is a generic term used for a diverse group of fodder grasses identified for their small coarse grains (Weber 1998). Millets are the staple food crops that supply a major portion of calories and protein to large segments of populations in the semiarid tropical regions of Africa and Asia (O'Kennedy et al. 2006; Mundada et al. 2019). Small millets include barnyard millet (*Echinocloa frumentacea*), finger millet (*Eleusine coracana*), foxtail millet (*Setaria italica*), kodo millet (*Paspalum scrobiculatum*), little millet (*Panicum sumatrense*), and proso millet (*Panicum miliaceum*). About 97% of millet production takes place in developing countries and they are cultivated as an important staple crop in Asia and Africa (McDonough et al. 2000). Traditionally these grains are used in such diverse food types as leavened and unleavened flatbreads, porridges, steamed foods, and rice-like boiled products, and in alcoholic as well as nonalcoholic beverages (Murty and Kumar 1995). Besides food for human health, millets serve several other purposes such as stems are used as fodder, for construction of hut walls, fences, and thatches. Dry stems are also used in industries to produce brooms, mats, baskets, sunshades, etc. (Kumar et al. 2012).

Considering the nutritional parameters, millets are way ahead of wheat and rice. Some millets have more than fifty times of fiber than rice (Anonymous 2015). A wide array of nutritional composition and phytoconstituents, millets help against several chronic diseases like ischemic strokes, cardiovascular diseases, cancers, obesity, and type II diabetes (Jones et al. 2000; Jones 2006).

6.2 Millets and Abiotic Stresses

Crop growth and development are constantly influenced by environmental conditions (Egli and Bruening 2000). Drought, one of the most important environmental stresses that limit crop growth, development, and ultimately yield. In the present era, agricultural productivity is lost mostly due to scarcity of water. In many countries, more than 70% of the total fresh water is used for agriculture (Moumeni et al. 2011; Mundada et al. 2020). To meet the needs of a growing population, more food must be produced using less water (Moumeni et al. 2011). Plant growth and productivity are adversely affected by water stress. Therefore, the development of plants with increased survivability and growth during water stress is a major objective in the breeding of crop plants. Water deficit causes reduced carbon accumulation, tissue expansion, and reduced cell number. These processes involve a number of genes, enzymes, hormones, and metabolites (Skirycz and Inzé 2010; Tardieu and Tuberosa 2010). In fact, plants have evolved various molecular mechanisms to reduce their consumption of resources and adjust their growth to adapt to adverse environmental conditions (Skirycz and Inzé 2010; Nishiyama et al. 2011; Kim et al. 2017).

Different types of abiotic stresses are often amalgamated, and these stresses alone or in combination persuade various cellular damages. These stresses ultimately induce complex osmotic stresses in plants (Huang et al. 2012). Furthermore, all the types of abiotic stresses including osmotic stress induce oxidative damage and involve the formation of reactive oxygen species (ROS) in plant cells (Upadhyaya and Panda 2013; Mundada et al. 2020). Among the different types of abiotic stresses, drought and salt stress result in osmotic stress. Plants recognize stress stimuli through different mechanisms that activated the signaling pathways which include secondary messengers, hormones, transcription regulators, and signal transducers (Cvikrová et al. 2013; Danquah et al. 2014). A specific acclimation response to stress-induced stimuli is generated by these multiple signals that unite to modulate the stress-induced genes encoding proteins and enzymes involved in stress metabolism (Casaretto et al. 2016). However, plants are found to be susceptible to different abiotic stresses, especially osmotic stress at germination level and early developmental stages of the seedling (Ahire and Nikam 2011; Saha et al. 2016; Parvathi and Nataraja 2017). Understanding the molecular mechanism of plants' response under such stress conditions is of utmost importance for developing strategies for improvement of the plants' tolerance under stress conditions. Transcriptome provides detailed insights into the molecular mechanisms by which plant responds to their environment. The expression pattern of transcriptome is dynamic and reflects the level of expression of all genes in a particular cell, tissue, or organ at a particular stage of growth and development in an organism. It is useful in understanding the complex regulatory network associated with the adaptability and tolerance of plants to stress. Transcriptome sequencing has gained a lot of importance due to advancement and reduced cost of sequencing due to Next Generation Sequencing technology. Presently, transcriptome data analysis of plants is performed widely in various organisms under diverse environmental conditions.

6.3 Transcriptome Efforts in Small Millets

6.3.1 Finger Millet (*Eleusine coracana* L. Gaertn)

Finger millet is cultivated broadly in major parts of India and Africa and commonly known as "mandua" and ragi in India (Chandra et al. 2016). Nutritionally finger millet is comparable or even better than that of the rice, wheat, and other prominent cereals (Latha et al. 2005; Chandrashekar 2010). Finger millet contains carbohydrate (81.5%), dietary fiber (18–20%), starch (65–75%), protein (9.8%), fat (1–1.7%), minerals (2.7%), and crude fiber (4.3%), which are equivalent to other millets and cereal crops (Saleh et al. 2013). Finger millet was found to be nutritionally superior to barley, rye, and oats and able to fulfill the expected requirement of growing children and pregnant women (Satish et al. 2017). Finger millet continues to grow even in severe water-deficit and osmotic stress conditions and displays notable recovery on alleviation of stress (Govind et al. 2009). It is also observed that application of Si improves seed germination and growth attributes under PEG induced osmotic stress by enhancing the uptake of Si, and by upregulating the biosynthesis of fatty acids (Mundada et al. 2021a, b).

Researchers have studied transcriptomic analysis in finger millet under abiotic stress (Table 6.1). Rahman et al. (2014) studied the effect of salinity stress on contrasting genotypes in finger millet by RNA sequencing of 27.91 million reads. The transcriptome analysis revealed that, salinity stress led to upregulation of functional groups of genes like transporters, transcription factors, genes involved in cell signaling, osmotic homeostasis, and biosynthesis of compatible solutes in the stress tolerant cultivar Trichy 1. Whereas genes involved in flavonoid biosynthesis were downregulated in stress-tolerant cultivar Trichy 1. Hittalmani et al. (2017) reported that finger millet genome harbors large number of drought tolerance related genes. Transcriptome analysis of low moisture stress and nonstress samples revealed the identification of number of drought-induced candidate genes, which could be used in breeding programs for drought stress tolerance. From the same transcriptome data, presence of genes involved in ROS scavenging activity along with cis-acting regulatory elements such as AuxRE, DRE, GARE, G-box, GATA-box, MBS, MYBR, and W-box under drought stress was reported by Avashthi et al. (2020). Presence of unique drought-responsive genes such as serine threonine protein phosphatase 2A (*PP2A*), calcineurin B-like interacting protein kinase31 (*CIPK31*), farnesyl pyrophosphate synthase (*FPS*), and signal recognition particle receptor α

Table 6.1 Transcriptomic analysis in Finger millet (*Eleusine coracana* L. Gaertn.) under abiotic stress

Cultivar used	Type of stress	Sequencing technology used	Findings	References
CO 12 and Trichy 1	Salinity stress	IonProton platform	Upregulation of functional groups of genes like transporters, transcription factors, genes involved in cell signaling, osmotic homeostasis, and biosynthesis of compatible solutes in tolerant cultivars. Gene for flavonoid synthesis were downregulated in tolerant cultivar.	Rahman et al. (2014)
Finger millet cultivar ML-365	Drought	Illumina HiSeq4000 and NextSeq500	Number of drought-related TFs.	Hittalmani et al. (2017)
GPU-28	Drought	llumina NextSeq 500 (2× paired end (151 bp))	Activation of diverse drought stress signaling cascade genes such as serine threonine protein phosphatase 2A (*PP2A*), calcineurin B-like interacting protein kinase31 (*CIPK31*), farnesyl pyrophosphate synthase (*FPS*), and signal recognition particle receptor α (*SRPR α*). The basal regulatory genes such as TATA-binding protein (TBP)-associated factors (TAFs).	Parvathi et al. (2019)
GP-1 and GP-45	Drought	Data from Hittalmani et al. (2017)	Genes involved in ROS scavenging activity along with cis-acting regulatory elements such as AuxRE, DRE, GARE, G-box, GATA-box, MBS, MYBR, and W-box are involved in drought tolerance.	Avashthi et al. (2020)
–	Drought	The Illumina HiSeq™ X-Ten platform	80,602 differentially expressed genes (DEGs) were identified from transcriptome.	Li et al. (2021)
Local cultivar FM/ST/01	Drought	Illumina Novaseq 6000	Silicon-mediated retaliation of osmotic stress in finger millet by upregulation of fatty acid biosynthesis and JA.	Mundada et al. (2021a)

(*SRPR α*). The basal regulatory genes such as TATA-binding protein (TBP)-associated factors (TAFs) was reported under drought stress by Parvathi et al. (2019). Li et al. (2021) reported a total of 80,602 differentially expressed genes (DEGs) and 3009 differentially expressed proteins by an integrated study of transcriptome and proteome of finger millet under drought stress. Silicon-mediated retaliation of osmotic stress in finger millet by upregulation of fatty acid biosynthesis and jasmonic acid was reported using an integrated study of transcriptome and metabolome by Mundada et al. (2021a). The study reported the upregulation of 19,811 genes and downregulation of 7882 genes in presence of silicon under PEG-inducted drought stress. The genes involved in the glycolytic pathways were reported to be upregulated along with genes involved in catabolism of amino acids to acetyl CoA. The elevated pool of acetyl CoA was further diverted to fatty acid biosynthesis leading to accumulation of fatty acids and biosynthesis of jasmonic acid under stress conditions. The genes of TCA pathway and fatty acid catabolism were downregulated in presence of silicon.

6.3.2 Barnyard Millet (*Echinochloa esculenta* A. Braun)

Barnyard millet a primeval millet crop also known as Japanese millet is cultivated in temperate regions across the globe. It is widely cultivated in Asian countries such as China, India, Korea, and Japan. India is one of the largest producers of barnyard millet in terms of area (0.146 mha) as well as the production (0.147 mt). The average yield was reported to be 1034 kg/ha in the last 3 years (IIMR 2018). Nutritionally, barnyard millet is rich and superior to all other millets. Its grains are enriched with minerals such as calcium, iron, magnesium, and zinc along with dietary fiber, proteins, certain essential amino acids, fats, and vitamins (Saleh et al. 2013; Chandel et al. 2014). The dietary intake of barnyard millet has been reported to reduce the levels of serum cholesterol and triglycerides along with blood glucose, in rats due to presence of resistant starch (Kumari and Thayumanavan 1998).

Echinochloa species have a high degree of tolerance to various abiotic stresses (Gupta et al. 2009; Singh et al. 2010). Arthi et al. (2019) assessed 89 *Echinochloa* accessions for salinity stress tolerance and reported that MDU1, PRJ1, CO(Kv)2, VL 29, TNEf 204, TNEf 301, TNEf361, and TNEf364 exhibit better germination as compared to rice. Farmers preferably cultivate the *Echinochloa* species in drought and salinity prone conditions. The tolerance of *Echinochloa* species to adverse environmental conditions might be attributed to specialized organization of rhizosphere that can facilitate the uptake and release of oxygen (O_2) from their roots under stressful conditions. Higher water uptake efficiency (deep root) of barnyard millet (*E. utilis*) over other minor millets, including pearl millet was reported by Zegada-Lizarazu and Iijima (2005). Barnyard millet is reported to sustain and increase the leaf area index, water use efficiency, and dry matter production under drought as well as flooding conditions. Therefore, it is also worth investigating the mechanism behind the tolerance to drought and flooding stress in *Echinochloa* species.

The only transcriptome study was carried out in barnyard millet under abiotic stress by Jayakodi et al. (2019) (Table 6.2). They sequenced 97,065 transcripts, including 65,276 protein-coding transcripts, out of which more than 90% have been functionally annotated, and contained 31,789 long noncoding RNA (lncRNA) transcripts. They reported differential expression in 4159 protein-coding and 2258 lncRNA transcripts under drought stress. The further confirmation of differential gene expression observed in the transcriptome data was carried out using qRT-PCR

Table 6.2 Transcriptomic analysis in Barnyard millet (*Echinochloa frumentacea* L.) under abiotic stress

Cultivar used	Type of stress	Sequencing technology used	Findings	References
CO (KV) 2 variety	Drought stress	Illumina NextSeq 500 platform	Transcriptome analysis identified 4159 protein-coding and 2258 lncRNA transcripts in Indian barnyard millet that showed either up- or downregulated expression Genes encoding photosystem II P680 reaction center D1 protein, photoreceptor superfamily, photosystem II CP43 chlorophyll apoprotein, photosystem I P700 chlorophyll apoprotein A1, photosystem II PsbM protein, photosystem I subunit PsaO, and photosystem II 22-kDa protein were showed higher expression under drought stress condition in *E. frumentacea* than *E. crus-galli*	Jayakodi et al. (2019)

analysis under drought stress condition for the transcripts of different gene encoding photosystem II P680 reaction center D1 protein, photoreceptor superfamily, photosystem II CP43 chlorophyll apoprotein, photosystem I P700 chlorophyll a apoprotein A1, photosystem II PsbM protein, photosystem I subunit PsaO, and photosystem II 22-kDa. All these genes were upregulated in the tolerant cultivar.

6.3.3 Proso Millet (*Panicum miliaceum* L.)

Proso millet grows at a wide range of altitudes, with a short growth cycle of 6–12 weeks and requires little water for growth and development. It is grown in northern China, Republic of Korea, Mongolia, southeastern Russia, Pakistan, India, Afghanistan, and southern Europe. The cultivation area of proso millet is 0.82 m ha in Russia, 0.7–1.0 m ha in China (Wang et al. 2016), 0.5 m ha in India (Salini et al. 2010), and 0.20 m ha in the USA (Habiyaremye et al. 2017). Proso millet is enriched with minerals such as calcium, iron, magnesium, manganese, sodium, phosphorus, potassium, and zinc along with several essential amino acids viz. methionine, tryptophan, phenylalanine, and valine (Saleh et al. 2013). According to Devi et al. (2014) proso millet has more amount of protein per 100 g of grains, i.e., 11–14.0% as comparted to rice (7.5%) and wheat (14.4%).

Proso millet is affected by abiotic stresses, though they are generally considered well-adapted to abiotic stresses as compared to most other cereals (Dwivedi et al. 2012). Yue et al. (2016a) for the first time carried out the transcriptome sequencing and large-scale de novo assembly in proso millet. A total of 113,643 unigenes were obtained, of which 62,543 were functionally annotated. Furthermore, more than

35,000 SSRs and 406,000 SNP loci were identified which can be further used in diversity studies. Further, they reported the differential gene expression of four genes namely, unigene34608, unigene35973, unigene33484, and unigene41558 under heat, cold and salt stress. The unigene41558 putative encodes a CBL-interacting protein kinase 9 (CIPK9), which interacts with calcium sensor and plays important roles in low-KC stress. High expression levels of unigene41558 were observed under several stress treatments. Similarly, unigene33484 likely plays a role in osmoregulation in Broomcorn millet and unigene35973 is predicted to encode a zinc-finger protein gene ISAP1, which involved in regulating cold, dehydration, and salt tolerance in transgenic tobacco. From the same transcriptome data using computational prediction methods, 32 PmWRKY genes were reported which play a vital role in abiotic stress response (Yue et al. 2016b). Later from the 37 core proso millet cultivars, Zhang et al. (2019) carried out transcriptome sequencing in drought sensitive cultivar (Jinshu 6) and tolerant cultivar (Neimi 5) under PEG-induced drought stress (Table 6.3). They reported 833 and 2166 differentially expressed genes (DEGs) in Jinshu 6 and Neimi 5, respectively. They also reported the delayed transcriptional responses of the ROS scavenging system to simulated drought treatment and relatively easy recovery of the expression of photosynthesis-associated genes in tolerant cultivar Neimi 5.

6.3.4 Foxtail Millet (*Setaria italica* (L.) P. Beauv)

Foxtail millet has been identified as the major millet in terms of worldwide production, as it is the sixth-highest yielding grain (Saleh et al. 2013). In terms of its yielding ability, foxtail millet ranks fourth among all millets (Bala Ravi 2004). Foxtail millet contains a pertinent number of nutritional components, especially starch, protein, vitamins, and minerals (Muthamilarasan et al. 2016). Due to coarse nature of grains, the digestible portion constitutes about 79%, and the remaining undigestible part of the grain contains relatively high levels of fiber as well as some anti-nutritional components (Sharma et al. 2018; Zhu et al. 2018). Foxtail millet is also a good source of crude fiber, helps in the digestive process and helps to induce bowl movement, thus producing a laxative effect that is beneficial for a healthy digestive system (Verma et al. 2015).

The foxtail millet is used as a model cereal crop for study of stress biology and functional genomics due to its small size (~490 Mbp) (Bennetzen et al. 2012; Zhang et al. 2012). Foxtail millet is herbicide tolerant (Zhu et al. 2006), drought, and salt-tolerant crop (Krishnamurthy et al. 2014; Sudhakar et al. 2015). The tolerance to abiotic stresses is attributed to enhanced biochemical activities such as elevated levels of antioxidants, antioxidative enzymes such as SOD, catalases, peroxidases, synthesis of compatible solutes, and their stress-related proteins (Lata et al. 2011). Aldo-Keto reductases (AKRs) are a superfamily of NADPH-dependent oxidoreductases known to be cytosolic, and monomeric oxidoreductases acts on carbonyl metabolites (Bohren et al. 1989). Foxtail millet AKR1 is a promising stress-responsive gene that modulates and enhances stress tolerance in major crops

Table 6.3 Transcriptomic analysis in Proso millet (*Panicum miliaceum* L.) under abiotic stress

Cultivar used	Type of stress	Sequencing technology used	Findings	References
Yumi No. 2 and Yumi No. 3	Abiotic stress—cold, heat, and salt stress	Illumina Paired end sequencing technology	A total of 113,643 unigenes were obtained, of which 62,543 were functionally annotated. Furthermore, more than 35,000 SSRs and 406,000 SNP loci were identified, four DEGs—unigene34608, unigene 35,973, Unigene 33,484, and unigene 41,558 were studied.	Yue et al. (2016a)
Yumi No. 3	Salt, drought, cold and heat stress	–	A total of 32 PmWRKY genes were identified out of which expression levels of 22 PmWRKY genes varied significantly under one or more abiotic stress treatments, which could be defined as abiotic stress-responsive genes.	Yue et al. (2016b)
Drought-tolerant (Neimi 5) and drought-sensitive (Jinshu 6) cultivars	Drought stress	constructed a cDNA library using leaf mRNA and subsequently sequenced using Hiseq 4000 sequencing platform.	Reported 833 and 2166 differentially expressed genes (DEGs) in Jinshu 6 and Neimi 5, respectively, in response to PEG-induced drought stress	Zhang et al. (2019)

(Kirankumar et al. 2016). Qi et al. (2013) reported certain small interfering RNAs (si RNA) and noncoding RNAs (ncRNA) that are involved in regulation of drought stress in foxtail millet. Similarly, Wang et al. (2014) reported presence of SiLEA 14 a member of late embryogenesis abundant (LEA) proteins in foxtail millet. The LEA proteins play an important role in protection of higher plants against environmental stresses (Wang et al. 2014).

Lata et al. (2010) studied the changes in the transcriptome of tolerant cultivar Prasad at two different stages (early 0.5 Hr and late 6 Hr) upon dehydration and identified a total of 327 unique ESTs. These ESTs were classified into 11 different categories according to their putative functions. The plant response against dehydration stress was complex, representing major transcripts involved in metabolism (17%), stress (6%), signaling (6%), transcription regulation (8%), translation (2%), and proteolysis (5%). Among these the function of around 37% transcripts is hypothetical or unknown. Furthermore, the confirmation of these findings were done by performing qRT-PCR analysis of 9 randomly selected upregulated ($\geq$2.5 fold) genes. Puranik et al. (2011) carried out transcriptome analysis in two contrasting cultivars of foxtail millet differing in their ability of salinity stress tolerance. In the study, they constructed two suppression subtractive hybridization cDNA libraries (forward and reverse) were of these cultivars and identified 249 non-redundant ESTs by random EST sequencing. These ESTs were grouped into 11 functional categories. Out of these 63.9% (159) clones showed $\geq$twofold variation in expression pattern under salinity stress. Out of 159, around 115 (72.3%) were upregulated and 44 (27.7%) were downregulated. Interestingly 81 (51%) of these differentially expressed transcripts were reported for the first time from this study. Qi et al. (2013) reported differential expression of 2824 genes upon drought stress, out of which 48.23% were upregulated and 51.77% were downregulated. Genes encoding late embryogenesis abundant protein (LEA), dehydrin, heat shock protein (HSP), aquaporin, and phosphatase 2C (PP2C) were the most abundant among the upregulated genes. While genes encoding protein kinase family protein, ribosomal protein, NAD(P)-binding Rossmann-fold superfamily protein, histone superfamily protein, and GDSL-like lipase/acylhydrolase superfamily protein were the most abundant among the downregulated genes. Genes that encode members of the MYB transcription factor family, peroxidase superfamily, and members of cytochrome P450 family were differentially expressed. Shi et al. (2018) exposed the drought-resistant F1 hybrid, M79, and its parental lines E1and H1 to drought stress. A transcriptomic study using leaves collected 6 days after drought treatment revealed 3066, 1895, and 2148 differentially expressed genes (DEGs) in M79, E1 and H1 compared to the respective untreated controls. Transcriptomic study suggested that DEGs in M79 contributed to the formation of a regulatory network involving multiple biological processes and pathways, including photosynthesis, signal transduction, transcriptional regulation, redox regulation, hormonal signaling, and osmotic regulation (Table 6.4). Xu et al. (2019) performed transcriptome profiling in two foxtail millet genotypes Damaomao (DM) (Tolerant) and Hongnian (HN) (Sensitive) to uncover the molecular mechanisms of drought stress tolerance. The genes involved in hormone synthesis, proline, and soluble sugar synthesis, and reactive oxygen species (ROS) metabolism were affected due to stress and the expression profile of some the genes implicated in these pathways were different in two cultivars. Overall, a total of 296 TF-encoding genes in 50 TF families were detected in DEGs of DM1/DM2 and HN1/HN2, and most of these DEGs belong to bHLH, WRKY, NAC, ERF, and MYB family. All TFs of four of these families were found upregulated (90.8%), with only one gene downregulated in HN1/HN2 in response to drought stress. Chlorophyll metabolism and photosynthesis responses showed seven differentially expressed genes involved in the chlorophyll metabolism

Table 6.4 Transcriptomic analysis in Foxtail millet (*Setaria italica* L.) under abiotic stress

Cultivar used	Type of stress	Sequencing technology used	Findings	References
Prasad	Dehydration (Drought Stress)	ABI Sequencer	A total of 327 unique ESTs were identified from both libraries and were classified into 11 different categories according to their putative functions. The plant response against dehydration stress was complex, representing major transcripts involved in metabolism, stress, signaling, transcription regulation, translation, and proteolysis.	Lata et al. 2010
Prasad and Lepakshi	Salinity Stress	ABI Sequencer, Version No. 3770, Applied Biosystems	A total of 249 nonredundant ESTs were identified by random EST sequencing and grouped into 11 functional categories. Macroarray analysis of these clones showed that 159 (63.9%) were differentially expressed (≥twofold) in response to salinity stress, with 115 (72.3%) up and 44 (27.7%) downregulated. Data revealed that 81 (51%) of the 159 differentially expressed transcripts found in foxtail millet have not been reported in previous studies.	Puranik et al. (2011)

(continued)

Table 6.4 (continued)

Cultivar used	Type of stress	Sequencing technology used	Findings	References
Yugu1	Drought stress	Illumina HiSeq 2000 sequencer (Illumina, USA)	A total of 2824 genes with drought-responsive expression patterns, among them 48.23% were up-regulated and 51.77% were down-regulated	Qi et al. (2013)
E1 (maternal line) H1 (paternal line) and their F1 hybrid M79, a drought-resistant variety	Drought stress	HiSeq X-Ten (Illumina, 140 USA)	3066, 1895, and 2148 DEGs were identified after drought treatment in M79, E1, and H1, respectively, with 1404, 1116, and 1328 upregulated genes and 1662, 779, and 820 downregulated genes in corresponding genotypes.	Shi et al. (2018)
Damaomao (DM; number: 00002348; drought resistance index (DRI) in 2016 regional trial, DRI: 1.08) and drought-sensitive Hongnian (HN; number: 00000183; DRI: 0.03)	Drought stress	Illumina HiSeq-TM 4000 System (Illumina Inc., San Diego, CA, USA)	Two genotypes possessed high conservatism in some vital biological pathways which respond to drought stress, involving hormone synthesis, proline, and soluble sugar synthesis, and reactive oxygen species (ROS) metabolism. However, some genes related to these pathways showed differential expression profiles in two cultivars.	Xu et al. (2019)
Jigu 16	Drought Stress	Illumina System	Drought stress-responsive genes were related to different metabolic processes, such as photosynthesis, signal transduction, phenylpropanoid	Qin et al. (2020)

(continued)

Table 6.4 (continued)

Cultivar used	Type of stress	Sequencing technology used	Findings	References
			biosynthesis, starch and sucrose metabolism, and osmotic adjustment. Furthermore, the D1-pyrroline-5-carboxylate synthetase genes, SiP5CS1 and SiP5CS2, were remarkably upregulated in foxtail millet under drought stress conditions.	
Jingu 20	PEG Stress (Osmotic)	Illumina Hiseq 2500	The transcriptional studies revealed that more DEGs were identified during the further increase in water uptake period (phase III) than during the rapid initial uptake period (phase I) and the plateau period (phase II) under PEG stress.	Yu et al. (2020)
Yugu2 and An04	Salinity Stress	Illumina Hiseq platform X	Transcriptome analysis revealed 8887 and 12,249 DEGs in Yugu2 and An04 in response to salinity, respectively. Out of these, 3149 genes were overlapped between two varieties. These salinity-responsive genes indicated that ion transport, redox homeostasis, phytohormone metabolism, signaling, and secondary metabolism were enriched in Yugu2 by GO and KEGG analysis.	Pan et al. (2020)

pathway were identified, of which two transcripts were upregulated and five were downregulated. In the foxtail millet transcriptomes of DM and HN, one superoxide dismutase and 13 peroxidase genes, found only in HN, were downregulated. In contrast, seven ascorbate-glutathione (GSHAsA) related genes (five upregulated and two downregulated) were identified under the 24 h drought stress. Three DEGs were found in proline metabolism, including ornithine aminotransferase (one upregulated in HN2/HN1), pyroline-5-carboxylate synthetase (two upregulated in DM2/DM1 and HN2/HN1), and proline dehydrogenase (one downregulated in HN2/HN1). Qin et al. (2020) carried out transcriptome analysis of drought-tolerant cultivar Jigu 16 under drought stress followed by rehydration. Their findings revealed that, drought stress-responsive genes were related to number of metabolic processes, such as photosynthesis, phenylpropanoid biosynthesis, signal transduction, starch and sucrose metabolism, and osmotic adjustment. Furthermore, the D1-pyrroline-5-carboxylate synthetase genes (SiP5CS1 and SiP5CS2) were remarkably upregulated in foxtail millet under drought stress conditions. Yu et al. (2020) screened eight different cultivars of foxtail millets and identified Jingu 20 as drought-tolerant cultivar. Seeds of tolerant cultivar were germinated at three different phases, i.e., for 2 h (phase I), 8 h (phase II), and 14 h (phase III) separately and the transcriptome sequencing at after treatment of PEG for 1 and 3 h for three phases independently. The transcriptional studies revealed that more differentially expressed genes were identified during the further increase in water uptake period (phase III) than during the rapid initial uptake period (phase I) and the plateau period (phase II) under PEG stress. Pan et al. (2020) reported 8887 and 12,249 differentially expressed genes in two cultivars of foxtail millet (Yugu2 and An04) differing in salinity stress. Out of these, 3149 genes were overlapped between two varieties. The cultivar Yugu2 by GO and KEGG analysis revealed that salinity-responsive genes were involved in ion transport, redox homeostasis, phytohormone metabolism, signaling, and secondary metabolism.

6.3.5 Kodo Millet (*Paspalum scrobiculatum* L.)

Kodo millet is one of the hardiest crops among the small millets, and it is grown primarily in India. It is common in upland rice regions in India, Indonesia, Philippines, Thailand, and Vietnam, and also present in Bangladesh and Myanmar (Galinato et al. 1999). It is a very hardy, drought-tolerant crop and can survive on marginal soils where other crops may not survive and can yield 450–900 kg/ha (Heuzé et al. 2012). Kodo millet is a balanced and staple food grain of tribal and poor populations. Kodo millet is the major source of protein and minerals in the daily diets of people residing in remote rural areas (Ahmad and Yadava 1996). Kodo millet is useful for diabetic and cardiovascular diseases as it contains a high proportion of complex carbohydrates and dietary fibers. The dietary fibers help in the prevention of constipation and complex carbohydrates act by slow release of glucose to the bloodstream. This leads to reduction in blood glucose levels among diabetic patients (Yadav et al. 2013). It was reported by Kothai and Roselin (2020) that germination and seedling growth of kodo millet seeds were significantly affected by salinity stress induced by various concentrations of NaCl and seawater. However, to date, the study of abiotic stress at transcriptomic level is not carried out in kodo millet.

6.3.6 Little Millet (*Panicum sumatrense* Roth. Ex Roemer and Schultes)

Little millet is one among the minor millets grown to a limited extent all over India, up to altitudes of 2100 m. It is regarded as an orphan crop due to its restricted cultivation in specific regions and low consumption. This crop is resistant to adverse agro-climatic conditions, but it is sensitive to terminal drought (Tadele 2016). Little millet being rich in cholesterol, is known to increase good cholesterol in the body. It is beneficial for kids to strengthen their body. The complex carbohydrates present in little millet are digested slowly thereby releasing the glucose slowly in the bloodstream, which is advantageous for diabetic patients (Bhat et al. 2018). It is a rich source of minerals such as phosphorous (220 mg/100 g) and iron (9.3 mg/100 g). It is especially good for people has low body mass. A few recipes, which can be prepared using little millet, are dosa, idli, pongal, kichadi, etc. (NIN 2007).

Das et al. (2020) carried out the transcriptome sequencing for the first time in little millet under drought and salt stress conditions (Table 6.5). The study of ~230 million reads from control and treated tissues revealed 86,814 unigenes. The sequencing of transcripts was carried out separately from leaf and root tissue and revealed a different mode of regulation upon abiotic stress. The study identified presence of 61 common differentially expressed genes in both tissues under drought and salinity stress, which was further validated by qRT-PCR. The analysis revealed that, the genes for transcription factor WRKY and ABC transporters exhibited more expression under drought stress. The transcriptomic data was also used to determine genic SSR markers relevant to abiotic stress tolerance in little millet.

Table 6.5 Transcriptomic analysis in little millet (*Panicum sumatrense* Roth. Ex Roemer and Schultes) under abiotic stress

Cultivar used	Type of stress	Sequencing technology used	Findings	References
OLM-20	Drought and salinity	Illumina HiSeq2500	RNA-Seq generated ~230 million reads from control and treated tissues, which were assembled into 86,614 unigene.	Das et al. (2020)

6.4 Conclusion and Future Perspectives

Next generation sequencing and bioinformatics tools have changed the scenario of gene expression studies dramatically in the last decade. The transcriptome studies provide a valuable resource for identification of differentially expressed genes, generation of molecular markers like microsatellites and single nucleotide polymorphism. Transcriptome sequencing is a relatively cheaper technique making it an obvious choice in the studies, where whole genome is not affordable. The technique along with cost efficacy yield functionally relevant information and generate a large amount of genomic resource in less time. Small millets are nutritionally enriched and are relatively hardy crops capable of growing under diverse environmental conditions. They play a pivotal role in nutritional food security and health benefits. Despite these benefits very few researchers have carried out the transcriptomic studies in small millet indicating that there is a huge amount of niche for further studies in small millets as the information on genome is obscure in majority of the small millets. Among small millets, transcriptomic studies are available majorly in finger millet (*Eleusine coracana* L. Gaertn.) and foxtail millet (*Setaria italica*). The transcriptomic studies in future will add to the identification of candidate genes tolerant to abiotic stresses for genetic breeding studies, development of nutritionally rich and resistant cultivars, development of molecular markers like SSR, SNP's, etc. Along with transcriptomics various tools and techniques such as genomics, metabolomics, ionmics, and phenomics have been devised to allow the understanding of genetic makeup in depth, their signaling cascade, and their adaptability under stress conditions. Diverse studies of these omics tools along with integrated approaches will be more beneficial in the coming future to decipher the molecular basis of abiotic stress tolerance in small millets.

Acknowledgments Authors are also grateful to Yashavantrao Chavan Institute of Science, Satara for financial support under the DBT-STAR college scheme. Financial assistance to the faculty under self-funded project and faculty exchange program under RUSA is acknowledged for providing facilities.

References

Ahire ML, Nikam TD (2011) Differential response of brinjal (*Solanum melongena* Linn.) varieties to salinity stress in relation to seed germination and osmolytes accumulation. Seed Sci Biotechnol 5:29–35

Ahmad MS, Yadava HS (1996) Assessment of productivity and economics of small millets in Madhya Pradesh. Crop Res Hisar 12:391–394

Anonymous (2015). https://milletindia.org/wp-content/uploads/2015/07/Milletsfutureoffoodandfarming.pdf

Arthi N, Rajagopal B, Geethanjali S, Nirmalakumari A, Senthil N (2019) Screening of barnyard millet (*Echinochloa frumentacea*) germplasm for salinity tolerance. Electron J Plant Breed 10: 659–666

Avashthi H, Pathak RK, Gaur VS, Singh S, Gupta VK, Ramteke PW, Kumar A (2020) Comparative analysis of ROS-scavenging gene families in finger millet, rice, sorghum, and foxtail millet revealed potential targets for antioxidant activity and drought tolerance improvement. Netw Model Anal Health Inf Bioinf 9:1–23

Bennetzen JL, Schmutz J, Wang H, Percifield R, Hawkins J, Pontaroli AC, Estep M, Feng L, Vaughn JN, Grimwood J, Jenkins J, Barry K, Lindquist E, Hellsten U, Deshpande S, Wang X, Wu X, Mitros T, Triplett J, Yang X, Ye CY, Mauro-Herrera M, Wang L, Li P, Sharma M, Sharma R, Ronald PC, Panaud O, Kellogg EA, Brutnell PTP, Doust AN, Tuskan GA, Rokhsar D, Devos KM (2012) Reference genome sequence of the model plant *Setaria*. Nat Biotechnol 30:555–561

Bhat S, Nandini C, Tippeswamy V (2018) Significance of small millets in nutrition and health-a review. Asian J Dairy Food Res 37:35–40

Bohren KM, Bullock B, Wermuth B, Gabbay KM (1989) The aldo-keto reductase superfamily: cdnas and deduced amino acid sequences of human aldehyde and aldose reductase. J Biol Chem 264:9547–9551

Bouis HE (2000) Enrichment of food staples through plant breeding: a new strategy for fighting micronutrient malnutrition. Nutrition 16:701–704

Casaretto JA, El-Kereamy A, Zeng B, Stiegelmeyer SM, Chen X, Bi YM, Rothstein SJ (2016) Expression of OsMYB55 in maize activates stress-responsive genes and enhances heat and drought tolerance. BMC Genomics 17:1–15

Chandel G, Meena RK, Dubey M, Kumar M (2014) Nutritional properties of minor millets: neglected cereals with potentials to combat malnutrition. Curr Sci 107:1109–1111

Chandra D, Chandra S, Sharma AK (2016) Review of Finger millet (*Eleusine coracana* (L.) Gaertn): a powerhouse of health benefiting nutrients. Food Sci Hum Wellness 5:149–155

Chandrashekar A (2010) Finger Millet: *Eleusine coracana*. In: Taylor SL (ed) Advances in food and nutrition research. Academic Press, Burlington, pp 215–262

Cvikrová M, Gemperlová L, Martincová O, Vanková R (2013) Effect of drought and combined drought and heat stress on polyamine metabolism in proline-over-producing tobacco plants. Plant Physiol Biochem 73:7–15

Danquah A, de Zelicourt A, Colcombet J, Hirt H (2014) The role of ABA and MAPK signalling pathways in plant abiotic stress responses. Biotechnol Adv 32:40–52

Das RR, Pradhan S, Parida A (2020) De-novo transcriptome analysis unveils differentially expressed genes regulating drought and salt stress response in Panicum sumatrense. Sci Rep 10:1–14

Devi PB, Vijayabharathi R, Sathyabama S, Malleshi NG, Priyadarisini VB (2014) Health benefits of finger millet (*Eleusine coracana* L.) polyphenols and dietary fiber: a review. J Food Sci Technol 51:1021–1040

Dwivedi S, Upadhyaya H, Senthilvel S, Hash C, Fukunaga K, Diao X, Santra D, Baltensperge D, Prasad M (2012) Millets: genetic and genomic resources. In: Janick J (ed) Plant Breeding Reviews. John Wiley and Sons, Inc., Hoboken, pp 247–374

Egli DB, Bruening WP (2000) Potential of early-maturing soybean cultivars in late plantings. Agron J 92:532–537

Galinato MI, Moody K, Piggin CM (1999) Upland rice weeds of South and Southeast Asia. International Rice Research Institute (IRRI), Philippines pp 108–122

Govind G, Thamme Gowda HV, Kalaiarasi PJ, Iyer DR, Muthappa SK, Nese S, Makarla UK (2009) Identification and functional validation of a unique set of drought induced genes preferentially expressed in response to gradual water stress in peanut. Mol Genet Genomics 281:591–605

Gupta A, Mahajan V, Kumar M, Gupta HS (2009) Biodiversity in the barnyard millet (*Echinochloa frumentacea* Link, Poaceae) germplasm in India. Genet Resour Crop Evol 56(6):883–889

Habiyaremye C, Matanguihan JB, D'Alpoim GJ, Ganjyal GM, Whiteman MR, Kidwell KK, Murphy KM (2017) Proso millet (*Panicum miliaceum* L.) and its potential for cultivation in the Pacific Northwest, US: a review. Front Plant Sci 7:1961

Heuzé V, Tran G, Giger-Reverdin S (2012) Scrobic (*Paspalum scrobiculatum*) forage and grain. Feedipedia.org. http://www.feedipedia.org/node/401

Hittalmani S, Mahesh HB, Shirke MD, Biradar H, Uday G, Aruna YR, Lohithaswa HC, Mohanrao A (2017) Genome and transcriptome sequence of finger millet (*Eleusine coracana* (L.) Gaertn.) provides insights into drought tolerance and nutraceutical properties. BMC Genomics 18:1–16

Huang J, Sun S, Xu D, Lan H, Sun H, Wang Z, Bao Y, Wang J, Tang H, Zhang H (2012) A TFIIIA-type zinc finger protein confers multiple abiotic stress tolerances in transgenic rice (*Oryza sativa* L.). Plant Mol Biol 80:337–350

IIMR (2018) Annual report 2017–18. Indian Institute of Millets Research, Hyderabad

Jayakodi M, Madheswaran M, Adhimoolam K, Perumal S, Manickam D, Kandasamy T et al (2019) Transcriptomes of Indian barnyard millet and barnyard grass reveal putative genes involved in drought adaptation and micronutrient accumulation. Acta Physiol Plant 41:1–11

Jones JM (2006) Grain-based foods and health. Cereal Foods World 51:108

Jones PJ, Raeini-Sarjaz M, Ntanios FY, Vanstone CA, Feng JY, Parsons WE (2000) Modulation of plasma lipid levels and cholesterol kinetics by phytosterol versus phytostanol esters. J Lipid Res 41:697–705

Kim YH, Khan AL, Waqas M, Lee IJ (2017) Silicon regulates antioxidant activities of crop plants under abiotic-induced oxidative stress: a review. Front Plant Sci 8:510

Kirankumar TV, Madhusudhan KV, Nareshkumar A, Kiranmai K, Lokesh U, Venkatesh B, Sudhakar C (2016) Expression analysis of Aldo-Keto Reductase 1 (AKR1) in foxtail millet (*Setaria italica* L.) subjected to abiotic stresses. Am J Plant Sci 7:500–509

Kothai VN, Roselin RM (2020) Evaluation of salinity stress effects on seed germination and seedling growth and estimation of protein contents in Kodo millet (*Paspalum scrobiculatum* L.). J Stress Physiol Biochem 16:70–81

Krishnamurthy L, Upadhyaya HD, Gowda CLL, Kashiwagi J, Purushothaman R, Singh S, Vadez V (2014) Large variation for salinity tolerance in the core collection of foxtail millet (*Setaria italica* (L.) P. Beauv.) germplasm. Crop Pasture Sci 65:353–361

Kumar A, Arya RK, Kumar S, Kumar D, Kumar S, Panchta R (2012) Advances in pearl millet fodder yield and quality improvement through breeding and management practices. Forage Res 38:1–14

Kumar A, Tomer V, Kaur A, Kumar V, Gupta K (2018) Millets: a solution to agrarian and nutritional challenges. Agric Food Secur 7:1–15

Kumari SK, Thayumanavan B (1998) Characterization of starches of proso, foxtail, barnyard, kodo, and little millets. Plant Foods Hum Nutr 53:47–56

Lata C, Jha S, Dixit V, Sreenivasulu N, Prasad M (2011) Differential antioxidative responses to dehydration-induced oxidative stress in core set of foxtail millet cultivars [*Setaria italica* (L.)]. Protoplasma 248:817–828

Lata C, Sahu PP, Prasad M (2010) Comparative transcriptome analysis of differentially expressed genes in foxtail millet (Setaria italica L.) during dehydration stress. Biochem Biophys Res Commun 393(4):720–727

Latha AM, Rao KV, Reddy VD (2005) Production of transgenic plants resistant to leaf blast disease in finger millet (*Eleusine coracana* (L.) Gaertn.). Plant Sci 169:657–667

Li J, Wang Y, Wang L, Zhu J, Deng J, Tang R, Chen G (2021) Integration of transcriptomic and proteomic analyses for finger millet [*Eleusine coracana* (L.) Gaertn.] in response to drought stress. PLoS One 16:e0247181

McDonough CM, Rooney LW, Serna-Saldivar SO (2000) The millets. WORLD 36(797):28–791

Moumeni A, Satoh K, Kondoh H, Asano T, Hosaka A, Venuprasad R, Serraj R, Kumar A, Leung H, Kikuchi S (2011) Comparative analysis of root transcriptome profiles of two pairs of drought-tolerant and susceptible rice near-isogenic lines under different drought stress. BMC Plant Biol 11:1–17

Mundada PS, Ahire ML, Umdale SD, Barmukh RB, Nikam TD, Pable AA, Deshmukh RK, Barvkar VT (2021b) Characterization of influx and efflux silicon transporters and understanding their role in the osmotic stress tolerance in finger millet (*Eleusine coracana* (L.) Gaertn.). Plant Physiol Biochem 162:677–689

Mundada PS, Barvkar VT, Umdale SD, Kumar SA, Nikam TD, Ahire ML (2021a) An insight into the role of silicon on retaliation to osmotic stress in finger millet (*Eleusine coracana* (L.) Gaertn). J Hazard Mater 403:124078
Mundada PS, Nikam TD, Kumar SA, Umdale SD, Ahire ML (2020) Morpho-physiological and biochemical responses of finger millet (*Eleusine coracana* (L.) Gaertn.) genotypes to PEG-induced osmotic stress. Biocatal Agric Biotechnol 23:101488
Mundada PS, Umdale SD, Nikam TD, Ahire ML (2019) Genetic diversity using RAPD markers, mineral composition and their correlation in selected local landraces of finger millet [*Eleusine coracana* (L.) Gaertn.]. Vegetos 32:1–10
Murty DS, Kumar KA (1995) Traditional uses of sorghum and millets. Sorghum and millets: Chemistry and technology, 221
Muthamilarasan M, Dhaka A, Yadav R, Prasad M (2016) Exploration of millet models for developing nutrient rich graminaceous crops. Plant Sci 242:89–97
Nutritive value of Indian Foods, NIN (2007) Millet in your meals. http://www.sahajasamrudha.org/
Nishiyama R, Watanabe Y, Fujita Y, Le DT, Kojima M, Werner T, Vankova R, Yamaguchi-Shinozaki K, Shinozaki K, Kakimoto T, Sakakibara H, Schmülling T, Tran LSP (2011) Analysis of cytokinin mutants and regulation of cytokinin metabolic genes reveals important regulatory roles of cytokinins in drought, salt and abscisic acid responses, and abscisic acid biosynthesis. Plant Cell 23:2169–2183
O'Kennedy MM, Grootboom A, Shewry PR (2006) Harnessing sorghum and millet biotechnology for food and health. J Cereal Sci 44:224–235
Pan J, Li Z, Dai S, Ding H, Wang Q, Li X, Ding G, Wang P, Guan Y, Liu W (2020) Integrative analyses of transcriptomics and metabolomics upon seed germination of foxtail millet in response to salinity. Sci Rep 10(1):1–6
Parvathi MS, Nataraja KN (2017) Discovery of stress responsive TATA-box binding protein associated Factor6 (TAF6) from finger millet (*Eleusine coracana* (L.) Gaertn). J Plant Biol 60:335–342
Parvathi MS, Nataraja KN, Reddy YN, Naika MB, Gowda MC (2019) Transcriptome analysis of finger millet (*Eleusine coracana* (L.) Gaertn.) reveals unique drought responsive genes. J Genet 98:46
Puranik S, Jha S, Srivastava PS, Sreenivasulu N, Prasad M (2011) Comparative transcriptome analysis of contrasting foxtail millet cultivars in response to short-term salinity stress. J Plant Physiol 168(3):280–287
Qi X, Xie S, Liu Y, Yi F, Yu J (2013) Genome-wide annotation of genes and noncoding RNAs of foxtail millet in response to simulated drought stress by deep sequencing. Plant Mol Biol 83: 459–473
Qin L, Chen E, Li F, Yu X, Liu Z, Yang Y, Wang R, Zhang H, Wang H, Liu B, Guan YA (2020) Genome-wide gene expression profiles analysis reveal novel insights into drought stress in foxtail millet (*Setaria italica* L.). Int J Mol Sci 21(22):8520
Rahman H, Jagadeeshselvam N, Valarmathi R, Sachin B, Sasikala R, Senthil N et al (2014) Transcriptome analysis of salinity responsiveness in contrasting genotypes of finger millet (*Eleusine coracana* L.) through RNA-sequencing. Plant Mol Biol 85:485–503
Rao SV, Raju MVLN, Panda AK, Saharia P, Sunder GS (2011) Effect of supplementing betaine on performance, carcass traits and immune responses in broiler chicken fed diets containing different concentrations of methionine. Asian Australas J Anim Sci 24:662–669
Ravi SB (2004) Neglected millets that save the poor from starvation. Leisa India 6:1–8
Saha B, Mishra S, Awasthi JP, Sahoo L, Panda SK (2016) Enhanced drought and salinity tolerance in transgenic mustard [*Brassica juncea* (L.) Czern & Coss.] overexpressing Arabidopsis group 4 late embryogenesis abundant gene (AtLEA4-1). Environ Exp Bot 128:99–111
Saleh ASM, Zhang Q, Chen J, Shen Q (2013) Millet grains: nutritional quality, processing, and potential health benefits. Compr Rev Food Sci Food Saf 12:281–295
Salini K, Nirmalakumari A, Muthiah AR, Senthil N (2010) Evaluation of proso millet (*Panicum miliaceum* L.) germplasm collections. Electron J Plant Breed 1:489–499

Satish L, Ceasar SA, Ramesh M (2017) Improved agrobacterium-mediated transformation and direct plant regeneration in four cultivars of finger millet (*Eleusine coracana* (L.) Gaertn.). Plant Cell Tissue Organ Cult 131:547–565

Sharma S, Saxena DC, Riar CS (2018) Changes in the GABA and polyphenols contents of foxtail millet on germination and their relationship with in vitro antioxidant activity. Food Chem 245: 863–870

Shi W, Cheng J, Wen X, Wang J, Shi G, Yao J, Guo J (2018) Transcriptomic studies reveal a key metabolic pathway contributing to a well-maintained photosynthetic system under drought stress in foxtail millet (*Setaria italica* L.). Peer J 6:e4752

Singh P, Raghuvanshi RS (2012) Finger millet for food and nutritional security. African J Food Sci 6:77–84

Singh KP, Mishra HN, Saha S (2010) Moisture-dependent properties of barnyard millet grain and kernel. J Food Eng 96:598–606. https://doi.org/10.1016/j.jfoodeng.2009.09.007

Skirycz A, Inzé D (2010) More from less: plant growth under limited water. Curr Opin Biotechnol 21:197–203

Sudhakar C, Veeranagamallaiah G, Nareshkumar A, Sudhakarbabu O, Sivakumar M, Pandurangaiah M, Kiranmai K, Lokesh U (2015) Polyamine metabolism influences antioxidant defense mechanism in foxtail millet (*Setaria italica* L.) cultivars with different salinity tolerance. Plant Cell Rep 34:141–156

Tadele Z (2016) Drought adaptation in millets. In: Shanker AK, Shanker C (eds) Abiotic and biotic stress in plants – recent advances and future perspectives. IntechOpen. https://doi.org/10.5772/61929

Tardieu F, Tuberosa R (2010) Dissection and modelling of abiotic stress tolerance in plants. Curr Opin Plant Biol 13:206–212

Upadhyaya H, Panda SK (2013) Abiotic stress responses in tea [*Camellia sinensis* L. (O) Kuntze]: an overview. Rev Agric Sci 1:1–10

Verma S, Srivastava S, Tiwari N (2015) Comparative study on nutritional and sensory quality of barnyard and foxtail millet food products with traditional rice products. J Food Sci Technol 52: 5147–5155

Wang M, Li P, Li C, Pan Y, Jiang X, Zhu D, Zhao Q, Yu J (2014) SiLEA14, a novel atypical LEA protein, confers abiotic stress resistance in foxtail millet. BMC Plant Biol 14:1–16

Wang R, Hunt HV, Qiao Z, Wang L, Han Y (2016) Diversity and cultivation of broomcorn millet (*Panicum miliaceum* L.) in China: a review. Econ Bot 70:332–342

Weber SA (1998) Out of Africa: the initial impact of millets in South Asia. Curr Anthropol 39:267–274

Xu BQ, Gao XL, Gao JF, Jing LI, Pu YANG, Feng BL (2019) Transcriptome profiling using RNA-seq to provide insights into foxtail millet seedling tolerance to short-term water deficit stress induced by PEG-6000. J Integr Agric 18:2457–2471

Yadav N, Chaudhary K, Singh A, Gupta A (2013) Evaluation of hypoglycemic properties of kodo millet based food products in healthy subjects. Int J Pharm 3:14–20

Yu A, Zhao J, Wang Z, Cheng K, Zhang P, Tian G, Liu X, Guo E, Du Y, Wang Y (2020) Transcriptome and metabolite analysis reveal the drought tolerance of foxtail millet significantly correlated with phenylpropanoids-related pathways during germination process under PEG stress. BMC Plant Biol 20(1):1–7

Yue H, Wang L, Liu H, Yue W, Du X, Song W, Nie X (2016a) De novo assembly and characterization of the transcriptome of broomcorn millet (*Panicum miliaceum* L.) for gene discovery and marker development. Front Plant Sci 7:1083

Yue H, Wang M, Liu S, Du X, Song W, Nie X (2016b) Transcriptome-wide identification and expression profiles of the WRKY transcription factor family in Broomcorn millet (*Panicum miliaceum* L.). BMC Genomics 17:1–11

Zegada-Lizarazu W, Iijima M (2005) Deep root water uptake ability and water use efficiency of pearl millet in comparison to other millet species. Plant Prod Sci 8:454–460

Zhang G, Liu X, Quan Z, Cheng S, Xu X, Pan S, Xie M, Zeng P, Yue Z, Wang W, Tao Y, Bian C, Han C, Xia Q, Peng X, Cao R, Yang X, Zhan D, Hu J, Zhang Y, Li H, Li H, Li N, Wang J, Wang C, Wang R, Guo T, Cai Y, Liu C, Xiang H, Shi Q, Huang P, Chen Q, Li Y, Wang J, Zhao Z, Wang J (2012) Genome sequence of foxtail millet (*Setaria italica*) provides insights into grass evolution and biofuel potential. Nat Biotechnol 30:549–554

Zhang Y, Gao X, Li J, Gong X, Yang P, Gao J, Wang P, Feng B (2019) Comparative analysis of proso millet (*Panicum miliaceum* L.) leaf transcriptomes for insight into drought tolerance mechanisms. BMC Plant Biol 19:1–17

Zhu XL, Zhang L, Chen Q, Wan J, Yang GF (2006) Interactions of aryloxyphenoxypropionic acids with sensitive and resistant acetyl-coenzyme a carboxylase by homology modeling and molecular dynamic simulations. J Chem Inf Model 46:1819–1826. https://doi.org/10.1021/ci0600307

Zhu Y, Chu J, Lu Z, Lv F, Bie X, Zhang C, Zhao H (2018) Physicochemical and functional properties of dietary fiber from foxtail millet (*Setaria italic*) bran. J Cereal Sci 79:456–461

Role of Proteomics in Understanding the Abiotic Stress Tolerance in Minor Millets

7

A. Karthikeyan, V. G. Renganathan, and N. Senthil

Abstract

Minor millets are an underutilized group of cereal crops, and their grains are nutritionally superior to cereals. These millets have a unique ability to survive under adverse environmental conditions where the cultivation of major cereals is either not popular or fails to produce considerable yield. Although minor millets possess tolerance to environmental stresses than cereals, their yield is limited by various abiotic stresses (i.e., drought, heat, and salinity), which need to be boosted to enhance their utilization as a food crop. Abiotic stress tolerance in plants is a complex phenomenon involving changes in the biochemical, molecular, and physiological processes. These changes include modifications in the genomic and proteomic complements that are related to the plant stress tolerance mechanisms. The field of "omics" has progressed in recent years and offered a set of techniques to understand how plants respond to abiotic stresses. Proteins are involved in many cellular processes; therefore, proteomics has the edge over other omics techniques to study the stress tolerance mechanism in plants. Over the decade, several reports detailed the abiotic stress tolerance in minor millets

A. Karthikeyan
Subtropical Horticulture Research Institute, Jeju National University, Jeju, South Korea

Department of Biotechnology, Centre of Excellence in Innovation, Agricultural College and Research Institute, Tamil Nadu Agricultural University, Madurai, Tamil Nadu, India

V. G. Renganathan
Department of Biotechnology, Centre of Excellence in Innovation, Agricultural College and Research Institute, Tamil Nadu Agricultural University, Madurai, Tamil Nadu, India

N. Senthil (✉)
Department of Plant Molecular Biology and Bioinformatics, Center for Plant Molecular Biology and Biotechnology, Tamil Nadu Agricultural University, Coimbatore, Tamil Nadu, India
e-mail: senthil_natesan@tnau.ac.in

R. N. Pudake et al. (eds.), *Omics of Climate Resilient Small Millets*,
https://doi.org/10.1007/978-981-19-3907-5_7

using proteomics approaches. This chapter discusses how proteomics can help researchers better understand the abiotic stress tolerance in minor millets.

Keywords

Abiotic stress · Minor millets · Proteomics · Underutilized crops · MALDI-TOF · 2D-GE

7.1 Introduction

Millets are small-seeded annual C4 Panicoid grasses referred to as "nutricereals or nutraceutical crops" due to their functional and nutraceutical properties. The nutritional content of millets is comparable to or better than cereals, with high amounts of carbohydrates, proteins, minerals, vitamins, and essential amino acids (Dwivedi et al. 2012). Also, millets are well adapted to adverse conditions (i.e., limited rainfall, poor soil fertility, and difficult land terrain) and require minimum inputs, where major cereals are failed to succeed. It makes them an eye-catching crop for marginal farming environments (Johnson et al. 2019; Rawat et al. 2021). Millets are distinguished by their appearances, morphological characteristics, growth stages, maturity, and grains. The secondary group of millets contains foxtail millet (*Setaria italica*), little millet (*Panicum sumatrense*), kodo millet (*Paspalum scrobiculatum*), proso millet (*Panicum miliaceum*), finger millet (*Eleusine coracana*), and barnyard millet (*Echinochloa* spp.) are popularly known as the minor millets (Padulosi et al. 2015). They have a short growing season with a crop duration of 70–90 days (Vali Pasha et al. 2018). Minor millets offer a nutritious diet and livelihood security to millions of households, particularly small and marginal farmers, and residents of rainfed areas in Asia and sub-Saharan Africa. Recently global status of minor millets cultivation was reviewed by Vetriventhan et al. (2020). So far, the minor millets production and cultivation area details were not well documented and primarily detailed together with millets.

With abrupt climate changes, plants are seriously affected by abiotic stresses and crop yield could be compromised by more than 50%. Although, in general, millets respond to abiotic stresses (i.e., drought, salinity, waterlogging, cold, and heat) better than cereals, additional development is required to build them more resilient to unparalleled effects of climate change and associated environmental stresses (Durairaj et al. 2019; Muthamilarasan and Prasad 2021). Introgression of traits related to tolerance in millets will increase their yield in marginal environments and will assist in alleviating the future food scarcity caused by climate change. Many conventional and molecular methods are available to elucidate the tolerance mechanism and increase stress tolerance and crop productivity. The breeding program is integrated with genomic techniques to reach new heights in molecular breeding that allow for the development of crop varieties tolerant to abiotic stresses (Deshmukh et al. 2014; Visioni et al. 2019).

In particular, advanced genomics approaches assist in getting the whole genome sequence (WGS) and provide the knowledge of gene composition that facilitates crop improvement. Nevertheless, WGS information alone is not very useful to understand in what way plant and environment interact, and not all open reading frames (ORF) match up to functional genes (Ribeiro et al. 2013). Proteomics is a potent technique for studying and understanding plant stress responses. In the post-genomics era, proteomics has developed as a complementary technique to others, including transcriptomics and metabolomics, thanks to advances in MS-based technology and fast-track progress of quantitative approaches. There are several reports on using proteomics to understand the abiotic stress tolerance in minor millets (Veeranagamallaiah et al. 2008; Pan et al. 2018; Li et al. 2021). This chapter discusses the minor millets' responses to abiotic stress and the latest research developments in proteomics to understand the abiotic stress tolerance in minor millets.

7.2 Morpho-Physio and Biochemical Responses of Minor Millets to Abiotic Stresses

Plants deal with abiotic stressors that desiccate the cells, leading to a series of events, i.e., growth, sterility, seed abortion, reduced grain filling, and ultimately loss of productivity. However, millet's exceptional tolerance to abiotic stress has developed various acclimation, avoidance, and adaptive strategies to deal with it compared to other staple cereals (Seerat et al. 2021; Numan et al. 2021). Mechanisms of tolerance involve a complex phenomenon regulated at the physiological, cellular, and molecular levels, resulting in adaptive changes in morphological, physiological, and biochemical processes. Tolerance is achieved via alterations in plant structure, growth rate, tissue osmotic potential, and antioxidant defenses (Tadele 2016). Morphological adaptations in plants include the following characteristics: small leaf area, leaf folding, thickened cell wall, increased length, and dense root system, decreased shoot lengths, high tillering, adjusted flowering time, and short life cycle (Tadele 2016). The physio-biochemical process enforces the plants to lower the osmotic potential in the cytoplasm, leading to leaf turgor pressure during moisture stress. Besides that, the low photosynthetic and stomatal conductance rates and root respiration also help them escape from water stress. Osmotic potential is regulated by a high accumulation of osmoprotectants (compatible solutes) such as proline, glycine betaine, and organic acids, leading to cell turgor maintenance. These osmoprotectants also enhance the activity of antioxidant enzymes and given that protection to reactive oxygen species (ROS).

A better insight into those morphological and physio-biochemical responses in millets could be used to select tolerant and sensitive genotypes against various abiotic stressors. Plants species vary with the manifesting strategies to cope with any moisture stress through changes in phenotypic traits. Subsequently, the degree of tolerance/susceptibility also varies between species and within the species level. For instance, the anatomical traits including root, shoot length, and leaf area plays important roles in millet plant's drought tolerance (Balsamo et al. 2006; Tyagi et al. 2017, 2021). In a study by Balsamo et al. (2006) have reported that, the leaf anatomical changes were observed between drought-tolerant and susceptible millet plants, where the midrib and lamina vascular bundles had robust bundle sheath extensions and fibrous regions in the former than the case of later. Moreover, extensive lignification of the bundle sheath extensions in the tolerant species makes the lamina more stable in prolonged drought conditions, thus preventing increased mechanical shear stress on the more delicate mesophyll cells. The study also concluded that the leaf tensile strength grouped the three plants with different levels of tolerance to drought, where the drought-tolerant species had higher tensile strength values than the moderately tolerant (medium), and this value was very low in the case of susceptible species, indicated a positive correlation between leaf tensile strength and drought tolerance. In another study, Ayele et al. (2001) reported that an extensive, deeper, and broader root system provided drought stress tolerance in plants. Similarly, the increased root length and decreased shoot length also reported small millets, more particularly in little millet (Ajithkumar and Panneerselvam 2014) and finger millet (Jamra et al. 2020; Tyagi et al. 2017, 2021). Increased root length and decreased shoot length can be attributed to the felicitation of increased water absorption from the moisture deficit soils and increased partitioning of carbohydrates to roots than shoots (Prasad et al. 2004). In a study, PEG-induced drought stress improves the nutrient use efficiency in drought-tolerant genotype by partitioning more biomass to root, resulting in a higher root–shoot (R/S) ratio in finger millet (Zhang et al. 2018). Tavakol and Pakniyat (2007) and Boldaji et al. (2012) also reported that the drought-tolerant varieties have an instinctive allocation of root biomass than drought-susceptible varieties. Finger millet also had characteristics features of metal toxicity tolerance. The nickel (Ni) accumulation in the roots helps the plant mitigate the effects of heavy metals (Gupta et al. 2017). Similarly, in foxtail millet, phosphorus adaptation leads to lateral root proliferation by increasing root number, density, and length, thus enlarging the root absorptive surface area (Nadeem et al. 2020). Generally, waterlogging stress affects roots, stomatal conductance, and photosynthesis, eventually leading due to inhibition of gas diffusion and accumulation of toxic compounds (Linkemer et al. 1998). To cope with this situation, the finger millet and other millets follow these two strategies: the development of adventitious roots and the other by solubilized sugar and enhanced activity of nitrogen reductase in the shoots (Kulkarni and Chavan 2014; Cannarozzi et al. 2018).

On the other hand, the osmotic adjustment helps maintain photosynthesis by maintaining the pressure potential of cells when leaf water potential falls. For instance, Karyudi and Fletcher (2003) reported that during moisture stress among *S. italica* species, the high osmoregulation group produced 1.58 times more grain yield and harvest index than the low osmoregulation group. In contrary to that report, Morgan (1984) and Matsuura et al. (2012) claimed that the osmotic adjustment of the leaf alone did not discriminate the drought tolerance among four millets (*P. miliaceum, P. sumatrense, S. italica,* and *S. glauca*). Among which the higher osmotic adjustment millet species *S. glauca* and *P. sumatrense* showed low

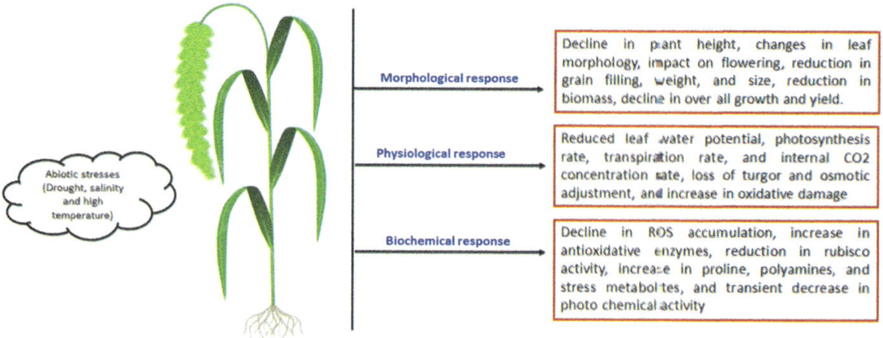

Fig. 7.1 Morpho-physio and biochemical responses of minor millets to abiotic stresses

productivity compared to *P. miliaceum* under water deficit. This suggests that deeper root growth of *P. miliaceum* compared to others may be more important than an osmotic adjustment in sustaining photosynthesis by maintaining leaf turgor (Matsuura et al. 2012). The relative water content (RWC) is another important physiological criterion in discriminating the tolerant and susceptible genotypes in many crop species. RWC level affects the physiology of plants, leading to growth inhibition, impaired photosynthesis, and biochemical changes that ultimately lead to disruption of ion homeostasis and generation of ROS, etc. (Wojtyla et al. 2020). Salinity stress reduces the RWC in crop plants due to the osmotic stress in roots restricting the water absorption and leading to dehydration under high salt (Makarana et al. 2019). Salinization and alkalization in foxtail and proso millet, although lead to the reduction of plant dry weight, relative growth rate (RGR), net assimilation rate (NAR), leaf area ratio (LAR), RWC, and nitrogen, the detrimental effects were greater on former than later (Islam et al. 2011). Further, it revealed that the tolerant proso millet species were high in chlorophyll compared to foxtail millet under saline conditions. In finger millet, the salt-tolerant varieties had lower shoot Na+/K+ ratios and higher leaf carbohydrate contents than susceptible genotypes; it suggested that the ion regulation and carbohydrate metabolism led to salt tolerance in finger millet (Vijayalakshmi et al. 2014) (Fig. 7.1).

ROS are secondary messengers in cell signaling at low concentrations, and however, it damages the cells when the concentration is high during stress conditions (Bartoli et al. 2013). Moisture stress increases ROS generation, which has a negative impact on cellular structure and metabolism. Fortunately, the ROS-mediated damage in plants depended on homeostasis between ROS and antioxidant levels at a time (Mirzaee et al. 2013). Several enzymes (SOD, CAT, POD) and nonenzymatic antioxidant systems like ascorbate and glutathione are involved in many defense responses and are important in ROS scavenging under abiotic stresses in plants (Sharma et al. 2012). Superoxide radicals (O^{2-}) get converted to hydrogen peroxide (H_2O_2) by SOD and utilizing different electron donors, H_2O_2 is reduced to water by POD; ascorbate reduces H_2O_2 to water, and CAT breaks down H_2O_2 into oxygen and water (Khan et al. 2019). The increased proline, ROS, and antioxidants concentration was observed in tolerant genotypes of foxtail millet, finger millet, and little millet (Jamra et al. 2020; Mundada et al. 2020; Aidoo et al. 2016; Bartwal et al. 2016). The higher proline content showed a positive correlation with increased glutamine synthetase and P5C reductase activities in drought-tolerant foxtail millet genotypes (Veeranagamallaiah et al. 2007). Likewise, the antioxidant enzymes activity was increased under drought stress to help the little millet cope with excessive ROS formation. Drought-tolerant little millet has a lower chlorophyll degradation ratio and a higher accumulation of compatible solutes such as proline, glycine betaine, and free amino acid (Ajithkumar and Panneerselvam 2014). Hormones including abscisic acid (ABA) and jasmonic acid (JA) regulate the complex signal transduction network in a plant grown under stress conditions. Both hormones help to regulate stomatal closure, which is the major way for plants to save water through the gaseous exchange in their leaves. As a result, it maintains the balance of plant development and defense response and enabling the plants to adapt to adverse conditions (Acharya and Assmann 2009). Drought-tolerant foxtail millet showed a modest rate of decrease in RCW and chlorophyll and higher soluble sugar and proline contents and a considerable upsurge of various stress hormones ABA and JA (Xu et al. 2019).

7.3 The Need for Proteomics Research to Understand the Abiotic Stress Tolerance

Advances in omics approaches have been game-changers for enhancing crop production under abiotic stress. Combining omics approaches with convention breeding methods led to a pronounced increase in our knowledge of understanding the abiotic stress tolerance mechanism. Abiotic stress tolerance is a complex phenomenon involving the physiological, biochemical, and molecular processes that are triggered by the interaction of multiple genes. Transcriptomics and microarray approaches brought considerable benefit to plant biologists to discover and characterize several abiotic stress-responsive genes. In particular, advances in next-generation sequencing technologies (NGS) such as RNA-sequencing have now outdated the former microarray technologies. In proso millet, using tolerant and sensitive cultivar's important gene resources for improving the drought tolerance was described by comparative transcriptome analysis (Zhang et al. 2019). In another research, Das et al. (2020) detailed the genes and molecular mechanisms of drought and salinity stress tolerance in little millet using transcriptomics analysis. Although the transcriptomic approach is useful for discovering the genes related to tolerance to abiotic stress, research at the proteome level is necessary to know the functional gene expression profiles. Also, the existing evidence has proven that the alterations that occurred in gene expression levels do not always relate to protein level (Gygi et al. 1999; Kosova et al. 2013). The main reason is that cells have posttranscriptional and posttranslational regulation mechanisms. Moreover, the research aiming at transcriptomics alone is insufficient to understand the stress-responsive mechanism in plants due to the coupled effects of precise temporal and spatial genes expression patterns. All the transcripts could not be translated into proteins in particular cells. As a result, if our studies solely depend on transcriptomics, such posttranscriptional regulation will be missed (Krishnamurthy et al. 2018). In addition to the modifications, proteins in cells are dynamic that comes to be directed for proteolysis. Thus, studying the gene expression at the protein level is needed to thoroughly understand the mechanism associated with tolerance to abiotic stress (Sobhanian et al. 2011).

7.4 Potential and Advances of Proteomics to Improve the Abiotic Stress Tolerance

The information of major proteins that are involved in the growth and development of the plant is important. These proteins regulate the physiological and biochemical pathways to uphold cellular homeostasis in a specific environment. Proteomics is known as a versatile molecular tool that is extensively used for analyzing protein expression and its characterization in plants under adverse environmental conditions (Ghosh and Xu 2014). The current scope of proteomics is much wider than originally defined. It creates the opportunity to investigate the structure, function, abundance, and interactions of proteins at a particular time point. Even though genomics-related research investigations are useful to researchers in understanding what is possible theoretically, proteomics unveils the key players for controlling particular cellular processes. Unlike the genome, which is static in nature, the proteome has dynamic capabilities (Eldakak et al. 2013). Therefore, proteomics is an edge over other omics approaches for studying plant stress tolerance mechanisms. Proteomics can identify translational and posttranslational regulations, in addition to supplementing transcriptome level changes, bringing new understandings into plant responses to abiotic stress (Hu et al. 2015). In recent decades, scientists have been able to study the plant proteome responses to environmental stress in a complex way due to the upswing of high-throughput proteomics approaches. Vanderschuren et al. 2013 and Ghosh and Xu 2014 described the general workflow for proteomic experiments in plants (Fig. 7.2). Proteomics analysis is carried out using (i) 2-DE or paired gel-free shotgun liquid chromatography-tandem mass spectrometry (LC-MS/MS) platforms for protein separation and identification. (ii) Through protein mapping and characterization of posttranslational modifications and protein–protein interactions unraveling the functions of protein and their networks of metabolic and signaling pathways in plants, (iii) Bioinformatics approaches and the usage of databases for model and non-model plant species. In recent times, the gel-free protein separation techniques, and next-generation proteomics techniques, including isobaric tags for relative and absolute quantitation (iTRAQ), multidimensional protein identification technology (MudPIT), isotope-coded affinity tags (ICATs), and targeted mass tags (TMTs), were extensively applied for studying the plant development and metabolic strategies in abiotic stress adaptation using descriptive and comparative proteomics analyses. There have been several proteomic studies in different crops, including millets under abiotic stress using above-mentioned

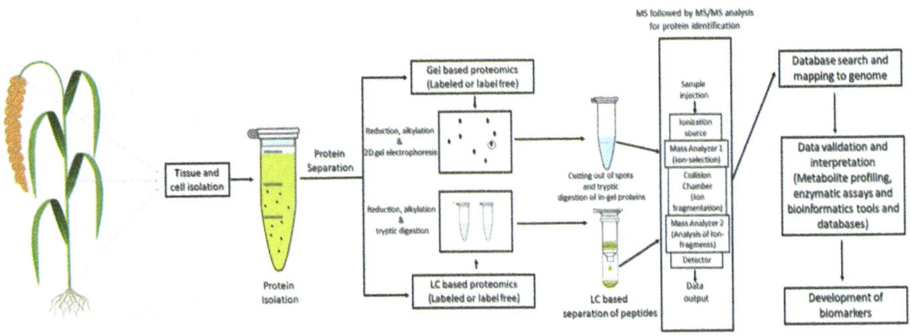

Fig. 7.2 Workflow of proteomic experiments in plants

methods (Ghatak et al. 2016; Mustafa and Komatsu 2021; Punia et al. 2020; Xu et al. 2020).

7.5 Proteomics of Minor Millets Grown under Abiotic Stress

Proteomics research in small millets is not well developed, and limited studies are only conducted. Although some progress in proteome research has been made in foxtail millet, it is still far behind the other major millet and cereal crops. Here, we discussed the latest research developments on proteomics in understanding the abiotic stress tolerance in minor millets. Drought is the key abiotic stress impacting plant growth and yield around the world, and it is becoming more widespread as a result of global climate change. Foxtail millet is considered to be a model crop for abiotic stress tolerance studies due to its genome size, less repetitive DNA, crop duration, and inbreeding nature, as well as better performance under those conditions. Pan et al. (2018) integrated TMT and LC-MS/MS-based proteomic approaches to quantify the differentially expressed proteins (DEPs) after subjecting foxtail millet seedlings to drought stress. A total of 2474 DEPs were detected; among them, 321 proteins (252 upregulated and 69 downregulated) related to drought response were identified. Many proteins found in this study have earlier been reported to involve in drought or other abiotic stresses.

For instance, stress and defense (LEA, ASR, ns LTPs, and aquaporin), photosynthesis (Rubisco, LH CII, and PEPC), signaling (Protein kinases and phosphatases), carbon metabolism (GAPDH, phosphoglycerate kinase, and fructokinase), protein synthesis and processing (ribosomal protein, elongation factor, HSPs, and protease), metabolism-related proteins (polyamine metabolism, lignin, and GB biosynthesis) and transcription (RNA helicases, PPR, and WD 40). In another study, foxtail millet landrace "Huangjinmiao" grown at control and drought and harvested grains were subjected to 2-DE combined MALDI-TOF/TOF analysis to elucidate the foxtail millet response to drought stress at a proteomic level (Li et al. 2019). There were 104 DEPs that were identified, among which 57 were upregulated and

47 downregulated proteins. The DEPs were related to the pathway of glycolysis/gluconeogenesis (Fructose-bisphosphate aldolase cytoplasmic isozyme-like), energy metabolism (ATP synthase F1 beta subunit), tricarboxylic acid (Putative aconitate hydrase), starch and sucrose metabolism (1, 4, alpha glucan branching enzyme 2 and glucose-1-phosphate adenyltransferase), protein folding and degradation (Protein disulfide isomerase and trypsin inhibitor), and globulin-1 S allele like (Storage proteins). Comparative proteomic profiling of contrasting drought response foxtail millet cultivars Longgu 6 (tolerance) and Huangjinmiao (Susceptible) under various drought treatment groups analyzed by 2-DE coupled with tandem MS. A total of 83 protein spots were detected in the two groups (W1 vs W2 and W0 vs W2), all of which were linked to various biochemical and metabolic pathways, including response to stress and defense, and chaperones, storage protein, amino acid, and protein metabolisms, and energy metabolism (Xu et al. 2020). Recently, Li and coworkers detailed the finger millet response to drought stress by combining transcriptomics and proteomics analyses. RNA seq and iTRAQ approaches were combined to study the finger millet grown at control, drought stress (19), and rehydration (36 hours) conditions. There were 80,602 DEGs and 3009 DEPs found at the transcriptomic and proteomic levels. The study's main finding is that under drought stress, 113 DEGs-DEPs had different levels of expression, and network analysis revealed that two DEGs-DEPs (BAMY2, barley K-amylase 2, and ISA1, isoamylase-type starch debranching enzyme) play a key regulatory role (Li et al. 2021).

Salinity is another major abiotic stress that limits crop production and the existence of additional salt in the environment, causing trouble in metabolic processes that generally manifest in nutrient imbalance, reduced nutrient uptake, and inhibited plant growth to osmotic stress and ultimately poor productivity. Foxtail millet cultivar Prasad (Salt tolerant) response to NaCl stress at proteome level analyzed by 2D gel electrophoresis. It revealed that 29 proteins (i.e., carbohydrate and nucleotide metabolisms, cell wall biogenesis, energy and lipid metabolisms, nitrogen metabolism, photosynthesis, and signal transduction) that upregulated or downregulated in response to NaCl stress (Veeranagamallaiah et al. 2008). There are no other proteomics reports available in minor millet's response to salinity stress. Also, no efforts have been made to explore the abiotic stress responses at proteomics levels in little millet, kodo millet, proso millet, and barnyard millet. Therefore, a significant gap in the literature as no proteins have been discovered with an association with abiotic stress tolerance mechanism in these millets. But the comparative study on protein profiling in proso millet can be helpful in future research (Roy et al. 2017). Finally, the studies discussed in this section provide the knowledge of proteins related stress response and adaptation, as well as a regulatory network of various metabolic pathways under stress stimulation in minor millets. Also, it will give light on the drought stress tolerance mechanisms of minor millets, as well as crop breeding and innovation. The details of proteomics for abiotic stress tolerance in minor millets are summarized in Table 7.1.

Table 7.1 Proteomics for abiotic stresses in minor millets

S. no.	Species	Plant part	Stress	Techniques used	Major findings	References
1	Foxtail millet	Seedlings	Salinity	2-DE MALDI-TOF/TOF MS	29 proteins (upregulated or downregulated) responded to NaCl stress	Veeranagamallaiah et al. (2008)
2	Foxtail millet	Seedlings	Drought	TMT-LC-MS/MS	A total of 2474 proteins were identified, among them, 321 drought-responsive proteins (252 upregulated and 69 downregulated proteins) were identified.	Pan et al. (2018)
3	Foxtail millet	Grains	Drought	2-DE MALDI-TOF/TOF	104 proteins responded to DROUGHT stress, among which, 57 upregulated and 47 downregulated proteins.	Li et al. (2019)
4	Foxtail millet	Grains	Drought	2-DE MALDI-TOF/TOFT M	A total of 83 protein spots detected among the different stress groups (W1 vs W2 and W0 vs W2), linked to biochemical and metabolic pathways, response to stress and defense.	Xu et al. (2020)
5	Foxtail millet	Seedlings	Salinity	Phosphoproteomic analysis	The differentially regulated phosphoproteins (DRPPs) that were identified in this study were mainly involved in signal transduction, regulation of gene expression, translation, ion transport, and metabolism processes.	Pan et al. (2021)
5	Finger millet	Leaves	Drought	TMT-iTRAQ coupled with transcriptomics	The major findings from this study shows that 113 DEGs-DEPs had different levels of expression under drought treatment, and network analysis revealed that two DEGs-DEPs (BAMY2, barley K-amylase 2 and ISA1, isoamylase-type starch debranching enzyme) play a key regulatory role.	Li et al. (2021)

7.6 Conclusion and Future Perspectives

Abiotic stresses are one of the major factors that greatly influence crop production worldwide. Although minor millets responded well to abiotic stresses, supplementary research is required to build them more resilient crops to abiotic stresses. Because minor millets safeguard the livelihood and nutrition security in people of Asia and sub-Saharan Africa. In these circumstances, plant biologists require to concentrate on the accurate mechanism underlying the tolerance at various growth stages of plants. Minor millets have good market value owing to their nutritional value and are a potential candidate for proteomic research. They possess an innate potential to modify their gene expression, by this means remodeling the proteome. Proteomics is gaining momentum in all three levels via studying the proteins at cellular and subcellular levels, developmental and structural levels, and genetic and physiological levels. It is a very useful approach to investigating the correlation between gene expression and the protein profile of a plant. But the progress of proteomics research in minor millets is slow due to the lack of genomic information.

To date, proteomics research could not be delivered more data related to the abiotic stress tolerance potential of small millets. There are few reports published on proteomics for abiotic stress tolerance in minor millets, including foxtail millet and finger millet, and no reports are available on little millet, kodo millet, proso millet, and barnyard millet. Hence, there is an immense need to study the abiotic stress response at the protein level in these millets using comparative proteomics. Moreover, the development of a proteome map could permit the rapid evaluation of minor millet cultivars, mutants, and transgenic lines. Also, the detailed studies related to physiology could provide a useful proteome reference map. The research outcome from millet proteomics will assist in understanding the role of proteins and facilitate the cloning of genes in the future.

Besides these research gaps, there is also a need for research in protein–protein and protein–ligand associations and interdisciplinary research of proteomics together with transcriptomics and metabolomics useful to determine the communicating pathways among proteins, transcripts, and metabolites involved in tolerance mechanisms to abiotic stress. The detailed information regarding the expression of abiotic stress tempted new marker proteins that could further allow us to develop genetically engineered stress-tolerant millet cultivars. Although, protein biomarkers discovered by proteomics will be efficiently used in breeding programs. The progress of stress-tolerant and high-yielding cultivars is important for increasing minor millets production. Proteomic research will accelerate marker-assisted genetic enhancement studies in minor millets, which will lead to the development of high-yielding stress-tolerant genotypes under abiotic stresses.

Conflicts of Interest All authors have read and agreed to the published version of the chapter and the authors declare that they have no conflict of interest.

References

Acharya B, Assmann S (2009) Hormone interactions in stomatal function. Plant Mol Biol 69:451–462

Aidoo MK, Bdolach E, Fait A, Lazarovitch N, Rachmilevitch S (2016) Tolerance to high soil temperature in foxtail millet (*Setaria italica* L.) is related to shoot and root growth and metabolism. Plant Physiol Biochem 106:73–81. https://doi.org/10.1016/j.plaphy.2016.04.038

Ajithkumar IP, Panneerselvam R (2014) ROS scavenging system, osmotic maintenance, pigment and growth status of *Panicum sumatrense* Roth. under drought stress. Cell Biochem Biophys 68(3):587–595. https://doi.org/10.1007/s12013-013-9746-x

Ayele M, Blum A, Nguyen HT (2001) Diversity for osmotic adjustment and root depth in TEF [*Eragrostis* tef (Zucc) Trotter]. Euphytica 121:237–249. https://doi.org/10.1023/A:1012099914738

Balsamo RA, Willigen CV, Bauer AM, Farrant J (2006) Drought tolerance of selected Eragrostis species correlates with leaf tensile properties. Ann Bot 97(6):985–991. https://doi.org/10.1093/aob/mcl068

Bartoli CG, Casalongué CA, Simontacchi M, Marquez-Garcia B, Foyer CH (2013) Interactions between hormone and redox signalling pathways in the control of growth and cross tolerance to stress. Environ Exp Bot 94:73–88

Bartwal A, Pande A, Sharma P, Arora S (2016) Intervarietal variations in various oxidative stress markers and antioxidant potential of finger millet (*Eleusine coracana* L.) subjected to drought stress. J Environ Biol 37(4):517

Boldaji SH, Khavari-Nejad RA, Sajedi RH, Fahimi H, Saadatmand S (2012) Water availability effects on antioxidant enzyme activities, lipid peroxidation, and reducing sugar contents of alfalfa (*Medicago sativa* L.). Acta Physiol Plant 34(3):1177–1186

Cannarozzi G, Weichert A, Schnell M, Ruiz C, Bossard S, Blösch R, Plaza-Wüthrich S, Chanyalew S, Assefa K, Tadele Z (2018) Waterlogging affects plant morphology and the expression of key genes in tef (*Eragrostis tef*). Plant Direct 2:e00056. https://doi.org/10.1002/pld3.56

Das RR, Pradhan S, Parida A (2020) *De-novo* transcriptome analysis unveils differentially expressed genes regulating drought and salt stress response in *Panicum sumatrense*. Sci Rep 10:21251

Deshmukh R, Sonah H, Patil G, Chen W, Prince S, Mutava R, Vuong T, Valliyodan B, Nguyen HT (2014) Integrating omic approaches for abiotic stress tolerance in soybean. Front Plant Sci 5:244

Durairaj M, Gurumurthy G, Nachimuthu V, Muniappan K, Balasubramanian S (2019) Dehulled small millets: the promising nutricereals for improving the nutrition of children. Matern Child Nutr 15:e12791

Dwivedi S, Upadhyaya H, Senthilvel S, Hash C, Fukunaga K, Diao X (2012) Millets: genetic and genomic resources. Plant Breed Rev 35:247–375

Eldakak M, Milad SIM, Nawar AI, Rohila JS (2013) Proteomics: a biotechnology tool for crop improvement. Front Plant Sci 4:35

Ghatak A, Chaturvedi P, Nagler M, Roustan V, Lyon D, Bachmann G, Postl W, Schröfl A, Desai N, Varshney RK, Weckwerth W (2016) Comprehensive tissue-specific proteome analysis of drought stress responses in Pennisetum glaucum (L.) R. Br. (Pearl millet). J Proteome 143:122–135

Ghosh D, Xu J (2014) Abiotic stress responses in plant roots: a proteomics perspective. Front Plant Sci 5:6

Gupta SM, Arora S, Mirza N, Pande A, Lata C, Puranik S, Kumar J, Kumar A (2017) Finger millet: a "certain" crop for an "uncertain" future and a solution to food insecurity and hidden hunger under stressful environments. Front Plant Sci 8:643

Gygi SP, Rochon Y, Franza BR, Aebersold R (1999) Correlation between protein and mRNA abundance in yeast. Mol Cell Biol 19(3):1720–1730

Hu J, Rampitsch C, Bykova NV (2015) Advances in plant proteomics toward improvement of crop productivity and stress resistance. Front Plant Sci 6:209

Islam MS, Akhter MM, El Sabagh A, Liu LY, Nguyen NT et al (2011) Comparative studies on growth and physiological responses to saline and alkaline stresses of Foxtail millet (*'Setaria italica'* L.) and Proso millet (*'Panicum miliaceum'* L.). Aust J Crop Sci 5(10):1269

Jamra G, Shah P, Agarwal A, Sharma D, Kumar A (2020) Elucidating the physio-morphological and biochemical responses towards peg-induced drought stress in finger millet genotypes. Int J Curr Microbiol Appl Sci 9(7):1672–1687

Johnson M, Deshpande S, Vetriventhan M, Upadhyaya HD, Wallace JG (2019) Genome-wide population structure analyses of three minor millets: kodo millet, little millet, and proso millet. Plant Genome 12:190021

Karyudi, Fletcher RJ (2003) Osmoregulation in birdseed millet under conditions of water stress II. Variation in F3 lines of *Setaria italica* and its relationship to plant morphology and yield. Euphytica 132:191–197. https://doi.org/10.1023/A:1024620413781

Khan MN, Zhang J, Luo T, Liu J, Ni F, Rizwan M, Fahad S, Hu L (2019) Morpho-physiological and biochemical responses of tolerant and sensitive rapeseed cultivars to drought stress during early seedling growth stage. Acta Physiol Plant 41(2):25

Kosova K, Prail IT, Vitamvas P (2013) Protein contribution to plant salinity response and tolerance acquisition. Int J Mol Sci 14(4):6757–6789

Krishnamurthy P, Qingsong L, Kumar PP (2018) Proteomics perspectives in post-genomic era for producing salinity stress-tolerant crops. In: Kumar V, Wani S, Suprasanna P, Tran LS (eds) Salinity responses and tolerance in plants, vol 2. Springer, Cham

Kulkarni S, Chavan P (2014) Study of effect of waterlogging on root anatomy of ragi and rice. Am J Plant Physiol 9:46–51. https://doi.org/10.3923/ajpp.2014.46.51

Li J, Li X, Yang Q, Luo Y, Gong X, Zhang W, Hu Y, Yang T, Dong K, Feng B (2019) Proteomic changes in the grains of foxtail millet (*Setaria italica* (L.) Beau) under drought stress. Span J Agric Res 17:802

Li J, Wang Y, Wang L, Zhu J, Deng J, Tang R et al (2021) Integration of transcriptomic and proteomic analyses for finger millet [*Eleusine coracana* (L.) Gaertn.] in response to drought stress. PLoS One 16(2):e0247181

Linkemer G, Board JE, Musgrave ME (1998) Waterlogging effects on growth and yield components in late-planted soybean. Crop Sci 38:1576–1584. https://doi.org/10.2135/cropsci1998.0011183X003800060028x

Makarana G, Kumar A, Yadav RK, Kumar R, Soni PG et al (2019) Effect of saline water irrigations on physiological, biochemical and yield attributes of dual purpose pearl millet (*Pennisetum glaucum*) varieties. Indian J Agric Sci 89(4):624–633

Matsuura A, Tsuji W, Ping A, Shinobu I, Kouhei M (2012) Effect of pre- and post-heading water deficit on growth and grain yield of four millets. Plant Prod Sci 15(4) 323–331

Mirzaee M, Moeini A, Ghanati F (2013) Effects of drought stress on the lipid peroxidation and antioxidant enzyme activities in two canola (*Brassica napus* L.) cultivars. J Agric Sci Technol 15(3):593–602

Morgan JM (1984) Osmoregulation and water stress in higher plants. Annu Rev Plant Physiol 35: 299–319

Mundada PS, Nikam TD, Kumar SA, Umdale SD, Ahire ML (2020) Morpho-physiological and biochemical responses of finger millet (*Eleusine coracana* (L.) Gaertn.) genotypes to PEG-induced osmotic stress. Biocatal Agric Biotechnol 23:101488. https://doi.org/10.1016/j.bcab.2019.101488

Mustafa G, Komatsu S (2021) Plant proteomic research for improvement of food crop under stresses: a review. Mol Omics 17:860–880

Muthamilarasan M, Prasad M (2021) Small millets for enduring food security amidst pandemics. Trends Plant Sci 26:33–40

Nadeem F, Ahmad Z, Ul Hassan M, Ruifeng W, Diao X et al (2020) Adaptation of foxtail millet (*Setaria italica* L.) to abiotic stresses: a special perspective of responses to nitrogen and phosphate limitations. Front Plant Sci 11:187. https://doi.org/10.3389/fpls.2020.00187
Numan M, Serba DD, Ligaba-Osena A (2021) Alternative strategies for multi-stress tolerance and yield improvement in millets. Genes (Basel) 12(5):739. https://doi.org/10.3390/genes12050739
Padulosi S, Mal B, King OI, Gotor E (2015) Minor millets as a central element for sustainably enhanced incomes, empowerment, and nutrition in rural India. Sustainability 7:8904–8933
Pan J, Li Z, Wang Q, Garrell AK, Liu M, Guan Y, Zhou W, Liu W (2018) Comparative proteomic investigation of drought responses in foxtail millet. BMC Plant Biol 18(1):315
Pan J, Li Z, Wang Q, Guan Y, Li X, Huangfu Y, Meng F, Li J, Dai S, Liu W (2021) Phosphoproteomic profiling reveals early salt-responsive mechanisms in two foxtail millet cultivars. Front Plant Sci 2021:1934
Prasad PVV, Staggenborg SA, Ristic Z (2004) Impacts of drought and/or heat stress on physiological, developmental, growth, and yield processes of crop plants. Adv Agric Syst Model Series 1 1:301–355
Punia H, Tokas J, Bhadu S, Mohanty AK, Rawat P, Malik A, Satpal (2020) Proteome dynamics and transcriptome profiling in sorghum (*Sorghum bicolor*) under salt stress. 3 Biotech 10(9):412
Rawat L, Karnatak AK, Bisht TS, Kukreti A (2021) Minor millets: profile and ethnobotanical scenario. In: Kumar A, Tripathi MK, Joshi D, Kumar V (eds) Millets and millet technology. Springer, Singapore
Ribeiro M, Nunes-Miranda JD, Branlard G, Carrillo JM, Rodriguez-Quijano M, Igrejas G (2013) One hundred years of grain omics: identifying the glutens that feed the world. J Proteome Res 12:4702–4716
Roy SK, Kwon SJ, Yu JH, Sarker K, Cho SW, Moon YJ, Jung TW, Park CH, Woo SH (2017) Comparison of protein profiles of proso millet (*Panicum miliaceum*) seeds of various korean cultivars. Korean J Crop Sci 62(1):40–50
Seerat S, Mushtaq NU, Shah WH, Rasool A, Hakeem KR et al (2021) Morpho-physiological, biochemical and molecular adaptation of millets to abiotic stresses: a review. Phyton-Int J Exp Bot 90(5):1363–1385
Sharma P, Jha AB, Dubey RS, Pessarakli M (2012) Reactive oxygen species, oxidative damage, and antioxidative defense mechanism in plants under stressful conditions. J Bot 2012:217037. https://doi.org/10.1155/2012/217037
Sobhanian H, Aghaei K, Komatsu S (2011) Changes in the plant proteome resulting from salt stress: toward the creation of salt-tolerant crops? J Proteome 74(8):1323–1337
Tadele Z (2016) Drought adaptation in millets. In: Shanker A, Shanker C (eds) Abiotic and biotic stress in plants: recent advances and future perspectives. InTech, Rijeka, pp 639–662
Tavakol E, Pakniyat H (2007) Evaluation of some drought resistance criteria at seedling stage in wheat (*Triticum aestivum* L.) cultivars. Pak J Biol Sci 10(7):1113–1117
Tyagi J, Shrivastava N, Sharma AK, Varma A, Pudake R (2021) Effect of *Rhizophagus intraradices* on growth and physiological performance of Finger Millet (*Eleusine coracana* L.) under drought stress. Plant Sci Today 8(4):912–923
Tyagi J, Varma A, Pudake RN (2017) Evaluation of comparative effects of arbuscular mycorrhiza (*Rhizophagus intraradices*) and endophyte (*Piriformospora indica*) association with finger millet (*Eleusine coracana*) under drought stress. Eur J Soil Biol 81:1-0
Vali Pasha K, Ratnavathi CV, Ajani J, Raju D, Manoj Kumar S, Beedu SR (2018) Proximate, mineral composition and antioxidant activity of traditional small millets cultivated and consumed in Rayalaseema region of South India. J Sci Food Agric 98(2):652–660
Vanderschuren H, Lentz E, Zainuddin I, Gruissem W (2013) Proteomics of model and crop plant species: status, current limitations and strategic advances for crop improvement. J Proteome 20(93):5–19
Veeranagamallaiah G, Chandraobulreddy P, Jyothsnakumari G, Sudhakar C (2007) Glutamine synthetase expression and pyrroline-5-carboxylate reductase activity influence proline

accumulation in two cultivars of foxtail millet (Setaria italica L.) with differential salt sensitivity. Environ Exp Bot 60:239–244. https://doi.org/10.1016/j.envexpbot.2006.10.012

Veeranagamallaiah G, Jyothsnakumari G, Thippeswamy M, Reddy PCO, Surabhi GK, Sriranganayakulu G et al (2008) Proteomic analysis of salt stress responses in foxtail millet (*Setaria italica* L. cv. Prasad) seedlings. Plant Sci 175:631–641

Vetriventhan M, Azevedo VCR, Upadhyaya HD (2020) Genetic and genomic resources, and breeding for accelerating improvement of small millets: current status and future interventions. Nucleus 63:217–239

Vijayalakshmi D, Ashok SK, Raveendran M (2014) Screening for salinity stress tolerance in rice and finger millet genotypes using shoot Na+/K+ ratio and leaf carbohydrate contents as key physiological traits. Indian J Plant Physiol 19(2):156–160. https://doi.org/10.1007/s40502-014-0090-y

Visioni A, al-Abdallat A, Elenien JA (2019) Genomics and molecular breeding for improving tolerance to abiotic stress in barley *(Hordeum vulgare* L.). In: Rajpal VR, Sehgal D, Kumar A et al (eds) Genomics assisted breeding of crops for abiotic stress tolerance, vol II. Springer, Cham, pp 49–68

Wojtyla Ł, Paluch-Lubawa E, Sobieszczuk Nowicka E, Garnczarska M (2020) Drought stress memory and subsequent drought stress tolerance in plants. In: Priming-mediated stress and cross-stress tolerance in crop plants. Academic Press, pp 115–131

Xu B, Gao X, Dong K, Xia L, Yang P, Yang T, Feng B (2020) Grain protein content comparison and proteomic analysis of foxtail millet (*Setaria italica*) seed response to different drought stress levels. Acta Physiol Plant 42:20

Xu BQ, Gao XL, Gao JF, Jing LI, Pu YA et al (2019) Transcriptome profiling using RNA-seq to provide insights into foxtail millet seedling tolerance to short-term water deficit stress induced by PEG-6000. J Integr Agric 18(11):2457–2471

Zhang H, Li Y, Zhu JK (2018) Developing naturally stress-resistant crops for a sustainable agriculture. Nat Plants 4(12):989–996. https://doi.org/10.1038/s41477-018-0309-4

Zhang Y, Gao X, Li J, Gong X, Yang P, Gao J, Wang P, Feng B (2019) Comparative analysis of proso millet (*Panicum miliaceum*) leaf transcriptomes for insight into drought tolerance mechanisms. BMC Plant Biol 19:397

Plant Regeneration and Transgenic Approaches for the Development of Abiotic Stress-Tolerant Small Millets

8

Muthukrishnan Arun, Nandakumar Vidya,
Krishnagowdu Saravanan, Jayachandran Halka,
Kumaresan Kowsalya, and Jaganathan Sakthi Yazhini Preetha

Abstract

Small millets, a member of the Poaceae family grows well in Asia, Africa, and some regions of Europe. These millets are a great source of protein, therefore, it is highly used as a source of food for humans, animals, and birds. Despite the presence of higher content of protein, it is also rich in other essential nutrients such as vitamins, minerals, and, dietary fibers. These millets exhibit biological properties such as being anti-inflammatory, anti-cancerous, and lowering cholesterol and glucose levels in the body. Even though millets exhibit significant biological functions, their consumption and production are declining globally. This may be due to easy availability and easier cooking methods of other prominent cereals such as rice, wheat, and maize. Small millets are also widely used as a potent source for the production of starch and alcohol which is escalating the demand for these millets. The production of small millets is mainly affected due to biotic and abiotic factors. Biotic factors include fungal, bacterial, and viral infections whereas drought, salinity, waterlogging, and lodging are the abiotic factors that greatly affect their production and yield Reliable and robust methods of plant regeneration, identification of novel functional genes responsible for abiotic stress tolerance, and introduction of new traits to small millets by establishing *Agrobacterium*-mediated transformation have paved the way for the development of abiotic stress-tolerant millets. This chapter highlights different biotic and abiotic factors that inhibit the growth of small millets, various plant regeneration methods, transformation studies, potential genes for abiotic stress tolerance, and transgenic approaches for the production of improved abiotic stress-tolerant millets.

M. Arun (✉) · N. Vidya · K. Saravanan · J. Halka · K. Kowsalya · J. S. Y. Preetha
Department of Biotechnology, Bharathiar University, Coimbatore, Tamil Nadu, India
e-mail: arun@buc.edu.in

R. N. Pudake et al. (eds.), *Omics of Climate Resilient Small Millets*,
https://doi.org/10.1007/978-981-19-3907-5_8

Keywords

Small millets · Biotic stress · Abiotic stress · Plant regeneration · Transformation · Stress tolerance

8.1 Introduction

The small millets belong to the Poaceae family, widely grown in arid and semiarid regions of the world such as Asia, Africa, and some parts of Europe (Rao 1989). The presence of small seeds is one of the distinguished features of this group of millets and is highly used as a source of food for humans and feed for animals and birds. It is ranked sixth among other economically important crops in the world and can serve as a source of cheap protein source to unprivileged communities across the globe. These millets play a major role in boosting the economy of developing and underdeveloped countries and also in dealing with food security (Asharani et al. 2010). Many reports suggest that these millets are highly tolerant to a range of environmental stresses and need fewer inputs in the form of fertilizers, manure, and regular irrigation. One of the major attractions of millets is that they can grow well in a shorter duration and their harvesting time usually range between 70 and 80 days (Singh et al. 2020). This group of small millets consists of eight species namely finger millet (*Eleusine coracana*), foxtail millet (*Setaria italica*), proso millet (*Panicum miliaceum*), little millet (*Panicum sumatrense),* barnyard millet (*Echinochloa esculenta),* kodo millet (*Paspalum scrobiculatum*), tef millet (*Eragrostis tef*), and fonio millet (*Digitaria exilis*) (Upadhyaya et al. 2008). Finger millet is a highly productive crop that can withstand drastic changes in climatic conditions. It can grow in less fertile soil and can also grow well without any supplementations (Gull et al. 2014). Moreover, this millet shows a high resistance toward storage pests and has a very long shelf life of around 10 years, making it available throughout the year and hence, it got another name as "Famine crop" (Mgonja et al. 2007). Foxtail millet is tolerant to drought and is used as a model crop in studies related to bioenergy grasses namely switchgrass and pearl millets (Zhang et al. 2017b). This millet has high efficiency in using water than other millets and is used as a model crop to study the drought mechanism due to its special physical characteristics such as thicker cell walls, denser roots, and smaller leaves (Satyavathi et al. 2019). Proso millet is a summer annual crop that can resist drought (Herdrich 2001; Sateesh 2010; Changmei and Dorothy 2014) and requires a warmer climate for its development (Habiyaremye et al. 2017). Similarly, the barnyard millet is also reported to have the ability to survive in adverse environmental conditions with limited inputs and can resist both abiotic and biotic stresses (Renganathan et al. 2020). Furthermore, kodo millet, tef millet, and fonio millet are drought stress-resistant varieties and provide a sustainable source of food for poor people in Asian and African countries. Even now, tef millet is considered a staple food in Ethiopia and Eritrea. This extraordinary character of millets toward tolerating abiotic stresses should be logically applied to develop new varieties of crops that are very susceptible to inconsistent climatic conditions.

Millets are known as the powerhouse of nutrients and hence, it is also called nutri-cereals. The nutritional quality of these millets is far more superior than staple crops such as wheat and rice as they contain abundant amounts of minerals such as magnesium, iron, phosphorous, and potassium, and contain substantial amounts of proteins, essential amino acids (cysteine and methionine), vitamins, polyphenols, dietary fiber along with a higher fat content than rice, maize, and sorghum (Obilana and Manyasa 2002; Devi et al. 2014). Proso millet and foxtail millet contains the highest protein content. Finger millet is found to have the highest calcium content (300–350 mg 100 g^{-1}) among all other millets and is around 10 times greater than staple crops like rice and wheat (Malathi et al. 2016). Furthermore, little millet and barnyard millet are considered to be very rich in iron where as dietary fiber is relatively higher in barnyard millet and kodo millet. In general, these nutri-cereals are also considered rich sources of niacin, vitamin B complex, and folic acid (Hulse et al. 1980; Pathak 2013). The presence of several bioactive compounds in millets such as lignans, flavonoids, beta-glucan, phytate, fiber, sterols, inulin, and pigments contributes positively to health (Rao 2003; Kamara et al. 2009; Amir et al. 2014). The most frequently seen flavonoids in millets are catechin, quercetin, luteolin, orientin, apigenin, isoorientin, vitexin, myricetin, isovitexin, daidzein, sponarin, violanthin, lucenin-1, and tricin (Chandrasekara and Shahidi 2011; Pradeep and Sreerama 2017; Pathak et al. 2018). Several phenolic compounds are also reported to be present in millets namely gallic acid, polymeric tannins, protocatechuic acid, gentisic acid, caffeic acid, vanillic acid, syringic acid, ferulic acid, para-coumaric acid, trans-cinnamic acid, and 5-n-alkylresorcinols (Shobana et al. 2009; Siwela et al. 2010; Pradeep and Guha 2011; Bellato et al. 2013; Pradeep and Sreerama 2017). The nutritional value of small millets is presented in Table 8.1 (Hassan et al. 2021).

The presence of highly potent nutraceutical compounds makes the millet a versatile crop in providing enormous health benefits such as lowering cholesterol, decreasing glucose levels, inhibiting the growth of cancerous cells, and alleviating the risk of anti-inflammatory bowel disorders (Nithiyanantham et al. 2019). The small millets are gluten-free, therefore, it becomes an ideal substitute for people who are suffering from Celiac disease, a disease often triggered by ingesting gluten, which is mostly present in wheat and other cereals (Saleh et al. 2013). Higher content of magnesium in small millets is found to be very effective in reducing migraine and cardiac arrest (Nithiyanantham et al. 2019). Phytic acid present in these millets is considered one of the factors responsible for lowering cholesterol (Coulibaly et al. 2010). Finger millet has proven roles in reducing the blood glucose level, decreasing the risk of gastrointestinal, and inflammatory bowel disorders. This millet is also considered a potent antioxidant (Muthamilarasan et al. 2016). Foxtail millet acts as an anti-hyperglycemic and anti-lipidemic agent during diabetic conditions which might be due to the presence of alkaloids and glycosides (Sireesha et al. 2011). Consumption of proso millet is very effective to alleviate diabetes and cardiac diseases by lowering blood glucose levels and increasing HDL (High-density

Table 8.1 Nutritional composition of small millets

Millets	Carbohydrates (g)	Protein (g)	Ash (g)	Total fat (g)	Dietary fiber(g)	Energy (Kcal)	Total Amino acid (mg) [a]	Vitamins (mg)/100g				Minerals (mg)/100g		
								A	B1	B2	B3	Ca	Fe	P
Barnyard millet	65.5	6.2	6.1	2.2	9.8	307	2436	–	0.3	0.09	–	20	5.0	280
Kodo millet	65.9	8.3	3.0	1.4	9.0	309	1824	–	0.2	0.10	–	27	0.5	188
Little millet	67.0	7.7	4.7	4.7	7.6	341	2474	–	0.41	0.28	–	17	9.3	220
Foxtail millet	60.9	12.3	2.9	4.3	8.0	331	2957	–	0.6	0.10	0.99	31	2.8	290
Finger millet	72.0	7.3	3.6	1.3	3.6	328	2606	6.0	0.3	0.10	1.4	344	3.9	283
Proso millet	70.4	12.5	5.0	1.1	2.2	341	2766	–	0.41	0.28	4.54	14	0.8	206

[a]Isoleucine, leucine, lysine, methionine, phenylalanine, histidine, threonine, tryptophan, valine

lipoprotein) cholesterol respectively (Park et al. 2008). Furthermore, proso millet also acts as an antioxidant by scavenging free radicals during D-galactosamine and lipopolysaccharide-induced liver injury (Ito et al. 2007). Like other millets, barnyard millet also helps in lowering glucose and lipid levels (Ugare et al. 2014). Therefore, these millets can be undoubtedly used as a substitute for staple crops in developing and underdeveloped countries for providing nutrition.

Asian and African countries account for 96.89% of total millet production and these countries have the maximum area under cultivation (30,697,888 hectares) (Chandra et al. 2021). Similarly, Asian and African countries have the maximum area under cultivation (30,697,888 hectares) followed by Europe (344,533 hectares), America (166,260 hectares), and Australia (35,750 hectares) (Chandra et al. 2021). For the past two decades, the significance of millets as a staple food, particularly in Asian countries has been decreasing due to various elements such as increasing incomes, ascending trends in urbanization, and government policies supporting the production and consumption of fine cereals. Nevertheless, the same aspects are provoking the demand and usage of these crops as feed and for the production of starch and alcohol on an industrial scale (Thakur and Sharma 2018).

Even though millets have many beneficial properties for promoting good health, the production rate of small millets in India is witnessing a downward trend and this may be due to two major reasons such as a decrease in demand for small millets and the stationary phase of millets' production in India. Crops such as rice, maize, and corn have taken dominance over small millets, especially in the rural belt of India. This might be due to various issues such as government subsidies given for fine cereals (rice, wheat, maize, etc.), more availability of fine cereals, easier access toward processed food made from these cereals and requires less cooking time. These factors adversely affect the production and consumption of small millets among the Indian population. There was an eightfold decrease (53.35 lakh ha to 6.82 lakh ha) in the area under cultivation for small millets from 1955 to 1956 to 2013–2014. Consequently, a fourfold reduction in its production was observed during this time frame. When compared to the yield of other popular crops, only a slight increase in the yield of small millets was observed (Perumal 2017; Thakur and Sharma 2018). The total area under cultivation for small millets in India is 24.04 lakh hectares. The largest small millet-producing state is Karnataka (56.7%) followed by Uttarakhand (11.4%), and Tamil Nadu (8%) and other states constitute only less than one-fifth of the total small millets' production. The production of small millets in different states of India is given in Fig. 8.1. Even though small millets exhibit numerous nutritional properties, its commercial importance is gradually decreasing. Hence, more focus should be given to this category of neglected crops. This chapter focuses on different biotic and abiotic factors that inhibit the growth of small millets, various plant regeneration methods, transformation studies, potential genes for abiotic stress tolerance, and transgenic approaches for the production of improved abiotic stress-tolerant millets.

Fig 8.1 Production of small millets in India

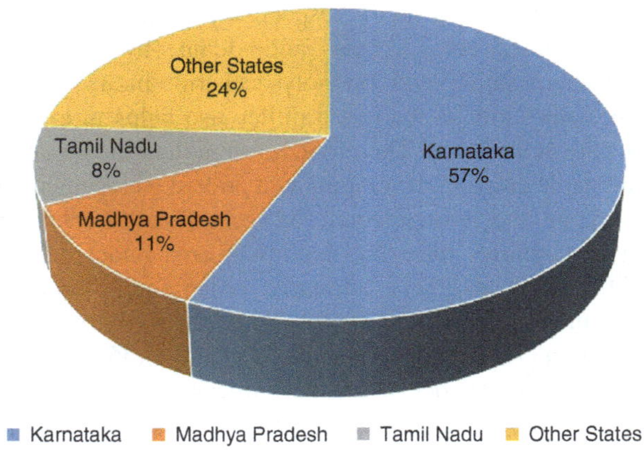

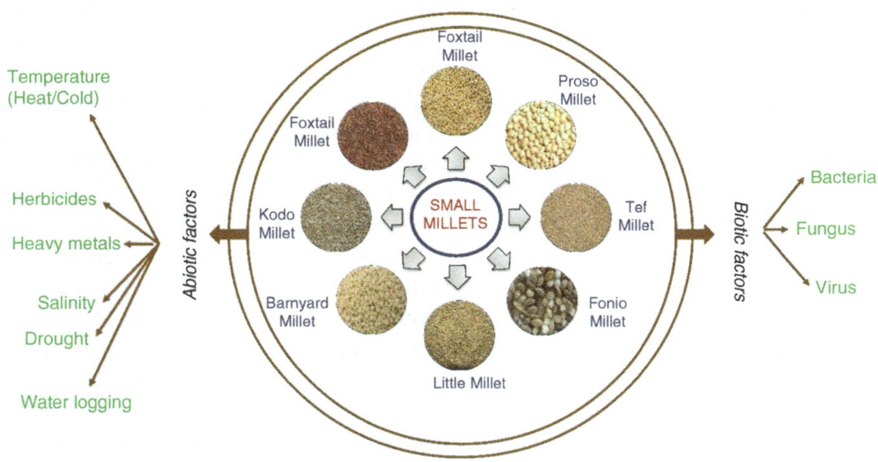

Fig. 8.2 Factors that hinder the production of millets

8.2 Factors That Hinder the Production of Small Millets

The biotic and abiotic stresses are the factors that hinder the production and yield of small millets. However, small millets effectively tackle the damage caused due to both biotic and abiotic stresses, but under susceptible conditions, some diseases severely affect these crops in various ecological regions. Biotic and abiotic factors that are hindering the production of small millets are summarized in Fig. 8.2.

Fungal, bacterial, and viral pathogens are the biotic factors that adversely affect the growth and production of small millets. Fungal diseases are more dominant in small millets than bacterial and viral diseases. This disease affects its vegetative and reproductive parts as well as its roots and stalks. The smuts and rust diseases affect all types of small millet and decrease its yield and production. Rust is a fungal disease that affects finger millet, foxtail millet, little millet, kodo millet, and tef millet by *Uromyces eragrostidis*, *Uromyces setariae-italicae*, *Uromyces linearis*, *Puccinia substriata*, and *Uromyces eragrostidis*, respectively. Moreover, this disease causes less damage to the crops and hence leads to less economic loss. On the other hand, smut disease affects all millets and it drastically affects the productivity of high-yielding varieties. Smut disease is commonly seen in foxtail millet and is caused by *Ustilago crameri*. Other types of fungal diseases namely blast, brown spot, leaf blight, crazy top, udbatta, foot rot, sheath rot, and banded sheath blight are specific to a few millets (Nagaraja and Das 2016). One of the major diseases that hinder the production of small millet is the blast disease caused by *Pyricularia grisea*. This disease gets affected during the rainy season and causes an average loss of yield from 28% to 36%, at times it may shoot up to 90% in endemic places (Rao 1990; Nagaraja and Mantur 2007). Blast disease is commonly seen in finger millet, foxtail millet, little millet, barnyard millet, and proso millet. Another fungal disease that causes drastic damage to yield and economic loss is seedling blight and leaf blight disease, which is caused by *Drechslera nodulosum*. It is reported that green ear disease or crazy downy dew disease caused by *Sclerophthora macrospora* in finger millet can decrease their yield by 50%. Udbatta is a panicle disease caused by the fungus *Ephelis oryzae*, which affects the panicle of the millet and makes it sterile leading to decreased grain production. This is commonly seen in kodo millet, foxtail millet, and little millet. Banded sheath blight is another fungal disease caused by *Rhizoctonia solani*, which causes lesions in the leaves of all the small millets and reduces their yield (Nagaraja and Das 2016). Bacterial diseases also affect small millets and are of less importance. Though bacterial diseases occur occasionally on small millets, they can sometimes impose severe threats to their growth and productivity. Bacterial leaf spot, bacterial blight, and bacterial leaf streak are common bacterial diseases seen in small millets. Bacterial leaf spot disease characterized by spots is caused by *Xanthomonas eleusinae* mainly affects the leaves of small millets and bacterial blight caused by *Xanthomonas axonopodis* causes premature wilting leading to yield loss (Desai et al. 1965). Furthermore, bacterial leaf streaks seen in finger millet caused by *Pseudomonas eleusinae* led to drooping of leaves (Billimoria and Hegde 1971). Viral diseases also infect small millets and their incidence increases during favorable climatic conditions. They occur on an irregular basis and are mostly not economically important. Ragi severe mosaic, ragi mottle streak, and ragi streak diseases are viral diseases that infect small millets. Ragi mottle streak, a viral disease transmitted through two types of jassids (*Cicadulina bipunctella* and *Cicadulina chinai*) causes a loss of yield ranging from 50% to 100% (Maramorosch et al. 1977). On the other hand, Ragi streak viral disease cause a decrease in grain weight, leading to a reduction in the overall quality of the grains (Nagaraju et al. 1982).

Even though biotic stress factors cause yield loss in small millets but, abiotic stress factors negatively affect the yield of small millets more than biotic stresses. Abiotic stresses poorly affect growth and yield recovery and also trigger various physical and biological modifications in small millets. Drought stress, heat stress, waterlogging stress, lodging effect (Numan et al. 2021), and salt stress are the major types of abiotic factors that negatively affect the production of small millets. Even though millets are given the status of "neglected crops," many scientific studies have been carried out on finger millet and foxtail millet related to abiotic stress tolerance. Only limited data is available on barnyard millet, kodo millet, proso millet, tef millet, little finger, and fonio millet. Research has to be expanded further to other small millets to fully understand the effect of abiotic factors and to identify potent genes that are contributing to abiotic stress tolerance.

Water dearth is one of the primary reasons for the loss of crops in arid and semiarid regions. Drought decreases both the quality and quantity of small millets. Under drought, foxtail millet, little millet, and proso millet have shown a significant decrease in yield before their flowering stage. Moreover, drought caused a complete loss of yield in finger millets just after 4 weeks of sowing (Maqsood and Ali 2007; Tadele 2016). The approximate yield loss caused due to drought in tef millet was 40% (Wondewosen et al. 2012; Mengistu and Mekonnen 2012). Nevertheless, the productivity of these millets gets reduced due to drought stress but, the nutritional qualities such as protein and mineral content of the grains also get reduced (Ravindran 1991; Saleh et al. 2013).

Even though most of the small millets are resistant to heat stress but still can induce many biological and molecular modifications in plants. Heat stress can modify cellular processes such as photosynthesis and respiration, thereby creating a drastic effect on the rate of millet production. A yield loss of 75% and 84% in finger millet (Opole et al. 2018) and foxtail (Aidoo et al. 2016), respectively, was observed. Divya et al. (2019) reported that a combined heat and drought stress caused severe yield loss in foxtail millet. Moreover, there was a reduction in chlorophyll fluorescence, chlorophyll stability index, relative water content, rate of photosynthesis, and nitrate reductase activity which reduced the overall yield. It has been reported that high temperature decreases the transportation of electrons, disrupts the function of photosystem (PS) II, and enhances reactive oxygen species (ROS) accumulation (Ray et al. 2019). Heat dehydrates the reproductive parts and results in plant sterility, seed abortion, decreased number of seeds, and reduced grain filling period (Asthir 2015).

Waterlogging is the major reason for a lower yield of millets in areas where there is a higher rate of precipitation. Waterlogging inhibits gas diffusion and triggers the buildup of toxic compounds by closing soil pores. This action hinders the growth of roots, fluctuates the rate of photosynthesis, and interferes with stomatal conductance. Around 16% of grain loss was recorded in proso millets which were subjected to waterlogging for 14 days (Linkemer et al. 1998).

Lodging is not uncommon among small millets. Lodging targets both the stem and the roots of millets. It is the bending of stem or dislocation of roots due to excessive force exerted during watering, rain, wind, or their combinations. This excessive bending of the stem causes breakage, which ultimately leads to yield loss (Cannarozzi et al. 2018). The shoot system of tef millet is fragile in nature making it less resistant to root lodging (Berry et al. 2004). A loss of yield was observed in foxtail millet due to lodging (Tian et al. 2010) and excessive application of fertilizers also caused lodging in finger millet (Opole 2012). Even though millets are generally resistant to lodging, new strategies for developing resistant varieties should be established to increase the overall productivity and yield.

Salt stress is seen in arid and semiarid places and causes impaired growth in plants by generating free radicals which result in improper functioning of cellular activities. Salt stress adversely affects cell division, seed germination, excessive buildup of sodium ions, nutrient imbalance, and stomatal closure in plants. which overall affects the yield of crops. It also activates oxidative stress and triggers the generation of reactive oxygen species resulting in disruption of cell integrity and cell death (Polash et al. 2019). Small millets are also susceptible to higher concentrations of salt. Salt stress decreased the tolerance index of shoots and root, biomass, relative water content, and photosynthetic pigments in a dose-dependent manner during the NaCl treatment (50–200 Mm) in foxtail millet (Rasool et al. 2020). A significant reduction in seed germination, length of root and shoot, photosynthetic pigments (chlorophyll a, chlorophyll b), and protein content in foxtail millet and proso millets were observed under higher concentrations of NaCl (Shah et al. 2020). Foxtail millet and proso millet showed a significant reduction in yield and other parameters related to yield under higher concentrations of salt (Kafi et al. 2009).

8.3 Mechanism of Abiotic Stress Tolerance in Small Millets

During stress conditions, plants become susceptible to damages caused by various abiotic stress factors, still, they exhibit many biological mechanisms to overcome these stresses. Small millets exhibit many stress tolerance mechanisms to overcome and survive drought. These tolerance mechanisms are classified into four types namely (1) drought avoidance: the ability of the plants to withstand the balance of water during stress to evade water deficiency in plant tissues, (2) drought tolerance: the ability of the plants to produce biomass by surviving in reduced water potential, (3) drought escape: plants mature early to avoid drought, and (4) drought recovery: plants produce some amount of yield after recovering from a sporadic drought (Numan et al. 2021; Tyagi et al. 2017). Many physiological changes such as stomatal conductance, osmotic adjustment, and cell membrane stability also take place inside the cells of small millets to overcome the drought stress (Kusaka et al. 2005). Osmotic adjustment permits the leaves of small millets to maintain their turgor pressure which enables the absorption of water from dry soil (Debieu et al. 2018; Tyagi and Pudake 2017). Finger millet survives drought stress through modifications in the root system (Ramakrishnan et al. 2017). A better root system will enhance the uptake of nutrients from the soil, increase soil–microbe interactions, provide support to the shoot system, and improve drought tolerance (Comas et al. 2013). Furthermore, tef millet produces stronger roots to combat drought stress (Ayele et al. 2001; Debieu et al. 2018).

Heat tolerance mechanisms specific for small millets have not been reported yet. Plants exhibit different mechanisms to survive heat stress which mainly include maintaining membrane stability, scavenging reactive oxygen species (ROS), production of antioxidants and compatible solutes, stimulation of mitogen-activated protein kinase (MAPK), and calcium-dependent protein kinase (CDPK) cascades. Plants also activate chaperone signaling and transcriptional activation to cope up with the heat stress (Wahid et al. 2007).

Several mechanisms are shown by small millets to overcome waterlogging stress. These millets generally show hypoxia or anoxia during waterlogging and in such situations, plants carry out anaerobic respiration to withstand this condition. Finger millet carries out anaerobic respiration during reduced oxygen levels to survive this condition (Hossain and Uddin 2011). Another mechanism to overcome waterlogging is by maintaining root growth (Erdmann and Wiedenroth 1986; Daugherty and Musgrave 1994; Huang et al. 1994, 1997). Adventitious root formation is also another strategy shown by waterlogging tolerant plants to overcome this stress (Matsuura et al. 2016). Finger millet has been reported to have the ability to produce adventitious roots during waterlogging stress (Ni et al. 2019). Shoots of tef millet show an increased nitrogen reductase activity during this stress (Sairam et al. 2008), which causes some deterioration of plants. Waterlogging tolerant plants produce air containing spongy tissues known as aerenchyma. The formation of aerenchyma is a feature that permits roots to grow in waterlogged soils, by letting more internal oxygen diffusion from shoots to the root tip and also to the rhizosphere (Thomson et al. 1990; Visser et al. 1996; Jackson and Armstrong 1999).

Reduction of plant height using genetic engineering and application of chemicals are the commonly used scientific methods to overcome the effects of lodging (Shivhare and Lata 2017). The time of sowing the seeds, tilling of the soil, increasing the space between each row, and decreasing the number of seedlings in each row are some of the common agricultural practices to avoid lodging effects (van Delden et al. 2010; Piñera-Chavez et al. 2016; Cannarozzi et al. 2018). One of the effective methods to tolerate the lodging effect is to inhibit the synthesis of plant growth hormones such as gibberellic acid (van Delden et al. 2010). The application of paclobutrazol reduced both the plant height and lodging effect in finger millet and tef millet (Desta and Amare 2021). Kodo millet mutants having tolerance to lodging effect could be generated by gamma radiation or ethyl methane sulfonate (EMS) (Jency et al. 2020). A mutant variety of tef millet produced by the application of EMS showed good tolerance to lodging effect (Jöst et al. 2015).

Small millets belong to a group of plants called glycophytes and can tolerate only an average concentration of salt. These millets exhibit both enzymatic and nonenzymatic defense mechanisms to overcome high salt stress (Ul Mushtaq et al. 2021).

8.4 Plant Regeneration Methods in Small Millets

Plant regeneration experiments in small millets started during the late twentieth century and were mainly done in kodo, finger, and proso millets (Rangan 1974, 1976). Even though numerous reports are available on finger millet, foxtail millet, kodo millet, and proso millet, however, tef millet and fonio millet are still understudied. Although many scientific reports are available about the in vitro regeneration methods, it is challenging to get the desired result due to many governing factors such as the type of explant, media composition, and physical parameters such as temperature, humidity, light, etc. Therefore, a robust understanding of these growth factors as well as the small millets is necessary to establish successful protocols. Plant regeneration is the final output of plant tissue culture technique and it occurs through organogenesis, somatic embryogenesis, and a combination of both. Organogenesis involves the formation of whole plants directly from different types of explants whereas, in somatic embryogenesis, somatic embryos are produced from the totipotent cells which undergo embryogenic pathways to produce whole plants. Somatic embryogenesis is considered the most preferred method for plant regeneration among small millets (Kumar et al. 2001). Plant regeneration studies in small millets are summarized in Table 8.2. and discussed in detail as follows.

8.4.1 Finger Millet

Mature seeds, mesocotyl, leaf base, roots, immature inflorescence, and shoot tips were used as the explants for regenerating finger millets. Among all these explants, mature seeds were widely used as the potent explant for its regeneration. Plant growth hormones such as 2,4-Dichlorophenoxyacetic acid (2,4-D), picloram, 2,4,5-trichlorophenoxyacetic acid, and 1-naphthaleneacetic acid (NAA) have shown to be effective for callus formation. The majority of the work highlights the synergistic effect of both auxins and cytokinins on callus induction and shoot regeneration. Auxin such as 2,4-D at different concentrations (1.5 mg l^{-1}, 2.0 mg l^{-1}, 2.5 mg l^{-1}, 3.0 mg l^{-1}, and 4.0 mg l^{-1}) in combination with minute quantities of cytokinin (0.5 mg l^{-1} kinetin or 0.5 mg l^{-1} 6-benzylaminopurine (BAP) induced the formation of callus from the aforementioned explants. Moreover, lower concentrations of hormones such as 2,4-D, NAA, gibberellic acid (GA_3), and kinetin promoted higher shoot regeneration from seed-derived calli (Mohanty et al. 1985; Wakizuka and Yamaguchi 1987; Pius et al. 1994; Poddar et al. 1997; Kumar et al. 2001; Kothari-Chajer et al. 2008; Sharma et al. 2011; Suman and Kalpana 2012; Jagga-Chugh et al. 2012; Yemets et al. 2013). Eapen and George (1990) and Pius et al. (1999) induced callus from seeds by augmenting 4.0 mg l^{-1} picloram and 0.5 mg l^{-1} kinetin in Murashige and Skoog (MS) media. 2.0 mg l^{-1} 2,4-D and 0.5 mg l^{-1} BAP had a positive effect on callus induction (Nethra et al. 2009; Jayasudha et al. 2014a; Hema et al. 2014; Pande et al. 2015). Interestingly MS basal media also supported the shoot regeneration (Eapen and George 1990; Pius et al. 1999; Gupta et al. 2001) and root induction (Jayasudha et al. 2014a; Pande et al. 2015) form calli produced from mature seeds. A combination of auxin and cytokinin such as 2–2.5 mg l^{-1} 2,4-D and 0.5 mg l^{-1} kinetin induced calli derived from seeds (Dosad and Chawla 2015a, b; Satish et al. 2016a, b). Anju et al. (2016) reported that 2 mg l^{-1} 2, 4-D and 0.5 mg l^{-1} BAP promoted callus induction, whereas 1 mg l^{-1} GA_3 and half-strength MS basal media promoted shoot regeneration and root induction, respectively. Latha et al. (2005) and Antony Ceasar and Ignacimuthu (2008) advocated the usage of 2,4-D and kinetin in inducing calli from shoot tips.

Table 8.2 Plant regeneration studies in small millets

Millet type	Explant	Processes	Growth regulators	References
Finger millet (*Eleucine coracana*)	Mature seed Mature embryo and epicotyl Leaf and root explant and mature seed	Somatic embryogenesis Plant regeneration Somatic embryogenesis Plant regeneration Root formation Callus induction Somatic embryogenesis	2, 4-D alone or with KIN GA_3, BA or NAA alone or KIN; IAA 2, 4-D BA or KIN IBA; BA 2, 4-D or 3, 6-D or dicamba 2, 4-D or 3, 6-D or dicamba; ABA, BA; KIN	Sivadas et al. (1990), Poddar et al. (1997), Gupta et al. (2001), Kothari et al. (2004), Kothari-Chajer et al. (2008), Nethra et al. (2009) and Sharma et al. (2011). Patil et al. (2009) Bekele et al. (1995) and Mekbib et al. (1997)
Foxtail millet (*Setaria italica*)	Immature inflorescence, mature embryo, mature seed, and shoot tip	Somatic embryogenesis Plant regeneration	2, 4-D alone or with KIN or BA NAA with BA or KIN or 2, 4-D with KIN	Xu et al. (1984), Rao et al. (1988), Reddy and Vaidyanath (1990), Osuna-Avila et al. (1995), Liu et al. (2005), Qin et al. (2008) and Wang et al. (2011).
Proso millet (*Panicum millaceum*)	Immature and mature embryo, mature seed, immature inflorescence, mesocotyl, shoot tip, and leaf and stem segments Immature inflorescence, immature and mature seed; young leaf base and mesocotyl	Somatic embryogenesis Plant regeneration Somatic embryogenesis Shoot regeneration Root formation	2, 4-D alone or with KIN 2, 4-D or NAA Picloram or 2, 4-D alone or with TDZ; KIN NAA alone or with BA PAA	Rangan (1974), Bajaj et al. (1981), Heyser and Nabors (1982), Rangan and Vasil (1983) and Heyser (1984). Rangan (1976), Nayak and Sen (1989, 1991), Vikrant and Rashid (2001, 2002a, b, 2003), Kaur and Kothari (2004) and Kothari-Chajer et al. (2008).
Barnyard millet (*Echinochloa crusgalli*)	Mature seed	Somatic embryogenesis	2, 4-D	Gupta et al. (2001).
Kodo millet (Paspalum scrobiculatum)	Shoot tip	Somatic embryogenesis Plant regeneration Root formation	2, 4-D alone or with KIN TDZ alone or BA; NAA IBA	Arockiasamy et al. (2001) and Ceasar and Ignacimuthu (2010).

(continued)

Table 8.2 (continued)

Millet type	Explant	Processes	Growth regulators	References
Foxtail millet (*Setaria italica*)	Immature inflorescence; mature embryo; mature seed and shoot tip	Somatic embryogenesis Plant regeneration	2, 4-D alone or with KIN or BA NAA with BA or KIN or 2, 4-D with KIN	Xu et al. (1984), Rao et al. (1988), Reddy and Vaidyanath (1990), Osuna-Avila et al. (1995), Liu et al. (2005), Qin et al. (2008) and Wang et al. (2011)
Proso millet (*Panicum millaceum*)	Immature and mature embryo, mature seed, immature inflorescence, mesocotyl, shoot tip and leaf, and stem segment Immature inflorescence, immature and mature seed; young leaf base and mesocotyl	Somatic embryogenesis Plant regeneration Somatic embryogenesis Shoot regeneration Root formation	2, 4-D alone or with KIN 2, 4-D or NAA Picloram or 2, 4-D alone or with TDZ; KIN NAA alone or with BA PAA	Rangan (1974), Bajaj et al. (1981), Heyser and Nabors (1982), Rangan and Vasil (1983) and Heyser (1984). Rangan (1976), Nayak and Sen (1989, 1991), Vikrant and Rashid (2001, 2002a, b, 2003), Kaur and Kothari (2004) and Kothari-Chajer et al. (2008).
Barnyard millet (*Echinochloa crusgalli*)	Mature seed	Somatic embryogenesis	2, 4-D	Gupta et al. (2001).
Kodo millet (Paspalum scrobiculatum)	Shoot tip	Somatic embryogenesis Plant regeneration Root formation	2, 4-D alone or with KIN TDZ alone or BA; NAA IBA	Arockiasamy et al. (2001) and Ceasar and Ignacimuthu (2010).
Foxtail millet (*Setaria italica*)	Immature inflorescence; mature embryo; mature seed and shoot tip	Somatic embryogenesis Plant regeneration	2, 4-D alone or with KIN or BA NAA with BA or KIN or 2, 4-D with KIN	Xu et al. (1984), Rao et al. (1988), Reddy and Vaidyanath (1990), Osuna-Avila et al. (1995), Liu et al. (2005), Qin et al. (2008) and Wang et al. (2011).
Tef (*Eragrostis tef*)	Mature seed Immature spikelet and panicle segment Immature anther and embryo	Somatic embryogenesis Embryo promotion Plant regeneration Gynogenic tissue	2, 4-D followed by TIBA 2, 4-D; KIN followed by IAA; BA GA_3 2, 4-D; BA	Assefa et al. (1998). Gugsa et al. (2006) Tadesse et al. (2009) and Gugsa and Kumlehn (2011).

(continued)

Table 8.2 (continued)

Millet type	Explant	Processes	Growth regulators	References
		induction Somatic embryogenesis Plant regeneration	2, 4-D BA alone or with NAA	
Fonio (*Digitaria exilis*)	Stem segment	Somatic embryogenesis Shoot development	2, 4-D BA; GA_3	Ntui et al. (2010).

2, 4-D 2, 4-Dichlorophenoxyacetic acid, *KIN* kinetin, *GA_3* Gibberellic acid, *BA* Benzyl adenine, *NAA* Naphthalene acetic acid, *IAA* Indole-3-acetic acid, *IBA* Indole butyric acid, *ABA* Abscisic acid, *TIBA* Triiodobenzoic acid, *TDZ* Thidiazuron, *PAA* Phenylacetic acid

Furthermore, cytokinins such as BAP, thidiazuron (TDZ), and kinetin had a stimulatory effect on shoot regeneration. Alex et al. (2018) stated that shoot tip-derived callus induction was highest in MS medium supplemented with 2.0 mg l^{-1} 2,4-D plus 1.0 mg l^{-1} BAP and higher root induction was observed in GBK-043050 variety in media augmented with 1.0 mg l^{-1} BAP plus 0.25 mg l^{-1} NAA whereas higher shoot regeneration was observed in GBK-043137 variety in 1.75 mg l^{-1} BAP medium. Recently, Jamra et al. (2021) reported that GP-45 genotype showed 98% callus induction derived from seed in MS medium supplemented with 2.0 mg l^{-1} 2, 4-D and 2.5 mg l^{-1} BAP and 73.7% of shoot regeneration in MS medium supplemented with 0.5 mg l^{-1} 2, 4-D and 2.5 mg l^{-1} BAP. Many reports suggest the potency of complex compounds such as amino acids, inorganic compounds, casein hydrolysate, and vitamins for an efficient regeneration in finger millet. Mohanty et al. (1985) stated that 500 mg l^{-1} casein hydrolysate induced callus. Eapen and George (1990) demonstrated that proline (100 mg l^{-1}), glutamine (800 mg l^{-1}), and casein (250 mg l^{-1}) improved regeneration. Similarly, Anjaneyulu et al. (2011) stated 97% of callus induction when growth regulators were used in combination with proline (500 mg l^{-1}), casein hydrolysate (300 mg l^{-1}), and tryptophan (250 mg l^{-1}). On the other hand, Yemets et al. (2013) stated that by supplementing 100 mg l^{-1} glutamine, 200 mg l^{-1} casein hydrolysate, and 500 mg l^{-1} proline did not markedly improve the frequency of callus induction, however, there was a significant improvement in the quality of callus. Furthermore, augmenting basal MS medium with 500 mg l^{-1} proline, and 100 mg l^{-1} glutamine produced around 30% of regeneration, whereas no regeneration was observed on basal MS media lacking proline and glutamine. Satish et al. (2015) stated that a higher concentration of casein (500 mg $l^{-1}$and 750 mg l^{-1}) produced the highest induction of embryogenic callus and regeneration of shoots. Shoot apical meristem (SAM) of finger millet (CO 9 genotype) when placed in MS medium containing 3.0 mg l^{-1} BAP, kinetin 2.0mg l^{-1}, 300 mg l^{-1}proline, 400 mg l^{-1} casein hydrolysate, 3.0 mg l^{-1} glycine, and 5 % coconut water produced a higher shoot frequency of 95.89%, number of shoots per explant (46.52), and shoot length (10.8 cm) (Atul Babu et al. 2018).

Inorganic nutrients are an essential part of the MS medium that greatly affects the morphogenic potency of plants. Kothari et al. (2004) stated that a fivefold increase in copper sulfate ($CuSO_4$) in the media enhanced shoot regeneration efficiency by four times. Correspondingly, Sharma et al. (2011) reported that a tenfold increase in $CuSO_4$ was found effective in enhancing calli formation and recovery of plantlets. Kothari-Chajer et al. (2008) observed a dose-dependent effect of $CuSO_4$ on regeneration pattern whereas the growth medium lacking these micronutrients did not regenerate. Silver nitrate ($AgNO_3$) is a well-known inhibitor of ethylene henceforth supplementing $AgNO_3$ to the culture media greatly improved the callus induction, and formation of friable embryogenic callus, shoots, roots, and somatic embryogenesis in several plants (Benson 2000; Kumar et al. 2009). Kothari-Chajer et al. (2008) reported that adding $AgNO_3$ enhanced the growth of callus in finger millet. Furthermore, a two or threefold increase in plant regeneration was observed when 1.0 mg l^{-1} $AgNO_3$. Sharma et al. (2011) advocated that the addition of 5mg l^{-1} to 10 mg l^{-1} $AgNO_3$ had a stimulatory effect on the formation of embryogenic calli and plant regeneration.

8.4.2 Foxtail Millet

Compared to finger millets, there are only some reports about plant regeneration methods in foxtail millets. Different explants such as leaf base, mesocotyl, immature inflorescence, seed, and young inflorescence are used to induce callus in foxtail millet (Dosad and Chawla 2016). Shoot tip (Osuna-Avila et al. 1995) and immature glumes (Reddy and Vaidyanath 1990) are also used as explants for regeneration in foxtail. Linsmaier and Skoog (LS) medium with 4 mg l^{-1} kinetin and 1 mg l^{-1} indole acetic acid (IAA) showed the highest shoot regeneration rate (47%) from callus (Reddy and Vaidyanath 1990). Mostly, 2,4-D is used as the plant hormone for the induction of callus along with cytokinins such as kinetin and NAA (Rout et al. 1998; Satish et al. 2016a, b; Wang et al. 2016). Moreover, 2,4-D alone is also used for callus induction from immature influorescence (Wang et al. 2011). Rout et al. (1998) reported that 3.0 mg l^{-1} 2,4-D, and 0.5 mg l^{-1} kinetin induced callus from leaf base and mesocotyl regions. Moreover, 1.0 mg l^{-1} kinetin, 1.0 mg l^{-1} BAP, and 0.5 mg l^{-1} 2,4-D promoted shoot regeneration and MS basal medium alone initiated root formation. Interestingly half-strength MS basal media also promoted root formation from shoots regenerated from immature influorescence (Wang et al. 2011) and seeds (Satish et al. 2016a, b). Inorganic nutrients play a vital role in plant regeneration. Wang et al. (2011) also stated that inclusion of 5 mg l^{-1} to 10 mg l^{-1} $AgNO_3$ had a stimulatory effect on the formation of embryogenic calli and plant regeneration.

8.4.3 Proso Millet

Very limited data is available on plant regeneration studies in proso millet. This millet has immense potential and more regeneration studies should be carried out. Nabors et al. (1983) stated that 1.0 mg l^{-1} 2,4,5-Trichlorophenoxyacetic acid (2,4,5-T) alone induced both non-embryogenic and embryogenic calli that were derived from seeds cultured on MS medium with 2,4-D and BAP. Jain et al. (2001) reported that calluses having somatic embryos from immature embryos and inflorescences developed well on MS medium with 2.5 mg l^{-1} 2,4-D, 1 mg l^{-1} kinetin, and somatic embryos from mature embryos grew well on MS medium with 0.5 mg l^{-1} 2,4-D and 0.5 mg l^{-1} BAP. In addition, plant regeneration was witnessed when calli were placed on MS medium devoid of plant hormones and further, abundant root induction was observed on a half-strength MS medium containing 2% sucrose.

8.4.4 Barnyard Millet

Mature seeds are taken as explants to study plant regeneration in this type of millet. Two genotypes (CO2 and PRJ1) of barnyard millet were used for plant regeneration through somatic embryogenesis by using mature seeds as explants and CO2 genotype responded better for callus induction, shoot and root formation. Furthermore, the percentage of somatic embryogenic induction was 76% (CO2) and 69% (PRJ1), respectively, when cultured on MS medium supplemented with 5 mg l^{-1} 2,4-D whereas maximum shoot induction (75.4%) was induced in MS medium containing 0.5 mg l^{-1} BAP and the maximum root induction (90.3%) was induced in half-strength MS medium lacking plant growth hormones (Rajak et al. 2018). In another study by Dosad and Chawla (2015b), maximum callus induction (41.1 %) was observed on MS medium supplemented with 3.5 mg l^{-1} 2, 4-D. In addition, maximum shoot induction was produced (84.6 %) on calli that were grown on MS medium containing 1 mg l^{-1} NAA, and maximum rooting was also observed in the same medium composition.

8.4.5 Kodo Millet

Immature inflorescence, seeds, immature embryos, and shoot tips were taken as explants and different combinations of plant growth hormones were tested to optimize the best protocol for plant regeneration. N6 medium is used in most of the regeneration studies in kodo millet. N6 medium with and higher concentration of 2,4-D (22 mg l^{-1}) initiated callus from seeds and N6 medium without any plant hormones produced shoots from the callus (Vikrant and Rashid 2003). In addition, N6 medium with and lower concentration of 2,4-D (2.0 mg l^{-1}) induced callus from immature fluorescence and MS medium devoid of plant hormones produced shoots from the callus (Vikrant and Rashid 2002a). Somatic embryogenesis developed from immature inflorescence-derived calli which were induced in N6 medium containing

1.0 mg l^{-1} 2,4-D. Moreover, MS medium supplemented with 5% charcoal initiated shoots from calli (Vikrant and Rashid 2001). Correspondingly, the same explant was used to induce embryogenic callus in MS medium having 1 mg l^{-1} picloram and 1 mg l^{-1} kinetin. Moreover, lower concentrations of cytokinins promoted shoot formation (NAA—0.5 mg l^{-1} and BA—1 mg l^{-1}) and phenyl acetic acid (0.5 mg l^{-1}) in MS medium initiated root formation (Kaur and Kothari 2004). Shoot formation was promoted in MS medium containing 1 mg l^{-1} 2,4-D and 2.5 mg l^{-1} TDZ from seeds and root formation was induced in MS basal medium (Vikrant and Rashid 2002b). A combination of 2.0 mg l^{-1} 2,4-D and 0.5 mg l^{-1} kinetin initiated shoot tip-derived callus and 1.0 mg l^{-1} TDZ promoted both shoots and roots (Ceasar and Ignacimuthu 2010).

8.4.6 Tef Millet

Mature seeds, immature spikelet, immature anther, embryo, and panickle segment were employed as the explants for plant regeneration studies in tef millet. Most of the plant regeneration studies pertaining to tef are reported via somatic embryogenesis, embryo, and gynogenic tissue formation were induced by using auxins such as 2,4-D in combination with various other hormones such as indole acetic acid (IAA), BA, and 2,3,5-Triiodobenzoic Acid (TIBA) (Assefa et al. 1998; Gugsa et al. 2006; Tadesse et al. 2009; Gugsa and Kumlehn 2011). Though MS is commonly used in tef regeneration, N6 medium was preferred for regenerating tef from explants such as immature spikelets, panicle segments, and anthers (Plaza-Wüthrich and Tadele 2013). Medium containing KBP minerals has been proved to be beneficial for successful regeneration of tef from immature zygotic embryos of size 0.2–0.35 mm is used as explant and optimized results were observed by culturing these explants on a medium containing 9.2–13.8 μM 2,4-D, KBP minerals, 6 mM glutamine with 0.5% Phytagel (Gugsa and Kumlehn 2011).

8.4.7 Fonio Millet

The plant regeneration studies on fonio are also very limited. Stem segment is used as the explant for somatic embryogenesis and shoot development by using 2,4-D and a combination of BA and GA_3, respectively (Ntui et al. 2010).

8.5 Transformation Studies in Small Millets

In small millets, both biolistic and *Agrobacterium*-mediated transformation studies have been carried out. Even though the biolistic method for gene delivery system is independent of the type of species and genotype; still the chances for chimeral plantlets, incorporation of multiple copies of the gene, and tissue damage due to particle bombardment method are very frequent. On the contrary,

Agrobacterium-mediated system allows more accurate and stable integration of relatively smaller copies of a gene. This method also allows the transfer of larger DNA fragments. Moreover, *Agrobacterium* transformation system is highly cultivar specific. There are many governing factors that decide the success rate of *Agrobacterium*-mediated transformation such as the type of explant, *Agrobacterium* strain, composition of co-cultivation medium, duration of co-cultivation, selectable markers, promoters, and plant tissue culture conditions. A summary of transformation studies in small millets is summarized in Table 8.3, and discussed below.

8.5.1 Finger Millet

Despite the nutritional richness of finger millet, transgenic studies for genetic improvement are still insufficient. Earlier reports on transformation in finger millet were done by the biolistic method. Gupta et al. (2001) reported that the *gus* gene expression in callus derived from leaves and seeds via the biolistic method of transformation can be highly driven by *ubiquitin1* gene promoter. Even though there was a higher expression of transgene but, stable transgene expression was not obtained. Subsequently, Latha et al. (2005) reported the successful stable expression of *pin* gene driven by cauliflower mosaic virus 35S (CaMV35S) promoter by biolistic transformation in embryogenic callus and enhanced the resistance against leaf blast disease. Following the method described by Latha et al. (2005), and Mahalakshmi et al. (2006) used shoot tip-derived embryogenic callus for developing salt-tolerant transgenic finger millet lines by transferring serine-rich protein (*PcSrp*) encoding gene from *Porteresia coarctata* into finger millet driven by rice Actin-1 promoter. In another study by Sharma et al. (2011) several physical and chemical parameters for an enhanced transformation were evaluated and concluded that a rupture disk pressure of 1,100 psi with a distance of 3 cm from rupture disk to macrocarrier, microprojectile travel distance of 12 cm, 1.0 μm-sized gold particles provided maximum transient GUS expression and transformation efficiency. Moreover, osmotic treatment of callus with 0.4 M sorbitol also enhanced transformation efficiency of 45.3%. In another study, Antony Ceasar and Ignacimuthu (2011) employed shoot apices from genotype GPU 45 for callusing, somatic embryogenesis, and *Agrobacterium*-mediated transformation. The transformation efficiency of 19% was achieved with pCAMBIA1301:*gus* lines and 3.8% was attained for stable transformation. Later, Ignacimuthu and Ceasar (2012) developed transgenic finger millets that expressed rice chitinase (pHyg-*chi.11*) gene driven by maize *ubiquitin1* promoter by using the optimized transformation protocol. Furthermore, the transgenic plants showed an increased resistance toward leaf blast disease. Babu et al. (2012) employed *Agrobacterium*-mediated method to transfer *PDH45* gene into the seed of finger millet to produce salt-tolerant variety. Further, the seed-derived callus was co-cultivated with *Agrobacterium* containing pCAMBIA binary vector with *PDH45* gene, *nptII*, *hptII*, and *gus* reporter gene under the control of CaMV 35S promoter. To carry out a study with a similar objective, tolerance, Anjaneyulu et al. (2014) employed *Sorghum bicolor* (*SbVPPase*) gene under the control of CaMV35S

Table 8.3 An overview of genetic transformation studies in economically important millet species

Millet species	Explant used for transformation	Transformation mode	Vector	Promoter selectable marker/ reporter	References
Finger millet (*Eleucine coracana*)	Callus	Biolistic	NA	*CaMV35s:uidA, Act 1:uidA, Uq1:uidA,RbcS:uidA,Ft 1: uidA*	Gupta et al. (2001)
	Embryonic callus	Biolistic	pBar35S,pPur	*CaMV35s:bar, CaMV35s: uidA*	Latha et al. (2005)
	Shoot apical meristem	*Agrobacterium* mediated	pCAMBIA 1301	*CaMV35s:gus, CaMV35s:hpt*	Antony Ceasar and Ignacimuthu (2011)
	Mature seed	*Agrobacterium* mediated	pBI-121, pCNL-56, p35S-GUS-INT	*CaMV35s:npt, CaMV35s: uidA*	Sharma et al. 2011
	Shoot apical meristem	*Agrobacterium* mediated	pCAMBIA 1301	*Ubi1:Chi11,CaMV35s:hpt*	Ignacimuthu and Ceasar (2012)
	Mature seed	*Agrobacterium* mediated	pCAMBIA 1301	*CaMV35s:hpt, CaMV35s: SbVPPase*	Anjaneyulu et al. (2014)
	Mature seed	*Agrobacterium* mediated	pCAMBIA 1380, pCAMBIA 1301	*CaMV35s:hpt, CaMV35s: uidA*	Hema et al. (2014)
Foxtail millet (*Setaria italica*)	Immature inflorescence-derived embryonic calli	*Agrobacterium* mediated	pBI-121	*CaMV35s:nptII*	Liu et al. (2005)
	Immature inflorescence-derived embryonic calli	*Agrobacterium* mediated	pBIZm13Si401	*Zm13-Si401*	Qin et al. (2008)
	Immature inflorescence-derived embryonic calli	Biolistic	pROKf40s, pROKf40and pROKf40i	*CaMV35s:hpt*	Liu et al. (2009)
	Immature inflorescence-derived embryonic calli	*Agrobacterium* mediated	pCAMBIA 2300,pSB	*CaMV35s:nptII, CaMV35s: hpt,CaMV35s:uidA*	Wang et al. (2011)

(continued)

Table 8.3 (continued)

Millet species	Explant used for transformation	Transformation mode	Vector	Promoter selectable marker/ reporter	References
	Immature inflorescence-derived embryonic calli	*Agrobacterium* mediated	pCOU	*Ubi:hpt*	Wang et al. (2014)
	Immature inflorescence-derived embryonic calli	*Agrobacterium* mediated	Modified pCOU	*Ubi:hpt*	Li et al. (2014)
	Immature inflorescence-derived embryonic calli	*Agrobacterium* mediated	pCAMBIA 2300	pF128:uidA, p19Z:uidA, *CaMV35s:uidA*	Pan et al. (2015)
	Immature inflorescence-derived embryonic calli	*Agrobacterium* mediated	pCAMBIA 2300, pCOU	*CaMV35s:nptII,Ubi:hpt*	Pan et al. (2016)
	Immature inflorescence-derived embryonic calli	Agrobacterium mediated	Modified pCOU	*Ubi:uidA*	Li et al. (2017)
	Shoot apices	*Agrobacterium* mediated	*pFGC-SiPHT1:2, pFGC-SiPHT1:3 or pFGC-SiPHT1:4*	*CaMV35s:hpt*	Ceasar et al. (2017)
Barnyard millet (*Echinochloa crusgalli*)	Embryonic callus and leaf segment	Biolistic	NA	*CaMV35s:uidA, ZmUbi, OsAct,RbcS:uidA,ppcA-L-Ft*	Gupta et al. (2001)
Kodo millet (*Paspalum scrobiculatum*)	Callus derived from seeds	*Agrobacterium* mediated	pCNL56, pCAMBIA 2300	*CaMV35s:gusA:nptII*	Bhatt et al. (2021)
Fonio millet (*Digitaria exilis (L)*)	Callus derived from stem segments	*Agrobacterium* mediated	pIG121-Hm	*hpt: nptII :gusA*	Otang ntui et al. (2017)

Tef millet (*Eragrostis tef* (Zucc.))	Immature embryo	*Agrobacterium* mediated	pGPTV-Kan harboring PcGA2ox1	*CaMV35s:uidA:nptII*	Gebre et al. (2013)

Ubi1 maize ubiquitin promoter, *CaMV35S* cauliflower mosaic virus 35S promoter, *Act1* rice actin gene promoter, *bar* or *pat* phosphinothricin acetyltransferase, *hptII* or *hpt* hygromycin phosphotransferase gene, *nptII* neomycin Phosphotransferase II gene, *GUS* or *uid* A β-glucuronidase gene, *Ft Ft5* promoter of C4 isoform of phosphoenolpyruvate carboxylase gene from *Flaveria trinervia*, *Zm13 Zea mays* 13 promoter, *ZmUbi*maize ubiquitin, *OsAct*rice actin, *RbcS:uidA*rice small subunit of ribulose 1,5-biphospate carboxylase, *ppcA-L-Ft Flaveria trinervia* phosphoenol pyruvate carboxylase, PcGA2ox1 *Phaseolus coccineus*, NA not available

promoter was transformed into embryogenic calli and Hema et al. (2014) employed pCAMBIA1301 to introduce the mannitol-1-phosphate dehydrogenase (*mtlD*) gene to produce multiple abiotic stress-tolerant variety of finger millets.

8.5.2 Foxtail Millet

Transformation studies in foxtail millet were carried out by both biolistic-mediated (Dong and Duan 1999, 2000; Liu et al. 2009) and *Agrobacterium tumefaciens*-mediated approaches (Liu et al. 2005, 2007; Wang et al. 2011). Only a few reports are available in foxtail millet related to biolistic mediated transformation. Diao et al. (1999) reported for the first time the significance of several parameters that affect this method of transforming genes into immature fluorescence-derived embryogenic callus. Later, Dong and Duan (1999, 2000) developed transgenic foxtail millets by targeting pollen grains and influorescence by microprojectile bombardment method. However, lower transformation efficiency was obtained through these studies. Overexpression and RNAi-induced mutant millets expressing *SiPf40* were developed by Liu et al. (2009) where floret-derived embryogenic calli were used for transformation via bombardment method. However, the procedure of bombarding the desired gene to callus was done according to O'Kennedy et al. (2004). The first report on *Agrobacterium*-mediated method in foxtail millet was done by Liu et al. (2005) who reported a transformation efficiency of about 6.6%. Later, Liu et al. (2007) optimized important parameters that govern the transformation efficiency such as the type of explant, *Agrobacterium* strain, concentration of acetosyringone, and duration of co-cultivation. Further, it was concluded that the influorecence can be used as the best explant whereas the *Agrobacterium* strain LBA4404 with low cell density with inoculation time ranging from 30 to 40 mins, the optimal acetosyringone concentration in the medium was determined as 0.1 mmol l^{-1}, and duration for co-cultivation for 48 h.

Later, Qin et al. (2008) adopted *Agrobacterium*-mediated transformation method as described by Liu et al. (2005) and transformed callus derived from panicle with *Agrobacterium* strain LBA4404 containing an expression cassette pBIN19:*Zm13: Si401*. Further, Wang et al. (2011) enhanced the foxtail millet transformation procedure by optimizing various factors such as *Agrobacterium* strain, age of the callus, co-cultivation media, and other transformation conditions and the standardized protocol was efficiently used for obtaining *SBgLR* transgenic lines with a transformation efficiency of 5.5%. Furthermore, it was concluded that the highest regeneration was obtained in a 25-day-old callus that is induced from inflorescences (0.5–1.0 cm long) with *Agrobacterium* strain LBA4404 where the temperature for co-cultivation was 22 °C and dithiothreitol concentration was found to be 0.15 g l^{-1}; DTT in both the infection as well as in the co-cultivation media. Subsequently, Wang et al. (2014) employed the protocol described by Qin et al. (2008) and Wang et al. (2011) to carry out an overexpression study where *SiLEA1A* overexpressed millets showed better growth under salt and drought conditions. Furthermore, Li et al. (2014, 2017) assessed the abiotic stress-tolerant functions of

SiARDP and *SiASR4* genes in foxtail millet using the method of Wang et al. (2011). Interestingly, Pan et al. (2015) also used the transformation method done by Qin et al. (2008) and Wang et al. (2011) to identify a novel seed-specific promoter namely *F128* in foxtail millet which is more active than CaMV35S (constitutive promoter), and 19Z promoter (seed-specific promoter from maize) in transgenic foxtail, maize, and *Arabidopsis*. Pan et al. (2016) carried out overexpression of lipid transfer protein gene (*SiLTP*) and also identified the function of this gene against salt and drought stress via earlier reported transformation protocols. Transformation efficiency of 10% was achieved when shoot apices were used as explants by Ceasar et al. (2017) for developing RNAi plantlets for three members (SiPHT1:2, SiPHT1:3, and SiPHT1:4) belonging to the PHT1 phosphate transporter family. Moreover, the silenced plants showed a substantial decrease in the inorganic phosphate content of shoots and root tissues.

8.5.3 Barnyard Millet

There is only one report available on transformation in barnyard millet which was done by Gupta et al. (2001) where the *gus* gene expression in callus derived from leaves and seeds via the biolistic method of transformation was highly driven by *ubiquitin1* gene promoter.

8.5.4 Kodo Millet

Considering the nutritional significance of kodo millets, recently Bhatt et al. (2021) reported a proficient, highly repeatable, and an optimized protocol for *Agrobacterium*-mediated transformation in five different varieties of kodo millets (RK 15, GPUK-3, RBK-155, RK-390-25, and TNAU 86). The following transformation conditions significantly increased the transformation efficiency in kodo millet which includes cell densities of 0.8 for EHA 105 strain and 1.0 for LBA 4404 strain, the inclusion of antinecrotic mixture AM3, supplementing surfactant such as 0.5% pluronic acid F-68, infection time of 20 minutes, 200 μM acetosyringone for EHA 105 strain and 250 μM for LBA4404 strain, co-cultivation temperature of 24 °C with GPUK-3 variety as the best responding variety among the above-mentioned varieties. Moreover, higher transformation efficiency was obtained by using *Agrobacterium* strain EHA105 harboring plasmids pCNL 56 and *Agrobacterium* strain LBA4404 harboring plasmids pCNL driven by CaMV35s promoter in callus derived from mature seeds.

8.5.5 Tef Millet

The transformation studies done in tef millet are very few. Successful genetic transformation for integrating a transgene depends upon efficient plant regeneration protocols and finally the detection and expression of the inserted gene. Embryogenic callus derived from immature embryos from tef cultivar DZ-01-196 was transformed using *Agrobacterium* strain LBA4404. Furthermore, the embryogenic callus was transformed with PcGA2ox that inactivates GA_3, driven by CaMV35S promoter and the callus was grown in co-cultivation medium (CCM) followed by selection in KBP and regeneration (K4NB) media (Gebre et al. 2013).

8.5.6 Fonio Millet

Genetic studies in fonio millet are always stagnant as there is no efficient transformation protocol and this situation has badly affected crop improvement research in fonio millet. Valentine et al. (2017) developed a protocol for transforming two genotypes (Agyong and Churiwe) of fonio millet. Stem-derived callus which is 21 days old was co-cultivated for 3 days with *Agrobacterium tumefaciens* strain EHA101 that carries the plasmid pIG121-Hm containing hygromycin phosphotransferase (*hpt*), neomycin phosphotransferase II (*nptII*), and β-glucuronidase (*gus*) as reporter genes. Later, the selection was done on hygromycin B medium, and plants were regenerated on MS medium supplemented with 0.5 mg l^{-1} BA and 0.1 mg l^{-1} GA_3 with a transformation efficiency of 2.1 % in "Agyong" and 2.7% "Churiwe" genotypes. This study has paved a path for crop improvement studies in fonio millet.

8.6 Transgenic Approaches to Produce Abiotic Stress-Tolerant Small Millets

Despite the nutritional value of millets, research in genetic engineering through a transgenic approach is very limited in millets. The current focus of research on this crop is to improve the production of stress-tolerant and high-yielding varieties of millets (Saxena et al. 2018; Lata 2015). Some meaningful research has been carried out to improve abiotic stress tolerance in small millets.

8.6.1 Finger Millet

Very few studies are related to the production of abiotic stress-tolerant finger millets through the transformation technique. For the past two decades, research on small millets is gaining momentum and the majority of the work related to abiotic stress tolerance on finger millets was carried out to increase salt tolerance. The transgenic finger millet produced through particle-inflow-gun method expressing serine-rich protein (*PcSrp*) gene from *Porteresia coarctata* under rice actin-1 promoter showed an increased salt tolerance by maintaining ion homeostasis (Mahalakshmi et al. 2006). *Agrobacterium*-mediated transformation of *PDH45* gene from pea increased considerable salt tolerance in finger millet (Babu et al. 2012). A similar method of transformation was used to introduce vacuolar H^+-pyrophosphatase (*SbVPPase*) gene from Sorghum in finger millet to improve salt tolerance. Furthermore, the transgenic millet showed a drastic increase in chlorophyll content, and enzymatic activities (Anjaneyulu et al. 2014). Furthermore, two genes namely Na^+/H^+ antiporter of *Pennisetum glaucum* (*PgNHX1*) and H^+-pyrophosphatase (*AVP1*) from *Arabidopsis thaliana*, were transformed using *Agrobacterium tumefaciens* strain EHA105 to improve salt tolerance (300 μM NaCl) in finger millet by increasing enzymatic activities (Jayasudha et al. 2014b). Herbicide resistance of millets is one of the important agronomic traits which has to be improved along with other characteristics such as increased yield, biotic and abiotic stress tolerances. Bayer et al. (2014) developed transgenic finger millet with a mutant αtubulin gene having resistance to dinitroanilines by both *Agrobacterium*-mediated transformation and bioballistic transformation methods.

8.6.2 Foxtail Millet

Mannitol, a vital osmolyte, scavenges reactive oxygen species (ROS) and confers tolerance to various abiotic stresses in plants (Seckin et al. 2009). *Agrobacterium*-mediated transformation of finger millet expressing mannitol-1-phosphate dehydrogenase (*mtlD*) isolated from *Escherichia coli* imparted multiple stress tolerance such as drought, salinity, and oxidative stress compared to the control. Moreover, the transgenic plants showed improved growth, higher chlorophyll retention, and osmotic adjustment (Hema et al. 2014). Another study was done by overexpressing *SiLEA14* in foxtail millet which showed a significant increase in multiple stress tolerance namely salinity, drought, and osmotic stress that increased the production of free proline, sugars, and osmotic protectants (Wang et al. 2014).

8.7 Functionally Characterized Abiotic Stress Resistance Genes in Small Millets

When a plant is exposed to abiotic stress, several genes get either upregulated or downregulated. This leads to increased biosynthesis of numerous proteins and metabolites, which may cater to plant resistance toward abiotic stresses. Though many genes have been functionally characterized in small millets, especially in foxtail millet and finger millet, but few scientific reports are only available about transgenic finger millet in tolerating abiotic stresses. Genes that confer the abiotic stress tolerance in small millets are presented in Table 8.4.

Many genes in foxtail millet have been functionally characterized for various types of abiotic stress. Genes such as *PHGPX* (Sreenivasulu et al. 2004) and aldose reductase (Veeranagamallaiah et al. 2009) contribute to combat salt-induced oxidative damage in foxtail millet. 12-Oxophytodienoic acid reductase 1 encoded by *SiORP1* gene in foxtail millet showed a positive response toward drought stress and was upregulated due to osmotic stress (Zhang et al. 2007). Jia et al. (2007)

Table 8.4 Genes that confer abiotic stress tolerance in small millets

Name of the gene	Name of the millet	Type of stress tolerance	References
EcNAC1	Finger millet	Abiotic stress tolerance	Ramegowda et al. (2012)
Ec-apx1	Finger millet	Drought stress	Bhatt et al. (2013)
Metallothionein	Finger millet	Drought stress	Parvathi et al. (2013)
Farnesylated protein *ATFP6*	Finger millet	Drought stress	Parvathi et al. (2013)
Farnesyl pyrophosphate synthase	Finger millet	Drought stress	Parvathi et al. (2013)
Protein phosphatase 2A	Finger millet	Drought stress	Parvathi et al. (2013)
RISBZ4	Finger millet	Drought and salt stress	Parvathi et al. (2013)
EcDehydrin7	Finger millet	Abiotic stress tolerances	Singh et al. (2014)
Monodehydroascorbate reductase	Finger millet	Drought, salt, and UV radiation stress stress	Sudan et al. (2015)
NAC 67	Finger millet	Drought and salt stress stress	Rahman et al. (2016)
PHGPX	Foxtail millet	Salt tolerance	Sreenivasulu et al. (2004)
Aldose reductase	Foxtail millet	Salt tolerance	Veeranagamallaiah et al. (2009)
SiORP1	Foxtail millet	Drought stress	Zhang et al. (2007)
psbA	Foxtail millet	Atrazine stress	Jia et al. (2007)
Glutamine synthetase Pyrroline-5-carboxylate reductase 12-oxophytodienoic acid reductase (*OPR1*)	Foxtail millet	Drought tolerance	Zhang et al. (2007)
DNAj	Foxtail millet	Drought and heat tolerance	Wang et al. (2009)
Si69	Foxtail millet	Aluminum tolerance	Zhao et al. (2009)
SiPLDa1	Foxtail millet	Drought stress	Peng et al. (2010)
Acetyl-CoA carboxylase	Foxtail millet	Resistance to sethoxydim herbicide	Dong et al. (2011)
Dehydration-responsive element-binding protein 2 (*DREB2*)	Foxtail millet	Dehydration tolerance	Lata et al. (2011)

(continued)

Table 8.4 (continued)

Name of the gene	Name of the millet	Type of stress tolerance	References
NAC transcription factor	Foxtail millet	Salt stress tolerance	Puranik et al. (2011)
WD-40	Foxtail millet	Associated with dehydration stress-responsive pathway	Mishra et al. (2014)
SiREM6	Foxtail millet	Salt tolerance	Yue et al. (2014)
C2H2 type of zinc finger transcription factors (TFs)	Foxtail millet	Salt, dehydration, and cold stress	Muthamilarasan et al. (2014)
SiALDH	Foxtail millet	Salt tolerance	Zhu et al. (2014)
ABA-responsive DRE-binding protein (*ARDP*)	Foxtail millet	Tolerance to salt and drought stresses	Li et al. (2014)
Autophagy-related gene (*ATG*)	Foxtail millet	Tolerance to nitrogen starvation and drought stresses	Li et al. (2015)
Nuclear factor-Y (*SiNF-YA1*, *SiNFYB8*) genes	Foxtail millet	Drought and salt tolerance	Feng et al. 2015
SiWRKY	Foxtail millet	Dehydration, salt, and hormone stress	Muthamilarasan et al. (2015)
Argonaute protein 1 encoding gene	Foxtail millet	Regulation of stress responses	Liu et al. (2016)
Abscisic acid stress ripening gene (*ASR*)	Foxtail millet	Tolerance to drought and oxidative stresses	Feng et al. (2016)
SiDof	Foxtail millet	Drought stress	Zhang et al. (2017a, b)
SiASR4	Foxtail millet	Salinity and drought tolerance	Li et al. (2017)
SiPIP	Foxtail millet	Dehydration stress	Singh et al. (2019)
SiWLM2b	Foxtail millet	Drought stress	Yang et al. (2019)
CIPK	Foxtail millet	Abiotic stress	Zhao et al. (2019)
EcDREB2A	Foxtail millet	Heat stress	Singh et al. (2021)

reported that atrazine resistance in foxtail millet was imparted by *psbA* gene which encodes photosystem II D1 protein and this resistance is due to a single amino acid change (glycine instead of serine at 264th position) in the resistant variety. Foxtail *Si69* gene resistance to aluminum was transformed to *Arabidopsis* under CaMV35s promoter and the transgenic *Arabidopsis* showed a significant increase in resistance to aluminum by decreased lipid peroxidation and root damage. Interestingly, the protein which is encoded by *Si69* in foxtail millet shares homology to all other aluminum-induced proteins present in other species (Zhao et al. 2009). Further, DNAj gene in foxtail millet conferred drought and heat tolerance and was tested in four wheat cultivars through the pollen-tube pathway (Wang et al. 2009). This study has paved the mode for creating drought-tolerant wheat lines expressing foxtail millet genes. Phospholipase D also plays an important role in drought tolerance in plants. The *SiPLDα1* overexpression in *Arabidopsis* improved drought resistance in transformed plants which showed increased biomass and relative water content (RWC), reduced electrolytic leakage, and improved survival percentages compared to the control (Peng et al. 2010). In plants, acetyl-coenzyme A carboxylase (*ACCase*) is involved in fatty acid metabolism and catalyzes the carboxylation of acetyl-CoA to malonyl-CoA. Acetyl-coenzyme A carboxylases are important factors for plant sensitivity toward a group of herbicides such as cyclohexanediones (CHDs) and aryloxyphenoxy propionates (APPs). Foxtail *ACCase* resistance to sethoxydim was transformed to maize under maize ubiquitin promoter and the transgenic maize showed a significant increase in resistance to sethoxydim. In addition, the transgenic maize seeds had an increased oil content of about 24% to 65% (Dong et al. 2011). Dehydration-responsive-element-binding (*DREB*) genes in plants code for many transcription factors and signal transduction activities to carry out a myriad of metabolic activities. Dehydration-responsive-element-binding gene in foxtail millet (*SiDREB2*) has been found to have high expression levels during dehydration and salinity stress. This indicates its potential role to tolerate the aforementioned stress factors (Lata et al. 2011). Several transcription factors in plants play their role as mediators for stress tolerance. One of the transcription factors in plants that have this ability is NAC transcription factors. Puranik et al. (2011) reported that salinity and dehydration stresses upregulated the expression of *SiNAC* gene in foxtail millet. Hence, this gene is probably involved in abiotic stress responses.

The transcription factor, *EcNAC1* from finger millet conferred polyethylene glycol and mannitol induced osmotic stress and salt stress tolerance in tobacco through *Agrobacterium*-mediated transformation (Ramegowda et al. 2012). Ascorbate peroxidase (Ec-apx1) was isolated from finger millet and functionally characterized in both drought-resistant and drought-susceptible varieties of finger millet. Moreover, the expression of *Ec-apx1* was noticeably higher in the drought-tolerant variety of finger millet which confirms its role in tolerating drought stress (Bhatt et al. 2013). Plants' response toward drought is controlled by the synergistic action of several genes (Sehgal et al. 2012). Metallothionein, farnesylated protein ATFP6, protein phosphatase 2A, *RISBZ4*, and farnesyl pyrophosphate synthase genes identified in finger millet had a strong role in tolerating drought stress in this millet. Furthermore, the expression of the above-mentioned genes was high during moderate and severe drought stress (Parvathi et al. 2013). C_2H_2zinc-type transcription factor has a vital role in stress tolerance and controlling various signaling mechanisms. C_2H_2 transcription factor from foxtail millet (*SiC₂H₂*) has the potential to provoke tolerance against salinity, dehydration, and cold stress (Muthamilarasan et al. 2014). Dehydrin genes are usually expressed during abiotic stress and overexpression of *EcDehydrin7* gene from finger millet conferred a higher tolerance to drought and heat in transgenic tobacco (Singh et al. 2014). Late embryogenesis abundant (LEA) proteins are generally involved in imparting protection to higher plants from abiotic stresses. A novel LEA gene, *SiLEA14*, from foxtail millet was expressed in *Arabidopsis* and the transgenic plant showed a significant increase in stress tolerance against salinity and osmotic stress (Wang et al. 2014). WD40 proteins generally boost the protein–protein interactions and also enhance the activity of proteins. *SiWD40* was functionally characterized in foxtail millet and there was a significant expression of *SiWD40* gene during different stresses such as salinity, dehydration, abscisic acid, and cold stress (Mishra et al. 2014). The remorin proteins (REMs) play a crucial role during the abiotic stress response. Furthermore, overexpression of *SiREM6* conferred high salinity tolerance in transgenic *Arabidopsis* (Yue et al. 2014). Aldehyde dehydrogenase genes (*ALDH*) from foxtail millet, such as *SiALDH7B1*, *SiALDH12A1*, and *SiALDH18B2* were upregulated during osmotic stress, cold, hydrogen peroxide (H_2O_2), and abscisic acid (ABA). In addition, introducing *SiALDH2B2*, *SiALDH10A2*, *SiALDH5F1*, *SiALDH22A1*, and *SiALDH3E2* into *Escherichia coli* (*E. coli*) improved salt tolerance (Zhu et al. 2014). The DREB (dehydration-responsive element binding)-type transcription factors regulate the expression of stress-inducible genes by binding the DRE/CRT cis-elements in promoter regions. The abscisic acid (ABA)-responsive DREB-binding protein gene (*SiARDP*) was cloned from foxtail millet and the expression of *SiARDP* increased after drought, high salt, and at low temperature in foxtail millet seedlings. In addition, the expression of *SiARDP* in *Arabidopsis* increased drought and salt tolerance during seed germination and seedling development (Li et al. 2014).

The monodehydroascorbate reductase gene (*mdar*) has a profound effect in combating abiotic stresses in plants due to its ability to scavenge reactive oxygen species (ROS). Sudan et al. (2015)studied the role of the monodehydroascorbate reductase gene (*mdar*) in tolerating different abiotic stresses such as drought, salinity, and UV radiation. Furthermore, *mdar* had a profound effect in combating abiotic stresses in finger millet by scavenging ROS. Nuclear factors are yet another important factor for abiotic stress tolerance in plants and overexpression of *SiNF-YA1* increased drought and salt tolerance by stimulating stress-related genes such as *NtERD10* and *NtCAT1* in *Nicotiana tabacum* (tobacco). Another nuclear factor in foxtail millet, *SiNF-YB8* regulated the expression of *NtSOD, NtPOD, NtLEA5*, and *NtERD10* in *Nicotiana tabacam.* Furthermore, both transformation experiments proved that the transgenic plants were able to maintain relative water content (RWC), chlorophyll content, activities of superoxide dismutase (SOD), peroxidase (POD), catalase (CAT), and malondialdehyde (MDA) enzymes under stress conditions (Feng et al. 2015). Several studies show the WRKY transcription factors get upregulated during abiotic stresses such as drought, heat, cold, and salinity. Muthamilarasan et al. (2015) advocated that *SiWRKY066* and *SiWRKY082* are potent genes that can mediate both drought and salinity stress tolerance in foxtail millet and these genes can be further transformed to other crops for increasing the resistance toward abiotic stresses. Autophagy is a biological mechanism commonly seen in eukaryotes for the recycling of nutrients inside the cell. This process is involved during low nitrogen stress in *Arabidopsis*. An autophagy-related gene, *SiATG8a,* from foxtail millet was transformed to *Arabidopsis* and the transgenic plant showed high resistance to drought and nitrogen starvation (Li et al. 2015).

The NAC proteins in plants are vital for their growth and are also reported to play a crucial role during stress tolerance. Overexpression of *EcNAC67*, a transcription factor from finger millet conferred drought and salt tolerance in rice through *Agrobacterium*-mediated transformation (Rahman et al. 2016). SET domain genes regulate the transcription of genes that are involved in diverse physiological and developmental processes. Overexpression of *SiSET14* from foxtail millet conferred cold stress in yeast. Even though the expression of *SiSET14* was higher during late stages in types of stresses (salinity, dehydration, and cold stress) but transgenic yeast harboring *SiSET14* showed a better rate of proliferation during cold stress than other mentioned stresses (Yadav et al. 2016). Argonaute protein 1 encoding gene (AGO1) gene regulates the plant developmental processes and abiotic stress tolerance. The activity of *SiAGOI* was established by mutating this gene (*SiAGOIb*) and it was observed that mutated foxtail millet lines showed a lower degree of growth and drought tolerance compared to wild type (Liu et al. 2016). Abscisic acid stress ripening protein also plays a crucial role in abiotic stress tolerance. Overexpression of *SiASR1* gene from foxtail millet positively regulated drought and oxidative stress in transgenic tobacco. Moreover, this overexpression of *SiASR1* mitigated hydrogen peroxide accumulation and increased the activities of antioxidant enzymes (Feng et al. 2016). Lipid transfer proteins (LTPs), a group of small molecular weight cysteine-rich soluble proteins are involved in the development of flowers and seeds, deposition of cuticular wax also plays a crucial role during abiotic stress responses. The LTP gene (*SiLTP*) from foxtail millet mediated drought and salinity tolerance in transgenic *Arabidopsis* (Pan et al. 2016). Zhang et al. 2017a identified 35 Dof genes and out of these *SiDof7* and *SiDof15* were found to involve in the drought stress signaling mechanism. Li et al. (2017) demonstrated that transformation of *Arabidopsis* with *SiASR4* gene and overexpression of *SiASR4* in foxtail millet revealed increased tolerance to drought and salt stresses. Furthermore, genes associated with stress and ROS scavenging activity were triggered in transgenic plants. A novel endoplasmic reticulum (ER) membrane-tethered bZIP transcription factor from finger millet, *EcbZIP17* was transformed to tobacco plants and the transgenic plants exhibited better growth and yield. Moreover, the transgenic plants showed a better plant architecture under the influence of abiotic stresses (250 mM NaCl, 10% PEG6000, 400 mM mannitol) (Ramakrishna et al. 2018). Cis-element analysis showed that CBL-Interacting protein kinase (CIPK) in foxtail millet can confer abiotic stress tolerance (Zhao et al. 2019). Aquaporins are proteins that play a crucial role in biological and molecular functions in plants. The small basic intrinsic proteins, *SiPIP3;1* and *SiSIP1;1* were differentially expressed in two cultivars of foxtail millet in response to stress treatments. Overexpression of these genes in yeast also confirmed its role to tolerate dehydration as well as salt stress (Singh et al. 2019). Many studies have been reported about LIM proteins which are key factors in carrying out various cellular activities such as controlling gene expression, production of the cytoskeleton, signal transduction, and regulation of metabolic activities. Overexpression of *SiWLIM2b* from foxtail millet increased drought tolerance in transformed rice, and the transgenic rice had a greater rate of survival, higher RWC, and reduced cell damage than the control (Yang et al. 2019). Recently, overexpression of *EcDREB2A* transcription factor from finger millet in tobacco increased tolerance to heat stress through ROS scavenging activity (Singh et al. 2021). The above-mentioned studies demonstrate that these genes are a potential candidate for improving agronomic traits in crops, especially in those crops which are highly susceptible to abiotic stresses. However, much literature is available about foxtail millet and finger millet, but other small millets are yet to be explored to reveal more potential genes for both biotic and abiotic stress-tolerant studies.

8.8 Conclusion and Future Perspective

Small millets are nutritionally rich crops and can serve as a key source of sustenance, especially for the economically backward community in developing and underdeveloped countries. The development and improvement of stress-tolerant millets is still a challenging task due to inadequate genomic resources. The availability of the genome sequence of foxtail millet, finger millet will be highly beneficial to the upcoming research pertaining to biotic and abiotic stress tolerance studies in minor millets. The candidate genes for stress tolerance isolated from small millets can be introduced in other crops which are very susceptible to environmental stresses. CRISPR/ Cas9-based genome editing tool can also be applied extensively in this crop to enhance its natural capacity of tolerance to a wide range of stress conditions. Research on other neglected minor millets such as proso millet, barnyard millet, little millet, kodo millet, tef millet, and fonio millet, which are also equally nutritional should be given priority. Currently, only some genes have been used for the transgenic approach for the development of stress tolerance transgenic millet. More potent genes have to be identified, characterized, and transformed to unravel the full potential of small millets in maintaining food security.

References

Aidoo MK, Bdolach E, Fait A et al (2016) Tolerance to high soil temperature in foxtail millet (*Setaria italica* L.) is related to shoot and root growth and metabolism. Plant Physiol Biochem 106:73–81. https://doi.org/10.1016/j.plaphy.2016.04.038

Alex N, Cecilia M, Mathew N et al (2018) Efficient plant regeneration protocol for finger millet [*Eleusine coracana* (L.) Gaertn.] via somatic embryogenesis. African J Biotechnol 17:660–667. https://doi.org/10.5897/AJB2018.16452

Amir G, Romee J, Gulzar A et al (2014) Significance of finger millet in nutrition, health and value added products: A review. J Environ Sci Comput Sci Eng Technol 3:1601–1608

Anjaneyulu E, Hemalatha S, Raj SB, Balaji M (2011) Callus induction and plant regeneration in finger millet (Eleusine coracana L.). Libyan Agric Res Cent J Int 2:57–61

Anjaneyulu E, Reddy PS, Sunita MS et al (2014) Salt tolerance and activity of antioxidative enzymes of transgenic finger millet overexpressing a vacuolar H+-pyrophosphatase gene (SbVPPase) from *Sorghum bicolor*. J Plant Physiol 171:789–798. https://doi.org/10.1016/j.jplph.2014.02.001

Anju C, Rabindran R, Velazhahan R, Ravikesavan R (2016) Callusing and regeneration in finger millet [*Eleusine coracana* (L.) Gaertn.]. Res J Agric Sci 7:324–329

Antony Ceasar S, Ignacimuthu S (2008) Efficient somatic embryogenesis and plant regeneration from shoot apex explants of different Indian genotypes of finger millet (*Eleusine coracana* (L.) Gaertn.). Vitr Cell Dev Biol - Plant 44:427–435. https://doi.org/10.1007/s11627-008-9153-y

Antony Ceasar S, Ignacimuthu S (2011) *Agrobacterium*-mediated transformation of finger millet (Eleusine coracana (L.) Gaertn.) using shoot apex explants. Plant Cell Rep 30:1759–1770. https://doi.org/10.1007/s00299-011-1084-0

Arockiasamy S, Prakash S, Ignacimuthu S (2001) High regenerative nature of *Paspalum scrobiculatum* L, an important millet crop. Curr Sci 80:496–498

Asharani VT, Jayadeep A, Malleshi NG (2010) Natural Antioxidants in Edible Flours of Selected Small Millets. Int J Food Prop 13:41–50. https://doi.org/10.1080/10942910802163105

Assefa K, Gaj M, Maluszynski M (1998) Somatic embryogenesis and plant regeneration in callus culture of tef, *Eragrostis tef* (Zucc.). Plant Cell Rep 18:154–158

Asthir B (2015) Mechanisms of heat tolerance in crop plants. Biol Plant 59:620–628. https://doi.org/10.1007/s10535-015-0539-5

Atul Babu G, Vinoth A, Ravindhran R (2018) Direct shoot regeneration and genetic fidelity analysis in finger millet using ISSR markers. Plant Cell, Tissue Organ Cult 132:157–164. https://doi.org/10.1007/s11240-017-1319-z

Ayele M, Blum A, Nguyen HT (2001) Diversity for osmotic adjustment and root depth in TEF [*Eragrostis tef* (Zucc) Trotter]. Euphytica 121:237–249. https://doi.org/10.1023/A:1012099914738

Babu A, Geetha K, Manjunatha V, Shankar A (2012) An efficient high throughput plant regeneration and transformation protocol for production of transgenics tolerant to salt in finger millet. Int J Crop Improv 3:6–20

Bajaj YPS, Sidhu BS, Dubey VK (1981) Regeneration of genetically diverse plants from tissue cultures of forage grass — Panicum sps. Euphytica 30:135–140. https://doi.org/10.1007/BF00033669

Bayer GY, Yemets AI, Blume YB (2014) Obtaining the transgenic lines of finger millet *Eleusine coracana* (L.). with dinitroaniline resistance. Cytol Genet 48:139–144. https://doi.org/10.3103/S0095452714030025

Bekele E, Klock G, Zimmermann U (1995) Somatic embryogenesis and plant regeneration from leaf and root explants and from seeds of *Eragrostis tef* (Gramineae). Hereditas 123:183–189

Bellato S, Ciccoritti R, Del Frate V et al (2013) Influence of genotype and environment on the content of 5-n alkylresorcinols, total phenols and on the antiradical activity of whole durum wheat grains. J Cereal Sci 57:162–169. https://doi.org/10.1016/j.jcs.2012.11.003

Benson EE (2000) Sepecial symposium: In vitro plant recalcitrance in vitro plant recalcitrance: An introduction. Vitr Cell Dev Biol - Plant 36:141–148. https://doi.org/10.1007/s11627-000-0029-z

Berry PM, Sterling M, Spink JH et al (2004) Understanding and Reducing Lodging in Cereals. Adv Agron 84:217–271

Bhatt D, Saxena SC, Jain S et al (2013) Cloning, expression and functional validation of drought inducible ascorbate peroxidase (Ec-apx1) from *Eleusine coracana*. Mol Biol Rep 40:1155–1165. https://doi.org/10.1007/s11033-012-2157-z

Bhatt R, Asopa PP, Jain R et al (2021) Optimization of *Agrobacterium* Mediated Genetic Transformation in *Paspalum scrobiculatum* L. (Kodo Millet). Agronomy 11:1104. https://doi.org/10.3390/agronomy11061104

Billimoria K, Hegde R (1971) A new bacterial disease of ragi, *Eleusine coracana* (Linn.) Gaertn in Mysore state. Curr Sci 40:611–612

Cannarozzi G, Weichert A, Schnell M et al (2018) Waterlogging affects plant morphology and the expression of key genes in tef (*Eragrostis tef*). Plant Direct 2:e00056. https://doi.org/10.1002/pld3.56

Ceasar SA, Baker A, Ignacimuthu S (2017) Functional characterization of the PHT1 family transporters of foxtail millet with development of a novel *Agrobacterium*-mediated transformation procedure. Sci Rep 7:14064. https://doi.org/10.1038/s41598-017-14447-0

Ceasar SA, Ignacimuthu S (2010) Effects of cytokinins, carbohydrates and amino acids on induction and maturation of somatic embryos in kodo millet (*Paspalum scorbiculatum* Linn.). Plant Cell, Tissue Organ Cult 102:153–162. https://doi.org/10.1007/s11240-010-9716-6

Chandra AK, Chandora R, Sood S, Malhotra N (2021) Global production, demand, and supply. In: Millets and Pseudo Cereals. Elsevier, pp 7–18

Chandrasekara A, Shahidi F (2011) Determination of antioxidant activity in free and hydrolyzed fractions of millet grains and characterization of their phenolic profiles by HPLC-DAD-ESI-MSn. J Funct Foods 3:144–158. https://doi.org/10.1016/j.jff.2011.03.007

Changmei S, Dorothy J (2014) Millet-the frugal grain. Int J Sci Res Rev 3:75–90

Comas LH, Becker SR, Cruz VMV et al (2013) Root traits contributing to plant productivity under drought. Front Plant Sci 4. https://doi.org/10.3389/fpls.2013.00442

Coulibaly A, Kouakou B, Chen J (2010) Phytic acid in cereal grains: Structure, healthy or harmful ways to reduce phytic acid in cereal grains and their effects on nutritional quality. Am J Plant Nutr Fertil Technol 1:1–22. https://doi.org/10.3923/ajpnft.2011.1.22

Daugherty CJ, Musgrave ME (1994) Characterization of populations of rapid-cycling *Brassica rapa* L. selected for differential waterlogging tolerance. J Exp Bot 45:385–392. https://doi.org/10.1093/jxb/45.3.385

Debieu M, Sine B, Passot S et al (2018) Response to early drought stress and identification of QTLs controlling biomass production under drought in pearl millet. PLoS One 13:e0201635. https://doi.org/10.1371/journal.pone.0201635

van Delden SH, Vos J, Ennos AR, Stomph TJ (2010) Analysing lodging of the panicle bearing cereal teff (Eragrostis tef). New Phytol 186:696–707. https://doi.org/10.1111/j.1469-8137.2010.03224.x

Desai S, Thirumalachar M, Patel M (1965) Bacterial blight disease of *Eleusine caracana* Gaertn. Indian Phytopath 28:384–386

Desta B, Amare G (2021) Paclobutrazol as a plant growth regulator. Chem Biol Technol Agric 8:1. https://doi.org/10.1186/s40538-020-00199-z

Devi PB, Vijayabharathi R, Sathyabama S et al (2014) Health benefits of finger millet (Eleusine coracana L.) polyphenols and dietary fiber: a review. J Food Sci Technol 51:1021–1040. https://doi.org/10.1007/s13197-011-0584-9

Diao X, Chen Z, Duan S et al (1999) Factors influencing foxtail millet embryogenic calli transformation by particle bombardment. Acta Agric Boreali Sin 14:31–36

Divya K, Kalarani M, Jeyakumar P et al (2019) Screening for drought and heat tolerance in foxtail millet by physiological and biochemical indices. Pharma Innov J 8:665–669

Dong Y, Duan S (1999) Establishment of embryogenic cell suspension culture and plant regeneration of millet and gene transfer. J Basic Sci Eng 7:34–40

Dong Y, Duan S (2000) Production of transgenic millet plants via particle bombardment. Acta Bot Boreal-Occident Sin 20:175–178

Dong Z, Zhao H, He J et al (2011) Overexpression of a foxtail millet Acetyl-CoA carboxylase gene in maize increases sethoxydim resistance and oil content. African J Biotechnol 10:3986–3995

Dosad S, Chawla HS (2015a) In vitro plant regeneration from mature seeds of finger millet (*Eleusine coracana*) through somatic embryogenesis. Indian J Plant Physiol 20:360–367. https://doi.org/10.1007/s40502-015-0191-2

Dosad S, Chawla HS (2015b) Optimization of different doses of growth regulators for in vitro regeneration of *Echinochloa frumentacea* Roxb. from caryopsis. Indian J Plant Physiol 20:339–344. https://doi.org/10.1007/s40502-015-0184-1

Dosad S, Chawla HS (2016) In vitro plant regeneration and transformation studies in millets: current status and future prospects. Indian J Plant Physiol 21:239–254. https://doi.org/10.1007/s40502-016-0240-5

Eapen S, George L (1990) Influence of phytohormones, carbohydrates, aminoacids, growth supplements and antibiotics on somatic embryogenesis and plant differentiation in finger millet. Plant Cell Tissue Organ Cult 22:87–93. https://doi.org/10.1007/BF00043683

Erdmann B, Wiedenroth EM (1986) (1986) Changes in the root system of wheat seedlings following root anaerobiosis II. Morphology and anatomy of evolution forms. Ann Botany 58(5):607–616

Feng Z-J, He G-H, Zheng W-J et al (2015) Foxtail millet NF-Y families: genome-wide survey and evolution analyses identified two functional genes important in abiotic stresses. Front Plant Sci 6. https://doi.org/10.3389/fpls.2015.01142

Feng Z-J, Xu Z-S, Sun J et al (2016) Investigation of the ASR family in foxtail millet and the role of ASR1 in drought/oxidative stress tolerance. Plant Cell Rep 35:115–128. https://doi.org/10.1007/s00299-015-1873-y

Gebre E, Gugsa L, Schlüter U, Kunert K (2013) Transformation of tef by *Agrobacterium* through immature embryo regeneration system for inducing semi-dwarfism. South African J Bot 87:9–17. https://doi.org/10.1016/j.sajb.2013.03.004

Gugsa L, Kumlehn J (2011) Somatic embryogenesis and massive shoot regeneration from immature embryo explants of tef. Biotechnol Res Int 2011:1–7. https://doi.org/10.4061/2011/309731

Gugsa L, Sarial AK, Lörz H, Kumlehn J (2006) Gynogenic plant regeneration from unpollinated flower explants of *Eragrostis tef* (Zuccagni) Trotter. Plant Cell Rep 25:1287–1293. https://doi.org/10.1007/s00299-006-0200-z

Gull A, Jan R, Nayik GA et al (2014) Significance of finger millet in nutrition, health and value added products: a review. Magnesium (mg) 3:1601–1608

Gupta P, Raghuvanshi S, K Tyagi a (2001) Assessment of the efficiency of various gene promoters via biolistics in leaf and regenerating seed callus of millets, *Eleusine coracana* and *Echinochloa crusgalli.* Plant Biotechnol 18:275–282. https://doi.org/10.5511/plantbiotechnology.18.275

Habiyaremye C, Matanguihan JB, D'Alpoim Guedes J, Ganjyal GM, Whiteman MR, Kidwell KK, Murphy KM (2017) Proso millet (Panicum miliaceum L.) and its potential for cultivation in the Pacific Northwest, US: A review. Front Plant Sci 7:1961

Hassan ZM, Sebola NA, Mabelebele M (2021) The nutritional use of millet grain for food and feed: a review. Agric and Food Secur. 16(10):1–4

Hema R, Vemanna RS, Sreeramulu S et al (2014) Stable expression of mtlD gene imparts multiple stress tolerance in finger millet. PLoS One 9:e99110. https://doi.org/10.1371/journal.pone.0099110

Herdrich N (2001) Grower experiences with millet in eastern washington. Washington State University, Pullman,WA, pp 1997–1999

Heyser J (1984) Callus and shoot regeneration from protoplasts of proso millet (*Panicum miliaceum* L.). Z. Z Pflanzenphysiol Bd 113:293–299

Heyser JW, Nabors MW (1982) Regeneration of proso millet from embryogenic calli derived from various plant parts 1. Crop Sci 22:1070–1074. https://doi.org/10.2135/cropsci1982.0011183X002200050043x

Hossain A, Uddin S (2011) Mechanisms of waterlogging tolerance in wheat: Morphological and metabolic adaptations under hypoxia or anoxia. Aust J Crop Sci 5:1094–1101

Huang B, Johnson JW, Box JE, NeSmith DS (1997) Root characteristics and hormone activity of wheat in response to hypoxia and ethylene. Crop Sci 37:812–818. https://doi.org/10.2135/cropsci1997.0011183X003700030020x

Huang B, Johnson JW, NeSmith DS, Bridges DC (1994) Root and shoot growth of wheat genotypes in response to hypoxia and subsequent resumption of aeration. Crop Sci 34:1538–1544. https://doi.org/10.2135/cropsci1994.0011183X003400060023x

Hulse J, Laing E, Pearson O (1980) Sorghum and the millets: their composition and nutritive value. Academic Press, pp 187–193

Ignacimuthu S, Ceasar SA (2012) Development of transgenic finger millet (*Eleusine coracana* (L.) Gaertn.) resistant to leaf blast disease. J Biosci 37:135–147. https://doi.org/10.1007/s12038-011-9178-y

Ito K, Ozasa H, Noda Y et al (2007) Effects of free radical scavenger on acute liver injury induced by d-galactosamine and lipopolysaccharide in rats. Hepatol Res 38(2):194–201. https://doi.org/10.1111/j.1872-034X.2007.00252.x

Jackson M, Armstrong W (1999) Formation of aerenchyma and the processes of plant ventilation in relation to soil flooding and submergence. Plant Biol 1:274–287. https://doi.org/10.1055/s-2007-978516

Jagga-Chugh S, Kachhwaha S, Sharma M et al (2012) Optimization of factors influencing microprojectile bombardment-mediated genetic transformation of seed-derived callus and regeneration of transgenic plants in *Eleusine coracana* (L.) Gaertn. Plant Cell Tissue Organ Cult 109:401–410. https://doi.org/10.1007/s11240-011-0104-7

Jain S, Varshney A, Kothari SL (2001) Embryogenic callus induction and efficient plant regeneration in proso millet. Cereal Res Commun 29:313–320. https://doi.org/10.1007/BF03543676

Jamra G, Shah P, Agarwal A et al (2021) Endogenous phytonutrient, phytochemical, and phytohormone levels modulate in-vitro callus induction and plant regeneration in finger millet (*Eleusine coracana*) genotypes. Plant Biosyst - An Int J Deal with all Asp Plant Biol:1–10. https://doi.org/10.1080/11263504.2021.1918779

Jayasudha B, Sushma A, Prashantkumar H, Sashidha V (2014a) An efficient in-vitro agrobacterium -mediated transformation protocol for raising salinity tolerant transgenic plants in fingermillet [*Eleusine coracana* (L.) Gaertn.]. Plant Arch 14:823–829

Jayasudha B, Sushma A, Prashantkumar H, Sashidhar V (2014b) An efficient in-vitro Agrobacterium-mediated transformation protocol for raising salinity tolerant transgenic finger millet (*Eleusine coracana* (L.) Gaertn). Plant Arch 14823–829 14:823–829

Jency JP, Rajasekaran R, Singh RK et al (2020) Induced mutagenesis enhances lodging resistance and photosynthetic efficiency of kodomillet (*Paspalum Scrobiculatum*). Agronomy 10:227. https://doi.org/10.3390/agronomy10020227

Jia X, Yuan J, Shi Y et al (2007) A Ser–Gly substitution in plastid-encoded photosystem II D1 protein is responsible for atrazine resistance in foxtail millet (Setaria italica). Plant Growth Regul 52:81–89. https://doi.org/10.1007/s10725-007-9181-3

Jöst M, Esfeld K, Burian A et al (2015) Semi-dwarfism and lodging tolerance in tef (Eragrostis tef) is linked to a mutation in the α-Tubulin 1 gene. J Exp Bot 66:933–944. https://doi.org/10.1093/jxb/eru452

Kafi M, Zamani G, Ghoraishi G (2009) Relative salt tolerance of south Khorasan millets. Desert 14: 63–70

Kamara MT, Huiming Z, Kexue Z et al (2009) Comparative study of chemical composition and physicochemical properties of two varieties of defatted foxtail millet flour grown in China. Am J Food Technol 4:255–267. https://doi.org/10.3923/ajft.2009.255.267

Kaur P, Kothari SL (2004) In vitro culture of kodo millet: Influence of 2,4-D and picloram in combination with kinetin on callus initiation and regeneration. Plant Cell Tissue Organ Cult 77: 73–79. https://doi.org/10.1023/B:TICU.0000016505.20448.44

Kothari SL, Agarwal K, Kumar S (2004) Inorganic nutrient manipulation for highly improved in vitro plant regeneration in finger millet—*Eleusine coracana* (L.) Gaertn. Vitr Cell Dev Biol - Plant 40:515–519. https://doi.org/10.1079/IVP2004564

Kothari-Chajer A, Sharma M, Kachhwaha S, Kothari SL (2008) Micronutrient optimization results into highly improved in vitro plant regeneration in kodo (Paspalum scrobiculatum L.) and finger (*Eleusine coracana* (L.) Gaertn.) millets. Plant Cell Tissue Organ Cult 94:105–112. https://doi.org/10.1007/s11240-008-9392-y

Kumar S, Agrawal K, Kothari S (2001) In vitro induction and enlargement of apical domes and formation of multiple shoots in finger millet, Eleusine coracana (L.) Gaertn and crowfoot grass, *Eleusine indica* (L.) Gaertn. Curr Sci 18:1482–1485

Kumar V, Parvatam G, Ravishankar GA (2009) AgNO3 - a potential regulator of ethylene activity and plant growth modulator. Electron J Biotechnol 12:8–9. https://doi.org/10.2225/vol12-issue2-fulltext-1

Kusaka M, Lalusin AG, Fujimura T (2005) The maintenance of growth and turgor in pearl millet (*Pennisetum glaucum* [L.] Leeke) cultivars with different root structures and osmo-regulation under drought stress. Plant Sci 168:1–14. https://doi.org/10.1016/j.plantsci.2004.06.021

Lata C (2015) Advances in omics for enhancing abiotic stress tolerance in millets. Proc Indian Natl Sci Acad 81. https://doi.org/10.16943/ptinsa/2015/v81i2/48095

Lata C, Bhutty S, Bahadur RP et al (2011) Association of an SNP in a novel DREB2-like gene SiDREB2 with stress tolerance in foxtail millet [*Setaria italica* (L.)]. J Exp Bot 62:3387–3401. https://doi.org/10.1093/jxb/err016

Latha AM, Rao KV, Reddy VD (2005) Production of transgenic plants resistant to leaf blast disease in finger millet (Eleusine coracana (L.) Gaertn.). Plant Sci 169:657–667. https://doi.org/10.1016/j.plantsci.2005.05.009

Li C, Yue J, Wu X et al (2014) An ABA-responsive DRE-binding protein gene from Setaria italica, SiARDP, the target gene of SiAREB, plays a critical role under drought stress. J Exp Bot 65: 5415–5427. https://doi.org/10.1093/jxb/eru302

Li J, Dong Y, Li C et al (2017) SiASR4, the target gene of SiARDP from Setaria italica, improves abiotic stress adaption in plants. Front Plant Sci 7. https://doi.org/10.3389/fpls.2016.02053

Li W, Chen M, Zhong L et al (2015) Overexpression of the autophagy-related gene SiATG8a from foxtail millet (*Setaria italica* L.) confers tolerance to both nitrogen starvation and drought stress in *Arabidopsis*. Biochem Biophys Res Commun 468:800–806. https://doi.org/10.1016/j.bbrc.2015.11.035

Linkemer G, Board JE, Musgrave ME (1998) Waterlogging effects on growth and yield components in late-planted soybean. Crop Sci 38:1576–1584. https://doi.org/10.2135/cropsci1998.0011183X003800060028x

Liu X, Tang S, Jia G et al (2016) The C-terminal motif of SiAGO1b is required for the regulation of growth, development and stress responses in foxtail millet (*Setaria italica* (L.) P. Beauv). J Exp Bot 67:3237–3249. https://doi.org/10.1093/jxb/erw135

Liu Y, Feng X, Xu Y et al (2009) Overexpression of millet ZIP-like gene (SiPf40) affects lateral bud outgrowth in tobacco and millet. Plant Physiol Biochem 47:1051–1060. https://doi.org/10.1016/j.plaphy.2009.08.007

Liu Y, Yu J, Ao G, Zhao Q (2007) Factors influencing *Agrobacterium*- mediated transformation of foxtail millet (Setaria italica). Chin J Biochem Mol Biol 23:531–536

Liu Y, Yu J, Zhao Q et al (2005) Genetic transformation of millet (*Setaria italica)* by Agrobacterium-mediated. Chin J Agr Biotechnol 13:32–37

Mahalakshmi S, Christopher GSB, Reddy TP et al (2006) Isolation of a cDNA clone (PcSrp) encoding serine-rich-protein from *Porteresia coarctata* T. and its expression in yeast and finger millet (Eleusine coracana L.) affording salt tolerance. Planta 224:347–359. https://doi.org/10.1007/s00425-005-0218-4

Malathi B, Appaji C, Rajender Reddy G et al (2016) Growth pattern of millets in India. Indian. J Agric Res 50:382–386. https://doi.org/10.18805/ijare.v50i4.11257

Maqsood M, Ali S (2007) Effects of drought on growth, development, radiation use efficiency and yield of finger millet (*Eleucine coracana*). Pak J Bot 39:123

Maramorosch K, Govindu H, Kondo F (1977) Rhabdo virus particles associated with mosaic disease of naturally infected *Eleusine coracana* (finger millet) in Karnataka state (Mysore) South India. Plant Dis Rep 61:1029–1031

Matsuura A, An P, Murata K, Inanaga S (2016) Effect of pre- and post-heading waterlogging on growth and grain yield of four millets. Plant Prod Sci 19:348–359. https://doi.org/10.1080/1343943X.2016.1146907

Mekbib F, Mantell SH, Buchanan-Wollaston V (1997) Callus induction and in vitro regeneration of tef [*Eragrostis tef* (Zucc.) Trotter] from leaf. J Plant Physiol 151:368–372. https://doi.org/10.1016/S0176-1617(97)80267-2

Mengistu D, Mekonnen LS (2012) Integrated agronomic crop managements to improve tef productivity under terminal drought. In: Water Stress. InTech, pp 235–254

Mgonja MA, Lenne JM, Manyasa E, Sreenivasaprasad S (2007) Finger millet blast management in East Africa. Creating opportunities for improving production and utilization of finger millet. In: Proceedings of the First International Finger Millet Stakeholder Workshop, Projects R8030 & R8445 UK Department for Inter, p 196

Mishra AK, Muthamilarasan M, Khan Y et al (2014) Genome-wide investigation and expression analyses of WD40 protein family in the model plant foxtail millet (*Setariaitalica* L.). PLoS One 9:e86852. https://doi.org/10.1371/journal.pone.0086852

Mohanty BD, Gupta SD, Ghosh PD (1985) Callus initiation and plant regeneration in ragi (*Eleusine coracana* Gaertn.). Plant Cell Tissue Organ Cult 5:147–150. https://doi.org/10.1007/BF00040311

Muthamilarasan M, Bonthala VS, Khandelwal R et al (2015) Global analysis of WRKY transcription factor superfamily in Setaria identifies potential candidates involved in abiotic stress signaling. Front Plant Sci 6. https://doi.org/10.3389/fpls.2015.00910

Muthamilarasan M, Bonthala VS, Mishra AK et al (2014) C2H2 type of zinc finger transcription factors in foxtail millet define response to abiotic stresses. Funct Integr Genomics 14:531–543. https://doi.org/10.1007/s10142-014-0383-2

Muthamilarasan M, Dhaka A, Yadav R, Prasad M (2016) Exploration of millet models for developing nutrient rich graminaceous crops. Plant Sci 242:89–97. https://doi.org/10.1016/j.plantsci.2015.08.023

Nabors MW, Heyser JW, Dykes TA, DeMott KJ (1983) Long-duration, high-frequency plant regeneration from cereal tissue cultures. Planta 157:385–391. https://doi.org/10.1007/BF00397195

Nagaraja A, Das IK (2016) Disease resistance in pearl millet and small millets. In: Biotic Stress Resistance in Millets. Elsevier, pp 69–104

Nagaraja A, Mantur SG (2007) Screening of Eleusine coracana germplasm for blast resistance. J Mycopathol Res 45:66–68

Nagaraju VS, Reddy H, Lucy CK (1982) Ragi streak a leaf hopper transmitted virus disease in Karnataka. Mysore J Agric Sci 16:301–305

Nayak P, Sen S (1989) Plant regeneration through somatic embryogenesis from suspension cultures of a minor millet, Paspalum scrobiculatum. Plant Cell Rep 8:296–299

Nayak P, Sen SK (1991) Plant regeneration through somatic embryogenesis from suspension culture-derived protoplasts of *Paspalum scrobiculatum* L. Plant Cell Rep 10. https://doi.org/10.1007/BF00193160

Nethra N, Gowda R, Gowda PHR (2009) Influence of culture medium on callus proliferation and morphogenesis in finger millet.InNew approaches to plant breeding of orphan crops in Africa. In: Proceedings of an International Conference, Bern, Switzerland, 19-21 September 2007, pp 167–178

Ni X-L, Gui M-Y, Tan L-L et al (2019) Programmed cell death and aerenchyma formation in water-logged sunflower stems and its promotion by Ethylene and ROS. Front Plant Sci 9. https://doi.org/10.3389/fpls.2018.01928

Nithiyanantham S, Kalaiselvi P, Mahomoodally MF et al (2019) Nutritional and functional roles of millets—A review. J Food Biochem 43. https://doi.org/10.1111/jfbc.12859

Ntui VO, Azadi P, Supaporn H, Mii M (2010) Plant regeneration from stem segment-derived friable callus of "Fonio" (*Digitaria exilis* (L.) Stapf.). Sci Hortic (Amsterdam) 125:494–499. https://doi.org/10.1016/j.scienta.2010.04.017

Numan M, Serba DD, Ligaba-Osena A (2021) Alternative strategies for multi-stress tolerance and yield improvement in Millets. Genes (Basel) 12:739. https://doi.org/10.3390/genes12050739

O'Kennedy MM, Burger JT, Botha FC (2004) Pearl millet transformation system using the positive selectable marker gene phosphomannose isomerase. Plant Cell Rep 22:684–690. https://doi.org/10.1007/s00299-003-0746-y

Obilana AB, Manyasa E (2002) In: Belt PS (ed) Pseudo cereals and less common cereals: Grain properties and utilization potential. Springer-Verlag, New York, pp 177–217

Opole R. (2012) Effect of environmental stress and management on grain and biomass yield of Finger Millet (*Eleusine coracana* (L.) Gaertn.) Kansas State University, PhD Thesis.

Opole RA, Prasad PV, Djanaguiraman M et al (2018) Thresholds, sensitive stages and genetic variability of finger millet to high temperature stress. J Agron Crop Sci 204:477–492. https://doi.org/10.1111/jac.12279

Osuna-Avila P, Nava-Cedillo A, Jofre-Garfias AE, Cabrera-Ponce JL (1995) Plant regeneration from shoot apex explants of foxtail millet. Plant Cell Tissue Organ Cult 40:33–35. https://doi.org/10.1007/BF00041115

Pan Y, Li J, Jiao L et al (2016) A non-specific *Setaria italica* lipid transfer protein gene plays a critical role under abiotic stress. Front Plant Sci 7. https://doi.org/10.3389/fpls.2016.01752

Pan Y, Ma X, Liang H et al (2015) Spatial and temporal activity of the foxtail millet (*Setaria italica*) seed-specific promoter pF128. Planta 241:57–67. https://doi.org/10.1007/s00425-014-2164-5

Pande A, Dosad S, Chawla HS, Arora S (2015) In-vitro organogenesis and plant regeneration from seed-derived callus cultures of finger millet (*Eleusine coracana*). Brazilian J Bot 38:19–23. https://doi.org/10.1007/s40415-014-0102-1

Park K-O, Ito Y, Nagasawa T et al (2008) Effects of dietary korean proso-millet protein on plasma adiponectin, hdl cholesterol, insulin levels, and gene expression in obese type 2 diabetic mice. Biosci Biotechnol Biochem 72:2918–2925. https://doi.org/10.1271/bbb.80395

Parvathi MS, Nataraja KN, Yashoda BK et al (2013) Expression analysis of stress responsive pathway genes linked to drought hardiness in an adapted crop, finger millet (*Eleusine coracana*). J Plant Biochem Biotechnol 22:193–201. https://doi.org/10.1007/s13562-012-0135-0

Pathak H (2013) Role of millets in nutritional security of India. National Academy of Agricultural Sciences, New Delhi, pp 1–16

Pathak RK, Gupta A, Shukla R, Baunthiyal M (2018) Identification of new drug-like compounds from millets as Xanthine oxidoreductase inhibitors for treatment of Hyperuricemia: A molecular docking and simulation study. Comput Biol Chem 76:32–41. https://doi.org/10.1016/j.compbiolchem.2018.05.015

Patil S, Sawardekar S, Bhave S et al (2009) (2009). Development of somaclones and their genetic diversity analysis through RAPD in finger millet (*Eleusine coracana* L. Gaertn.), Indian J. Genet., 69: 132-139. Indian J Genet 69:132–139

Peng Y, Zhang J, Cao G et al (2010) Overexpression of a PLDα1 gene from Setaria italica enhances the sensitivity of *Arabidopsis* to abscisic acid and improves its drought tolerance. Plant Cell Rep 29:793–802. https://doi.org/10.1007/s00299-010-0865-1

Perumal A (2017) Production and consumption of minor millets in India- A structural break analysis. Ann Agric Res New Ser 38:1–8

Piñera-Chavez FJ, Berry PM, Foulkes MJ et al (2016) Avoiding lodging in irrigated spring wheat. I. Stem and root structural requirements. F Crop Res 196:325–336. https://doi.org/10.1016/j.fcr.2016.06.009

Pius J, Eapen S, George L et al (1999) Performance of plants regenerated through somatic embryogenesis in finger millet (*Eleusine coracana* Gaertn.). Trop Agric gesentch £xtension 2: 87–90

Pius J, George L, Eapen S, Rao PS (1994) Influence of genotype and phytohormones on somatic embryogenesis and plant regeneration in Finger millet. Proc Indian natninal Sci Acad 60:53–56

Plaza-Wüthrich S, Tadele Z (2013) Regeneration and transformation studies on Tef. In: Achievements and prospects of Tef improvement, pp 67–79

Poddar K, Vishnoi R, Kothari S (1997) Plant regeneration from embryogenic callus of finger millet Eleusine coracana (L.) Gaertn. on higher concentrations of NH4NO3 as a replacement of NAA in the medium. Plant Sci 129:101–106. https://doi.org/10.1016/S0168-9452(97)00183-0

Polash MA, Sakil M, Hossain M (2019) Plants responses and their physiological and biochemical defense mechanisms against salinity: A review. Trop Plant Res 6:250–274. https://doi.org/10.22271/tpr.2019.v6.i2.35

Pradeep PM, Sreerama YN (2017) Soluble and bound phenolics of two different millet genera and their milled fractions: Comparative evaluation of antioxidant properties and inhibitory effects on starch hydrolysing enzyme activities. J Funct Foods 35:682–693. https://doi.org/10.1016/j.jff.2017.06.033

Pradeep SR, Guha M (2011) Effect of processing methods on the nutraceutical and antioxidant properties of little millet (*Panicum sumatrense*) extracts. Food Chem 126:1643–1647. https://doi.org/10.1016/j.foodchem.2010.12.047

Puranik S, Bahadur RP, Srivastava PS, Prasad M (2011) Molecular cloning and characterization of a membrane associated NAC family gene, SiNAC from Foxtail Millet [*Setaria italica* (L.) P. Beauv.]. Mol Biotechnol 49:138–150. https://doi.org/10.1007/s12033-011-9385-7

Qin FF, Zhao Q, Ao GM, Yu JJ (2008) Co-suppression of Si401, a maize pollen specific Zm401 homologous gene, results in aberrant anther development in foxtail millet. Euphytica 163:103–111. https://doi.org/10.1007/s10681-007-9610-4

Rahman H, Ramanathan V, Nallathambi J et al (2016) Over-expression of a NAC 67 transcription factor from finger millet (*Eleusine coracana* L.) confers tolerance against salinity and drought stress in rice. BMC Biotechnol 16:35. https://doi.org/10.1186/s12896-016-0261-1

Rajak K, Sprae S, Rathour S et al (2018) To Study the Indirect Plant Regeneration of Two Cultivars in Barnyard Millet using Different Combination of Plant Growth Regulators and Compared between Superior Genotype for in-vitro Culture. Int J Curr Microbiol Appl Sci 51:2055–2061

Ramakrishna C, Singh S, Raghavendrarao S et al (2018) The membrane tethered transcription factor EcbZIP17 from finger millet promotes plant growth and enhances tolerance to abiotic stresses. Sci Rep 8:2148. https://doi.org/10.1038/s41598-018-19766-4

Ramakrishnan M, Ceasar SA, Vinod KK et al (2017) Identification of putative QTLs for seedling stage phosphorus starvation response in finger millet (*Eleusine coracana* L. Gaertn.) by association mapping and cross species synteny analysis. PLoS One 12:e0183261. https://doi.org/10.1371/journal.pone.0183261

Ramegowda V, Senthil-Kumar M, Nataraja KN et al (2012) Expression of a Finger Millet Transcription Factor, EcNAC1, in Tobacco Confers Abiotic Stress-Tolerance. PLoS One 7:e40397. https://doi.org/10.1371/journal.pone.0040397

Rangan TS (1974) Morphogenic investigations on tissue cultures of *Panicum miliaceum*. Zeitschrift für Pflanzenphysiologie 72:456–459. https://doi.org/10.1016/S0044-328X(74)80066-8

Rangan TS (1976) Growth and Plantlet Regeneration in Tissue Cultures of Some Indian Millets: *Paspalum scrobiculatum* L., *Eleusine coracana*Gaertn. and *Pennisetum typhoideum*Pers. Zeitschrift für Pflanzenphysiologie 78:208–216. https://doi.org/10.1016/S0044-328X(73)80003-0

Rangan TS, Vasil IK (1983) Somatic embryogenesis and plant regeneration in tissue cultures of Panicum miliaceum L. and Panicum miliare Lamk. Zeitschrift für Pflanzenphysiologie 109(1):49–53

Rao AM, Kishor PBK, Reddy LA, Vaidyanath K (1988) Callus induction and high frequency plant regeneration in Italian millet (*Setaria italica*). Plant Cell Rep 7:557–559. https://doi.org/10.1007/BF00272756

Rao AN (1990) Estimates of losses in finger millet (*Eleusine coracana*) due to blast disease (Pyricularia grisea). J Agric Sci 24:57–60

Rao B (2003) Bioactive phytochemicals in Indian foods and their potential in health promotion and disease prevention. Asia Pac J Clin Nutr 12:9–22

Rao M (1989) Small Millets in Global Agriculture.The Small Millets: Their Importance, Present Status and Outlook. Oxford and IBH Publishing Co Pvt, Delhi, pp 9–12

Rasool A, Shah WH, Tahir I et al (2020) Exogenous application of selenium (Se) mitigates NaCl stress in proso and foxtail millets by improving their growth, physiology and biochemical parameters. Acta Physiol Plant 42:116. https://doi.org/10.1007/s11738-020-03109-w

Ravindran G (1991) Studies on millets: Proximate composition, mineral composition, and phytate and oxalate contents. Food Chem 39:99–107. https://doi.org/10.1016/0308-8146(91)90088-6

Ray DK, West PC, Clark M et al (2019) Climate change has likely already affected global food production. PLoS One 14:e0217148. https://doi.org/10.1371/journal.pone.0217148

Reddy LA, Vaidyanath K (1990) Callus formation and regeneration in two induced mutants of foxtail millet (*Setaria italica*). J Genet Breed 44:133–138

Renganathan VG, Vanniarajan C, Karthikeyan A, Ramalingam J (2020) Barnyard Millet for Food and Nutritional Security: Current Status and Future Research Direction. Front Genet 11. https://doi.org/10.3389/fgene.2020.00500

Rout GR, Samantaray S, Das P (1998) In vitro selection and characterization of Ni-tolerant callus lines of *Setaria italica* L. Acta Physiol Plant 20:269–275. https://doi.org/10.1007/s11738-998-0058-5

Sairam RK, Kumutha D, Ezhilmathi K et al (2008) Physiology and biochemistry of waterlogging tolerance in plants. Biol Plant 52:401–412. https://doi.org/10.1007/s10535-008-0084-6

Saleh ASM, Zhang Q, Chen J, Shen Q (2013) Millet Grains: Nutritional Quality, Processing, and Potential Health Benefits. Compr Rev Food Sci Food Saf 12:281–295. https://doi.org/10.1111/1541-4337.12012

Sateesh P (2010) Millets: future of food and farming. Millet Network of India Deccan Development Society FIAN, Hyderabad, pp 2–9

Satish L, Ceasar SA, Shilpha J et al (2015) Direct plant regeneration from in vitro-derived shoot apical meristems of finger millet (*Eleusine coracana* (L.) Gaertn.). Vitr Cell Dev Biol - Plant 51:192–200. https://doi.org/10.1007/s11627-015-9672-2

Satish L, Rathinapriya P, Ceasar SA et al (2016a) Effects of cefotaxime, amino acids and carbon source on somatic embryogenesis and plant regeneration in four Indian genotypes of foxtail millet (*Setaria italica* L.). Vitr Cell Dev Biol - Plant 52:140–153. https://doi.org/10.1007/s11627-015-9724-7

Satish L, Rency AS, Rathinapriya P et al (2016b) Influence of plant growth regulators and spermidine on somatic embryogenesis and plant regeneration in four Indian genotypes of finger millet (*Eleusine coracana* (L.) Gaertn). Plant Cell, Tissue Organ Cult 124:15–31. https://doi.org/10.1007/s11240-015-0870-8

Satyavathi C, Solanki RK, Kakani KC et al (2019) Genomics assisted breeding for abiotic stress tolerance in millets. In: Genomics assisted breeding of crops for abiotic stress tolerance, vol II, pp 241–255

Saxena R, Vanga SK, Wang J, Orsat V, Raghavan V (2018) Millets for food security in the context of climate change: A review. Sustainability 10(7):2228

Seckin B, Sekmen AH, Türkan İ (2009) An Enhancing Effect of Exogenous Mannitol on the Antioxidant Enzyme Activities in Roots of Wheat Under Salt Stress. J Plant Growth Regul 28:12–20. https://doi.org/10.1007/s00344-008-9068-1

Sehgal D, Rajaram V, Armstead IP et al (2012) Integration of gene-based markers in a pearl millet genetic map for identification of candidate genes underlying drought tolerance quantitative trait loci. BMC Plant Biol 12:9. https://doi.org/10.1186/1471-2229-12-9

Shah WH, Rasool A, Tahir I, Rehman RU (2020) Exogenously applied selenium (Se) mitigates the impact of salt stress in *Setaria italica* L. and Panicum miliaceum L. Nucl 63:327–339. https://doi.org/10.1007/s13237-020-00326-z

Sharma M, Kothari-Chajer A, Jagga-Chugh S, Kothari SL (2011) Factors influencing *Agrobacterium tumefaciens*-mediated genetic transformation of *Eleusine coracana* (L.) Gaertn. Plant Cell, Tissue Organ Cult 105:93–104. https://doi.org/10.1007/s11240-010-9846-x

Shivhare R, Lata C (2017) Exploration of Genetic and Genomic Resources for Abiotic and Biotic Stress Tolerance in Pearl Millet. Front Plant Sci 7. https://doi.org/10.3389/fpls.2016.02069

Shobana S, Sreerama YN, Malleshi NG (2009) Composition and enzyme inhibitory properties of finger millet (*Eleusine coracana* L.) seed coat phenolics: Mode of inhibition of α-glucosidase and pancreatic amylase. Food Chem 115:1268–1273. https://doi.org/10.1016/j.foodchem.2009.01.042

Singh A, Kumar M, Shamim M (2020) Importance of minor millets (Nutri Cereals) for nutrition purpose in present scenario. Int J Chem Stud 8:3109–3113. https://doi.org/10.22271/chemi.2020.v8.i1au.9226

Singh RK, Shweta S, Muthamilarasan M et al (2019) Study on aquaporins of *Setaria italica* suggests the involvement of SiPIP3;1 and SiSIP1;1 in abiotic stress response. Funct Integr Genomics 19:587–596. https://doi.org/10.1007/s10142-018-00653-0

Singh S, Chopperla R, Shingote P et al (2021) Overexpression of EcDREB2A transcription factor from finger millet in tobacco enhances tolerance to heat stress through ROS scavenging. J Biotechnol 336:10–24. https://doi.org/10.1016/j.jbiotec.2021.06.013

Singh UM, Chandra M, Shankhdhar SC, Kumar A (2014) Transcriptome Wide Identification and Validation of Calcium Sensor Gene Family in the Developing Spikes of Finger Millet Genotypes for Elucidating Its Role in Grain Calcium Accumulation. PLoS One 9:e103963. https://doi.org/10.1371/journal.pone.0103963

Sireesha Y, Kasetti RB, Nabi SA et al (2011) Antihyperglycemic and hypolipidemic activities of Setaria italica seeds in STZ diabetic rats. Pathophysiology 18:159–164. https://doi.org/10.1016/j.pathophys.2010.08.003

Sivadas P, Kothari SL, Chandra N (1990) High frequency embryoid and plantlet formation from tissue cultures of the Finger millet-*Eleusine coracana* (L.) Gaertn. Plant Cell Rep 9. https://doi.org/10.1007/BF00231557

Siwela M, Taylor JRN, de Milliano WAJ, Duodu KG (2010) Influence of phenolics in finger millet on grain and malt fungal load, and malt quality. Food Chem 121:443–449. https://doi.org/10.1016/j.foodchem.2009.12.062

Sreenivasulu N, Miranda M, Prakash HS et al (2004) Transcriptome changes in foxtail millet genotypes at high salinity: Identification and characterization of a PHGPX gene specifically up-regulated by NaCl in a salt-tolerant line. J Plant Physiol 161:467–477. https://doi.org/10.1078/0176-1617-01112

Sudan J, Negi B, Arora S (2015) Oxidative stress induced expression of monodehydroascorbate reductase gene in *Eleusine coracana*. Physiol Mol Biol Plants 21:551–558. https://doi.org/10.1007/s12298-015-0327-x

Suman K, Kalpana A (2012) Effects of heavy metals on *Eleusine coracana* (L.) Gaertn. Res Plant Biol 2:43–54

Tadele Z (2016) Drought adaptation in millets. InTech, London

Tadesse A, Tefera H, Guzmann M et al (2009) Advances in haploid production in higher plants. In: Tournaev A (ed) Androgenesis, pp 274–283

Thakur S, Sharma HO (2018) Trend and growth of small millets production in Madhya Pradesh as compared to India. Int J Agric Sci 10:4983–4986. https://doi.org/10.9735/0975-3710.10.1.4983-4986

Thomson CJ, Armstrong W, Waters I, Greenway H (1990) Aerenchyma formation and associated oxygen movement in seminal and nodal roots of wheat. Plant, Cell Environ 13:395–403. https://doi.org/10.1111/j.1365-3040.1990.tb02144.x

Tian B, Wang J, Zhang L et al (2010) Assessment of resistance to lodging of landrace and improved cultivars in foxtail millet. Euphytica 172:295–302. https://doi.org/10.1007/s10681-009-9999-z

Tyagi J, Pudake RN (2017) Spectrophotometric assays to evaluate the rhizospheric microbes mediated drought tolerance in plants. In: Modern tools and techniques to understand microbes. Springer, Cham, pp 413–429

Tyagi J, Varma A, Pudake RN (2017) Evaluation of comparative effects of arbuscular mycorrhiza (Rhizophagus intraradices) and endophyte (Piriformospora indica) association with finger millet (Eleusine coracana) under drought stress. Eur J Soil Biol 81:1–10

Ugare R, Chimmad B, Naik R et al (2014) Glycemic index and significance of barnyard millet (Echinochloa frumentacae) in type II diabetics. J Food Sci Technol 51:392–395. https://doi.org/10.1007/s13197-011-0516-8

Ul Mushtaq N, Saleem S, Rasool A et al (2021) Salt stress threshold in millets: Perspective on cultivation on marginal lands for biomass. Phyton (B Aires) 90:51–64. https://doi.org/10.32604/phyton.2020.012163

Upadhyaya H, Gowda CL, Reddy VG, Sube S (2008) Diversity of small millets germplasm in genebank at ICRISAT. In: 5th International Symposium on New Crops and Uses: their role in a rapidly changing world, 3-4 September, 2007. University of Southampton, Southampton

Valentine ON, Edak AU, Ikuo N, Masahiro M (2017) *Agrobacterium*-mediated genetic transformation of Fonio (Digitaria exilis (L.) Stapf). African J Biotechnol 16:1302–1307. https://doi.org/10.5897/AJB2017.15903

Veeranagamallaiah G, Ranganayakulu GS, Thippeswamy M et al (2009) Aldose reductase expression contributes in sorbitol accumulation and 4-hydroxynon-2-enal detoxification in two foxtail millet (*Setaria italica* L.) cultivars with different salt stress tolerance. Plant Growth Regul 59: 137–143. https://doi.org/10.1007/s10725-009-9396-6

Vikrant, Rashid A (2001) Direct as well as indirect somatic embryogenesis from immature (unemerged) inflorescence of a minor millet Paspalum scrobiculatum L. Euphytica 120:167–172

Vikrant, Rashid A (2002a) Somatic embryogenesis from immature and mature embryos of a minor millet *Paspalum scrobiculatum* L. Plant Cell Tissue Organ Cult 69:71–77. https://doi.org/10.1023/A:1015048529095

Vikrant, Rashid A (2002b) Induction of multiple shoots by thidiazuron from caryopsis cultures of minor millet (*Paspalum scrobiculatum* L.) and its effect on the regeneration of embryogenic callus cultures. Plant Cell Rep 21:9–13. https://doi.org/10.1007/s00299-002-0466-8

Vikrant, Rashid A (2003) Somatic embryogenesis from mesocotyl and leaf-base segments of *Paspalum scrobiculatum* L., a minor millet. Vitr Cell Dev Biol - Plant 39:485–489. https://doi.org/10.1079/IVP2003457

Visser EJW, Blom CWPM, Voesenek LACJ (1996) Flooding-induced adventitious rooting in Rumex: morphology and development in an ecological perspective. Acta Bot Neerl 45:17–28

Wahid A, Gelani S, Ashraf M, Foolad M (2007) Heat tolerance in plants: An overview. Environ Exp Bot 61:199–223. https://doi.org/10.1016/j.envexpbot.2007.05.011

Wakizuka T, Yamaguchi T (1987) The Induction of Enlarged Apical Domes In Vitro and Multishoot Formation from Finger Millet (*Eleusine coracana*). Ann Bot 60:331–336. https://doi.org/10.1093/oxfordjournals.aob.a087452

Wang J, Nie J, Pattanaik S, Yuan L (2016) Efficient Agrobacterium-mediated transformation of *Artemisia annua* L. using young inflorescence. Vitr Cell Dev Biol - Plant 52:204–211. https://doi.org/10.1007/s11627-015-9744-3

Wang M, Li P, Li C et al (2014) SiLEA14, a novel atypical LEA protein, confers abiotic stress resistance in foxtail millet. BMC Plant Biol 14:290. https://doi.org/10.1186/s12870-014-0290-7

Wang YF, Zhang J, Cui RL, Li W, Zhi H, Li HQ, Diao XM (2009) Transformation of wheat with DNAj gene from Foxtail millet via pollen-tube pathway. Acta Agric Bor Sin 24(2):17–21

Wang MZ, Pan YL, Li C, Liu C, Zhao Q, Ao GM, Yu JJ (2011) Culturing of immature inflorescences and Agrobacterium-mediated transformation of foxtail millet (*Setaria italica*). African J Biotechnol 10. https://doi.org/10.5897/AJB10.2330

Wondewosen S, Alemayehu B, Hussen M (2012) Genetic variation for grain yield and yield related traits in tef [<i>*Eragrostis tef*</i> (Zucc.)Trotter] under moisture stress and non-stress environments. Am J Plant Sci 03:1041–1046. https://doi.org/10.4236/ajps.2012.38124

Xu Z, Wang D, Yang L, Wei Z (1984) Somatic embryogenesis and plant regeneration in cultured immature inflorescences of *Setaria italica*. Plant Cell Rep 3:149–150. https://doi.org/10.1007/BF00270210

Yadav CB, Muthamilarasan M, Dangi A et al (2016) Comprehensive analysis of SET domain gene family in foxtail millet identifies the putative role of SiSET14 in abiotic stress tolerance. Sci Rep 6:32621. https://doi.org/10.1038/srep32621

Yang R, Chen M, Sun J-C et al (2019) Genome-wide analysis of LIM family genes in Foxtail Millet (*Setaria italica* L.) and Characterization of the Role of SiWLIM2b in Drought Tolerance. Int J Mol Sci 20:1303. https://doi.org/10.3390/ijms20061303

Yemets AI, Bayer GY, Blume YB (2013) An Effective Procedure for In Vitro Culture of *Eleusine coracana* (L.) and Its Application. ISRN Bot 1–7. https://doi.org/10.1155/2013/853121

Yue J, Li C, Liu Y, Yu J (2014) A Remorin Gene SiREM6, the target gene of SiARDP, from Foxtail Millet (*Setaria italica*) promotes high salt tolerance in transgenic arabidopsis. PLoS One 9: e100772. https://doi.org/10.1371/journal.pone.0100772

Zhang J-P, Zhang J-P, Liu T-S et al (2007) Cloning and characterization of a putative 12-oxophytodienoic acid reductase cDNA induced by osmotic stress in roots of foxtail millet. DNA Seq 18:138–144. https://doi.org/10.1080/10425170601060764

Zhang L, Liu B, Zheng G et al (2017a) Genome-wide characterization of the SiDof gene family in foxtail millet (*Setaria italica*). Biosystems 151:27–33. https://doi.org/10.1016/j.biosystems.2016.11.007

Zhang L, Shu H, Zhang AY et al (2017b) Foxtail millet WRKY genes and drought stress. J Agric Sci 155:777–790. https://doi.org/10.1017/S0021859616000873

Zhao J, Yu A, Du Y et al (2019) Foxtail millet (*Setaria italica* (L.) P. Beauv) CIPKs are responsive to ABA and abiotic stresses. PLoS One 14:e0225091. https://doi.org/10.1371/journal.pone.0225091

Zhao L, Zhao Q, Ao G, Yu J (2009) The foxtail millet Si69 gene is a Wali7 (wheat aluminum-induced protein 7) homologue and may function in aluminum tolerance. Sci Bull 54:1697–1706. https://doi.org/10.1007/s11434-009-0238-8

Zhu C, Ming C, Zhao-shi X et al (2014) Characteristics and Expression Patterns of the Aldehyde Dehydrogenase (ALDH) Gene Superfamily of Foxtail Millet (*Setaria italica* L.). PLoS One 9: e101136. https://doi.org/10.1371/journal.pone.0101136

Mining Genes and Markers Across Minor Millets Using Comparative Genomics Approaches

9

Theivanayagam Maharajan, Stanislaus Antony Ceasar, Thumadath Palayullaparambil Ajeesh Krishna, and Savarimuthu Ignacimuthu

Abstract

Nutritionally important cereals of small millets are used as food crop for poor people around the world. They are mainly cultivated in arid and semi-arid regions of Asian and African countries. Despite the nutritional benefits, small millets remain an underutilized crop and have received very little attention from researchers as well as farmers around the world. In the last few decades, very little research efforts have been made to study the features of small millets. More concentrated research efforts are needed to characterize germplasm resources, developing mapping population and identify quantitative trait loci (QTL) or genes which may help to improve the growth and production of small millets under biotic and abiotic stresses. The annotated genome sequences are currently available for foxtail millet and finger millet. Draft genome sequences are also available for proso millet and barnyard millet. The availability of whole and draft genome sequences of these millets can be used for further studies, such as SNP identification, next-generation sequencing-based allele discovery, association and linkage map construction, identification of candidate genes for agronomically important traits, and marker-assisted breeding programs. Two millets (little millet and kodo millet) have no genome sequences. Research is needed in the future to develop genome sequence for these two millets to improve the growth and production. In this chapter, we discuss about the details on genome sequences of small millets, how to mine genes from draft genome sequence of millets with finger millet as example, and comparative genomic studies of small millets. This chapter will

T. Maharajan · S. A. Ceasar (✉) · T. P. A. Krishna
Department of Biosciences, Rajagiri College of Social Sciences, Kalamassery, Cochin, Kerala, India

S. Ignacimuthu
Xavier Research Foundation, St. Xavier's College, Palayamkottai, Tirunelveli, Tamil Nadu, India

R. N. Pudake et al. (eds.), *Omics of Climate Resilient Small Millets*,
https://doi.org/10.1007/978-981-19-3907-5_9

help understand the current progress of research work related to comparative genomics approaches and mining genes across small millets.

Keywords

Comparative genomics approach · Genome sequence · Mining genes · Molecular markers · Small millets

Abbreviations

CISP	Conserved intron scanning primer
EST-SSR	Expressed sequences tags-simple sequence repeat
PHT1	Phosphate transporter 1
QTL	Quantitative trait loci
SSR	Simple sequence repeat
ZIP	Zinc-regulated, iron-regulated transporter-like proteins

9.1 Introduction

Millets come under Paniceae, which is the largest tribe of the Family Poaceae with more than 1400 species. They are small-seeded food cereals with drought and pest resistance. Millets are often considered as the only cereal crops that require very less water (350–400 mm annual rainfall) for the production (Hasan et al. 2019). They can be easily cultivated within a minimum time with lesser effort and resources than the mainstream cereals like wheat and rice (Michaelraj and Shanmugam 2013). Among the millets, pearl millet (*Cenchrus americanus* (L.) Morrone) is considered as a major millet. The remaining six millets such as foxtail millet (*Setaria italica* (L.) P. Beauv), proso millet (*Panicum miliaceum* L.), finger millet (*Eleusine coracana* (L.) Gaertn), barnyard millet (*Echinochloa esculenta* A. Braun), kodo millet (*Paspalum scrobiculatum* L.), and little millet (*Panicum sumatrense* Roth ex Roem. & Schult) are considered as minor millets (Krishna et al. 2018). Among the cereals, millets rank sixth in terms of consumption and millets are consumed by more than 33% of the global population (Saleh et al. 2013). Millets possess high contents of nutrients like calcium, phosphorous, iron, and proteins (Annor et al. 2017; Devi et al. 2014; Ceasar et al. 2018; Maharajan et al. 2021). They are also rich in vitamins, minerals, and essential amino acids and are better than maize, rice, and sorghum with regard to the content of fatty acids (Vinoth and Ravindhran 2017; Devi et al. 2014; Gull et al. 2015). Millets minimize the risk of diseases like duodenal ulcers, anemia, constipation, and atopic dermatitis (Watanabe 1999; Nambiar et al. 2011). They also have blood clot inhibition property and other properties including anti-oxidant, anti-cancer, anti-inflammatory, and anti-fungal

(Dykes and Rooney 2007; Neelam et al. 2013) and serve as a desirable food source for diabetics (Bora et al. 2019). There has been a close link between millets and low incidence of cardiovascular, diabetes and certain cancer diseases (Radhika et al. 2011; Singh and Raghuvanshi 2012; Gong et al. 2018). Most of the millets are used as food (developing countries) and as animal feed (developed countries). Apart from being cheap, safe, and staple food among other cereal grains, millets could be used even after years of their harvest, which is a chief aspect of food security.

Millets are considered very important for food and nutritional securities of many developing Asian and African countries (Bora et al. 2019). According to the Consultative Group for International Agricultural Research (CGIAR) and Consortium's Climate change, Agriculture, and Food Security (CCAFS), the world agricultural production needs to increase by 70% to supply the food for nine billion people by 2050 (Wheeler and Von Braun 2013). Nearly two to three billion people may not suffer from poverty and malnutrition if the farmers cultivate drought and heat tolerant crops like millets (Amadou et al. 2011). Compared to other cereals, very limited research work have been conducted on small millets due to their cultivation and consumption in less developed countries. The availability of genomic resources in small millets are also limited in number as compared to mainstream cereals. In this chapter, we have discussed about the details on genome sequences of small millets, how to mine genes from draft genome sequence of millets with finger millet as an example (as it helps to mine any other important genes from other millets draft genome sequence), and comparative genomic studies of small millets.

9.2 Genome Sequences of Small Millets

Genome sequences of small millets are being released only in recent years. The genome sequences are available for four small millets till date viz. proso millet, finger millet, barnyard millet, and foxtail millet. Foxtail millet is the first small millet to have its whole genome sequenced independently by two research institutes, US Department of Energy Joint Genome Institute and Beijing Genomics Institute (Bennetzen et al. 2012; Zhang et al. 2012). The genome size of foxtail millet is around 515 Mb. Totally 38,801 genes were predicted from the genome sequence of foxtail millet by Zhang et al. (2012). Among these 30,579 genes (78.81%) were annotated and 8220 (21.19%) were unannotated. About 24,000 to 29,000 protein encoding genes were completely annotated by Bennetzen et al. (2012). A total of 63,286 expressed sequence tags (ESTs) were sequenced by Sanger ABI3730xl platform and another 1,217,156 by 454 FLX platform (Bennetzen et al. 2012). Two different research groups have released the draft genome sequence of finger millet. The first draft genome sequence of finger millet (genotype ML-365) was released in 2017 (Hittalmani et al. 2017). The genome assembly consisted of 525,759 scaffolds (>200 bp) with N50 length of 23.73 Kb, and the average scaffold length of 2275 bp (Hittalmani et al. 2017). A total of 85,243 genes (78,647 non-transposable elements and 6596 transposable elements related genes) were predicted in ML-365 genome based on de novo method of gene prediction using

Augustus. About 2866 drought responsive genes, 330 calcium transport and accumulation genes, and 146 C_4 pathway genes were also identified in the ML-365 genome of finger millet. In the subsequent year, Hatakeyama et al. (2017) released the draft genome assembly of finger millet (genotype PR-202) using a novel polyploidy genome assembly workflow. The genome size of finger millet was estimated to be 1.5 Gb and the assembled genome to be 1189 Mb which is estimated to cover 78.2% genome. The whole genome consisted of 2387 scaffolds with the N50 value of 905.318 Kb having maximum sequence length of 5 Mb. The FASTA file format of final scaffolds and annotation are publicly available at National Center for Biotechnology Information (NCBI) (Bio Sample number: SAMD00076255). They have also identified 62,348 genes among which nearly 91% genes are functionally annotated and 96.5% are found to be single copy genes (Hatakeyama et al. 2017). It is interesting to note that the completely annotated finger millet genome sequence (version 1.0 assembly) is currently available in the Phytozome database (https://phytozome-next.jgi.doe.gov/info/Ecoracana_v1_1). The complete annotated genome sequence was deposited in the Phytozome by School of Plant Sciences, Ecology and Evolutionary Biology, Arizona Genomics Institute on July, 2020. The genome assembly was generated by MECAT assembler (Xiao et al. 2017) and subsequently polished using QUIVER. The released genome assembly (version 1.0) contains 1110.3 Mb sequence, consisting of 674 contigs with a contig N50 of 15.3 Mb and a total of 97.2% of assembled bases in chromosomes.

Proso millet shotgun genome sequence was submitted during 2018 at NCBI by China Agricultural University, Beijing (ASM289544v2) (Zou et al. 2019). The bio sample and bio project number are SAMN08335224 and PRJNA429322, respectively. The genome length of proso millet was estimated to be 923 Mb (Table 9.1). They also identified protein-coding genes (55,930), microRNAs (339), transfer RNAs (1420), ribosomal RNAs (1640), and small nuclear RNAs (2302) from the genome sequence of proso millet. Guo et al. (2017) generated the draft genome sequence for barnyard grass *(Echinochloa crus-galli)* (Table 9.1). They have used STB08 genotype to generate the draft genome sequence for *Barnyard grass* by Illumina HiSeq 2000 system. The genome size of barnyard grass was estimated to be 1.27 Gb with a scaffold N50 length of 1.8 Mb. A total of 917 *cytochrome P450 monooxygenase*, 277 *glutathione S-transferase,* and 4945 differentially expressed genes were identified from the draft genome sequence of barnyard grass. They have also predicted 108,771 protein-coding genes, 785 microRNAs, and other non-coding RNAs from the genome sequence of barnyard grass. The draft genome of barnyard grass could be used as a reference genome for barnyard millet, like in bread wheat. The annotated (foxtail millet and finger millet) and draft (proso millet and barnyard millet) genome sequences will help to identify the genes and markers and to improve the growth and production of small millets under both biotic and abiotic stresses.

Table 9.1 Details on genome sequence of millets

Name of the millet	Name of the genotype	Year	Name of the Institute	Platform used for sequencing	N50 length of genome assembly	Predicted Genome size (Mb/Gb)	Total number of predicted genes	Total number of ESTs/ SSRs/SNPs predicted	References
Foxtail millet	Yugu1	2012	Department of Energy Joint Genome Institute, University of Georgia, USA	ABI3730xl capillary sequencer; 454 FLX platform; Illumina genome analyzer II platform	47.3 Mb	400.9 Mb	29,000	63,286 ESTs	Bennetzen et al. (2012)
Foxtail millet	Zhang gu	2012	Beijing Genomics Institute, Chinese Ministry of Agriculture, China	Illumina second-generation sequencer	380 Mb	423 Mb	38,801	542,322 SNPs	Zhang et al. (2012)
Finger millet	ML-365	2017	University of Agricultural Sciences, India	Illumina and SOLiD sequencing	23.73 Kb	1196 Mb	85,243	114,083 SSRs	Hittalmani et al. (2017)
	PR-202	2018	University of Zurich, Switzerland	Illumina NextSeq 500, MiSeq instrument (Illumina) and PacBio RS II system	905.318 Kb	1189 Mb	62,348	NA	Hatakeyama et al. (2017)
Proso millet	Landrace	2018	China Agricultural University, China	Illumina short-read coupled with Pac-bio long-read sequencing	369 kb	923 Mb	55,930	112,158 SSR	Zou et al. (2019)
Barnyard millet	STB08	2017	Chinese Academy of Agricultural Sciences, China	Illumina HiSeq platform	1.8 Mb	1.27 Gb	1,14,910	NA	Guo et al. (2017)

Abbreviations: *EST* Expressed sequence tag, *NA* Not applicable, *SSR* Simple sequence repeats, *SNP* Single nucleotide polymorphisms

Details on name of the genotypes and publishers, platform used for sequencing, N50 length of genome assembly, genome size, total number of predicted genes and molecular markers are included

9.3 Comparative Genomics Approaches in Small Millets

Comparative genomics approach is helpful to mine the genes and genomic resources of crops with limited or no genomic resources. In small millets, the comparative genomics approach has been used to mine genes and characterize the molecular markers and nutrient transporters for the improvement of their growth and yield under biotic and abiotic stresses. For example, molecular markers derived from the closely related cereals (including some millets) were successfully utilized to characterize the millet germplasm. Due to the non-availability of whole genome sequence in some millets, the genomes of maize, rice, sorghum, wheat, and foxtail millet have served as essential models to study the marker-based syntenic relationships. Several reports are available on the transferability of microsatellite markers from closely related cereals to small millets. Some researchers also found candidate genes in the genome sequences of various cereals having completely annotated genome sequence for quantitative trait loci (QTL) in small millets. Many reports are also available for the successful transfer of microsatellite markers from one millet to the other millet. In this section, we discuss all these works.

9.3.1 Analysis of Microsatellite Markers of Millets with Closely Related Cereals Through Comparative Genomic Approaches

Microsatellite markers are generally used for structural, functional, and comparative genomics studies, namely, variety identification, genetic diversity and phylogenetic relationships, construction of high-density genome maps, mapping of important genes, comparative genome mapping, and marker-assisted selection (Maharajan et al. 2018, 2021b; Krishna et al. 2020). Microsatellite markers have been developed for several crop plants including small millets and these are used to detect loci for related species. Comparative genomic studies were successfully applied in small millets for microsatellite marker analysis. In the first step, researchers start to amplify the microsatellite markers from one species to another closely related species. For example, cross-genome transferability of 128 simple sequence repeat (SSR) markers of two millets (101 SSR for finger millet and 26 for foxtail millet) were analyzed in little millet, proso millet, barnyard millet, and kodo millet (Table 9.2) (Krishna et al. 2018). Among these, 35 SSR markers (33 finger millet and 2 foxtail millet SSR) were showed 100% cross-genome transferability in all these millets (Krishna et al. 2018). Around 106 foxtail millet's expressed sequences tags SSR (EST-SSR) markers were used to analyze the cross-genome transferability of barnyard millet, finger millet, kodo millet, little millet, and proso millet (Kumari et al. 2013). Of the 106 EST-SSR markers assayed, the highest (90.6%) and lowest (80.2%) cross-genome transferability were observed in barnyard millet and finger millet, respectively (Table 9.2). The same group also used 236 SSR markers of finger millet for cross-genome amplification in kodo millet and barnyard millet. A total of 20 and 15 SSR markers were amplified in kodo millet and barnyard millet, respectively (Arya et al. 2014). In the same study, they have analyzed the cross-genome

Table 9.2 Transferability of microsatellite markers from various cereals to small millets

Crop Species	Marker type	Number of markers used	Crop species cross amplified	Number of markers amplified	Transferability (%)	References
Foxtail millet	EST-SSR	106	Proso millet	94	88.7	Kumari et al. (2013)
			Finger millet	86	80.2	
			Kodo millet	93	87.7	
			Little millet	95	89.6	
			Barnyard millet	96	90.6	
Foxtail millet	SSR	26	Proso millet	18	69.2	Krishna et al. (2018)
			Finger millet	8	30.7	
			Kodo millet	14	53.8	
			Little millet	18	69.2	
			Barnyard millet	17	65.3	
Finger millet	SSR	101	Proso millet	62	61.3	Krishna et al. (2018)
			Foxtail millet	53	52.4	
			Kodo millet	53	52.4	
			Little millet	54	53.4	
			Barnyard millet	56	55.4	
	SSR	236	Kodo millet	20	15.5	Arya et al. (2014)
			Barnyard millet	15	14.2	
	SSR	18	Barnyard millet	7	39.0	Babu et al. (2018a)
	SSR	32	Barnyard millet	29	91.0	Babu et al. (2018b)
Pearl millet	SSR	67	Kodo millet	10	22.2	Arya et al. (2014)
			Barnyard millet	10	22.2	
Rice	SSR	50	Finger millet	27	53.0	

(continued)

Table 9.2 (continued)

Crop Species	Marker type	Number of markers used	Crop species cross amplified	Number of markers amplified	Transferability (%)	References
						Wang et al. (2005)
	SSR	345	Finger millet	202	58.6	Babu et al. (2017)
	SSR	120	Barnyard millet	85	71.0	Babu et al. (2018b)
Wheat	SSR	50	Finger millet	36	65.6	Wang et al. (2005)
Maize	SSR	50	Finger millet	29	57.0	Wang et al. 2005
	SSR	46	Barnyard millet	32	70.0	Babu et al. 2018b
Sorghum	SSR	60	Finger millet	30	57.6	Wang et al. (2005)
Switch grass	SSR	548	Proso millet	339	62.0	Rajput et al. (2014)

Abbreviation: *EST-SSR* expressed sequences tags-simple sequence repeat, *SSR* simple sequence repeat

transferability in kodo millet and barnyard millet using 67 SSR markers of pearl millet. Of which, twenty SSR markers (10 for kodo millet and 10 for barnyard millet) were amplified (Table 9.2) (Arya et al. 2014). Rajput et al. (2014) developed 548 SSR markers from the genome sequences of switch grass and these all markers were characterized in eight genotypes of proso millet. Interestingly, 339 SSR markers out of 548 were amplified in all the genotypes of proso millet. A total of 210 SSR markers were developed from rice (50 SSR), wheat (50 SSR), maize (50 SSR), and sorghum (60 SSR) and these all markers were used to analyze the cross-genome transferability in finger millet. Among them, more than 100 SSR primers were amplified in finger millet (Wang et al. 2005). Cross-genome transferability analysis was performed in 24 genotypes of barnyard millet using 64 SSR markers derived from maize (46 SSR) and finger millet (18 SSR) (Babu et al. 2018a). Out of the 64 SSR, 39 SSR markers were amplified in all the genotypes of barnyard millet. All the identified markers can be used for further studies related to the genetic variation, population structure, and germplasm characterization of wild and cultivated population of millets.

9.3.2 Identification of Candidate Genes Targeting QTLs Reported in Small Millets

In recent years, many workers used comparative genomics approach to validate and find the genes related to QTL from the genome assembly of closely related cereals. We have provided a stepwise protocol on how to mine candidate genes using the QTL sequences at Phytozome (Fig. 9.1). Eleven putative candidate genes (*1,4-βGlucanase*, *Cytochrome P450 CYP2*, *cytokinin dehydrogenase* (*CKX*), *auxin response factor* (*ARF*), *ethylene responsive factor* (*ERF*), *minichromosome maintenance protein 1 agamous deficiens serum response factor* (*MADS*), *MADS box protein*, *zinc finger Cys2His2* (*ZF-C2H2_6*), *serine/ threonine protein kinase*, *pectin methylesterase inhibitor (PMEI),* and *calmodulin* (*CaM*) *binding protein*) were identified on the genome sequence of maize, foxtail millet, rice, *Panicum hallii*, *Brachypodium distachyon*, *Panicum virgatum,* and *Brachypodium stacei* using eight blast resistance QTL sequences of finger millet through comparative genomics analysis (Ramakrishnan et al. 2016). The same group further identified some putative candidate genes (*Cytochrome P450*, *OsPHT1*;*8*, *serine/threonine protein kinase* and PMEI) and transcription factors [*basic helix-loop-helix* (*bHLH*)] and *WRKY* from the genome assembly of closely related cereals based on the comparative genomics using the four QTL (UGEP13, 19, 68, and 90) of finger millet related to low phosphorus starvation response traits (Ramakrishnan et al. 2017). Very recently, we also found more than 10 candidate genes in different members of Poaceae family using four drought resistance traits QTL (UGEP7, 13, 16, and 95) of finger millet through the same comparative genomics (David et al. 2021). Puranik et al. (2020)

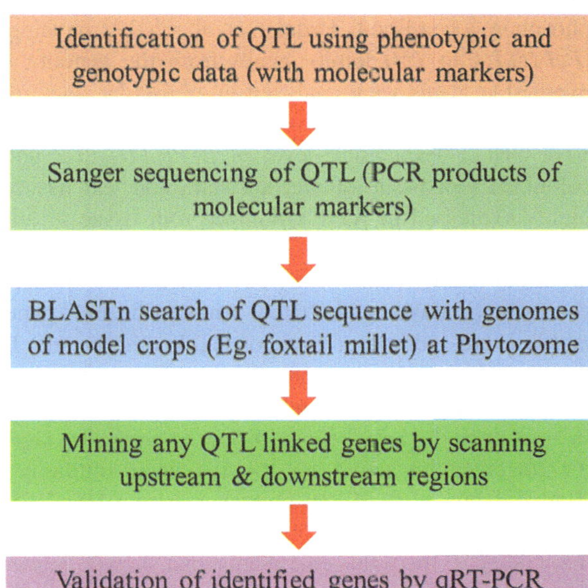

Fig. 9.1 Mining and validation of quantitative trait loci (QTL) by comparative genomics. Since the complete annotation is not yet available for some millets. PCR products for the identified QTL will be sequenced by Sanger sequencing. These sequences will be used for "BLASTn" search with genome of related model cereals (rice, maize, wheat, foxtail millet, sorghum, etc.) at Phytozome website to identify any candidate gene linked to QTL. This approach was successfully used by us (Ramakrishnan et al. 2016, 2017).

used comparative genomics to identify the metal ion binding, metal remobilization and detoxification candidate genes from the genome of various cereals using 25 QTL related to nutritional traits of finger millet. All the candidate genes were identified only through comparative genomics. Further research is needed to validate the candidate genes which help to improve the growth and development of small millets under biotic and abiotic stresses. No candidate genes have been yet identified by comparative genomics using the QTL of little millet, barnyard millet, and kodo millet. Because the study on QTL in little millet, barnyard millet, and kodo millet is comparatively less than other millets. So, some crucial research is needed to identify QTL on these millets, which will help to improve the growth and production of these millets under biotic and abiotic stresses.

9.3.3 Expression Analysis of Stress-Related Genes in Small Millets by Comparative Genomics

Like molecular markers, some candidate genes of small millets have been analyzed using the genic information of closely related model cereals. Twelve *PHT1* (*PHT1;1-1;12*) family genes of foxtail millet were used to analyze the expression pattern in leaf and root of all small millets under low and high phosphorus conditions (Maharajan et al. 2019). *PHT1;2* was expressed in leaf and root of all millets under low and high phosphorus conditions. Four (*PHT1;2, 1;3, 1;4,* and *1;9*) and three (*PHT1;3, 1;4,* and *1;6*) genes were expressed in root and leaf of foxtail millet, respectively, under low phosphorus condition. *PHT1;2, 1;3,* and *1;4* were expressed in leaf of finger millet and kodo millet under low phosphorus condition. In root of finger millet and proso millet, two genes (*PHT1;2* and *1;3*) were expressed under low phosphorus condition. Pudake et al. (2017) developed primers for four *PHT1* (*EcPHT1;1* to *1;4*) family genes from the genome sequences of rice and their expression levels were analyzed in arbuscular mycorrhiza fungi (AMF) colonized leaf and root tissues of three finger millet genotypes (ragi korchara, khairna, and VHC 3611). Expression level of *EcPHT1;1, 1;2* and *1;3* were higher in AMF colonized root tissue of ragi korchara and leaf tissue of khairna. *EcPHT1;4* was highly expressed in AMF colonized root tissue of all genotypes. They have also analyzed the expression level of these four *PHT1* family genes in leaf and root tissues of ragi korchara genotype of finger millet under low and normal phosphorus condition (Pudake et al. 2017). Among the four genes, *EcPHT1;1* and *1;2* were highly expressed in root tissue of ragi korchara under low phosphorus compared to the normal phosphorus condition. *EcPHT1;3* was found to be highly expressed in both leaf and root tissues under low phosphorus than normal phosphorus condition. In another study, basic leucine zipper transcription factor (*EcZIP17*) was developed for finger millet based on the sequence similarity of closely related species (Ramakrishna et al. 2018). Overexpression of *EcZIP17* in tobacco plants showed higher growth and yield of transgenic plants compared to the wild type. Compared to the wild type, the transgenic plants also showed higher germination rate, biomass, primary and secondary root formation and recovery rate under various abiotic stresses (salt, drought, and heat) (Ramakrishna et al. 2018). Like this, two zinc-regulated, iron-regulated transporter-like proteins (ZIP) family genes (*ZIP3* and *ZIP7*) were analyzed in leaf, stem, and spica of little millet under drought stress condition (Alagarasan et al. 2017). *ZIP3* was expressed in leaf, stem, and spica of little millet under drought stress. No other report yet available for comparative gene expression analysis in small millets under biotic and abiotic stresses. The complete and annotated genome sequence of finger millet and foxtail millet is available now. Also, draft genome sequences of barnyard millet and proso millet are also available at NCBI website. The available genome sequence will certainly help to identify the specific genes that help the growth and development of small millets under biotic and abiotic stresses. For example, 12 *SiPHT1* family genes were identified from genome sequences of foxtail millet and their expression level were identified in leaf, shoot, and root (under low and high phosphorus condition) (Ceasar et al. 2014; Roch et al. 2020) and leaf and roots (colonized by AMF) (Ceasar et al. 2014) of foxtail millet. In another study, seven *ZIP* family genes were developed from the genome sequences of foxtail millet and analyzed in shoot and root of foxtail millet under drought stress (Alagarasan et al. 2017). Therefore, these types of studies are required in all the small millets which help to improve the growth and development of other millets.

9.4 Mining of Genes from Genome Sequences of Millets

The complete annotated sequences are currently available for finger millet and foxtail millet. The genome sequences of remaining two millets (proso millet and barnyard millet) are not yet completely annotated. As like barnyard millet and proso millet, no genes have yet been reported in finger millet. We discuss a stepwise protocol on how to mine the phosphate transporter (*PHT1*) family gene from the draft genome sequence of finger millet at NCBI in this section. The same protocol can be used to mine any genes from the genome sequences of barnyard millet (draft genome of barnyard grass) and proso millet. The stepwise protocol will be used to mine genes from genome sequences of finger millet, barnyard millet, and proso millet having only draft genome sequence. This protocol is followed in the mining of finger millet *PHT1* genes using genome assembly (ASM218045v1) reported by Hittalmani et al. (2017) using genotype ML-365.

9.5 Steps

1. *Mining of PHT1 family genes for finger millet using the protein sequences of closely related cereals from the draft genome assembly of finger millet*: Foxtail millet and rice PHT1 family protein sequences are used for tblastn search with finger millet genome assembly (Fig. 9.2a, b) at NCBI.
2. *Prediction of genomic sequences for PHT1 family genes by tblastn*: The respective protein sequence is pasted in the query entry box (Enter accession number(s),

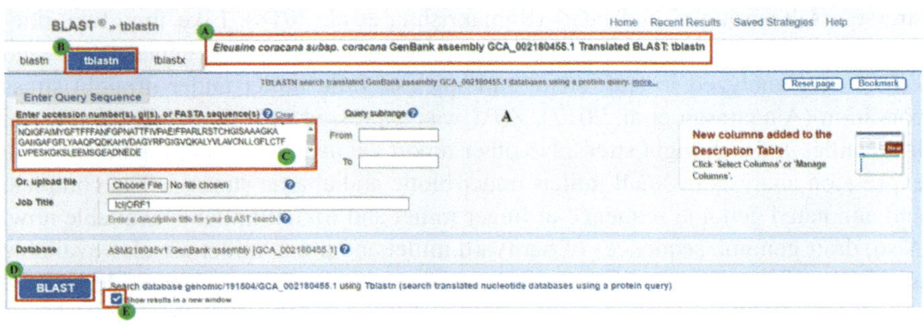

Fig. 9.2 Mining and prediction of *PHT1* family genes for finger millet from the draft genome assembly of finger millet using the protein sequences of closely related cereals. Draft genome assembly of finger millet was selected at NCBI (**a**). The "tblastn" option was selected for mining the PHT1 family protein sequences of foxtail millet with the draft genome assembly of finger millet (**b**). PHT1 family protein sequences of foxtail millet were pasted (fasta format) in the query entry box (**c**) and "blast" was clicked on (**d**). If we choose "show results in a new window" the results will appear in the new window (**e**), otherwise the results will appear in the same window

gi(s) or FASTA sequence(s)) (Fig. 9.2c) and "blast" option is clicked on (Fig. 9.2d). If we choose "show results in a new window" the results will appear in the new window (E), otherwise the results will appear in the same window.

3. *Collection of genomic sequences for PHT1 family genes by tblastn result*: After searching, the similar finger millet nucleotide sequences are displayed in a new window. The searching results are presented in Fig. 9.2. More number of these sequences (each sequences contains different scaffolds and ranges) are shown by tblastn. Selection was based on query cover (100%), except value (E) value = 0, start codon "ATG," scaffolds and sequencing range (Fig. 9.3a). Based on these components, the *PHT1* family gene nucleotide sequences are downloaded. Choose respective scaffold (e.g., 6389 after simply right click the selected scaffold) (Fig. 9.3b). After selecting the scaffold, click the scaffold; *PHT1* gene-related sequences are shown below in the same window (see in Fig. 9.4). After that, click graphics option, a sequence viewer will open in a new window (Fig. 9.4). Generally, in sequence viewer the excess ranges of sequences are shown (Fig. 9.5). So, after clicking graphics option, copy the sequences range in respective scaffold and the sequence ranges are pasted in find box (Fig. 9.5a). After click on find option then 5–10 seconds later the range of sequences only are shown. After that click download option (Fig. 9.5b). The download option containing various options are shown in Fig 9.5c. Choose "download" along with FASTA (visible range) option (Fig. 9.5d). Finally, the FASTA format of nucleotide sequences are retrieved in the respective box.
4. *Collection of protein sequences for PHT1 family genes by open reading frame (ORF) database*: The collected PHT1 family nucleotide sequences of finger millet were used for the retrieval of protein sequences from the ORF finder database (https://www.ncbi.nlm.nih.gov/orffinder/) and validated using the SMART BLAST option in the same window of ORF finder.

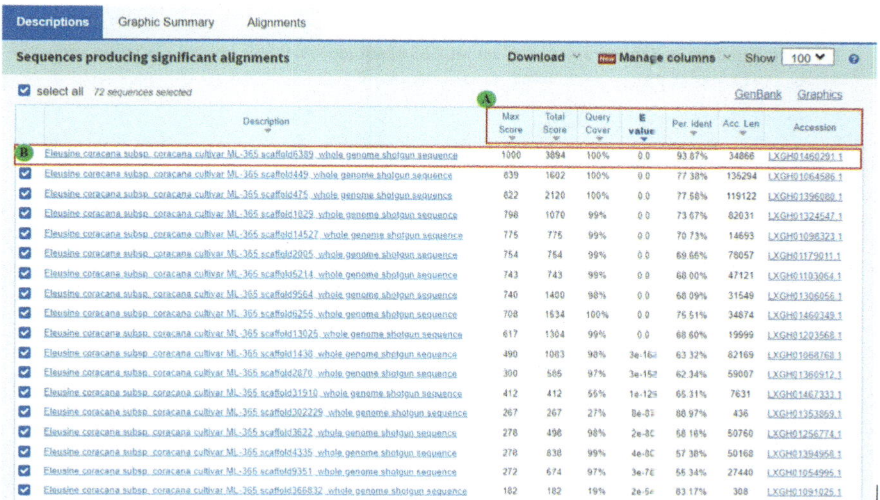

Description	Max Score	Total Score	Query Cover	E value	Per. ident	Acc. Len	Accession
Eleusine coracana subsp. coracana cultivar ML-365 scaffold6389, whole genome shotgun sequence	1000	3894	100%	0.0	93.87%	34866	LXGH01460291.1
Eleusine coracana subsp. coracana cultivar ML-365 scaffold449, whole genome shotgun sequence	639	1602	100%	0.0	77.38%	135294	LXGH01064586.1
Eleusine coracana subsp. coracana cultivar ML-365 scaffold475, whole genome shotgun sequence	622	2120	100%	0.0	77.58%	119122	LXGH01396080.1
Eleusine coracana subsp. coracana cultivar ML-365 scaffold1029, whole genome shotgun sequence	798	1070	99%	0.0	73.67%	82031	LXGH01324547.1
Eleusine coracana subsp. coracana cultivar ML-365 scaffold14527, whole genome shotgun sequence	775	775	99%	0.0	70.73%	14693	LXGH01098323.1
Eleusine coracana subsp. coracana cultivar ML-365 scaffold2005, whole genome shotgun sequence	754	764	99%	0.0	69.66%	78057	LXGH01179011.1
Eleusine coracana subsp. coracana cultivar ML-365 scaffold5214, whole genome shotgun sequence	743	743	99%	0.0	68.00%	47121	LXGH01103064.1
Eleusine coracana subsp. coracana cultivar ML-365 scaffold9564, whole genome shotgun sequence	740	1400	98%	0.0	68.09%	31549	LXGH01306056.1
Eleusine coracana subsp. coracana cultivar ML-365 scaffold6255, whole genome shotgun sequence	708	1634	100%	0.0	75.51%	34874	LXGH01460349.1
Eleusine coracana subsp. coracana cultivar ML-365 scaffold13025, whole genome shotgun sequence	617	1304	99%	0.0	68.60%	19999	LXGH01203568.1
Eleusine coracana subsp. coracana cultivar ML-365 scaffold1438, whole genome shotgun sequence	490	1083	98%	3e-16	63.32%	82169	LXGH01068768.1
Eleusine coracana subsp. coracana cultivar ML-365 scaffold2870, whole genome shotgun sequence	300	585	97%	3e-15	62.34%	59007	LXGH01360912.1
Eleusine coracana subsp. coracana cultivar ML-365 scaffold31910, whole genome shotgun sequence	412	412	56%	1e-12	65.31%	7631	LXGH01467333.1
Eleusine coracana subsp. coracana cultivar ML-365 scaffold302229, whole genome shotgun sequence	267	267	27%	8e-8	88.97%	436	LXGH01353869.1
Eleusine coracana subsp. coracana cultivar ML-365 scaffold3622, whole genome shotgun sequence	278	498	98%	2e-8	58.18%	50760	LXGH01256774.1
Eleusine coracana subsp. coracana cultivar ML-365 scaffold4335, whole genome shotgun sequence	278	838	99%	4e-8	57.38%	50168	LXGH01394968.1
Eleusine coracana subsp. coracana cultivar ML-365 scaffold9351, whole genome shotgun sequence	272	674	97%	3e-7	55.34%	27440	LXGH01054995.1
Eleusine coracana subsp. coracana cultivar ML-365 scaffold366832, whole genome shotgun sequence	182	182	19%	2e-5	83.17%	308	LXGH01091025.1

Fig. 9.3 Results on sequences producing significant alignments. After "tblastn," the similar finger millet nucleotide sequences were displayed (marked by rectangular box) The genomic sequences of PHT1 family genes were selected based on query cover (100%), except value (E) value = 0, scaffolds, etc. (A)

5. *Mining of genes by various comparative genomic approaches*: In recent years, the genome sequence of various millets has been published by different labs. Many computational tools such as Ensembl plants (https //plants.ensembl.org/index.html), Gramene (https://www.gramene.org/), PlantsDB (https://bio.tools/plantsdb), Phytozome (https://phytozome-next.jgi.doe.gov/), PLAZA (http://bioinformatics.psb.ugent.be/plaza/), GreenPhylDB (http://www.greenphyl.org/), SALAD (http://salad.dna.affrc.go.jp/), Plant Tribes (http://fgp.bio.psu.edu/tribedb/), and PlantgenIE.org (http://plantgenie.org/) are frequently used to conduct comparative genomic analysis in many plants. Ensembl plant is an integrated tool that helps to visualize, mine, and analyze the plant genomics data (Bolser et al. 2016). This tool provides genomic information on more than 90 sequenced plant species. Further, it provides an opportunity to conduct study related to comparative genomics (population structure, individual genetic variants, linkage mapping, gene models, and functional annotation) between small millets and other closely related cereals' genomic databases. Gramene allows to conduct comparative genomics studies across model plants and crops using phylogenetic structure and orthology-based predictions (Gupta et al. 2016). Gramene database contains well-annotated genome sequences, genetic and structural variation data, pathways and gene expression data for various plant species. Among the millets, details on foxtail millet genomic information is available in the Gramene database. This tool will enable to conduct comparative genomic analysis of small millets with other cereals, including foxtail millet. Phytozome is a large plant genomic portal of the Department of Energy's Joint Genome Institute (DOE-JGI)

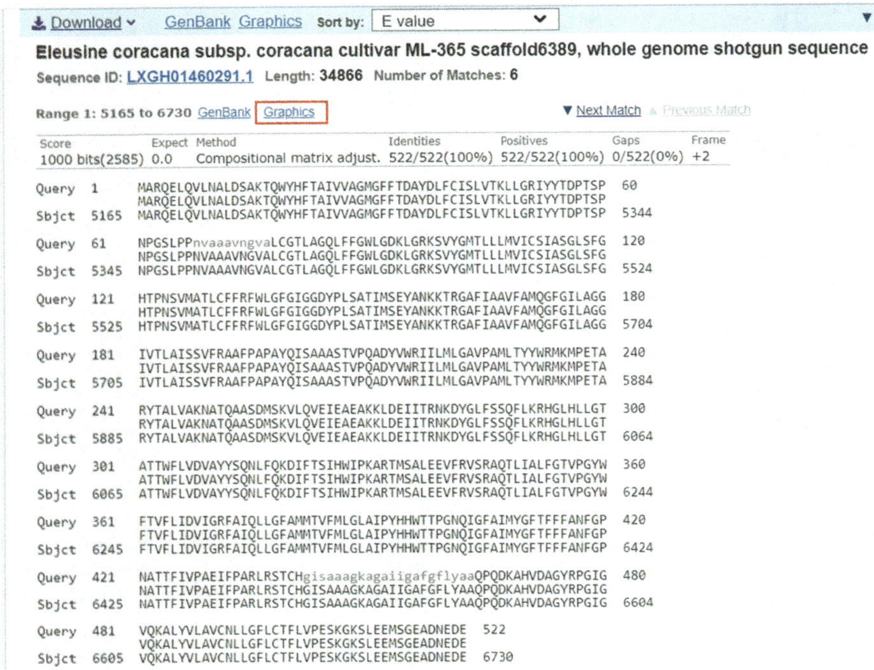

Fig. 9.4 Selection of scaffolds from sequences producing significant alignments. The best hit scaffold was selected from the sequences producing significant alignments. The graphics option is marked by rectangular box

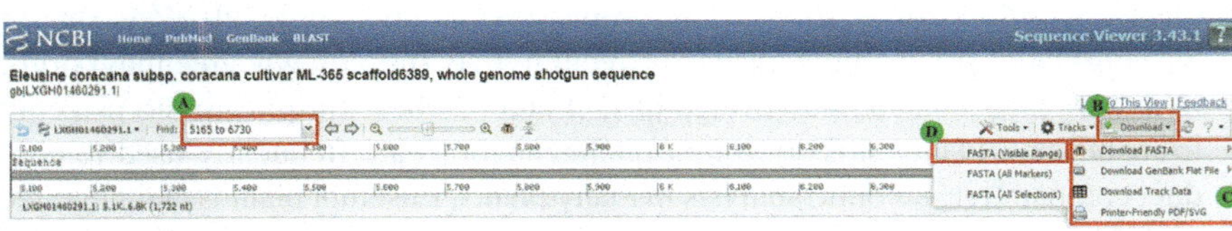

Fig. 9.5 Retrieval of nucleotide sequences for *PHT1* family genes. The sequence ranges of the respective scaffold were pasted in find box (**a**). Then download option was selected (**b**) after showing the range of sequences in sequence viewer (**b**). The download option contains various options which are shown in **c**. FASTA (visible range) option was chosen (**d**) then FASTA format nucleotide sequences were retrieved in the respective box

(Goodstein et al. 2012). It provides the broader plant science community a hub for accessing, visualizing, and analyzing plant genomes, as well as selected genomes and datasets that have been sequenced elsewhere. By integrating this large collection of plant genomes into a single resource and performing comprehensive and uniform annotation and analyses, Phytozome facilitates accurate and insightful comparative genomics studies. As of recent version (v13), Phytozome hosts 261 assembled and annotated genomes including foxtail millet and finger millet.

In Phytozome, collection of all genes have been annotated with eukaryotic ortholog groups of proteins (KOG), kyoto encyclopedia of genes and genomes (KEGG), enzyme, Pathway and the InterPro family of protein analysis tools. Search and visualization tools let users quickly find and analyze genes or genomic regions of interest. Other tools pave the way for identifying gene family copies (PLAZA), gene synteny (GoGe, PLAZA, PGSB), protein domains (PlantGDB), gene expression (PGSB), biomart (Phytozome, EnsemblPlants), intermine (Phytozome), gene ontology (GO) annotation (PLAZA and PGSB), and alternative splicing (plantgdb).

Seed shattering (detachment of fruit from the pedicel in cereals) is a major cause of yield loss and is essential for the propagation of their offspring (Dong and Wang 2015). In monocot plants, abscission layer in the pedicle is necessary for the seed shattering process (Fu et al. 2019). In rice, two genes such as, *Shattering* (*SH1*) and *SH4* play an important role in the formation of abscission layer during the early stages of flower development (Li et al. 2006; Yoon et al. 2014). In sorghum, seed shattering was controlled by *SH1* (Lin et al. 2012). Lodging reduces the plant height, head weight and length, culm diameter, stem length, internode length and diameter, number of tillers, peduncle length and diameter of cereals (Shah et al. 2019). It is one of the major factors limiting cereal productivity worldwide. Some genes associated with lodging tolerance were identified in rice and wheat. Several dwarfing genes such as *Rht1*, *2*, *4*, *5*, *8*, *9*, *2*, *13*, *14*, *15*, *16*, *18*, *19*, and *Rht-R107* were identified and they all are involved to reduce the plant height in wheat (Haque et al. 2011; Pearce et al. 2011; Shah et al. 2019). In rice, *semi-dwarf 1* (*sd1*) and polycomb genes (*OsCLF*, *OsEMF2b*, and *OsFIE2*) are responsible to reduce the plant height and they have played a key role to improve the lodging tolerance (Conrad et al. 2014; Zhong et al. 2018). Three genes (*prl5*, *SCM2*, and *SCM3*) were identified to be responsible to increase stem diameter, culm strength, and spikelet number under lodging tolerance (Ookawa et al. 2010; Yano et al. 2015). Lodging and shattering are also major problems limiting the growth and development of millets (Tian et al. 2018; Muthamilarasan and Prasad 2021). No genes associated with shattering and lodging in millets have yet been identified. Mining of lodging and shattering related genes from genome sequences of millets or through comparative genomic analysis may help to improve the growth and development of millet in the early stages of flower development.

9.6 Conclusion

Over the past decade, comparative genomic studies in millets helped to fill the gaps in the analysis of genes and markers in many millets that had limited or no genome sequences. There is no doubt that with the ongoing efforts, comparative genomic approaches in millets will continue to provide invaluable information for a better understanding of the adaptation of plants to their environment. Only few numbers of comparative genomics studies have been carried out on small millets compared to other cereals like rice, wheat, and maize. In millets, foxtail millet has been identified as model species for such studies since it has small and completely annotated genome. In future, comparative genomics using genome sequences of foxtail millet have a great potential in speeding up the development of genomic tools for breeding of little millet and kodo millet due to the non-availability of genome sequences for these millets. Ploidy level of the plant genome should be carefully considered when selecting the appropriate assembly method. The presence of two or more genes within the same nucleus may affect the accuracy of the assembly, making it difficult to distinguish between homologs. Genome sequences are used to detect genetic variations across different individual within a species. It is very difficult to detect genetic variations between different individual within a species in the genome of polyploid plants. Polyploidy (fusion of two or more genomes within one nucleus) can lead to high levels of heterozygosity which makes it more difficult to identify haplotypes. Compared to the sanger sequencing, the next-generation sequencing technology provides a way to generate genome sequences for polyploidy plant species. Most of the millets' genomes have been sequenced by the next-generation sequencing technology. But the annotation of genome sequences for millets is a major challenge because they come under the category of polyploidy species. However, genomes of two millets were annotated well and submitted on the Phytozome website. Availability of foxtail millet, finger millet, barnyard millet, and proso millet genome sequence will be enormously advantageous for comparative genomics, genome mapping, marker development, and molecular breeding of small millets and other cereals.

References

Alagarasan G, Dubey M, Aswathy KS, Chandel G (2017) Genome-wide identification of orthologous *ZIP* genes associated with zinc and iron translocation in *Setaria italica*. Front Plant Sci 8:775

Amadou I, Gbadamosi, Ole G (2011) Millet-based traditional processed foods and beverages-a review. Cereal Foods World 56:115–121

Annor GA, Tyl C, Marcone M, Ragaee S, Marti A (2017) Why do millets have slower starch and protein digestibility than other cereals? Trends Food SciTech 1:73–83

Arya L, Chauhan D, Yadav Y, Verma M (2014) Transferability of simple sequence repeat (SSR) markers developed in finger millet, and pearl millet to kodo millet and barnyard millet. In: Innovative approach in stem cell research, cancer biology and applied biotechnology, pp 60–64

Babu B, Rashmi C, Sood S (2018a) Cross transferability of finger millet and maize genomic SSR markers for genetic diversity and population structure analysis of barnyard millet. Indian J Genet Plant Breed 78:364–372

Babu BK, Joshi A, Sood S, Agrawal PK (2017) Identification of microsatellite markers for finger millet genomics application through cross transferability of rice genomic SSR markers. Indian J Genet 77:92–98

Babu BK, Sood S, Kumar D, Joshi A, Pattanayak A, Kant L, Upadhyaya HD (2018b) Cross-genera transferability of rice and finger millet genomic SSRs to barnyard millet (*Echinochloa* spp.). 3 Biotech 8:1–10

Bennetzen JL, Schmutz J, Wang H (2012) Reference genome sequence of the model plant *Setaria*. Nature Biotech 30:555–561

Bolser D, Staines DM, Pritchard E, Kersey P (2016) Ensembl plants: integrating tools for visualizing, mining, and analyzing plant genomics data. In: Edwards D (ed) Plant bioinformatics, Methods in molecular biology. Humana Press, New York, pp 115–140

Bora P, Ragaee S, Marcone M (2019) Characterisation of several types of millets as functional food ingredients. Int J Food Sci Nutr 70:1–11

Ceasar SA, Hodge A, Baker A, Baldwin SA (2014) Phosphate concentration and arbuscular mycorrhizal colonisation influence the growth, yield and expression of twelve PHT1 family phosphate transporters in foxtail millet (*Setaria italica*). PLoS One 9(9):e108459

Ceasar SA, Maharajan T, Ajeesh Krishna TP, Ramakrishnan M, Victor Roch G, Satish L, Ignacimuthu S (2018) Finger millet [*Eleusine coracana* (L.) Gaertn.] improvement: current status and future interventions of whole genome sequence. Front. Plant Sci 9:1054

Conrad LJ, Khanday I, Johnson C, Guiderdoni E, An G, Vijayraghavan U, Sundaresan V (2014) The polycomb group gene EMF 2B is essential for maintenance of floral meristem determinacy in rice. Plant J 80:883–894

David RHA, Ramakrishnan M, Maharajan T, Barathi Kannan K, Babu GA, Daniel MA, Ignacimuthu S (2021) Mining QTL and genes for root traits and biochemical parameters under vegetative drought in south Indian genotypes of finger millet (*Eleusine coracana* (L.) Gaertn) by association mapping and *in silico* comparative genomics. Biocatalysis Agril Biotech 32:101935

Devi PB, Vijayabharathi R, Sathyabama S, Malleshi NG, Priyadarisini VB (2014) Health benefits of finger millet (Eleusine coracana L.) polyphenols and dietary fiber: a review. J Food Sci Tech 51:1021–1040

Dong Y, Wang YZ (2015) Seed shattering: from models to crops. Front Plant Sci 6:476

Dykes L, Rooney L (2007) Phenolic compounds in cereal grains and their health benefits. Cereal Foods World 52:105–111

Fu Z, Song J, Zhao J, Jameso PE (2019) Identification and expression of genes associated with the abscission layer controlling seed shattering in Lolium perenne. AoB Plants 11:ply076

Gong L, Cao W, Chi H, Wang J, Zhang H, Liu J, Sun B (2018) Whole cereal grains and potential health effects: involvement of the gut microbiota. Food Res Int 103:84–102

Goodstein DM, Shu S, Howson R, Neupane R, Hayes RD, Fazo J, Rokhsar DS (2012) Phytozome: a comparative platform for green plant genomics. Nucl acids Res 40:D1178–D1186

Gull A, Prasad K, Kumar P (2015) Optimization and functionality of millet supplemented pasta. Food Sci Tech 35:626–632

Guo L, Qiu J, Ye C, Jin G, Mao L, Zhang H, Fan L (2017) *Echinochloa crus-galli* genome analysis provides insight into its adaptation and invasiveness as a weed. Nature Comm 8:1–10

Gupta P, Naithani S, Tello Ruiz MK, Chougule K, D'Eustachio P, Fabregat A, Jaiswal P (2016) Gramene database: navigating plant comparative genomics resources. Curr Plant Biol 7:10–15

Haque M, Martinek P, Watanabe N, Kuboyama T (2011) Genetic mapping of gibberellic acid-sensitive genes for semi-dwarfism in durum wheat. Cereal Res Comm 39:171–178

Hasan M, Maheshwari C, Garg NK, Kumar M (2019) Millets: Nutri-cereals. Biotech Express 6:18–21

Hatakeyama M, Aluri S, Balachadran MT, Sivarajan SR, Patrignani A, Grüter S, Poveda L, Shimizu-Inatsugi R, Baeten J, Francoijs KJ (2017) Multiple hybrid *de novo* genome assembly of finger millet, an orphan allotetraploid crop. DNA Res 25:39–47

Hittalmani S, Mahesh H, Shirke MD, Biradar H, Uday G, Aruna Y, Lohithaswa H, Mohanrao A (2017) Genome and transcriptome sequence of finger millet (*Eleusine coracana* (L.) Gaertn.) provides insights into drought tolerance and nutraceutical properties. BMC Genomics 18:1–16

Krishna TA, Maharajan T, David RHA, Ramakrishnan M, Ceasar SA, Duraipandiyan V, Ignacimuthu S (2018) Microsatellite markers of finger millet (*Eleusine coracana* (L.) Gaertn) and foxtail millet (*Setaria italica* (L.) Beauv) provide resources for cross-genome transferability and genetic diversity analyses in other millets. Biocatalysis Agrl Biotech 16:493–501

Krishna TA, Maharajan T, Roch GV, Ramakrishnan M, Ceasar SA, Ignacimuthu S (2020) Hybridization and hybrid detection through molecular markers in finger millet [*Eleusine coracana* (L.) Gaertn.]. J Crop Improv 34:335–355

Kumari K, Muthamilarasan M, Misra G, Gupta S, Subramanian A, Parida SK, Prasad M (2013) Development of eSSR-markers in *Setaria italica* and their applicability in studying genetic diversity, cross-transferability and comparative mapping in millet and non-millet species. PLoS One 8:e67742

Li C, Zhou A, Sang T (2006) Rice domestication by reducing shattering. Sci 311:1936–1939

Lin Z, Li X, Shannon LM, Yeh CT, Wang ML, Bai G, Yu J (2012) Parallel domestication of the Shattering1 genes in cereals. Nat Genet 44:720–724

Maharajan T, Antony Ceasar S, Ajeesh Krishna TP, Ignacimuthu S (2021) Finger millet [*Eleusine coracana* (L.) Gaertn]: An orphan crop with a potential to alleviate the calcium deficiency in the semi-arid tropics of Asia and Africa. Front Sust Food Syst 5:684447

Maharajan T, Ceasar SA, Ajeesh Krishna TP, Ramakrishnan M, Duraipandiyan V, Naif Abdulla AD, Ignacimuthu S (2018) Utilization of molecular markers for improving the phosphorus efficiency in crop plants. Plant Breed 137:10–26

Maharajan T, Ceasar SA, Krishna TPA, Ignacimuthu S (2019) Phosphate supply influenced the growth, yield and expression of PHT1 family phosphate transporters in seven millets. Planta 250:1433–1448

Maharajan T, Roch GV, Ceasar SA (2021b) Recent advancements of molecular breeding and functional genomics for improving nitrogen-, phosphorus-and potassium-use efficiencies in wheat. In: Hossain MA et al (eds) Molecular breeding in wheat, maize and sorghum: strategies for improving abiotic stress tolerance and yield. CAB International, Wallingford, pp 170–196

Michaelraj P, Shanmugam A (2013) A study on millets based cultivation and consumption in India. Int J Marketing Finan Serv Manag Res 2:49–58

Muthamilarasan M, Prasad M (2021) Small millets for enduring food security amidst pandemics. Trends Plant Sci 26:33–40

Nambiar VS, Dhaduk J, Sareen N, Shahu T, Desai R (2011) Potential functional implications of pearl millet (*Pennisetum glaucum*) in health and disease. J Appl Pharmal Sci 10:62–67

Neelam Y, Kanchan C, Alka S, Alka G (2013) Evaluation of hypoglycemic properties of kodo millet based food products in healthy subjects. IOSR J Pharmacy 3:14–20

Ookawa T, Hobo T, Yano M, Murata K, Ando T, Miura H, Matsuoka M (2010) New approach for rice improvement using a pleiotropic QTL gene for lodging resistance and yield. Nat Commun 1:1–11

Pearce S, Saville R, Vaughan SP, Chandler PM, Wilhelm EP, Sparks CA, Thomas SG (2011) Molecular characterization of Rht-1 dwarfing genes in hexaploid wheat. Plant Physiol 157: 1820–1831

Pudake RN, Mehta CM, Mohanta TK, Sharma S, Varma A, Sharma AK (2017) Expression of four phosphate transporter genes from finger millet (Eleusine coracana L.) in response to mycorrhizal colonization and pi stress. 3 Biotech 7:17

Puranik S, Sahu PP, Beynon S, Srivastava RK, Sehgal D, Ojulong H, Yadav R (2020) Genome-wide association mapping and comparative genomics identifies genomic regions governing grain nutritional traits in finger millet (*Eleusine coracana* L. Gaertn). Plants, People, Planet 2(6): 649–662

Radhika G, Sathya RM, Ganesan A, Saroja R, Vijayalakshmi P, Sudha V, Mohan V (2011) Dietary profile of urban adult population in South India in the context of chronic disease epidemiology (CURES–68). Public Health Nutr 14:591–598

Rajput SG, Plyler-Harveson T, Santra DK (2014) Development and characterization of SSR markers in proso millet based on switchgrass genomics. Am J Plant Sci 5:175–186

Ramakrishna C, Singh S, Raghavendrarao S, Padaria JC, Mohanty S, Sharma TR, Solanke AU (2018) The membrane tethered transcription factor EcbZIP17 from finger millet promotes plant growth and enhances tolerance to abiotic stresses. Sci Rep 8:1–14

Ramakrishnan M, Antony Ceasar S, Duraipandiyan V, Vinod KK, Kalpana K, Al-Dhabi NA, Ignacimuthu S (2016) Tracing QTLs for leaf blast resistance and agronomic performance of finger millet (*Eleusine coracana* (L.) Gaertn.) genotypes through association mapping and *in silico* comparative genomics analyses. PLoS One 11(7):e0159264

Ramakrishnan M, Ceasar SA, Vinod KK, Duraipandiyan V, Ajeesh Krishna TP, Upadhyaya HD, Ignacimuthu S (2017) Identification of putative QTLs for seedling stage phosphorus starvation response in finger millet (*Eleusine coracana* L. Gaertn.) by association mapping and cross species synteny analysis. PLoS One 12(8):e0183261

Roch GV, Maharajan T, Krishna TA, Ignacimuthu S, Ceasar SA (2020) Expression of PHT1 family transporter genes contributes for low phosphate stress tolerance in foxtail millet (*Setaria italica*) genotypes. Planta 252(6):1–9

Saleh AS, Zhang Q, Chen J, Shen Q (2013) Millet grains: nutritional quality, processing, and potential health benefits. Compr Rev Food Sci Food Safety 12:281–295

Shah L, Yahya M, Shah SMA, Nadeem M, Ali A, Ali A, Ma C (2019) Improving lodging resistance: using wheat and rice as classical examples. Int J Mol Sci 20:4211

Singh P, Raghuvanshi RS (2012) Finger millet for food and nutritional security. Afr J Food Sci 6: 77–84

Tian B, Luan S, Zhang L, Liu Y, Zhang L, Li H (2018) Penalties in yield and yield associated traits caused by stem lodging at different developmental stages in summer and spring foxtail millet cultivars. Field Crops Res 217:104–112

Vinoth A, Ravindhran R (2017) Biofortification in millets: a sustainable approach for nutritional security. Front Plant Sci 8:29

Wang ML, Barkley NA, Yu JK, Dean RE, Newman ML, Sorrells ME, Pederson GA (2005) Transfer of simple sequence repeat (SSR) markers from major cereal crops to minor grass species for germplasm characterization and evaluation. Plant Genet Resour 3:45–57

Watanabe M (1999) Antioxidative phenolic compounds from Japanese barnyard millet (*Echinochloa utilis*) grains. J Agrl Food Chem 47:4500–4505

Wheeler T, Von Braun J (2013) Climate change impacts on global food security. Science 341:508–513

Xiao CL, Chen Y, Xie SQ et al (2017) MECAT: fast mapping, error correction, and *de novo* assembly for single-molecule sequencing reads. Nat Methods 14:1072–1074

Yano K, Ookawa T, Aya K, Ochiai Y, Hirasawa T, Ebitani T, Matsuoka M (2015) Isolation of a novel lodging resistance QTL gene involved in strigolactone signaling and its pyramiding with a QTL gene involved in another mechanism. Mol Plant 8:303–314

Yoon J, Cho LH, Kim SL, Choi H, Koh HJ, An G (2014) The BEL 1-type homeobox gene SH 5 induces seed shattering by enhancing abscission-zone development and inhibiting lignin biosynthesis. Plant J 79:717–728

Zhang G, Liu X, Quan Z, Cheng S, Xu X, Pan S, Wang J (2012) Genome sequence of foxtail millet (*Setaria italica*) provides insights into grass evolution and biofuel potential. Nat Biotech 30: 549–554

Zhong J, Peng Z, Peng Q, Cai Q, Peng W, Chen M, Yao J (2018) Regulation of plant height in rice by the Polycomb group genes OsEMF2b, OsFIE2 and OsCLF. Plant Sci 267:157–167

Zou C, Li L, Miki D, Li D, Tang Q, Xiao L, Zhang H (2019) The genome of broomcorn millet. Nat Commun 10:1–11

Improving the Nutrient-Use Efficiency in Millets by Genomics Approaches

10

Thumadath Palayullaparambil Ajeesh Krishna, Theivanayagam Maharajan, Savarimuthu Ignacimuthu, and Stanislaus Antony Ceasar

Abstract

Food demand and malnutrition are the major issues for developing countries. Millets are helpful in strengthening both food and nutritional security in the future. They are nutrient-rich crops compared to major cereals. Consumption of millet-based foods helps for collective health benefits of nutrients and prevents metabolic disorders. Millets are considered a staple food for millions of people. The millet production is constrained by both biotic and abiotic stresses leading to reduced grain yield. Soil nutrient deficiency is the major issue in the semi-arid region of Asia and Africa. It crucially affects crop production, especially the quality of the crop. It is one of the major abiotic constraints of millet production. Therefore, improving the nutrient-use efficiency in millet under limited nutrient soil conditions is a major priority area of research. Introduction of functional genomics and associated techniques are helping to understand the character of the crop genome. Genome-based techniques are used for the identification of valuable traits/candidate genes. This will accelerate crop designing (insertion of desirable traits) and the development of improved varieties through marker-assisted selection (MAS). It could help to reduce the application of synthetic fertilizer and reduce the cost of crop production. So far, very little effort is done in the improvement of nutrient-use efficiency in millets. In this chapter, we present the scope of functional genomics and molecular marker approaches for improving nutrient-use efficiency in millets and give the future direction for the need for

T. P. A. Krishna · T. Maharajan · S. A. Ceasar (✉)
Department of Biosciences, Rajagiri College of Social Sciences, Kalamassery, Cochin, Kerala, India

S. Ignacimuthu
Xavier Research Foundation, St. Xavier's College, Palayamkottai, Tirunelveli, Tamil Nadu, India

R. N. Pudake et al. (eds.), *Omics of Climate Resilient Small Millets*,
https://doi.org/10.1007/978-981-19-3907-5_10

millet improvement. This chapter could help researchers understand the importance of the implication of functional genomics for crop improvement.

Keywords

Millets · Genomics · Germplasm · Crop improvement · Nutrient-use efficiency

10.1 Introduction

Crop plants are the primary resources of foods that provide energy and nutrient to human beings. The millets are minor-cereal crops that are considered as high-energy yielding nutritious foods which help in preventing malnutrition (Nithiyanantham et al. 2019). "Millets" is a generic term that indicates coarse cereals. The millets are generally categorized into major and minor millets. Sorghum (*Sorghum bicolor* L.) and pearl millet (*Cenchrus americanus* (L.) Morrone) are the major millets. The minor millets include foxtail millet (*Setaria italica* (L.) Beauv), finger millet (*Eleusine coracana* (L.) Gaertn), proso millet (*Panicum miliaceum* L.), little millet (*Panicum sumatrense* Roth. ex Roem. and Schuktz), kodo millet (*Paspalum scrobiculatum* L.), and barnyard millet (*Echinochloa esculenta* (A. Braun) H. Scholz) (Arya et al. 2014). Millets rank sixth in terms of world agriculture production of cereal grains (Kumar et al. 2018). Millets are nutritionally superior to major crops like rice and wheat. They are important food and fodder crops in semi-arid regions. Millets are still a staple food for poor people in many regions of the world (Kumar et al. 2018; Krishna et al. 2018).

Soil fertility is the primary limiting factor, which influences crop production under intensive agricultural farming (Chowdary and Patra 2019). Plants require a proper balance of all nutrients for normal growth and optimum yield. Both macro- and micro-nutrients play a vital role in physiological and metabolic activities of plants (Krishna et al. 2017). The unavailability of any nutrient affects plant growth and reduces crop production (Krishna et al. 2020a). Nutrient deficiency is one of the major abiotic constraints affecting crop production and reduction in the quality of crops. Krithika and Balachandar (2016) reported that the rice grown on low fertility soils usually produces very less yield with poor nutritional quality (Krithika and Balachandar 2016). Similarly, nutrient deficiency majorly affects millet production (Maharajan et al. 2019; Thilakarathna and Raizada 2015). Millets are important crops for achieving food and nutritional security in the developing countries. Hence, agriculture scientists across the world need to improve millets through genome-based approaches. Innovative crop improvement technologies like precision breeding and genome editing will be helpful to increase nutrient-use efficiency in millets. Genomic studies seem to be helpful to understand the structural, behavioral, and functional aspects of the crop genome. Marker-assisted selection (MAS) is very attractive and helps to develop improved varieties through precision breeding (Krishna et al. 2020b). The genome-assisted tools help for the identification and selection of beneficial quantitative trait loci (QTL)/genes for millet improvement.

Therefore, it provides the opportunity to improve the millets for nutrient-use efficiency.

This chapter focuses on the application of genomics strategies in millets for improving their nutrient-use efficiency. The scope and importance of molecular marker approaches for improving the nutrient-use efficiency in millets is presented. Furthermore, the details of germplasm characterization and genome-assisted breeding (GAB) in millet for improving nutrient-use efficiency are discussed. This chapter will help agricultural scientists and researchers to understand the importance and application of genomics and molecular marker approaches for millet improvement.

10.2 Nutritional Importance of Millets

Achieving food and nutritional security will help us in improving the health status of the world's population. About half of the global population, especially those from Asia and Africa, depends on cereal crops for food. Next to cereals, millets are the primary source of energy in the semi-arid tropics. The trends of consumption of polished rice and processed food are calorie-dense but not mineral nutrient-dense (Srilekha et al. 2019). Such calorie-dense consumption pattern may lead to a variety of lifestyle-related disorders like obesity, diabetes, cancer, and cardiovascular diseases. Millets are nutritionally superior to rice and wheat (Shobana et al. 2013). About 80% of millets are used as food while the remaining 20% is used as animal fodder and preparation of brewage products (Saleh et al. 2013; Shivran 2016; Sindhu and Khetarpaul 2001). The millet grains are ground into flour and used for the preparation of cakes or porridges (Vinoth and Ravindhran 2017). Millets are considered as a future nutritional food crop for developing countries having the potential to help prevent malnutrition problems around the world.

Millet grains contain a high amount of proteins, essential amino acids, dietary fibers, minerals, and vitamins (Hegde et al. 2005; Shobana et al. 2013; Saleh et al. 2013). Proso millet is a rich source of protein (12.5%) while barnyard millet is rich in crude fiber (13.6%) and iron (186 mg/kg dry matter) (Saleh et al. 2013). The finger millet grains are a rich source of minerals such as calcium, magnesium, and potassium (Devi et al. 2014; Saleh et al. 2013). The calcium content in finger millet (344 mg) ensures healthy bones (Maharajan et al. 2021). Foxtail millet is also rich in protein (11%) and fat (4%) contents (Zhang and Liu 2015). The little millet contains high amount of energy (351.65 kcal) and carbohydrates (74.75%) (Srilekha et al. 2019). Kodo millet provides high content of magnesium (1.1 g/kg dry matter). It also contains essential amino acids such as lysine, threonine, valine, and sulfur-containing amino acids (Antony et al. 1996; Ravindran 1992). Millets are considered as multi-nutrient grains and are termed as nutri-millets or nutri-cereals. Consumption of millets helps in achieving collective health benefits of nutrients (reviewed by Amadou et al. 2013; Hassan et al. 2021; Deepak Taggelli and Thakur 2018) and prevents metabolic disorders.

10.3 Influence of Nutrients on Millet Growth and Production

Millet production is threatened by abiotic stresses associated with climatic change. Nutrient deficiency in soil is the major abiotic constraint affecting millet production. It is well-known fact that millets respond well to the nutrient application (Maharajan et al. 2019). The application of nitrogen (N) fertilizer (40 kg/ha) helps in increasing the grain yield and quality in finger millet (Rao et al. 1989). Maharajan et al. (2019) reported that adequate application of phosphorus (P) significantly increases the grain yield of all millets. Millet production is reduced due to low soil fertility, drought, and poor agricultural management (Buerkert et al. 2002; Gandah et al. 2003). Most of the agricultural soil in semi-arid regions of Asia are deficient in nutrients (Thilakarathna and Raizada 2015; Sahrawat et al. 2007; Masto et al. 2007). Seed germination is one of the major challenges for small-seeded crops like millets under low fertility soil seriously affecting the millet growth at different stages. For example, P deficiency significantly influences the size of the panicle and reduces the grain yield of millets (Maharajan et al. 2019). The nutrient deficiency affects millet production worldwide and this situation demands the researchers to improve the millet production under stressful conditions. Plant breeders, biotechnologists, and molecular biologists need to apply new agricultural techniques for improving the nutrient-use efficiency in millets. Phenomic and genomic approaches are used in recent years for improving crops through precision breeding.

Millet production may be vulnerable to global climate change because meteorological factors influence crop growth and development. Global climate change has significantly increased the temperature and altered the amounts of rainfall. Drought and high rainfall are climatic phenomena, which occur periodically in all climatic regions. The acquisition of mineral nutrients by plants from the soil is depending upon the availability of nutrients in the soil solution. The climatic change will affect mineral nutrient availability in the soil both directly and indirectly. Both drought and excess water impair the plant's nutritional status. For example, the meta-analysis showed that drought reduces the concentration of N and P in plant tissue (He and Dijkstra 2014; Bista et al. 2018), and many studies have shown that drought stress can reduce the nutrient uptake from soil solution (Cramer et al. 2009; Waraich et al. 2011; Ge et al. 2012; Sardans and Peñuelas 2012). In plants, nutrient availability and uptake are reduced during poor mineralization processes under drought conditions (Sanaullah et al. 2012). Also, drought affects the mobility (kinetics) of mineral nutrient (ions) in the soil solution, which reduces the absorption of minerals by the plant root systems (Bassirirad 2000). Drought is a main limiting factor for millet growth and development (Seghatoleslami et al. 2008; Debieu et al. 2018). Rezaei et al. (2014) reported that water availability is a main limiting factor in pearl millet production whenever nutrient deficiency occurs due to climatic change. The high rainfall may lead to waterlogging if drainage is limited due to soil structure. It also leads to the unavailability of nutrients to plants. Under the waterlogging, the nutrient content such as N, Fe, K, and P reduced, and higher Na concentration was observed in cotton plants (Milroy et al. 2009). It indicates the imbalance of ion uptake in plants under waterlogging condition. So, the climatic change will influence the soil's physical and chemical properties, which directly influence soil fertility. The nutrient deficiency also influences the millet production (Maharajan et al. 2019; Ceasar et al. 2020). But millets have several agro-morphological, physiological, biochemical, and molecular characteristics which confer better tolerance to climatic changes (environmental stresses) than other cereal crops (Bandyopadhyay et al. 2017). Millet germplasm characterization and genomic selection provide the opportunity for improving millets under various climatic conditions. The genomic approaches will help to improve the nutrient-use efficiency in plants under abiotic stress conditions.

10.4 Characterization of Millet Germplasm for Nutrient-Use Efficiency

Plant genetic resources are an important part of agricultural biodiversity that play a major role in developing improved varieties (Saha et al. 2016). The genetic variability existing in the millet germplasm provides an excellent opportunity for millet improvement (Saha et al. 2016; Ceasar et al. 2018; Dwivedi et al. 2012). A huge number of germplasm accessions are available for millets worldwide (Fig. 10.1). This could help in conservation and utilization of genetic resources for millets improvement. Millets are well known to possess unique features related to resilience to adverse conditions (Nadeem et al. 2020; Umesh et al. 2019). The genotypic variations of millet accessions for stress tolerance may help to protect and improve their agronomic performance under stressful conditions. Therefore, characterization of millets' germplasm is a very important step for the genetic improvement of millets. Screening of millets germplasm from different agro-ecological systems helps to identify their ability to produce a high yield in both sufficient and deficient nutrient conditions and to access the extent of genotypic variation among the genotype in the germplasm. This helps in the identification of

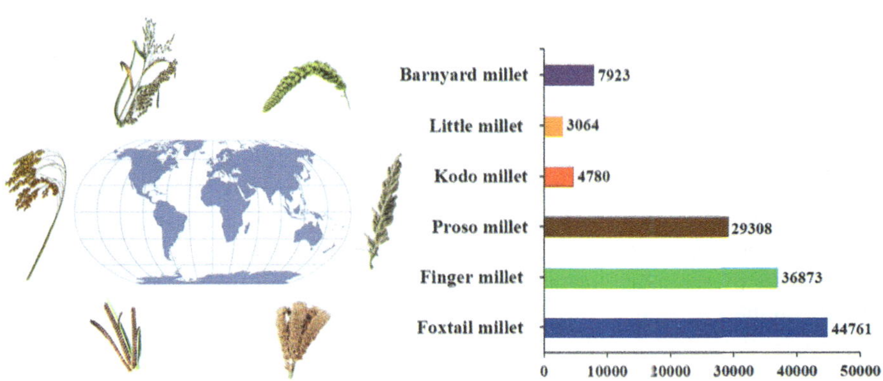

Fig. 10.1 Germplasm collection of millets. The figure represents that the total number of cultivated germplasm accessions of millets preserved globally in national and international genebanks (Vetriventhan et al. 2020)

morphological and physiological adaptation of the millets to low-nutrient conditions. Therefore, the first step for any genetic mapping and breeding approaches is the identification of existing variation available in the germplasm (Garnett et al. 2015).

So far, only a little effort has been made towards the characterization of millets germplasm for nutrient-use efficiency. Ramakrishnan et al. (2017) grew 128 genotypes of finger millet in low and high P conditions under greenhouse and categorized these into low-P-tolerant and low-P-susceptible genotypes. The genotypes such as GPU 45, IE 5201, IE 2871, IE 7320, GPU 66, HOSUR 1, TCUM 1, IE 2034, SVK 1, RAU 8, VR 708, and IE 3391 were found to be low-P-tolerant genotypes. Similarly, 79 cultivars of foxtail millet were used to assess their genetic response under low (0.2 mmol L^{-1}) and high (6 mmol L^{-1}) N application resulting in the identification of N-use efficient cultivars (Erying et al. 2019). Evaluation of the phenotypic response of 54 foxtail millet genotypes in low and high P conditions under greenhouse and field conditions by Ceasar et al. (2020) led to the identification of low-P-tolerant foxtail millet genotypes (ISe 1181, ISe 1655, ISe 783, and ISe 1892) in terms of grain yield. Furthermore, most of the studies were reported on morphological and agro-morphological characterization of millets in the field condition (Goswami et al. 2015; Dhanalakshmi et al. 2014). Therefore, researchers need to pay more attention towards evaluation of the millets in terms of plant nutrient aspects, which could be helpful in the selection of best breeding materials for nutrient-use efficiency.

10.5 Genomic Approaches for Improving Nutrient-Use Efficiency in Millets

The genomic information in the form of genetic maps, molecular markers, and genome sequences is very essential for the genetic improvement of cultivars through molecular marker-assisted breeding or genetic engineering (Fig. 10.2). Genome-based approaches have reduced the time and effort involved in the selection of breeding material for crop improvement. Nowadays, crop breeding has gradually changed from phenotype-based selection to genotype-based selection. Precision breeding is very attractive for the development of improved cultivars (Fig. 10.3). The identification of valuable QTL and candidate genes related to nutrient-use efficiency is the first step in marker-assisted breeding (MAB), which will help in the development of nutrient-use efficient millet varieties leading to a reduction in the application of synthetic fertilizers. Application of lesser quantity of fertilizers is the critical part of economic sustainability of agriculture and the use of nutrient-use efficient millets directly benefits the poor farmers.

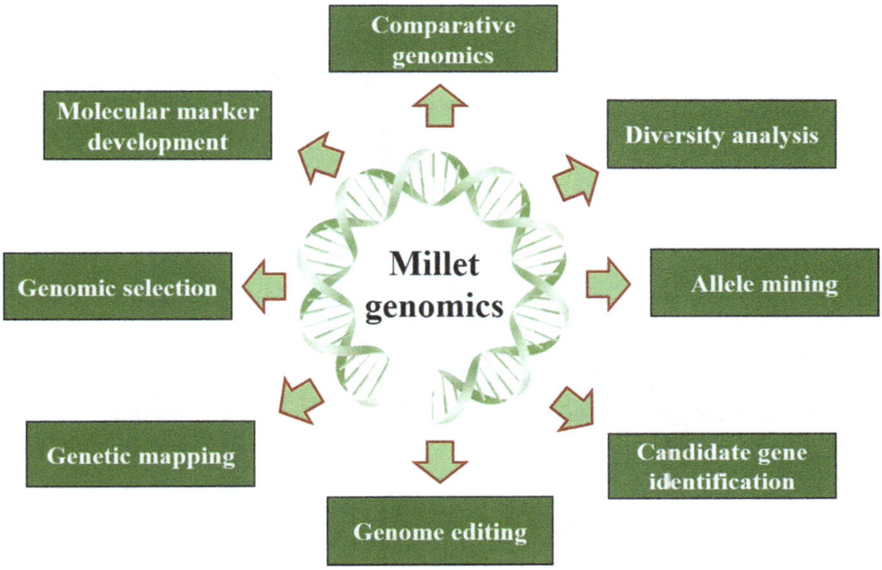

Fig. 10.2 Genomic approaches for millet improvement. This diagram shows that various applications of genomic resources for millet improvement

10.5.1 Molecular Marker-Assisted Breeding Approaches

Molecular markers provide a foundation for MAS, which is widely used in crop improvement programs. Molecular markers are popularly used for the identification of the QTL/genes responsible for grain yield and resistance to various biotic and abiotic stresses. In agriculture, MAB has enormous potential to improve the efficiency and precision of plant breeding through MAS. Non-interference by environment and accuracy in the selection (desirable traits) makes the MAS very attractive (Krishna et al. 2020b). In small millets, very little efforts reported towards the identification of valuable QTL/candidate genes for nutrient-use efficiency. About four putative QTL were identified in finger millet viz., *qLRDW.1*, *qLRDW.2*, *qHSDW.1*, and *qHRL.1*, which were associated with low-P-tolerance and were linked to the SSR markers UGEP19, UGEP68, UGEP13, and UGEP90, respectively (Ramakrishnan et al. 2017). These QTL could be used for the development of low-P-tolerant finger millet varieties in the future through molecular breeding. Many QTL have been identified in various crops on nutrient-use efficiency as compared to small millets. For example, QTL associated with nutrient-use efficiency traits were identified in rice (Jewel et al. 2019; Wang et al. 2014; Fang et al. 2015; Wei et al. 2012), maize (Li et al. 2015; Pestsova et al. 2016; Ribaut et al. 2007), barely (Han et al. 2016; Kindu et al. 2014; Karunarathne et al. 2020), and wheat (Safdar et al. 2020; Fan et al. 2018; Zhang et al. 2019). Therefore researchers need to focus on the identification of QTL/candidate genes for nutrient-use efficiency in millets, which

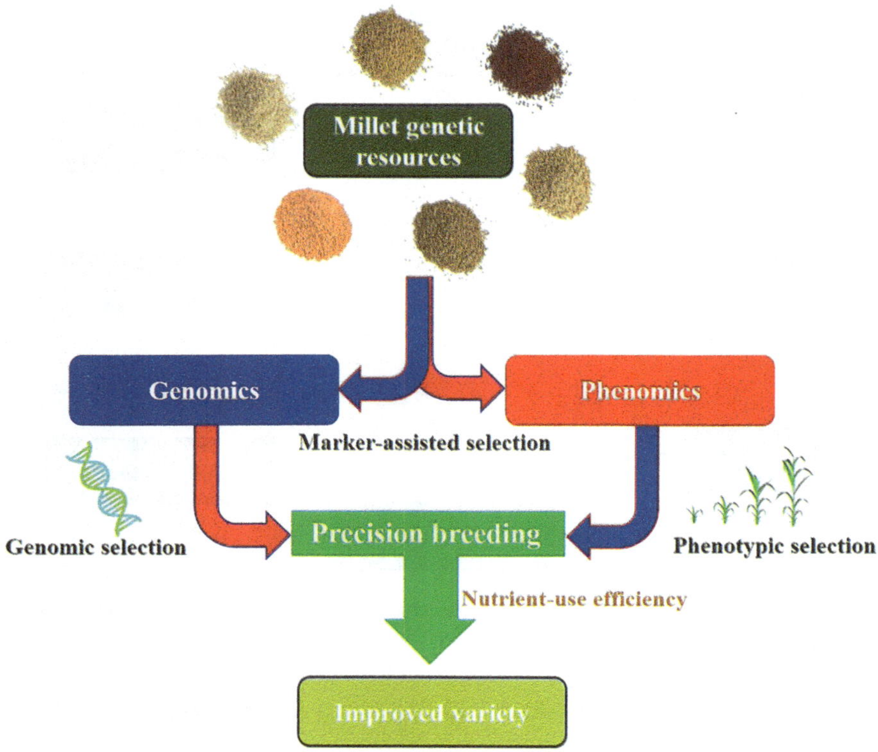

Fig. 10.3 Genomic approaches for millet improvement. This diagram shows that various applications of genomic resources for millet improvement

may be useful to increase millets production under low-nutrient conditions. Also, researchers need to use high-throughput phenomics facilities for the identification of valuable QTL. It will help for the selection of efficient breeding materials.

10.5.2 Functional Genomic Approaches

Understanding of the molecular mechanism of nutrient-use efficiency can facilitate the development of improved millets with efficient nutrient-use efficiency. The discovery of novel genes associated with nutrient-use efficiency is important for millet improvement under low-nutrient soil (Ceasar et al. 2014, 2017). The nutrient transporters are responsible for nutrient uptake, transport, and homeostasis in plants (Krishna et al. 2017). In millets, genes for a few nutrient transporter family members have been identified so far (Table 10.1). Recently, *EcZIP1* transporter was identified from the whole genome sequencing of finger millet. Analysis of the expression level of *EcZIP1* in the root, shoot, root-shoot zone, and flag leaf of six finger millet

Table 10.1 Details of nutrient transporter genes identified in millets

Name of the millet	Gene name	Functions/role	References
Finger millet	*EcPT1, EcPT2, EcPT3, and EcPT4*	Phosphate transport	Pudake et al. (2017)
	EcHNRT2, EcLNRT1, EcNADH-NR, EcGS, and *EcFd-GOGAT*	Nitrate uptake and assimilation	Gupta et al. (2013)
	EcCAX3	Calcium transport	Singh et al. (2015)
	CAX1, TPC1, CaMK1, and *CaMK2*	Calcium uptake, transport, and accumulation	Mirza et al. (2014)
Foxtail millet	*SiPHT1;1* to *SiPHT1;12*	Phosphate transport	Roch et al. (2020), Maharajan et al. (2019)
	SiHAK1	Potassium uptake and homeostasis maintenance	Zhang et al. (2018)
	SiPHT1;1 to *SiPHT1;4, SiPHT1;8* to *SiPHT1;12*	Phosphorus transport	Ahmad et al. (2018)
	SiAMT1.1 and *SiAMT1.3*	Ammonium transport	Ahmad et al. (2018)
	SiNRT1.1, SiNRT1.11, SiNRT1.12 SiNRT2.1 and *SiNAR2*	Nitrate transport	Ahmad et al. (2018), Nadeem et al. (2018)

genotypes (VHC 3582, IE 3618, IE 6240, VL 330, GE 724, and VHC 3893) revealed that the expression level was higher in the root, shoot, root-shoot zone, and flag leaf of GE 724 as compared to the other genotypes (Chandra et al. 2020). The study also revealed that the expression level of nutrient transporter depends upon the genetic variation of genotypes. Expression patterns of 12 *SiPHT1* family transporters (*SiPHT1;1 to SiPHT1;12*) analyzed by Roch et al. (2020) in 20 genotypes with contrasting characters of growth and P uptake responses under low and high P supply revealed genotype to genotype variation in the expression pattern of 12 *PHT1* transporter genes . Similarly, the analysis of expression pattern of *SiPHT1;1* to *SiPHT1;12* transporter genes under both low and high P supply (Maharajan et al. 2018) indicated differential expression patterns of *PHT1* family transporter genes in all seven millets. These differential expression patterns of *PHT1* family transporter family genes may be due to the genetic variation of the foxtail millet genotypes. This highlights the fact that the genetic variation among the genotypes may be contributing to the nutrient-use efficiency in foxtail millet. The *SiPHT1* family transporter genes play a main role of low-P-tolerance in foxtail millet and further characterization of genes will help in the improvement of P-use efficiency in foxtail millet and other cereals (Roch et al. 2020). Hence, the researchers need to focus on the identification of nutrient transporter genes in millets for the improvement of low-nutrient stress tolerance, which will help in understanding the genetic basis of molecular mechanism of low-nutrient tolerance. The functional characterization of nutrient transporter proteins may help in the improvement of crop nutrient-use efficiency (Krishna et al. 2017, 2020a). To date, very little information is available on the functional characterization of nutrient transporters in millets. In the recent past, the foxtail millet *SiPHT1* transporters were functionally characterized through yeast complementation assay and RNA interference (RNAi) technology (Ceasar et al. 2017) revealing the downregulation of *SiPHT1;2, SiPHT1;3*, and *SiPHT1;4* transporters in the transgenic foxtail millet leading to a significant reduction of total and inorganic P content in shoot and root tissue. The high-affinity potassium (HAK) family transporter plays a key role in uptake and transport of potassium and regulation of growth and development of plants under low-potassium condition (Chen et al. 2015; Yang et al. 2014). A few HAK family transporters have been identified and functionally characterized in small millets. For example, Zhang et al. (2018) reported that expression of *SiHAK1* restores the growth of potassium uptake-deficient strain of *Saccharomyces cerevisiae* (CY162; *trk1*△ *trk2*△) under low-potassium (<1 mM) conditions. The growth of mutant yeast indicated that the *SiHAK1* is a high-affinity transporter that helps in the uptake of potassium under low-potassium conditions. The mutant strain (CY162, *trk1*△ *trk2*△) also showed a perfect growth as compared to the empty vector transformants under high-potassium conditions, indicating that *SiHAK2* might facilitate potassium uptake in a high-potassium environment (Zhang et al. 2018). This kind of high-resolution study is needed for the functional characterization of nutrient transporters in millets. This could help in understanding the function of nutrient transporters under low-nutrient conditions.

10.5.3 Genomics-Assisted Breeding Approaches

Genome sequencing technology is helpful for accurate dissection of the genetic basis of phenotypic variation of the genotypes (Maharajan et al. 2021). Recent development in the next-generation sequencing (NGS) technology leads to the discovery of high-throughput molecular markers and multiplexed genotyping of germplasm to speed up MAB approaches. This technology plays a vital role in genome-wide association studies (GWAS) for crop improvement. For example, meta-QTL (MQTL) analysis identified the novel genomic loci associated with potassium-use efficiency in bread wheat (Safdar et al. 2020). The identified MQTL could be used for breeding and development of potassium stress-resistant bread wheat variety. Therefore, GWAS accelerates crop designing (insertion of desirable traits) and the development of improved varieties through MAB. GWAS has helped in the identification of beneficial traits or candidate genes and their alleles for nutrient-use efficiency in many crops such as rice (Pariasca-Tanaka et al. 2020; Wissuwa et al. 2015), wheat (Safdar et al. 2020; Cormier et al. 2014; Soumya et al. 2021), barely (Karunarathne et al. 2020), and maize (Sun et al. 2020). Very limited studies are available on GWAS in small millets as compared to other crops. In finger millet, GWAS was used for the identification of genomic region governing grain nutritional content (Puranik et al. 2020; Tiwari et al. 2020). No report is available on GWAS for nutrient-use efficiency in small millets under low-nutrient condition. Millets are a future nutritional food crop and their production is constrained by nutrient deficiency. High-resolution research like GWAS towards the identification of QTL in small millets for nutrient-use efficiency is essential for MAB. Therefore, researchers need to focus on GWAS in small millets for the improvement of their nutrient-use efficiency, which will help in reducing the application of synthetic fertilizers resulting in increase in the millet production.

10.6 Conclusion and Future Prospectus

Millets are important food and nutritional crops and are considered as crops for poor people. Millet production is reduced by both biotic and abiotic stresses. Improving nutrient-use efficiency in millet is crucial for sustainable crop growth and production under low-nutrient soils. The genetic improvement of millets is depended on the nature and extent of variation among the germplasms. Availability of huge germplasm for many millets provides the opportunity for genetic improvement in millets. Comprehensive screening of millet germplasm resources for nutrient-use efficiency is effective for improving millet under low-nutrient soils. Likewise, GAB plays a very crucial role in enhancing crop productivity in terms of the development of new varieties. The rapid advance in molecular marker technology followed by high-throughput phenomics and genomics provided opportunities to carry out efficient crop improvement. Many promising advanced phenomics facilities are available in these days for selecting better breeding materials. The advanced phenomics facilities like automation, imaging system, and efficient software will help in high-throughput phenotypic data collections. It may improve the efficiency of crop improvement through precision breeding. Also, the available genomic information of millets can be helpful for gene and allele discovery, association and linkage mapping, gene validation, and MAS. The NGS technology provides the opportunity for the GWAS of millets that helps for the genomic selection of desirable traits for crop improvement. Therefore, researchers need to focus on millet improvement related to nutrient-use efficiency. It could help reduce the application of synthetic fertilizers and improve crop and soil health. Consumption of millets also helps to strengthen human health and provides food and nutritional security.

References

Ahmad Z, Nadeem F, Wang R, Diao X, Han Y, Wang X, Li X (2018) A larger root system is coupled with contrasting expression patterns of phosphate and nitrate transporters in foxtail millet [*Setaria italica* (L.) Beauv.] under phosphate limitation. Front Plant Sci 9:1367

Amadou I, Gounga ME, Le G-W (2013) Millets: nutritional composition, some health benefits and processing—a review. Emirates J Food Agric 25(7):501–508

Antony U, Sripriya G, Chandra T (1996) Effect of fermentation on the primary nutrients in finger millet (*Eleusine coracana*). J Agric Food Chem 44(9):2616–2618

Arya L, Chauhan D, Yadav Y, Verma M (2014) Transferability of simple sequence repeat (SSR) markers developed in finger millet, and pearl millet to kodo millet and barnyard millet. Innov Appr Stem Cell Res Cancer Biol Appl Biotech:60–64

Bandyopadhyay T, Muthamilarasan M, Prasad M (2017) Millets for next generation climate-smart agriculture. Front Plant Sci 8:1266

Bassirirad H (2000) Kinetics of nutrient uptake by roots: responses to global change. New Phytol 147:155–169

Bista DR, Heckathorn SA, Jayawardena DM, Mishra S, Boldt JK (2018) Effects of drought on nutrient uptake and the levels of nutrient-uptake proteins in roots of drought-sensitive and-tolerant grasses. Plan Theory 7(2):28

Buerkert A, Piepho H-P, Bationo A (2002) Multi-site time-trend analysis of soil fertility management effects on crop production in sub-Saharan West Africa. Exp Agric 38(2):163–183

Ceasar SA, Hodge A, Baker A, Baldwin SA (2014) Phosphate concentration and arbuscular mycorrhizal colonisation influence the growth, yield and expression of twelve PHT1 family phosphate transporters in foxtail millet (*Setaria italica*). PLoS One 9(9):e108459

Ceasar SA, Baker A, Ignacimuthu S (2017) Functional characterization of the PHT1 family transporters of foxtail millet with development of a novel *Agrobacterium*-mediated transformation procedure. Sci Rep 7(1):1–16

Ceasar S, Maharajan T, Ajeesh Krishna T, Ramakrishnan M, Victor Roch G, Satish L, Ignacimuthu S (2018) Finger millet [*Eleusine coracana* (L.) Gaertn.] improvement: current status and future interventions of whole genome sequence. Front Plant Sci 9:1054

Ceasar SA, Ramakrishnan M, Vinod K, Roch GV, Upadhyaya HD, Baker A, Ignacimuthu S (2020) Phenotypic responses of foxtail millet (*Setaria italica*) genotypes to phosphate supply under greenhouse and natural field conditions. PLoS One 15(6):e0233896

Chandra AK, Pandey D, Tiwari A, Sharma D, Agarwal A, Sood S, Kumar A (2020) An omics study of iron and zinc homeostasis in finger millet: biofortified foods for micronutrient deficiency in an era of climate change? OMICS: J Integr Biol 24(12):688–705

Chen G, Hu Q, Luo L, Yang T, Zhang S, Hu Y, Yu L, Xu G (2015) Rice potassium transporter OsHAK1 is essential for maintaining potassium-mediated growth and functions in salt tolerance over low and high potassium concentration ranges. Plant Cell Environ 38(12):2747–2765

Chowdary KA, Patra BC (2019) Effect of micronutrient application with different sources of NPK on growth and yield of finger millet crop in red laterite zone. J Agric Sci Tech B 9:403–416

Cormier F, Le Gouis J, Dubreuil P, Lafarge S, Praud S (2014) A genome-wide identification of chromosomal regions determining nitrogen use-efficiency components in wheat (*Triticum aestivum* L.). Theor Appl Genet 127(12):2679–2693

Cramer MD, Hawkins HJ, Verboom GA (2009) The importance of nutritional regulation of plant water flux. Oecologia 161(1):15–24

Debieu M, Sine B, Passot S, Grondin A, Akata E, Gangashetty P, Vadez V, Gantet P, Fonceka D, Cournac L, Hash CT (2018) Response to early drought stress and identification of QTLs controlling biomass production under drought in pearl millet. PLoS One 13(10):e0201635

Deepak Taggelli R, Thakur V (2018) Minor millets-their potential health benefits and medicinal properties: a review. Int J Pure Appl Sci 6(1):1677–1681

Devi PB, Vijayabharathi R, Sathyabama S, Malleshi NG, Priyadarisini VB (2014) Health benefits of finger millet (*Eleusine coracana* L.) polyphenols and dietary fiber: a review. J Food Sci Tech 51(6):1021–1040

Dhanalakshmi T, Ramesh S, Upadhyaya H, Rao A, Gangappa E, Priyadarshini S (2014) Genetic variability for morpho-agronomic traits in core germplasm collections of finger millet (*Eleusine coracana* (L.) Gaertn.) based on third and fourth degree statistics and their origin. Int J Tropic Agric 32(1/2):239–242

Dwivedi SL, Upadhyaya HD, Senthilvel S, Hash CT, Fukunaga K, Diao X, Santra D, Baltensperge D, Prasad M (2012) Millets: genetic and genomic resources. Plant Breed Rev 35(1):247–375

Erying C, Ling Q, Yanbing Y, Huawen Z, Hailian W, Bin L, Shuting Y, Runfeng W, Yanan G (2019) Variability of nitrogen-use efficiency by foxtail millet cultivars at the seedling stage. Pesq Agrop Brasileira 55:1–9

Fan X, Zhang W, Zhang N, Chen M, Zheng S, Zhao C, Han J, Liu J, Zhang X, Song L (2018) Identification of QTL regions for seedling root traits and their effect on nitrogen-use efficiency in wheat (*Triticum aestivum* L.). Theor Appl Genet 131(12):2677–2698

Fang Y, Wu W, Zhang X, Jiang H, Lu W, Pan J, Hu J, Guo L, Zeng D, Xue D (2015) Identification of quantitative trait loci associated with tolerance to low potassium and related ions concentrations at seedling stage in rice (*Oryza sativa* L.). Plant Growth Reg 77(2):157–166

Gandah M, Bouma J, Brouwer J, Hiernaux P, Van Duivenbooden N (2003) Strategies to optimize allocation of limited nutrients to sandy soils of the Sahel: a case study from Niger, West Africa. Agric Ecosyst Environ 94(3):311–319

Garnett T, Plett D, Heuer S, Okamoto M (2015) Genetic approaches to enhancing nitrogen-use efficiency (NUE) in cereals: challenges and future directions. Funct Plant Biol 42(10):921–941

Ge TD, Sun NB, Bai LP, Tong CL, Sui FG (2012) Effects of drought stress on phosphorus and potassium uptake dynamics in summer maize (*Zea mays*) throughout the growth cycle. Acta Physiol Plant 34:2179–2186

Goswami A, Prasad B, Joshi V (2015) Characterization of finger millet [*Eleusine coracana* (L.) Gaertn.] germplasm for morphological parameters under field conditions. J Appl Nat Sci 7(2): 836–838

Gupta AK, Gaur VS, Gupta S, Kumar A (2013) Nitrate signals determine the sensing of nitrogen through differential expression of genes involved in nitrogen uptake and assimilation in finger millet. Funct Integr Genomics 13(2):179–190

Han M, Wong J, Su T, Beatty PH, Good AG (2016) Identification of nitrogen-use efficiency genes in barley: searching for QTLs controlling complex physiological traits. Front Plant Sci 7:1587

Hassan Z, Sebola N, Mabelebele M (2021) The nutritional use of millet grain for food and feed: a review. Agric Food Secur 10(1):1–14

He M, Dijkstra FA (2014) Drought effect on plant nitrogen and phosphorus: a meta-analysis. New Phytol 204(4):924–931

Hegde PS, Rajasekaran NS, Chandra T (2005) Effects of the antioxidant properties of millet species on oxidative stress and glycemic status in alloxan-induced rats. Nutr Res 25(12):1109–1120

Jewel ZA, Ali J, Mahender A, Hernandez J, Pang Y, Li Z (2019) Identification of quantitative trait loci associated with nutrient-use efficiency traits, using SNP markers in an early backcross population of rice (*Oryza sativa* L.). Int J Mol Sci 20(4):900

Karunarathne SD, Han Y, Zhang X-Q, Zhou G, Hill CB, Chen K, Angessa T, Li C (2020) Genome-wide association study and identification of candidate genes for nitrogen-use efficiency in barley (*Hordeum vulgare* L.). Front Plant Sci 11:1361

Kindu GA, Tang J, Yin X, Struik PC (2014) Quantitative trait locus analysis of nitrogen-use efficiency in barley (*Hordeum vulgare* L.). Euphytica 199(1):207–221

Krishna T, Ceasar S, Maharajan T, Ramakrishnan M, Duraipandiyan V, Al-Dhabi N, Ignacimuthu S (2017) Improving the zinc-use efficiency in plants: a review. SABRAO J Breed Genet 49(3): 1–22

Krishna TA, Maharajan T, David RHA, Ramakrishnan M, Ceasar SA, Duraipandiyan V, Roch GV, Ignacimuthu S (2018) Microsatellite markers of finger millet (*Eleusine coracana* (L.) Gaertn) and foxtail millet (*Setaria italica* (L.) Beauv) provide resources for cross-genome transferability and genetic diversity analyses in other millets. Biocatal Agric Biotechnol 16:493–501

Krishna T, Maharajan T, Victor Roch G, Ignacimuthu S, Antony Ceasar S (2020a) Structure, function, regulation and phylogenetic relationship of ZIP family transporters of plants. Front Plant Sci 11:662

Krishna T, Theivanayagam M, Roch GV, Duraipandiyan V, Ignacimuthu S (2020b) Microsatellite marker: importance and implications of cross-genome analysis for finger millet (*Eleusine coracana* (L.) Gaertn). Curr Biotech 9(3):160–170

Krithika S, Balachandar D (2016) Expression of zinc transporter genes in rice as influenced by zinc-solubilizing *Enterobacter cloacae* strain ZSB14. Front Plant Sci 7:446

Kumar A, Tomer V, Kaur A, Kumar V, Gupta K (2018) Millets: a solution to agrarian and nutritional challenges. Agric Food Secur 7(1):1–15

Li P, Chen F, Cai H, Liu J, Pan Q, Liu Z, Gu R, Mi G, Zhang F, Yuan L (2015) A genetic relationship between nitrogen-use efficiency and seedling root traits in maize as revealed by QTL analysis. J Exp Bot 66(11):3175–3188

Maharajan T, Ceasar SA, Ajeesh Krishna TP, Ramakrishnan M, Duraipandiyan V, Naif Abdulla AD, Ignacimuthu S (2018) Utilization of molecular markers for improving the phosphorus efficiency in crop plants. Plant Breed 137(1):10–26

Maharajan T, Ceasar SA, Krishna TPA, Ignacimuthu S (2019) Phosphate supply influenced the growth, yield and expression of PHT1 family phosphate transporters in seven millets. Planta 250(5):1433–1448

Maharajan T, Antony Ceasar S, Ajeesh Krishna T, Ignacimuthu S (2021) Finger millet [*Eleusine coracana* (L.) Gaertn]: an orphan crop with a potential to alleviate the calcium deficiency in the semi-arid tropics of Asia and Africa. Front Sust Food Syst 5:684447

Masto RE, Chhonkar PK, Singh D, Patra AK (2007) Soil quality response to long-term nutrient and crop management on a semi-arid inceptisol. Agric Ecosyst Environ 118(1–4):130–142

Milroy SP, Bange MP, Thongbai P (2009) Cotton leaf nutrient concentrations in response to waterlogging under field conditions. Field Crops Res 113(3):246–255

Mirza N, Taj G, Arora S, Kumar A (2014) Transcriptional expression analysis of genes involved in regulation of calcium translocation and storage in finger millet (*Eleusine coracana* L. Gartn.). Gene 550(2):171–179

Nadeem F, Ahmad Z, Wang R, Han J, Shen Q, Chang F, Diao X, Zhang F, Li X (2018) Foxtail millet [*Setaria italica* (L.) Beauv.] grown under low nitrogen shows a smaller root system, enhanced biomass accumulation, and nitrate transporter expression. Front Plant Sci 9:205

Nadeem F, Ahmad Z, Ul Hassan M, Wang R, Diao X, Li X (2020) Adaptation of foxtail millet (*Setaria italica* L.) to abiotic stresses: a special perspective of responses to nitrogen and phosphate limitations. Front Plant Sci 11:187

Nithiyanantham S, Kalaiselvi P, Mahomoodally MF, Zengin G, Abirami A, Srinivasan G (2019) Nutritional and functional roles of millets—a review. J Food Biochem 43(7):e12859

Pariasca-Tanaka J, Baertschi C, Wissuwa M (2020) Identification of loci through genome-wide association studies to improve tolerance to sulfur deficiency in rice. Front Plant Sci 10:1668

Pestsova E, Lichtblau D, Wever C, Presterl T, Bolduan T, Ouzunova M, Westhoff P (2016) QTL mapping of seedling root traits associated with nitrogen and water-use efficiency in maize. Euphytica 209(3):585–602

Pudake RN, Mehta CM, Mohanta TK, Sharma S, Varma A, Sharma AK (2017) Expression of four phosphate transporter genes from finger millet (*Eleusine coracana* L.) in response to mycorrhizal colonization and Pi stress. 3 Biotech 7(1):1–17

Puranik S, Sahu PP, Beynon S, Srivastava RK, Sehgal D, Ojulong H, Yadav R (2020) Genome-wide association mapping and comparative genomics identifies genomic regions governing grain nutritional traits in finger millet (*Eleusine coracana* L. Gaertn.). Plants People Planet 2(6): 649–662

Ramakrishnan M, Ceasar SA, Vinod K, Duraipandiyan V, Ajeesh Krishna T, Upadhyaya HD, Al-Dhabi N, Ignacimuthu S (2017) Identification of putative QTLs for seedling stage phosphorus starvation response in finger millet (*Eleusine coracana* L. Gaertn.) by association mapping and cross species synteny analysis. PLoS One 12(8):e0183261

Rao K, Rao C, Rao K (1989) Response of finger miller (*Eleusine coracana* L Gaertin) cultivars to nitrogen under rain-fed conditions. Int J Agron 34(3):302–306

Ravindran G (1992) Seed protein of millets: amino acid composition, proteinase inhibitors and in-vitro protein digestibility. Food Chem 44(1):13–17

Rezaei EE, Gaiser T, Siebert S, Sultan B, Ewert F (2014) Combined impacts of climate and nutrient fertilization on yields of pearl millet in Niger. Eur J Agron 1(55):77–88

Ribaut J-M, Fracheboud Y, Monneveux P, Banziger M, Vargas M, Jiang C (2007) Quantitative trait loci for yield and correlated traits under high and low soil nitrogen conditions in tropical maize. Mol Breed 20(1):15–29

Roch GV, Maharajan T, Krishna TA, Ignacimuthu S, Ceasar SA (2020) Expression of PHT1 family transporter genes contributes for low phosphate stress tolerance in foxtail millet (*Setaria italica*) genotypes. Planta 252(6):1–9

Safdar LB, Andleeb T, Latif S, Umer MJ, Tang M, Li X, Liu S, Quraishi UM (2020) Genome-wide association study and QTL meta-analysis identified novel genomic loci controlling potassium use efficiency and agronomic traits in bread wheat. Front Plant Sci 11:70

Saha D, Gowda MC, Arya L, Verma M, Bansal KC (2016) Genetic and genomic resources of small millets. Crit Rev Plant Sci 35(1):56–79

Sahrawat K, Wani S, Rego T, Pardhasaradhi G, Murthy K (2007) Widespread deficiencies of sulphur, boron and zinc in dryland soils of the Indian semi-arid tropics. Curr Sci 93:1428–1432

Saleh AS, Zhang Q, Chen J, Shen Q (2013) Millet grains: nutritional quality, processing, and potential health benefits. Compr Rev Food Sci Food Safety 12(3):281–295

Sanaullah M, Rumpel C, Charrier X, Chabbi A (2012) How does drought stress influence the decomposition of plant litter with contrasting quality in a grassland ecosystem? Plant Soil 352: 277–288

Sardans J, Peñuelas J (2012) The role of plants in the effects of global change on nutrient availability and stoichiometry in the plant-soil system. Plant Physiol 160(4):1741–1761

Seghatoleslami MJ, Kafi M, Majidi E (2008) Effect of drought stress at different growth stages on yield and water use efficiency of five proso millet (*Panicum miliaceum* L.) genotypes. Pak J Bot 40(4):1427–1432

Shivran A (2016) Biofortification for nutrient-rich millets. In: Biofortification of food crops. Springer, New York, pp 409–420

Shobana S, Krishnaswamy K, Sudha V, Malleshi N, Anjana R, Palaniappan L, Mohan V (2013) Finger millet (Ragi, *Eleusine coracana* L.): a review of its nutritional properties, processing, and plausible health benefits. Adv Food Nutr Res 69:1–39

Sindhu S, Khetarpaul N (2001) Probiotics: nutritional and health benefits. Life Sci Rep 3:1–10

Singh UM, Metwal M, Singh M, Taj G, Kumar A (2015) Identification and characterization of calcium transporter gene family in finger millet in relation to grain calcium content. Gene 566(1):37–46

Soumya PR, Burridge AJ, Singh N, Batra R, Pandey R, Kalia S, Rai V, Edwards KJ (2021) Population structure and genome-wide association studies in bread wheat for phosphorus efficiency traits using 35 K wheat Breeder's Affymetrix array. Sci Rep 11(1):1–17

Srilekha K, Kamalaja T, Maheswari KU, Rani RN (2019) Nutritional composition of little millet flour. Int Res J Pure Appl Chem 20(4):1–4

Sun X, Ren W, Wang P, Chen F, Yuan L, Pan Q, Mi G (2020) Evaluation of maize root growth and genome-wide association studies of root traits in response to low nitrogen supply at seedling emergence. Crop J 9(4):794–804

Thilakarathna MS, Raizada MN (2015) A review of nutrient management studies involving finger millet in the semi-arid tropics of Asia and Africa. Agronomy 5(3):262–290

Tiwari A, Sharma D, Sood S, Jaiswal JP, Pachauri SP, Ramteke PW, Kumar A (2020) Genome-wide association mapping for seed protein content in finger millet (*Eleusine coracana*) global collection through genotyping by sequencing. J Cereal Sci 91:102888

Umesh M, Angadi S, Gowda P, Ghimire R, Begna S (2019) Climate-resilient minor crops for food security. In: Agronomic crops. Springer, New York, pp 19–32

Vetriventhan M, Azevedo VC, Upadhyaya HD, Nirmalakumari A, Kane-Potaka J, Anitha S, Ceasar SA, Muthamilarasan M, Bhat BV, Hariprasanna K, Bellundagi A (2020) Genetic and genomic resources, and breeding for accelerating improvement of small millets: current status and future interventions. Nucleus 63:217–239

Vinoth A, Ravindhran R (2017) Biofortification in millets: a sustainable approach for nutritional security. Front Plant Sci 8:29

Wang K, Cui K, Liu G, Xie W, Yu H, Pan J, Huang J, Nie L, Shah F, Peng S (2014) Identification of quantitative trait loci for phosphorus-use efficiency traits in rice using a high density SNP map. BMC Genet 15(1):1–15

Waraich EA, Ahmad R, Ashraf MY (2011) Role of mineral nutrition in alleviation of drought stress in plants. Aust J Crop Sci 5(6):764–777

Wei D, Cui K, Pan J, Wang Q, Wang K, Zhang X, Xiang J, Nie L, Huang J (2012) Identification of quantitative trait loci for grain yield and its components in response to low nitrogen application in rice. Aust J Crop Sci 6(6):986–994

Wissuwa M, Kondo K, Fukuda T, Mori A, Rose MT, Pariasca-Tanaka J, Kretzschmar T, Haefele SM, Rose TJ (2015) Unmasking novel loci for internal phosphorus utilization efficiency in rice germplasm through genome-wide association analysis. PLoS One 10(4):e0124215

Yang T, Zhang S, Hu Y, Wu F, Hu Q, Chen G, Cai J, Wu T, Moran N, Yu L (2014) The role of a potassium transporter OsHAK5 in potassium acquisition and transport from roots to shoots in rice at low potassium supply levels. Plant Physiol 166(2):945–959

Zhang LZ, Liu RH (2015) Phenolic and carotenoid profiles and antiproliferative activity of foxtail millet. Food Chem 174:495–501

Zhang H, Xiao W, Yu W, Yao L, Li L, Wei J, Li R (2018) Foxtail millet SiHAK1 excites extreme high-affinity K^+ uptake to maintain K^+ homeostasis under low K^+ or salt stress. Plant Cell Rep 37(11):1533–1546

Zhang M, Gao M, Zheng H, Yuan Y, Zhou X, Guo Y, Zhang G, Zhao Y, Kong F, An Y (2019) QTL mapping for nitrogen-use efficiency and agronomic traits at the seedling and maturity stages in wheat. Mol Breed 39(5):1–17

11 Current Status of Bioinformatics Resources of Small Millets

Thiyagarajan Thulasinathan, Priyanka Jain, Arvind Kumar Yadav, Vishesh Kumar, Amitha Mithra Sevanthi, and Amolkumar U. Solanke

Abstract

Small millets belong to a Poaceae family that are widely cultivated and superior to other cereals for their nutritional quality. Genomic databases are designed for research purposes, it contains large repositories of DNA sequences, open reading frames (ORFs), intergenic sequences, and whole genome sequences of important crop species. Nuclear genome sequences for few millets like foxtail millet, proso millet, kodo millet, finger millet, little millet, Indian and Japanese barnyard millet, and Teff are available in the public domain. Gramene, millet genome databases, and phytozome databases are genomic resources for finger millet and foxtail millet. Foxtail millet database resources include transcription factor database and microRNA database. Besides that, Marker database and Transposable Elements-based Marker Database for foxtail millet are useful for marker-related studies in small millets. There are several areas of research still unexplored in small millets as genomic resources are meager compared to other major cereals. However, considering very high nutritional potential of these crops, there is a need to generate lot of database resources for small millets. In this chapter we have reviewed the information about all available databases related to small millets, and their potential application in improvement in yield and other traits.

Keywords

Small millets · Genome sequences · Markers · Transcription factor Databases

T. Thulasinathan · P. Jain · A. K. Yadav · V. Kumar · A. M. Sevanthi · A. U. Solanke (✉)
ICAR - National Institute for Plant Biotechnology, New Delhi, India
e-mail: amol.solanke@icar.gov.in

R. N. Pudake et al. (eds.), *Omics of Climate Resilient Small Millets*,
https://doi.org/10.1007/978-981-19-3907-5_11

11.1 Introduction

Small millets, also known as minor millets, belong to the grass family Poaceae. They include finger millet (*Eleusine coracana*), barnyard millet (*Echinochloa crus-galli*), foxtail millet (*Setaria italica*), proso millet (*Panicum miliaceum*), kodo millet (*Paspalum scrobiculatum*), fonio (*Digitaria exilis*), little millet (*Panicum sumatrense*), teff (*Eragrostis tef*), browntop millet (*Urochloa ramosa*), and guinea millet (*Brachiaria deflexa*). Small millets have different centers of origins. South Asian and Semi-Arid Tropics (SAT) of African countries are the native place for some of the small millets (Saha et al. 2016). Nutrition-wise, small millet grains possess micronutrients, especially iron and calcium. Apart from these, minor millets are rich in essential amino acids and high dietary fiber content (Devi et al. 2014; Shobana et al. 2013). They play an important role as a staple crop for poor people. With recent change in nutritional awareness, slowly minor millets are becoming nutritious food for urban people. Hence it is very essential to direct more research and development towards these minor millet crops. In the present chapter, we have discussed the status of genome sequences and available online databases of minor millets.

11.2 Genome Sequences Available in Minor Millets

Next-generation sequencing (NGS) can produce thousands of sequences in a fraction of time and is applicable to various crop plants without any information about their genetic maps or molecular markers. Advances in NGS have provided a lot of genomic sequences in different field crops (Varshney 2010; Varshney et al. 2009). Besides that, these sequence data could ease the generation of markers from the genome that helps in the identification of structure, diversity, evolution, variation in mapping sequence also for identifying marker tools to use in marker-assisted breeding for crop improvement. The noncoding and coding portions of the genome sequences can be identified by the genome sequencing and are involved in the growth regulation and development, and tolerance/resistance to the abiotic stresses. Increasing numbers of sequencing methods not only reduced the costs but also provide deeper and wide coverage and more reliability of genome sequences. In minor millets, nuclear genomes of foxtail millet, finger millet, proso millet, little millet, kodo millet, Japanese barnyard millet, Indian barnyard millet, and teff were already sequenced (Table 11.1) and complete genome sequences of chloroplast from proso millet, foxtail millet, barnyard millet, and little millet are also reported (Table 11.2).Genomic resources for finger millet, foxtail millet, barnyard millet, proso millet, little millet, and teff are present in National Center for Biotechnology Information (NCBI) database.

Foxtail millet is used as a reference for other C4 crops due to its many advantages including its short duration, small diploid genome size (423–510 Mb), and self-pollinating nature. Among small millets, the foxtail millet genome was first to be sequenced by researchers. Whole genome shotgun sequencing along with NGS was

Table 11.1 Nuclear genome sequence information of different minor millets

S. no	Common name	Scientific name	Ploidy level	Chromosome number	Genotype name	Genome size	No. of genes	ESTs available from NCBI	Sequence type	Reference
1	Finger millet	*E. coracana*	Tetraploid	$2n = 4x = 36$	ML-365	1196 Mb	85,243	1982	Draft genome	Hittalmani et al. (2017)
					PR202	1500 Mb	62,348		Draft genome	Hatakeyama et al. (2018)
2	Foxtail millet	*S. italica*	Diploid	$2n = 2x = 18$	Yugu1	510 Mb	24,000–29,000	66,051	Reference genome	Bennetzen et al. (2012)
					Zhang gu	423 Mb	38,801		Draft genome	Zhang et al. (2012)
4	Proso millet	*P. miliaceum*	Tetraploid	$2n = 4x = 36$	Landrace (Accession Number 00000390)	923 Mb	55,930	211	Draft genome	Zou et al. (2019)
					Longmi4	887.8 Mb	63,671		Draft genome	Shi et al. (2019)
5	Japanese barnyard millet	*E. crus-galli*	Hexaploid	$2n = 6x = 36$	STB08	1.27 Gb	108,771	–	Draft genome	Guo et al. (2017)
6	Teff	*E. tef*	Tetraploid	$2n = 4x = 40$	Tsedey (DZ-Cr-37)	672 Mb	28,113	–	Draft genome	Cannarozzi et al. (2014)

Table 11.2 Chloroplast genome sequence information of different minor millets

S. no	Crop	Scientific name	Genome size (bp)	No. of gene(s)	References
1.	Foxtail millet	*S. italica*	135,516	111	Wang and Gao (2016)
2.	Proso millet	*P. miliaceum*	139,929	132	Cao et al. (2017)
		P. miliaceum	139,826	108	Nie et al. (2018)
3.	Little millet Accession No. IT261894	*P. sumatrense*	139,384	125	Sebastin et al. (2018)
4.	Barnyard millet Accession No. IT230633	*E. esculenta*	139,851	111	Sebastin et al. (2019)
	Barnyard millet BTS02	*E. crus-galli*	139,800	–	Ye et al. (2014)
	Barnyard millet CO(KV)2	*E. frumentacea*	139,593	112	Perumal et al. (2016)

performed to a draft sequence of a foxtail millet cultivar "Zhang gu." Around 423 Mb (85%) from the total genome have been mapped, with 29% consisting of repeat elements, and 38,801 genes were identified and mapped on its total genome. The genome sequence contained a total of 16,903 contigs, and 439 scaffolds (Zhang et al. 2012a, b). "A10" and "Yugu1" were another foxtail millet cultivars that were sequenced by Bennetzen et al. (2012) using AB13730xl capillary sequencer, covering ~80% of the total genome with 5,736,559 of total reads. With 24,000–29,000 expressed genes identified and distributed on 9 chromosomes; the genome size for these sequences was computed to be 396.7 MB. The *Setaria viridis* ME034V cultivar is highly transformable, due to lack of genome sequence of this cultivar, that has reduced its usage. The genome sequencing was done using ultra long nanopore sequencing technology. Among all cultivars ME043V cultivar given the largest haploid genome size of 465.6 Mb from raw unprocessed sequence data, and 421.0 Mb from quality trimmed and organellar filtered sequence data. Around 37,908 protein coding genes and more than 300 K repetitive elements containing 46% of the genome, were identified through genome annotation studies (Thielen et al. 2020).

In another study, the whole genome sequencing of ML-365, a finger millet cultivar (a blast and drought tolerant) was completed using SOLiD (Sequencing by Oligonucleotide Ligation and Detection) and Illumina sequencing methods. The genome assembly consists of the average scaffold length of 2275 bp with 525,759 scaffolds having 23.73 Kb length of N50 (Hittalmani et al. 2017). Teff (*E. tef*) is a gluten-free cereal with high vitamin, protein, and mineral content. The genome of a teff accession "DZ-Cr-37" was sequenced by using Illumina HiSeq2000 and 454-FLX platforms, giving 44-fold coverage of the full genome. Forty Gigabit of the single-end and paired-end reads, covering 80% of the total genome were generated. From which 49,600 SSR markers from 672 Mb of the assembled genome were developed and additionally, 570 miRNAs, 1184 tRNAs, and 80 rRNAs were also detected (Cannarozzi et al. 2018).

Regardless of available NGS platforms, only draft sequences are available for most of the millets. To correct the mis-annotation and proper chromosomal assignments in genome sequences, the chromosomal level of genome assemblies and reannotation studies will be required to fill the gaps soon. However, the sequence information at the draft level can also be used by scientists for gene mining and genotyping at a large scale. In following paragraphs, we have discussed the available database information and its probable applications.

11.3 Online Genomic Resources and Databases Available for Small Millets

11.3.1 Gramene Portal

The Gramene Knowledgebase (https://www.gramene.org/) is a curated, freely available, integrated, open-source, database resource for model crops and plant species, specifically for comparative functional genomic studies. Gramene portal was created by partnership of Cold Spring Harbor Laboratory, Oregon State University, and EMBL-EBI (Tello-Ruiz et al. 2021). It provides free access to the information of 114 reference genomes with 123,064 gene family trees. Genomic information related to the Foxtail millet (*S. viridis, S. italica*) and Teff (*E. tef*) were present in this portal (Tello-Ruiz et al. 2016).

11.3.1.1 Gene and Genome Browsers

Gramene's web page (https://ensembl.gramene.org/genome_browser/index.html) has 114 genome sequences of different plant species, that can be accessed freely (https://ensembl.gramene.org/species.html; Accessed on March 28, 2022). An individual could choose the crop species and genome browser window of their interest which will show the tools like genome assembly, gene annotation (where we can find protein-coding and noncoding genes, splice variants, cDNA and protein sequences, noncoding RNAs.), comparative genomics (for studying homologs, gene trees, and whole genome alignments across multiple species), variation (for searching short sequence variants), and regulation (for microarray annotations) (Gupta et al. 2016).

11.3.1.2 Pathway Browser

Plant Reactome (http://plantreactome.gramene.org) is freely accessible database of plant metabolic and regulatory pathways. It gives access to various data analysis tools like video tutorials, quick search, user guide, data download, data model, database release summary, pathway browser, etc. By selecting the "browse pathways" or any link related to that pathway, researchers could access the default Pathway Browser for around 107 plant species (https://plantreactome.gramene.org/index.php?lang=en, Accessed on March 28, 2022). Data belonging to the selected entities can be downloaded in different file formats such as Word, PDF, SBML, and BioPAX (Gupta et al. 2016; Naithani et al. 2017, 2020).

11.3.1.3 Plant Gene Expression Atlas

From the Gramene search page researchers can download the Atlas of Plant Gene Expression (https://www.ebi.ac.uk/gxa/plant/experiments) was designed by people at European Molecular Biology Laboratory and European Bioinformatics Institute (EMBL-EBI). It is having transcriptomic data related to 982 experiments of 26 different plant species, which includes tissue-specific differential expression of genes at standard level (https://www.ebi.ac.uk/gxa/plant/experiments#; Accessed on March 28, 2022). This data related to differential expression can be seen on the website of Plant Gene Expression Atlas (Gupta et al. 2016). Transcriptome sequencing of Foxtail millet for different tissues is also available in this database, where the data for gene expression in leaf, root, stem, and tassel inflorescence is given (Zhang et al. 2012).

11.3.1.4 Analysis and Visualization Tools

Gramene has also many tools to analyze and visualize the processing of users uploaded and preloaded data. This genome browser provides several options for portraying different data types, including data on genomic variation [ESTs, structural variants, and single nucleotide polymorphisms (SNPs)]. Also, users can query different orthologs for multiple target species using nucleotide or amino acid sequences in blast tool. It also permits the users for uploading their derived data, including methylomes, long noncoding RNAs, transcriptomes, proteomes, etc. Users can also access and download the pre-analyzed data for few species, and the results of their genetic variants can be accessed online (Gupta et al. 2016; Naithani et al. 2017, 2020).

11.3.2 Millet Genome Database

The Millet genome database (https://db.cngb.org/datamart/plant/DATApla6/) is based on the data of the millet genome project developed by Beijing Genomics Institute (BGI) and Zhangjiakou Academy of Agricultural Sciences, China (CNGBdb; China National GeneBank Database). The database records the genotypic as well as phenotypic information of the millets. Besides, the database also applies machine learning to construct the genotype-phenotype model to promote molecular breeding (Fig. 11.1).

11.3.3 *Setaria italica* Genome Database (SiGDB)

SiGDB is derived from plantGDB genomics database. The main aim of SiGDB is to provide a suitable sequence-centered view of *foxtail* genome. SiGDB (https://www.plantgdb.org/SiGDB/) has been developed as a part of PlantGDB (http://www.

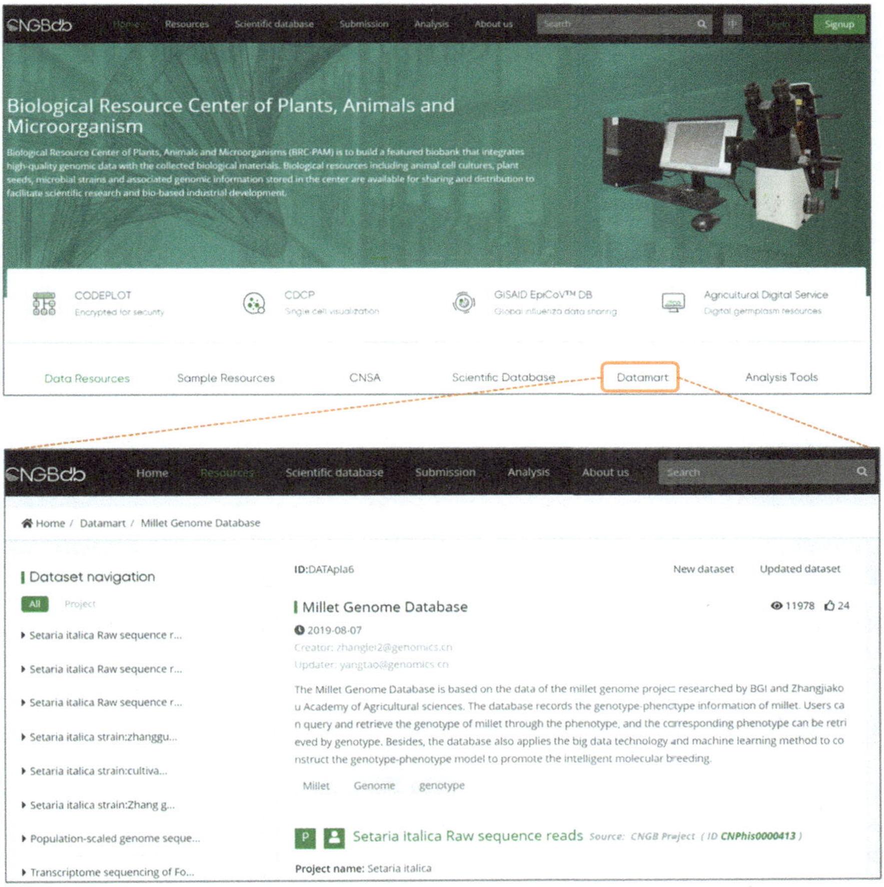

Fig. 11.1 Home page of Millet Genome Database (Accessed from China National Gene bank database on March 28, 2022)

plantgdb.org/). The database shows good quality of partial or full-length, spliced alignments for cDNA, EST, related species protein, PUT, and probe. Any sequence related to *Seteria italica* can be searched or downloaded using search/download links at SiGDB (Fig. 11.2).

Different tools available in this database are Blast SiGDB, Blast allGDBs, Gene seqer, and Genome Threader. Blast SiGDB tool utilizes Blastn, Blastp, tBlastn, and tBlastx to match a query sequence to SiGDB sequences, whereas blast allGDBs tool uses Blastn, Blastp, tBlastn, and tBlastx to match a query sequence to other crop sequences. Gene seqer tool has user specified transcript datasets spliced alignments to genomic DNA from any region in SiGDB, whereas genome threader tool is used to compute gene structure predictions, it has user specified protein datasets spliced alignments to genomic DNA from any region in SiGDB (Duvick et al. 2007).

Fig. 11.2 Home page of SiGDB (Accessed from https://www.plantgdb.org/SiGDB/ on March 28, 2022)

11.3.4 Foxtail Millet Marker Database (FmMDb)

The Foxtail millet Marker Database (FmMDb; http://www.nipgr.res.in/foxtail.html) is an online database to manage, retrieve, and visualize large number of markers with public open access in unrestricted manner. FmMDb database is the first to provide complete marker details to the plant researchers and breeders for improving elite cultivars of millet and grass species. It is the first database, which is completely designed for millet crop and other grass species for structural and comparative genomics studies. This database provides three different types of markers data, which includes genomic SSR, genic SSR, and intron length polymorphic (ILP) markers for plant researchers. Large number of DNA markers were identified from the genome sequence of foxtail millet after releasing of FmMDb to help in the breeding program for the advancement of foxtail and other small millets (Muthamilarasan et al. 2013).

11.3.5 Foxtail Millet MicroRNA Database (FmMiRNADb)

During the multiple biological processes in gene regulation, miRNAs play a major role in response to stress, through post-transcriptional control. FmMiRNADb (http://223.31.159.9/FmMiRNADb/index.html) is the first web-based database of foxtail microRNA (miRNA) sequences, its secondary structure and sequence

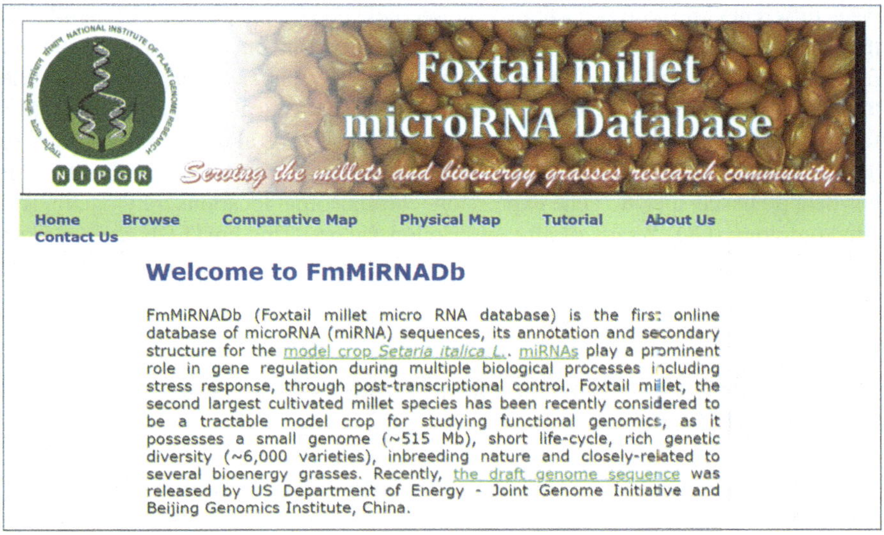

Fig. 11.3 Home page of FmMiRNADb. (Accessed from http://223.31.159.9/FmMiRNADb/index.html on March 28, 2022)

annotation (Khan et al. 2014). The browse section has start point for whole miRNA information, based on chromosomal location the miRNA data is sorted out in foxtail millet. By choosing individual chromosome number, it will show the presence of miRNA in the respective chromosome, whereas the view map will show the physical map of individual chromosome. The comparative map section will show the syntenic connection between physically mapped 355 miRNAs present over the 9 chromosomes in foxtail millet and their physical location on the chromosomes of closely related grass species genomes. It has also whole comparative mapping data, which is uploaded in the comparative map section for the researchers to download. The physical map section has physical map of all 123 miRNA-based markers on the 9 chromosomes in foxtail millet (Fig. 11.3).

11.3.6 Foxtail Millet Transcription Factor Database (FmTFDb)

Transcription factors are involved in controlling gene expression through binding to regulatory cis-elements in the promoters of particular genes. DNA binding transcription factors are especially involved in activation or repression of gene expression. Transcription factors are defined as multi-functional proteins, at the same time that may be involved in multiple signal transduction pathways during abiotic stresses in plants (Hrmova and Hussain 2021). Foxtail millet Transcription Factor Database (FmTFDb: http://59.163.192.91/FmTFDb/index.html) contains 2297 true transcription factors grouped into 55 different families. FmTFDb has three different approaches for browsing the database. The first approach is to browse TFs on the basis different chromosome, whereas the second approach is browsed by family. In this method, the transcription factors are separated out on the basis of different family. The third and last approach is to browse by physical map, in which the transcription factors are separated out on the basis of physical position of respective family. This database has BLAST search options and several other tools in order to identify gene ontology terms that are enriched and also to visualize physical map. It has data on tissue-specific gene expression, chromosomal locations, phylogeny, and gene ontology (GO) details of all transcription factors. This FmTFDb can act as an important resource for users involved in millet improvement program (Bonthala et al. 2014).

11.3.7 Foxtail Millet Transposable Elements-Based Marker Database (FmTEMDb)

Foxtail millet Transposable Elements-based Marker Database (FmTEMDb) (http://59.163.192.83/ltrdb/index.html) is an open resource database (Yadav et al. 2015). A total of 30,706 transposable elements (TE) were identified by a research group in foxtail millet. Further, all TE were classified as DNA transposons (24,386), full-length Gypsy type (1570), partial or solo Copia type (10,118), Long and Short-Interspersed Nuclear Elements (3659 and 53, respectively), partial or solo Gypsy type (23,293), and full-length Copia type (1038). They have also identified 20,278 TE-based markers. In millets for genotyping applications in large scale manner, six different types of markers were identified. In total 134 markers in Repeat Junction were also studied in 3 wild and 96 accessions of *Setaria italica,* and out of which 30 showed the polymorphism.

11.3.8 Phytozome Database

Phytozome is a portal of plant comparative genomics of joint genome institute. It is comparative platform for gene family data and plant genome and analysis. Phytozome gives the evolutionary record of each gene present in plant at the level of genome organization, sequence, gene structure, and gene family. Simultaneously, it provides access to complete plant genome functional annotations and sequences of all the plants available in the earth surface at joint genome institute. Ultimately phytozome is a strong comparative genomics tool, which assists to join model system with other economically and environmentally important plants (Goodstein et al. 2012). The following genomic resources are available in Phytozome: *Eleusine coracana* v1.1—finger millet, *Setaria italica* v2.2—foxtail millet, *Setaria viridis* v1.1—green foxtail millet, *Setaria viridis* v2.1—green foxtail millet (Available on https://phytozome-next.jgi.doe.gov/jbrowse/index.html on 28th March 2022) (Fig. 11.4).

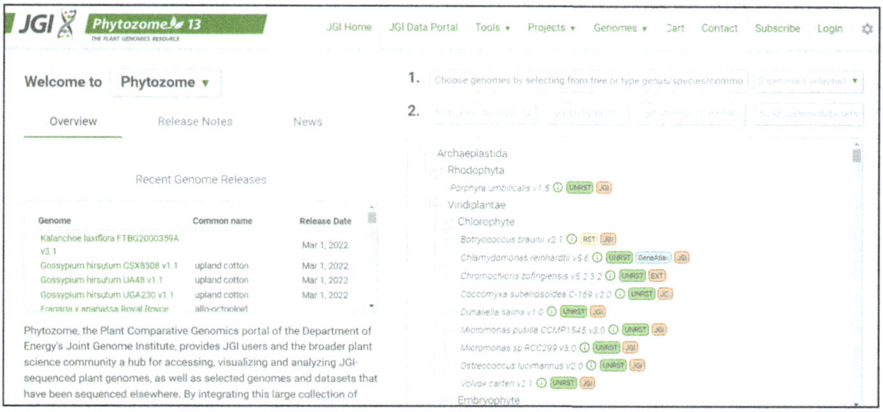

Fig. 11.4 Home page of phytozome. (Accessed from https://phytozome-next.jgi.doe.gov/jbrowse/index.html on 28th March 2022)

11.3.8.1 Text and Sequence Search

Gene families and genes can be obtained by using sequence similarity and keyword searches from phytozome database. User can also look for functional annotations from database (e.g., GO, KEGG, Panther, KOG, and PFAM), which obtains the whole set of all corresponding functional identifiers and later it will carry out another search for identifying important families possessing similar functions.

11.3.8.2 Gene Family and Gene Page Views

User can obtain elaborate information on each family and its component members from gene family view. Based on species, the gene in this family shows its individual family members. Separate links are provided to both a gene page view and a GBrowse view for each family member.

11.3.8.3 Data Access

Bulk data files are available in phytozome, that includes gene structure, transcript, coding sequence, genome assembly sequence, and peptide sequence in the FASTA format. Query-based data access can also be obtained by the phytozome's Phytomine and Biomart tools, whereas the bulk data sets can be derived through Joint Genome Institute genome portal.

11.3.9 Plant Metabolic Network Database

The following tools available in Plant metabolic network database can be accessed from https://plantcyc.org/content/pmn-tools.

11.3.9.1 Blast Tool

User can run a protein or gene of interest using PMN BLAST server to identify a homologous gene or proteins in species databases uploaded in plant metabolic network.

11.3.9.2 Metabolic Cluster View

It is web service for browsing, searching, and downloading metabolic gene clusters from several plant species. User can identify and see plant metabolic cluster from plant cluster finder.

11.4 Conclusion and Prospects

The minor millets are mostly cultivated by farmers for their high health boosting potentials, and nutritive value, able to display resistance to extreme environmental conditions especially drought stress and able to survive with low inputs. Based on availability of genomic resources and databases, which includes gene sequences, expressed sequences, microRNA sequences, transcription factors, molecular markers, and gene-based markers in minor millets up to date are utilized mostly in the cases of finger millet, foxtail millet, and proso millet. With the recent Whole Genome Sequencing (WGS) of finger millet, foxtail millet and proso millet varieties will act as a reference genome and serve as major function in relative analysis from genomic and genetic aspects. The versatile use of NGS tools for expressed sequences, genome and NGS derived SNP markers in minor millets, will also have significant contributions in genetic improvement of minor millets. Considering wider adaptation, good health benefits and agronomic characteristics of minor millet cultivars could be an alternative or additive millet to ensure food, feed, and nutritional security. With the help of online available database resources, now minor millets can be analyzed and enhanced in a better way. It is predicted that with the help of contemporary genetic and genomic resources along with its online databases, it can help us to improve the exploitation of this millet crops in a more effective manner.

References

Bennetzen JL, Schmutz J, Wang H, Percifield R, Hawkins J, Pontaroli AC, Estep M, Feng L, Vaughn JN, Grimwood J, Jenkins J (2012) Reference genome sequence of the model plant Setaria. Nat Biotechnol 30(6):555–561

Bonthala VS, Muthamilarasan M, Roy R, Prasad M (2014) FmTFDb: a foxtail millet transcription factors database for expediting functional genomics in millets. Mol Biol Rep 41(10):6343–6348

Cannarozzi G, Plaza-Wüthrich S, Esfeld K, Larti S, Wilson YS, Girma D, de Castro E, Chanyalew S, Blösch R, Farinelli L, Lyons E (2014) Genome and transcriptome sequencing identifies breeding targets in the orphan crop tef (*Eragrostis tef*). BMC Genomics 15(1):1–21

Cannarozzi G, Weichert A, Schnell M, Ruiz C, Bossard S, Blösch R, Plaza-Wüthrich S, Chanyalew S, Assefa K, Tadele Z (2018) Waterlogging affects plant morphology and the expression of key genes in tef (*Eragrostis tef*). Plant Direct 2(4):e00056

Cao X, Wang J, Wang H, Liu S, Chen L, Tian X, Qin H, Wang L, Na X, Qiao Z (2017) The complete chloroplast genome of *Panicum miliaceum*. Mitochondrial DNA B Resour 2(1):43–45

Devi PB, Vijayabharathi R, Sathyabama S, Malleshi NG, Priyadarisini VB (2014) Health benefits of finger millet (Eleusine coracana L.) polyphenols and dietary fiber: a review. J Food Sci Technol 51(6):1021–1040

Duvick J, Fu A, Muppirala U, Sabharwal M, Wilkerson MD, Lawrence CJ, Lushbough C, Brendel V (2007) PlantGDB: a resource for comparative plant genomics. Nucleic Acids Res 36 (Suppl_1):D959–D965

Goodstein DM, Shu SQ, Howson R, Neupane R, Hayes RD, Fazo J, Mitros T, Dirks W, Hellsten U, Putnam N, Rokhsar DS (2012) Phytozome: a comparative platform for green plant genomics. Nucleic Acids Res 40(D1):D1178–D1186

Guo L, Qiu J, Ye C, Jin G, Mao L, Zhang H, Yang X, Peng Q, Wang Y, Jia L, Lin Z (2017) Echinochloa crus-galli genome analysis provides insight into its adaptation and invasiveness as a weed. Nat Commun 8(1):1–10

Gupta P, Naithani S, Tello-Ruiz MK, Chougule K, D'Eustachio P, Fabregat A, Jiao Y, Keays M, Lee YK, Kumari S, Mulvaney J (2016) Gramene database: navigating plant comparative genomics resources. Curr Plant Biol 7:10–15

Hatakeyama M, Aluri S, Balachadran MT, Sivarajan SR, Patrignani A, Grüter S, Poveda L, Shimizu-Inatsugi R, Baeten J, Francoijs KJ, Nataraja KN (2018) Multiple hybrid de novo genome assembly of finger millet, an orphan allotetraploid crop. DNA Res 25(1):39–47

Hittalmani S, Mahesh HB, Shirke MD, Biradar H, Uday G, Aruna YR, Lohithaswa HC, Mohanrao A (2017) Genome and transcriptome sequence of finger millet (*Eleusine coracana* (L.) Gaertn.) provides insights into drought tolerance and nutraceutical properties. BMC Genomics 18(1):1–6

Hrmova M, Hussain SS (2021) Plant transcription factors involved in drought and associated stresses. Int J Mol Sci 22(11):5662

Khan Y, Yadav A, Bonthala VS, Muthamilarasan M, Yadav CB, Prasad M (2014) Comprehensive genome-wide identification and expression profiling of foxtail millet [Setaria italica (L.)] miRNAs in response to abiotic stress and development of miRNA database. Plant Cell Tissue Organ Cult 118(2):279–292

Muthamilarasan M, Misra G, Prasad M (2013) FmMDb: a versatile database of foxtail millet markers for millets and bioenergy grasses research. PLoS One 8(8):e71418

Naithani S, Preece J, D'Eustachio P, Gupta P, Amarasinghe V, Dharmawardhana PD, Wu G, Fabregat A, Elser JL, Weiser J, Keays M, Fuentes AM, Petryszak R, Stein LD, Ware D, Jaiswal P (2017) Plant Reactome: a resource for plant pathways and comparative analysis. Nucleic Acids Res 45(D1):D1029–D1039. https://doi.org/10.1093/nar/gkw932

Naithani S, Gupta P, Preece J, D'Eustachio P, Elser J, Garg P, Dikeman DA, Kiff J, Cook J, Olson A, Wei S, Tello-Ruiz MK, Mundo AF, Munoz-Pomer A, Mohammed S, Cheng T, Bolton E, Papatheodorou I, Stein L, Ware D, Jaiswal P (2020) Plant Reactome: a knowledgebase and resource for comparative pathway analysis. Nucleic Acids Res 48(D1): D1093–D1103. https://doi.org/10.1093/nar/gkz996

Nie X, Zhao X, Wang S, Zhang T, Li C, Liu H, Tong W, Guo Y (2018) Complete chloroplast genome sequence of broomcorn millet (*Panicum miliaceum* L.) and comparative analysis with other Panicoideae species. Agronomy 8(9):159

Perumal S, Jayakodi M, Kim DS, Yang TJ, Natesan S (2016) The complete chloroplast genome sequence of Indian barnyard millet, *Echinochloa frumentacea* (Poaceae). Mitochondrial DNA B Resour 1(1):79–80

Saha D, Gowda MC, Arya L, Verma M, Bansal KC (2016) Genetic and genomic resources of small millets. Crit Rev Plant Sci 35(1):56–79

Sebastin R, Lee GA, Lee KJ, Shin MJ, Cho GT, Lee JR, Ma KH, Chung JW (2018) The complete chloroplast genome sequences of little millet (*Panicum sumatrense* Roth ex Roem. and Schult.) (Poaceae). Mitochondrial DNA B Resour 3(2):719–720

Sebastin R, Lee KJ, Cho GT, Lee JR, Shin MJ, Kim SH, Lee GA, Chung JW, Hyun DY (2019) The complete chloroplast genome sequence of Japanese millet *Echinochloa esculenta* (A. braun) H. scholz (Poaceae). Mitochondrial DNA B Resour 4(1):1392–1393

Shi J, Ma X, Zhang J, Zhou Y, Liu M, Huang L, Sun S, Zhang X, Gao X, Zhan W, Li P (2019) Chromosome conformation capture resolved near complete genome assembly of broomcorn millet. Nat Commun 10(1):1–9

Shobana S, Krishnaswamy K, Sudha V, Malleshi NG, Anjana RM, Palaniappan L, Mohan V (2013) Finger millet (Ragi, Eleusine coracana L.): a review of its nutritional properties, processing, and plausible health benefits. Adv Food Nutr Res 69:1–39

Tello-Ruiz MK, Stein J, Wei S, Preece J, Olson A, Naithani S, Amarasinghe V, Dharmawardhana P, Jiao Y, Mulvaney J, Kumari S (2016) Gramene 2016: comparative plant genomics and pathway resources. Nucleic Acids Res 44(D1):D1133–D1140

Tello-Ruiz MK, Naithani S, Gupta P, Olson A, Wei S, Preece J, Jiao Y, Wang B, Chougule K, Garg P, Elser J, Kumari S, Kumar V, Contreras-Moreira B, Naamati G, George N, Cook J, Bolser DM, D'Eustachio P, Stein LD, Gupta A, Xu W, Regala J, Papatheodorou I, Kersey PJ, Flicek P, Taylor C, Jaiswal P, Ware D (2021) Gramene 2021: harnessing the power of comparative genomics and pathways for plant research. Nucleic Acids Res 49(D1):D1452–D1463. https://doi.org/10.1093/nar/gkaa979. PMID: 33170273

Thielen PM, Pendleton AL, Player RA, Bowden KV, Lawton TJ, Wisecaver JH (2020) Reference genome for the highly transformable Setaria viridis ME034V. G3 (Bethesda) 10(10):3467–3478

Varshney RK (2010) Gene-based marker systems in plants: high throughput approaches for marker discovery and genotyping. In: Molecular techniques in crop improvement. Springer, Dordrecht, pp 119–142

Varshney RK, Nayak SN, May GD, Jackson SA (2009) Next-generation sequencing technologies and their implications for crop genetics and breeding. Trends Biotechnol 27(9):522–530

Wang S, Gao LZ (2016) The complete chloroplast genome of an irreplaceable dietary and model crop, foxtail millet (Setaria italica). Mitochondrial DNA Part A 27(6):4442–4443

Yadav CB, Bonthala VS, Muthamilarasan M, Pandey G, Khan Y, Prasad M (2015) Genome-wide development of transposable elements-based markers in foxtail millet and construction of an integrated database. DNA Res 22(1):79–90. https://doi.org/10.1093/dnares/dsu039

Ye CY, Lin Z, Li G, Wang YY, Qiu J, Fu F, Zhang H, Chen L, Ye S, Song W, Jin G (2014) Echinochloa chloroplast genomes: insights into the evolution and taxonomic identification of two weedy species. PLoS One 9(11):e113657

Zhang G, Liu X, Quan Z, Cheng S, Xu X, Pan S, Xie M, Zeng P, Yue Z, Wang W, Tao Y (2012) Genome sequence of foxtail millet (*Setaria italica*) provides insights into grass evolution and biofuel potential. Nat Biotechnol 30(6):549–554

Advances in Omics for Enhancing Abiotic Stress Tolerance in Finger Millets

12

Sonam Singh, Mayuri D. Mahalle, Mareyam Mukhtar, Gitanjali Jiwani, Amitha Mithra Sevanthi, and Amolkumar U. Solanke

Abstract

Finger millet (*Eleusine coracana*) a Poaceae family member, is the sixth important food crop, which has very high nutritional value but can be grown in a variety of problem soils and marginal areas. Commonly known as "ragi," the crop is known for its tolerance to various abiotic stresses like drought, high temperature, alkaline/low pH, and saline soils. The cereal is not only a rich source of carbohydrates, proteins, and minerals but also highly enriched in amino acid methionine and calcium with certain medicinal properties. Development of molecular markers followed by that of a genetic map has accelerated research in finger millet improvement through genetic diversity studies, QTL mapping, association mapping, and marker-assisted selection. The recent release of two draft genome sequences of finger millet has given a major thrust to development of new age NGS-based markers like SNPs and DArT. The abundance of markers in ragi can serve as a platform for genomic selection in finger millet. On the other hand, lots of omics data has been generated from genomics, transcriptomics, proteomics, and a few metabolomics and ionomics studies in this crop. These data are available on various online databases and omics platforms to finger millet researchers. To understand the regulatory network for abiotic stress tolerance in finger millet, several functional genomics approaches were used leading to the identification of some potential candidate genes. These candidate genes can serve

S. Singh
ICAR-National Institute for Plant Biotechnology, New Delhi, India

Department of Horticulture, College of Agriculture and Life Science, Chungnam National University, Daejeon, Republic of Korea

M. D. Mahalle · M. Mukhtar · G. Jiwani · A. M. Sevanthi · A. U. Solanke (✉)
ICAR-National Institute for Plant Biotechnology, New Delhi, India
e-mail: amol.solanke@icar.gov.in

R. N. Pudake et al. (eds.), *Omics of Climate Resilient Small Millets*,
https://doi.org/10.1007/978-981-19-3907-5_12

for crop improvement program through transgenic and MAS approach. Transgenic approach has been used in finger millet for developing salt tolerance by overexpressing transgenes from other systems. However, the studies related to abiotic stress tolerance using mapping or transgenic approach are limited in finger millet. With the abundant genetic and genomic resources in finger millet, it offers a lot of opportunity to researchers for exploiting its unique characteristics, especially abiotic stress tolerance and nutritional properties.

Keywords

Ragi · Abiotic stress · Omics · Markers · Transgenic

12.1 Introduction

The steadily increasing human population (predicted to be 9.6 billion by 2050) is expected to fuel the global food requirement by additional 70% (FAO 2020). This coupled with reduction in cultivated land and drastic changes in the environment (Lesk et al. 2016) remain major limiting factors for world food security. By engineering tolerance to these environmental stresses, both biotic and abiotic, it is possible to realize the genetic potential of improved varieties and have sustainable production of crop plants. Globally, only 9% of the area is favorable for crop production with optimal conditions, whereas 91% is suffering from different stresses (Wahid et al. 2007; Godoy et al. 2021). Abiotic stresses alone account for greater than 50% yield loss in major crops (Bray 2000). Among the different abiotic stresses, high and low temperatures, drought, salinity, high/low pH soil, and radiation are the major ones adversely affecting the crop productivity. Under changing climatic conditions, high temperature has emerged as the main constraint for crop production, which accounts for about 20% loss, followed by drought (9%) and low temperature (7%) stress (Wang et al. 2007; Imran et al. 2021). The abiotic stress includes temperature (extremely high or low), drought, floods, etc. Apart from crop yield, these stresses also affect agricultural livestock, fisheries, and other produce. Hence, there is a need to either enrich the major cultivated crops with abiotic stress tolerant genes from primary/secondary/tertiary gene pools with stress tolerant genes or shift cultivation to climate resilient hardy crops.

Amidst various cultivated crop plants, millets are the most tolerant to various abiotic stresses. In Asia and Africa, widely recognized as semi-arid tropical environments, millets are cultivated as staple foods and are a rich source of proteins and calories (O'Kennedy et al. 2006; Puranik et al. 2017). Worldwide different kinds of millets are cultivated such as finger millet, foxtail millet, pearl millet, proso millet, barnyard millet, kodo millet, brown top millet, little millet, etc. Finger millet (*Eleusine coracana*) is the sixth foremost important crop and belongs to the cultivated *E. coracana* subsp. *coracana* from the Poaceae family (Mundada et al. 2021a). The word "Eleusine" comes from the Greek goddess of cereal crops "Eleusine"; and the word "finger millet" signifies the finger-like splitting arrangement of panicles. Finger millet is among the oldest cultivated crops and its domestication dates to ~5000 years. It originated in the tropical provinces of Africa and further distributed to Asia and South America (Gull et al. 2014). *E. coracana* comprises two subspecies, coracana ($2n = 4x = 36$) and Africana ($2n = 4x = 36$). The former is a cultivated species while the latter is a weedy species. The widely cultivated *E. coracana* phenotypically resembles the diploid, *E. indica* ($2n = 2x = 18$) as well as the tetraploid *E. coracana* subsp. *africana*. The weedy species are mainly grown in Africa while *E. indica* is found in Asia (Sood et al. 2016).

Finger millet is widely grown in about 25 countries of Asia and Africa. India leads the cultivation of finger millet growing it in a total of 1.58-million-hectare area. India is also the leading producer with 10.3 million tons of produce per annum and productivity of 1162 kg/ha contributing to more than 28% of the overall global production (FAO 2018). The yield potential of finger millet is remarkably high (>10 t/ha under optimum irrigated conditions) though mainly it is grown in dry and marginal lands. Finger millet solely accounts for 85% of the harvest produce of all the millets put together (Sakamma et al. 2018).

Tolerance of finger millet to a range of suboptimal growth conditions, such as drought, high temperature, alkaline/low pH/saline soil, and infestation by several pests and pathogens has made it suitable for dry farming as well as farming in steep topography (Chandrashekar 2010; Bhatt et al. 2013). Finger millet serves as a major source of carbohydrates, proteins, as well as minerals compared to any other cereal grains. It is an exclusively valuable source of amino acids such as methionine which is absent in the food diets of several millions of poor people (Tripathi and Platel 2010) especially tropical dryland regions where this crop food product serves as the prime food source. Compared to other crop plants, finger millet seeds have a much higher concentration of about tenfold of calcium (350 mg/100 g). Besides calcium, it has higher amounts of minerals like zinc, iron, potassium, and phosphorus (Puranik et al. 2020). Besides nutrition, finger millet has several medicinal benefits to humans as it is a diaphoretic, diuretic, and vermifuge. Pharmaceutical applications of finger millet are due to the presence of antioxidant, wound-healing, anti-ulcerative, and anti-microbial activity (Devi et al. 2014). It has efficient carbon assimilatory mechanism as it fixes carbon by C4 pathway. Given the hardy nature of finger millet and the climate change scenario, researchers have shifted their focus to finger millet for prospecting novel genes and alleles to provide tolerance under abiotic conditions. Thus, finger millet is a versatile crop and a valuable genetic resource to agriculturists. Overall, there are over and above 37,000 germplasm accessions conserved across the International GenBank of Asia, Africa, the USA, and European countries of which the leading countries are India, Kenya, Ethiopia, Uganda, and Zambia (Vetriventhan et al. 2016). Of these countries, India has the maximum collection in the world followed by Ethiopia (Dwivedi et al. 2012). This chapter reviews the upcoming research in omics technologies and subsequent gene characterization studies since the era of DNA markers with focus on abiotic stress tolerance in finger millet.

12.2 Development and Utilization of DNA Markers in Finger Millet

Molecular marker assisted breeding (MAB) is a routine in most of the cereal crops; however, it is not so in the small millet group crops including finger millet, at least till the last decade (Babu et al. 2013). Improving finger millet through classical recombination breeding is difficult due to its strictly self-pollinating nature, tetraploidy, and minute flower size which makes emasculation very tedious. Hence, development and utilization of molecular markers can be of immense help to identify the appropriate germplasm for breeding. Characterization and management of germplasm has become affordable, specific, and highly productive by means of identification of DNA polymorphism via genetic markers. This has led to a great impact on crop betterment and improvement.

Identification of germplasm tolerant to abiotic stress tolerance by assessing the genetic diversity patterns is vital for any drought breeding programs (Mahalle et al. 2021). The germplasm adapted to various environmental stresses possesses more allelic variation compared to their non-adapted counterparts. The germplasm having greater allele diversity is more suitable to serve as parents in breeding programs (Antony Ceasar et al. 2018). Initially, anonymous markers like Random amplified polymorphic DNA (RAPD), Inter simple sequence repeats (ISSR), and Amplified fragment length polymorphism (AFLP) were used in finger millet which was followed by more informative and locus specific markers like Restriction fragment length polymorphism (RFLP) and microsatellites (SSR), including both genomic and EST-SSRs. The first genetic map of finger millet was developed using SSR, EST, RFLP, and AFLP markers which covered all 18 chromosomes and spanning a length of 721 cM on A genome and 787 cM on B genome (Dida et al. 2007). These markers and the genetic map had accelerated the finger millet improvement by assisting genotypic classification and molecular breeding through diversity analysis, QTL mapping, association mapping, and marker-assisted selection (MAS).

12.2.1 Random Amplified Polymorphic DNA (RAPD)

Initially RAPD markers were the choice for studying the genetic diversity, cultivar identification and estimation of genetic relationships in the finger millet landraces and cultivars as they do not need any prior genomic information (Babu et al. 2007). The RAPD markers were used along with other markers to study 22 accessions from 5 different *Eleusine* species (Salimath et al. 1995). The study suggested that though RAPD markers are quicker and economic compared to other markers like RFLP and ISSR, it faces limitations in reproducibility in suboptimal degree of between-line and between-experiment (Salimath et al. 1995). Initial studies, however, showed that finger millet crop has very less molecular diversity. This could be a result of the narrow genetic base of the cultivated finger millet accessions. As the crop is grown and acclimatized to different environmental conditions in various parts of the world, higher genetic diversity can be found in geographically diverse accessions. Later studies showed the existence of such large genetic diversity in finger millet (Fakrudin et al. 2004; Das and Misra 2010; Kumari and Pande 2010; Panwar et al. 2010; Singh and Kumar 2010; Ramakrishnan et al. 2016a). An investigation is done with finger millet from southern India and non-Indian germplasm using 128 genotypes and 25 RAPD markers (Ramakrishnan et al. 2016a). The marker showed 76.48% polymorphism and it was able to divide the genotypes to their respective geographical locations. One more study was conducted using 17 RAPD markers and 11 genotypes where the RAPD markers assisted in efficient identification of germplasm. Thus, RAPD markers served as quick and economic markers in genetic diversity, population structure, and germplasm characterization studies in finger millet.

12.2.2 Inter Simple Sequence Repeat (ISSR)

ISSR markers also do not have a need for any prior sequence knowledge, similar to RAPD. The primers are designed based on simple repeat sequences like (CA) coupled with degenerate 3′ anchor sequences like $(AGC)_6TY$ or $(CA)_8RG$. The PCR product represents the region lying between the repeat sequences, resulting in the formation of multi-locus marker system which can aid in molecular diversity interpretations, fingerprinting, and genome mapping studies (Godwin et al. 1997). The ISSR marker system is proved to be effective in various studies due to its cost efficiency, simplicity, and reproducibility. A study done in finger millet suggested that ISSR markers are more promising than RAPD and RFLP in estimating the plant genetic diversity based upon the quantitative and qualitative data output generated (Salimath et al. 1995). The report also suggested that ISSR markers are suitable to differentiate the plants within and between the species. Another genetic diversity investigation was done using 80 accessions and 6 ISSR markers (Brhane et al. 2017). Though only six ISSR markers were used in the study, it revealed enough polymorphism to separate the accessions in different subgroups. Recently, 15 ISSR markers were used for the molecular characterization of 23 traditional varieties and hybrid derivatives of finger millet (Venkatesan et al. 2021). The material revealed high polymorphism with clear distinction of all the marker parameters. Thus, ISSR markers were more useful in determining the genotypic differences in finger millet cultivars.

Single primer amplification reaction (SPAR) markers are considered as an efficient system to detect molecular diversity as they are simple, economic, quick, easy to assay and most importantly, do not require prior sequence information (Fakrudin et al. 2004; Singh et al. 2014). Certain diversity-based research work in finger millet was conducted using other SPAR markers like DAMD (directed amplification of minisatellite region DNA) markers (Pandian et al. 2018).

12.2.3 Simple Sequence Repeats (SSRs)

Employing SSR markers to study millet genetic diversity can generate more useful molecular information about finger millets as SSRs have high polymorphic nature and show wide distribution within the genome (Lee et al. 2017). Dida et al. (2007) developed 82 genomic SSRs, especially by di- and trinucleotides SSR, from random genomic libraries of *Sal*I, *Hind*III, and *Pst*I, which were hybridized to 18,432 double-grid colonies and a mixture of di[(AG)15/(AC)15] and tri[(AAG)10/(AAC)10/(ATC)10, (AGC)10/(AGG)10,(CCG)10] probes. However, only 31of the 82 SSRs could be assigned loci on the finger millet genetic map. Subsequently they used 45 SSRs to assess genetic variation among 79 Asian and African millet genotypes. The investigation also explored the origin of the finger millet accessions and the results supported the theory that finger millet was cultivated in Africa earlier to its introduction in India. Later, SSRs became a regular feature in different genetic diversity studies as compared to other markers. Another group designed 12 SSR markers by sequencing of 815 clones from SSR enriched genomic DNA library (Lee et al. 2017).

Majority of the studies in finger millet are restricted to a small group of germplasm or local germplasm collections using a limited number of markers; only a very few have used large germplasm (De Villiers et al. 2015; Ramakrishnan et al. 2016b; Hiremath and Gowda 2018; Joshi et al. 2020). A study used a set of five SSR and nine RAPD markers to analyze local landraces from two districts of Nepal (Joshi et al. 2020) and found that marker-based studies are only useful if we can link them with the agro-morphological traits. Moreover, it was found that genetic distance generated by SSR markers is more compared to RAPD markers. Another diversity analysis conducted using 25 SSRs in 32 finger millet accessions also proved the efficiency of SSRs in identification of morphologically close and genetically distinct accessions (Hiremath and Gowda 2018). Another finding using 14 polymorphic genomic SSRs and 3 eSSRs showed greater extent of variation in African accessions compared to those of Indian origin (Arya et al. 2013). The South Indian genotypes were found to be grouped with African upland ones in this study. The findings suggested the geographical adaptability of the millet cultivars to African highland and lowland before it had moved to India. Also, the cultivars of finger millet from South India are morphologically similar to African lowland cultivars (Hilu and de Wet 1976). Another comparatively large-scale study was able to group 128 finger millet accessions based upon their geographical location using 87 genomic SSRs (Ramakrishnan et al. 2016b).

Very few GWAS studies are reported in finger millet that map QTL regions for abiotic stresses such as drought tolerance at vegetative stage and phosphorus deficit response at seedling stage (Ramakrishnan et al. 2017; David et al. 2021). The drought stress study employed 87 SSRs for identifying MTAs (marker-trait associations) for biochemical parameters associated with drought response. In the P-deficiency study, most of the genotypes studied showed adaptive response to phosphorus deficient conditions. Ten genotypes were shortlisted as low P tolerant and were screened further for root length and improved root hair initiation and long root hairs in P deficient soils. Two QTLs viz., *qLRDW.1* and *qLRDW.2* were reported for root dry weight during low P response which explained phenotypic variance above 10%. Two more QTLs were mapped for shoot dry weight (*qHSDW.1*) and root length (*qHRL.1*). These identified QTL regions also showed synteny with the grass genome which harbored genes like cytochrome P450, pectin methyl esterase inhibitor (*PMEI*) and phosphate transporters in its vicinity. As P deficiency in finger millets at seedling stage is detrimental during early growth and establishment, it is an important trait to breed for in order to develop climate resilient cultivars.

12.2.4 Expressed Sequence Tags (EST) SSRs (EST-SSRs)

With the lots of information on transcripts available in public databases in the current days, it has become possible to design EST-based SSRs, which are ideal markers for genetic applications as ESTs stand for expressed genes. The recurrence of microsatellites is remarkably more in ESTs than genomic DNA in various plant species (Morgante et al. 2002). EST-SSRs can be used for development of linkage map as well as comparative mapping due to their universal and evolutionarily conserved nature (Cordeiro et al. 2001; Schubert et al. 2001; Varshney et al. 2005). Out of the1740 ESTs available from the public domain, 31 EST-SSRs based on di-, tri-, tetra- and penta-nucleotide repeat sequences were designed (Arya et al. 2009). The ESTs were further tested on 13 elite germplasm from African and Indian origin. Seventeen of the 31 markers could be amplified in these accessions, 9 of which were polymorphic. These ESTs were also tested in pearl millet and 11 showed cross species transfer. Another group developed a set of 30 EST markers from the then available finger millet database on NCBI (Naga et al. 2012). A recent study developed 13 novel EST-SSRs from the enhanced genomic resources presently available in the public domain (Brhane et al. 2021). These ESTs were used to access the finger millet diversity in 55 landraces and 5 cultivars from Ethiopia. The report concluded that there is a poor association between the genetic makeup and geographic origin in a population of finger millet landraces. This suggests the impactful gene flow within and across geographic regions between and among the populations.

Overall, there are limited examples for molecular markers associated with abiotic stress tolerance in finger millets; and it has a vast scope for exploration in wild and cultivated accessions (Gupta et al. 2017). Drought tolerance in plants is regulated by various genes and transcription factors that allow them to grow in extreme environmental stresses (Pandian et al. 2018). From the draft whole genome sequencing data of finger millet, 85,243 genes were predicted that play a role in various cellular, molecular, and biological functions (Hittalmani et al. 2017). It is important to filter genes associated with drought tolerance from such a huge number of genes in order to dissect and engineer the regulatory network to enhance productivity under drought and select useful QTLs for crop breeding (Pandian and Ramesh 2019).

12.2.5 Next Generation Sequencing (NGS) Derived Markers in Finger Millet

Next generation sequencing technique enables unraveling of enormous genomic information needed for studying complex traits and map quantitative trait loci of importance across organisms. NGS techniques have made the identification of SNP marks simple and straightforward and thereby made other marker systems redundant and almost irrelevant. SNPs are abundant, reproducible, highly polymorphic, co-dominant, bi-allelic and ultimate DNA markers which are amenable to automated high-throughput genotyping; hence they are the most sought-after markers in the genomic era (Gimode et al. 2016). In case of finger millet SNP mining is comparatively tougher as it is an allotetraploid with low level of polymorphism resulting from a very high level of inbreeding. Also, there are high levels of homoeologous SNPs which occurs due to polymorphism among AA and BB sub-genomes of the same organism.

Diversity Array technology Sequencing (DArt) SNP Sequencing is appropriate for crops where limited genomic resources are available. DArT-Seq is a combination of DArT with genotyping-based sequencing (GBS) which has an edge as compared to other technologies. It has improved genome coverage and high reproducibility besides being cost effective. A total of 46,336 SNPs were deciphered as an outcome of genotyping by Dart seq platform in a study (Backiyalakshmi et al. 2021). This study indicated genetic variation and differentiation between regions, races, and biological status. These markers successfully delineated the population structure of Asian finger millet germplasm from the African ones. Genetic diversity was somewhat greater in Asian accessions as compared to their African counterparts because of integration of African alleles in the Asian gene pool via breeding. This study proves to be very useful for the breeders as it provides quality information about finger millet for development of high yielding varieties (Backiyalakshmi et al. 2021). Illumina and Roche 454 sequencing platforms have been used to identify 10,327 SSRs and 23,285 non-homoeologous SNP markers in finger millet and 101 SNP and 101 SSR markers were further tested for polymorphism in finger millet germplasm comprising of 30 wild and 59 cultivated accessions (Gimode et al. 2016). This study identified 49 and 80 polymorphic SSRs and SNPs, respectively. More importantly, the release of the draft whole genome sequencing of finger millet has given a major thrust to SNP markers in this nutritious, hardy, and poor man's cereal crop.

Genotyping by sequencing (GBS) technique has immense capacity to give large number of SNPs which would further help in identifying marker-trait associations to propagate genetic manipulations for generation of improved finger millet genotypes. GBS was conducted in a set of 113 genetically diverse finger millet accessions which yielded 23,000 SNPs and grouped the material into 3 distinct clusters with substantial admixture indicating free gene glow among South Asian and African germplasm (Kumar et al. 2016a, b). This SNP dataset was further used for identification of 109 marker-trait associations for 14 agro-morphological traits including grain yield using phenotype data from 2 different agro-ecological locations (Sharma et al. 2018). In yet another study from the same group, a total of 2977 genome-wide

SNPs were filtered from the GBS data and used for dissecting the genetic architecture of seed protein content (SPC) and its associated traits, days to maturity (DM), and grain yield (GY), phenotyped at two different locations (Tiwari et al. 2020). They employed General Linear Model, Mixed Linear Model (MLM), and Compressed MLM models for identifying MTAs which revealed 4–12 MTAs for each of the three traits. Downstream analysis of the associated SNPs and the genes revealed that aspartyl protease was the key player in determining SPC while ATP synthase was the key gene for GY and DM. In yet another study, GBS and Genome-wide association study (GWAS) were taken up to identify MTAs in 190 finger millet genotypes for 6 mineral (iron, magnesium, calcium, zinc, sodium, and potassium) and protein content. Using 169,365 SNPs their analysis unraveled 3 subpopulations and 34 significant and FDR-adjusted MTAs (Puranik et al. 2020). Downstream analysis of the 34 MTA associated genes revealed that 18 of them were associated with mineral transport, remobilization, and metal ion binding.

The availability of genome and a huge number of SNP markers can be a boon to practicing genomic selection (GS) in finger millet. Unlike biparental or association mapping panels, which are used to map QTLs, most of the drawbacks of MAS are solved by genomic selection (GS) and genomics-enabled prediction (GP), which allows for the quick selection of better genotypes. GS predicts genomic-estimated breeding values (GEBV) by using genome-wide dispersed DNA markers (Varshney et al. 2013). GS uses genotypic and phenotypic data of training population and builds a model which is further used to determine the GEBV of individuals in the test population, relying solely on their genotyping data. When compared to phenotype-based selection, it lowers the cost and resources required to generate a new crop variety (Crossa et al. 2017). Since there are several challenges in using GS and GP in small crops such as finger millet, the key outcomes achieved in certain important cereals imply that it might be used to generate better lines more rapidly, as well as increase the pace of genetic gain. Furthermore, GP models will be necessary for accessing the huge pool of undiscovered variety in the gene banks for finger millet (Crossa et al. 2017). Therefore, for using GS and GP in orphan crops like finger millet, comprehensive training populations and extensive marker maps are required. This will expedite the breeding process and enable the selection program for disease resistance, yield, and nutritional quality.

12.3 Different Omics Approaches in the Technological Era

Applied biology targets to study genetic control, cellular processes, and interactions with environmental changes, which necessitates extensive experiments in structural genetics and functional mechanisms. These substantial investigations are termed "omics" which are not only limited to genomics, transcriptomics, and proteomics but extends to epigenomics, metabolomics, metagenomics, and nutrigenomics (Langridge and Fleury 2011). These studies are primarily taken up using various high-throughput technologies that enable quantitative and/or qualitative identification of novel or known genes, transcription factors, proteins, primary and secondary metabolites, lipids, and other dynamic cellular constituents (Kato et al. 2011; Kulski 2016). However, they do demand comprehensive computational resources for the analysis and storage of high-throughput data generated in these omics studies. Consequently, various analysis servers, online databases, and several omics platforms have been developed over the past few years. Omics covers a broad interplanetary research fields, and for improvement of our present understanding of molecular mechanisms of traits, this calls for effective integration of several omics datasets.

Of late, the availability of genomic and genetic resources in finger millet has considerably improved which can act as a key to strengthening the introduction of a climate-smart crop plant breeding program (FAO 2015; Bandyopadhyay et al. 2017). Omics studies are greatly facilitated when the whole genome sequence (WGS) information is available in a crop species. Two independent whole genome sequence projects have been reported in two different varieties of finger millet (Hittalmani et al. 2017; Hatakeyama et al. 2018). Around 80–82% of finger millet genome has been covered by both groups, which is adequate owing to the occurrence of repetitive elements in grass genomes. The genome assembly of variety ML-365 has been elucidated using SOLiD and Illumina sequencing data (Hittalmani et al. 2017), whereas PR 202 variety has been sequenced using more recent technologies such as IlluminaMiseq, IlluminaHiSeq 4000, IlluminaNextSeq 500, and PacBio RSII system (Hatakeyama et al. 2018). Hence, the scaffold size of PR202 genome assembly was longer (>2.5 MB) in comparison to ML-365 (24 Kb). Around 62,348–85,243 genes were identified in the finger millet genome of which half of them were transposable elements containing large number of drought responsive genes and drought-related transcription factors (TFs). These two resources have provided a robust platform for millet researchers in identifying and validating genes involved in several regulatory pathways and nutrient transport. In the current section, we briefly describe the recent investigations on finger millet omics and pinpoint the development, progress, and challenges that lie ahead of their application in finger millet.

Finger millet genotype KNE755 and KNE796 was sequenced utilizing Illumina technologies and Roche454 by Gimode et al. (2016). The sequencing reads were assembled to 207 Mbp and comprise of 13% of finger millet genome. This report has led to the identification of 10,327 SSRs and 23,285 non-homoeologous SNPs.

12.3.1 Transcriptomics

The total amount of mRNA expressed by a particular cell or tissue or organism is referred as transcriptome and the study of this is called as "transcriptomics." RNA sequencing by NGS is more powerful, economical, and informative for transcriptomics studies in both non-model and model plant species, unlike microarray, as probes (gene-specific pre-synthesized oligonucleotides) and well-annotated genome are not pre-requisites for RNA-seq. Earlier, Roche 454 FLX was widely used for transcriptome sequencing because of its ability to generate relatively long reads which allows significantly de novo assembly compared to ABI SOLiD system and Illumina-Solexa platform (Guo et al. 2010; Sun et al. 2010). However, these platforms have become obsolete and only Illumina, PacBio, and Nanopore platforms are currently in vogue with Illumina being the forerunner (Sun et al. 2010). Since rice and finger millet are evolutionarily closer, and rice has huge RNA-seq resources, expression datasets of rice have been used for comparative genomic analysis in finger millets. For the non-model and C4 cereal crop plant like finger millet, with comparatively inadequate genomic information, RNA sequencing can unleash important information about the gene expression pattern of abiotic tolerance, and other nutritional trait elements (Kumar et al. 2016a, b).

Nageshbabu et al. (2013) reported miRNA profiling of finger millet under salinity and drought stress. Different drought responsive genes including protein phosphatase 2A, farnesylated protein ATFP6, metallothionein, farnesyl pyrophosphate synthase and RISBZ4, and their expression pattern were earlier reported by Parvathi et al. (2013) by expression profiling of drought stress response related genes. A subsequent study by Parvathi and Nataraja (2017) identified *TBP-Associated Factor6* (*EcTAF6*) gene in finger millet for drought stress by screening of cDNA library. Li et al. (2021) have also reported several drought responsive genes and their expression analysis in finger millet. Subsequently, novel drought responsive genes such as calcineurin B-like interacting protein kinase31 (CIPK31), signal recognition particle receptor α (SRPR α), serine threonine protein phosphatase 2A (PP2A), farnesyl pyrophosphate synthase (FPS), etc., were reported following RNA-seq of GPU28 variety of finger millet (Parvathi et al. 2019). Rahman et al. (2014) reported the transcriptome analysis of salinity responsive genes in contrasting genotypes using RNA-seq. Recently, differentially expressed genes have been identified under osmotic stresses after transcriptome profiling (Mundada et al. 2021a, b). Gupta et al. (2014) reported *EcDof1* and *EcDof2* transcription factors which were involved in carbon and nitrogen metabolism. Similar studies have been reported for drought, salinity, and osmotic stress responsive genes (Hittalmani et al. 2017; Ramegowda et al. 2017).

12.3.2 Proteomics

Proteomics is another major functional genomics tool that has tremendous ability in finding novel stress responsive proteins that can mitigate the adverse impact of abiotic stresses on field crops. To separate the complicated protein mixture, 2D gel electrophoresis is a highly recommended method used in proteomics. Mass spectrometry (MS) also acts as a feasible identification tool for various types of proteins (Harland 2005; Fuchs et al. 2005; Kussmann et al. 2006; Mariman 2006; Wang et al. 2006). Proteomics has seen a fresh revolution thanks to advances in MS systems, as it has evolved into a critical technique for studying protein–protein interactions and protein post-translational alterations, gaining understanding into biological processes (Shi et al. 2007). In a number of laboratories, LC-based proteome investigations are becoming more prevalent. However, due to the scarcity of genetic data, agricultural proteomics has been hindered significantly.

Various signaling cascades, ion transporters, and regulatory proteins are triggered due to stress, and understanding the proteins involved in the stress response processes can offer tremendous insight into the peculiar properties of millets which may be further used to increase crop yield. The importance of organ-specific proteome studies was highlighted by Komatsu and Hossain (2013) for detecting proteins that accumulate in distinct plant tissues in response to several stresses and thus might participate in stress responses in plants (Komatsu and Hossain 2013). A recent study by Li et al. (2021) provides a valuable resource for understanding the regulatory networks involved in finger millet under drought stress. This proteomics data unraveled many critical protein families, particularly thaumatin family and the glycosyl hydrolase family 17 (GHL17) that are key in the management of drought stress response, and are known to play multiple roles in the development of plants as well as responses to biotic stresses. The drought-related protein interaction networks discovered by Li et al. (2021) also identified 13 differentially expressed genes/proteins which participated in the response pathway towards drought stress. TAF6 which was earlier reported by Parvathi and Nataraja (2017) was validated by this proteome study. *BAMY2* (barley K-amylase 2) is a dextrinase inhibitory starch debranching enzyme found in barley. *ISA1* (isoamylase-type starch debranching enzyme) has also been reported to have a role in starch crystallization and biosynthesis. Both *BAMY2* and *ISA1* genes were observed to be considerably upregulated with respect to drought stress, thus limiting starch synthesis as well as dextrin, and boosting stress tolerance by limiting plant vegetative growth in finger millet (Li et al. 2021). Thus, response mechanism against drought can be explained with the help of proteomics data obtained from dehydration as well as rehydration of finger millet tissues. With the aid of additional investigation, these significant stress response genes might be exploited for functional research and genetic improvement of finger millet and other crops.

Jasmonic acid (JA) signaling pathway is activated by plants in exchange for a variety of developmental events with response to both abiotic and biotic stresses. The primary regulator of the plant Jasmonic acid signaling pathway, the Jasmonate ZIM-domain (JAZ) protein family, is involved in phytohormone crosstalk. Proteomics investigation conducted by Sen et al. (2016) discovered the in vivo interactions of finger millet protein JAZ (*Ec*JAZ) under methyl jasmonate treatment. The study discovered not just the JA signaling route, but also the signaling crosstalk of phytohormones in finger millet, which is a promising future crop. Antioxidants operate as redox buffers, interacting with a variety of cellular components to influence plant development and growth by influencing processes which range from cell division and cell elongation to senescence and then ultimately death. They work to reverse the stress-induced reactions in cells of plants. Agrawal et al. (2019) reported some natural proteins from finger millet which has tremendous antioxidant activity that might be used as a functional food component.

12.3.3 Ionomics

Nutritional composition is influenced by both the physiological state of the organism, i.e., differences in ion and mineral absorption, transport, and sequestration, as well as its genetic makeup. Hence, unraveling a plant's metabolic profile including ionic content, and phytochemicals can offer a functional explanation of its nutritional value (Sumner et al. 2003). In this context, metabolome profiling can be a useful phenotyping technique in developing biomarkers for dietary characteristics. Considering the fact that the nutritional value of finger millet is mainly due to its secondary metabolites such as polyphenols and mineral ions like calcium, a thorough metabolic profile analysis is necessary to select the optimal varieties for human consumption. Polyphenols in finger millet, represented primarily byproducts of benzoic acid, cinnamic acid, and quercetin, can be identified using high-performance liquid chromatography, nuclear magnetic resonance, and electrospray ionization mass spectrometry (Chandrasekara and Shahidi 2010; Banerjee et al. 2012). Consequently, applying this knowledge is critical for right biofortification of ragi.

Fungi, bacteria, viruses, and insects are among the top most biotic challenges to plants. Plants fight pathogens via a multi-tiered defensive mechanism mediated by ions like Ca^{2+}. Through the emergence of ionomics, mineral nutrients are now widely acknowledged to have a vital role in plant pathogen interactions. Minerals are necessary for boosting the development and tolerance of pest attacks by triggering the important enzymes which lead to the formation of metabolites like callose, glucosinolates, lignin, phytoalexins, and phenols. Ca^{2+} ions are crucial components of the cell wall as they contribute to lowering the severity of disease and susceptibility of the host by initiating a wide range of defensive cascades, such as an increasing the stiffness of the cell membrane (Dordas 2009). Furthermore, the activity of antioxidant enzymes such as peroxidase (POD), catalase (CAT), superoxide dismutase (SOD), and polyphenol oxidase (PPO), that scavenge reactive oxygen species (ROS) following pathogen infection, is significantly increased by Ca^{2+} content (Sun et al. 2020). This was observed in Ca^{2+} deficient plants, wherein several important metabolites like sugars and amino acids flowed out of the cytoplasm and into the apoplast, allowing growth of pathogens as well as disease progression (Clarkson 1996).

Previous research has shown that excessive Zn ion buildup is harmful to bacterial pathogens, notably, *Xanthomonas fastidiosa*, since Zn ions participate in starch production, protein structure, and plasma membrane protection from ROS (Cakmak 2000). Furthermore, zinc acts as a cofactor for numerous enzymes involved in hormone metabolism, cell reproduction, and cell membrane integrity (Singh et al. 2016). Likewise, during infection, both plants and pathogens compete for iron through accelerating their cellular activities. Chandra et al. (2020) used multiomics approach to reveal novel genes in *E. coracana* to get well acquainted about the regulatory networks and molecular mechanisms associated with zinc and iron homeostasis. The study discovered the potential role of 15 candidate genes that might be involved in iron and zinc homeostasis. In a nutshell, ions play a significant role in abiotic and biotic stress tolerance. The study of the ionome will undoubtedly be a way forward to explore and understand more about the interplay of ions leading to resistance or susceptibility of a disease.

12.3.4 Metabolomics

Metabolites give the energy which is required for the growth and metabolism of protoplasm as it is the basic element for various enzymatic and structural molecules. Metabolites not only take part in the determination of the physiological condition of the organism but also act as a bridge between genetic and phenotypic information. Metabolomics has a major role in the understanding of the phenotype-linked interrelated biological processes (Bino et al. 2004). Mostly all metabolomics work has been conducted on staple cereal plants and model crops; there are very few reports available for metabolic profiling in finger millet in abiotic stress treatment. One study on metabolite profiling was performed for Si-mediated osmotic stress tolerance using LC–QTOF–MS in finger millet (Mundada et al. 2021a). The transcriptomics data generated through this investigation revealed that the efflux and influx in Si transporters were increased by Si supplementation during stress conditions. This study indicates that uptake of Si is increased during stress in finger millet and its transport is controlled by Si transporters.

12.4 Reverse Genomics

Functional genomics tools like gene silencing, insertional mutagenesis, overexpression, targeted induced local lesions in the genome (TILLING), and gene editing have imparted a significant role in the improvement of our understanding of convoluted regulatory networks linked with various stress tolerance and acclimatization in plants. Various novel genes identified through numerous omics studies in finger millet are attractive candidates for crop improvement via transgenic or MAS approach in finger millet. This information is also critical for employing the reverse genetics strategies. From the previous sections in this chapter, it is evident that hundreds of such candidate genes and proteins have been identified in finger millet, but the major challenge remains the functional validation of these genes. One of the potential tools for discovering the function of a target gene is to overexpress it in the same or heterologous system under the control of a stress-inducible or constitutive promoter. Another significant strategy for discovering the function of a target gene is to silence/suppress its expression or knockout the gene. The available literature in these aspects is presented in detail in the following subsections.

12.4.1 Overexpressions (Transgenomics)

In the past two decades, a few attempts have been made for the development of transgenic finger millet for abiotic stress tolerance. Most of the attempts have been

Table 12.1 Genetic transformation in finger millet associated with abiotic stress

Transgene	Method of transformation	Outcome	Reference
PcSrp	Biolistic	Salt tolerance	Mahalakshmi et al. (2006)
PDH45	*Agrobacterium*-mediated gene transfer	Salt tolerance	Babu et al. (2012)
SbVPPase	*Agrobacterium*-mediated gene transfer	Salt tolerance	Anjaneyulu et al. (2014)
PgNHX1, *AVP1*	*Agrobacterium*-mediated gene transfer	Salt tolerance	Jayasudha et al. (2014)
mtlD	*Agrobacterium*-mediated gene transfer	Drought, oxidative, and salt tolerance	Hema et al. (2014)
OsSOS1	*Agrobacterium*-mediated gene transfer	Salt tolerance	Pushpa et al. (2020)

carried out to develop salt stress tolerant transgenic finger millet by overexpressing transgenes from other systems (Table 12.1).

One of the first attempts for the establishment of abiotic stress tolerant finger millet was described by Mahalakshmi et al. (2006). They used the biolistic gun to develop the salinity tolerance through a transformation of the serine-rich protein (*PcSrp)* gene from rice. Babu et al. (2012) carried out an in vitro plant regeneration and transformation for salt tolerance through the *PDH45* gene using Agrobacterium-mediated gene transfer. Anjaneyulu et al. (2014) reported increased salt tolerance in finger millet through the introduction of the vacuolar H+-pyrophosphatase (*SbVPPase*) gene from Sorghum via Agrobacterium-mediated gene transfer. Jayasudha et al. (2014) developed an efficient agrobacterium mediated transformation protocol for raising salt tolerant finger millet. They transformed a double gene construct of *AVP1* (from *Arabidopsis thaliana*) and *PgNHX1* (from *Pennisetum glaucum*) employing pCAMBIA 1301. This was the first reported study using double gene construct to enhance salt tolerance in finger millet. Recently, the *OsSOS1* gene from *O. sativa* was overexpressed using Agrobacterium-mediated gene transfer in finger millet (Pushpa et al. 2020). The overexpressing *OsSOS1* T1 transgenic lines showed high salt tolerance, enhanced root length, and seed germination plus reduced ROS compared to wild types. The results from this study demonstrated that *OsSOS* gene which is a Na^+ H^+ antiporter enhances salt tolerance probably due to SOS1 generated proton gradient. The gradient drives Na^+ H^+ antiporter causing excess Na^+ extrusion from cytosol to apoplast, thus ultimately reducing Na^+ toxicity inside cell. There is only one report on the development of the transgenic finger millet that enhanced tolerance to multiple abiotic stresses like drought, salt, and associated oxidative stress by overexpressing bacterial mannitol-1-phosphate dehydrogenase (*mtlD*) gene (Hema et al. 2014). As reported, finger millet is susceptible to these abiotic stresses especially during seed germination and early seedling development. Mannitol known as an osmolyte scavenges the hydroxyl radicals and minimizes the damage which abiotic stresses cause in plants. The T_1 and T_2 transgenic seedlings displayed better growth under drought and salinity stress

Table 12.2 Some genes from finger millet, associated with abiotic stress isolated and overexpressed in the model system

Transgene	Model system	Outcome	Reference
EcDREB2A	Tobacco	Heat tolerance	Singh et al. (2021)
EcCaM	*Arabidopsis*	Drought and salinity tolerance	Jamra et al. (2021)
Ecbzip	Tobacco	Multiple abiotic stress tolerance	Ramakrishna et al. (2018)
EcNAC67	Rice	Drought and salinity tolerance	Rahman et al. (2016)
EcbHLH57	Tobacco	Drought, oxidative, and salt tolerance	Babitha et al. (2015)
EcDehydrin7	Tobacco	Drought tolerance	Singh et al. (2014)
Ec-apx1	Bacterial system	Drought tolerance	Bhatt et al. (2013)
EcNAC1	Tobacco	Multiple abiotic stress tolerance	Ramegowda et al. (2012)

than the non-transformed wild type. Moreover, the plants showed chlorophyll retention and better osmotic adjustments under drought stress. But it was found that the overall effect on transgenic plants for abiotic stress tolerance was only marginal especially for osmotic stress when compared to other *mtlD* gene introduced in other plant species.

Though there are only a few reports on transgenic finger millet for abiotic stress tolerance, there are numerous studies where candidate genes from finger millets have been functionally validated in model plants such as tobacco or Arabidopsis (Table 12.2). NAC (*N*AM, *A*TAF, and *C*UC2) proteins known for their function in stress tolerance were isolated from subtracted cDNA library of drought adapted finger millet plants under drought stress (Ramegowda et al. 2012). From this, *EcNAC1* was transformed and expressed in tobacco plants and transgenic tobacco plants showed tolerance to osmotic and salinity stress. Expression of the *EcNAC67* gene was also shown to enhance tolerance against drought and salt in rice (Rahman et al. 2016). Overexpression of the *Ecbzip17* gene from the finger millet has been shown to increase tolerance against multiple abiotic stresses in tobacco (Ramakrishna et al. 2018). Overexpression of Calmodulin *EcCaM* from finger millet enhances tolerance against drought and salinity stress in Arabidopsis (Jamra et al. 2021). *EcDREB2A*, a novel A-2 type DREB gene was reported in finger millet which was subsequently demonstrated to be specific for heat stress tolerance (Singh et al. 2021).

12.4.2 Gene Silencing

Currently, gene silencing is an emerging reverse genetics tool for determining the function of the desired gene. However, no reports on genome editing have so far been available in finger millet. Virus-induced gene silencing (VIGS) via dsRNA is a successful gene knock-down genomics tool that is relatively faster and easier

(Baulcombe 2004). VIGS acts as a suppressor of targeted gene expression by virus vectors that harbor a target region of the host gene. In brief, it neither generates stable RNA interference (RNAi) nor produces knockout mutants nor can also be carried out in species that are not easy to transform (Scofield and Nelson 2009). Therefore, numerous VIGS vectors are available in crop plants but none of them have been used in finger millet for a long time. Parvathi and Nataraja (2017) reported for the first time, downregulation of a TATA-box Binding Protein Associated Factor6 (TAF6) from finger millet through VIGS method which led to developmentally defective and stress-sensitive phenotype in *Nicotiana benthamiana*. Other aspect of gene editing in small millets has been discussed in Chap. 13.

12.5 Conclusion and Future Prospects

Finger millet is an excellent choice as a high nutritional crop particularly for the economically weaker section in developing countries as it also shows tolerance against multiple abiotic stresses. For understanding the process of abiotic stress tolerance of finger millet, in the current genomics era, different molecular biology and biotechnology tools have been employed to date and substantial progress has been made though more can be done to exploit this poor man's cereal better for human welfare. The progress in genomics is impressive with two independent whole genome information published, following a couple of studies each from other major omics branches such as transcriptomic, metabolomics, proteomics, and ionomics are reported in the last 2–5 years both in the area of abiotic stress tolerance and mineral nutrition. At least 7–8 genes from finger millet have also been functionally characterized highlighting the growing importance of this crop among the researchers. Improvement in overall genomic resource and phenotyping can make the ground for genomic selection in finger millet so that genetic advancement of finger millet becomes feasible. The enhancement in these omics technologies coupled with functional genomics can serve as a boon for nutritional enhancement and stress management of other major food and horticultural crops.

References

Agrawal H, Joshi R, Gupta M (2019) Purification, identification, and characterization of two novel antioxidant peptides from finger millet (Eleusine coracana) protein hydrolysate. Food Res Int 120:697–707

Anjaneyulu E, Reddy PS, Sunita MS, Kishor PB, Meriga B (2014) Salt tolerance and activity of antioxidative enzymes of transgenic finger millet overexpressing a vacuolar H+-pyrophosphatase gene (SbVPPase) from Sorghum bicolor. J Plant Physiol 171(10):789–798

Antony Ceasar S, Maharajan T, Ajeesh Krishna TP, Ramakrishnan M, Victor Roch G, Satish L, Ignacimuthu S (2018) Finger millet [Eleusine coracana (L.) Gaertn.] improvement: current status and future interventions of Whole Genome Sequence. Front Plant Sci 9:1054

Arya L, Verma M, Gupta VK, Karihaloo JL (2009) Development of EST SSRs in finger millet (Eleusine coracana ssp. coracana) and their transferability to pearl millet (Pennisetum glaucum). J Plant Biochem Biotechnol 18(1):97–100

Arya L, Verma M, Gupta VK, Seetharam A (2013) Use of genomic and genic SSR markers for assessing genetic diversity and population structure in Indian and African finger millet (Eleusine coracana (L.) Gaertn.) germplasm. Plant Syst Evol 299(7):1395–1401

Babitha KC, Vemanna RS, Nataraja KN, Udayakumar M (2015) Overexpression of EcbHLH57 transcription factor from Eleusine coracana L. in tobacco confers tolerance to salt, oxidative and drought stress. PLoS One 10(9):e0137098

Babu BK, Senthil N, Gomez SM, Biji KR, Rajendraprasad NS, Kumar SS, Babu RC (2007) Assessment of genetic diversity among finger millet (Eleusine coracana (L.) Gaertn.) accessions using molecular markers. Genet Resour Crop Evol 54(2):399–404

Babu AG, Geetha KN, Manjunatha V, Shankar AG (2012) An efficient high throughput plant regeneration and transformation protocol for production of transgenics tolerant to salt in finger millet. Int J For Crop Improv 3(1):16–20

Babu BK, Singh UM, Yadav S, Kumar A (2013) Molecular marker technology for finger millet crop improvement—an under-utilized, food and nutritional security crop. Biotechnol Today 3(2):57–60

Backiyalakshmi C, Vetriventhan M, Deshpande S, Babu C, Allan V, Naresh D, Gupta R, Azevedo VC (2021) Genome-wide assessment of population structure and genetic diversity of the global finger millet germplasm panel conserved at the ICRISAT Genebank. Front Plant Sci 2021:1391

Bandyopadhyay T, Muthamilarasan M, Prasad M (2017) Millets for next generation climate-smart agriculture. Front Plant Sci 8:1266

Banerjee S, Sanjay KR, Chethan S, Malleshi NG (2012) Finger millet (Eleusine coracana) polyphenols: investigation of their antioxidant capacity and antimicrobial activity. Afr J Food Sci 6(13):362–374

Baulcombe D (2004) RNA silencing in plants. Nature 431(7006):356–363

Bhatt D, Saxena SC, Jain S, Dobriyal AK, Majee M, Arora S (2013) Cloning, expression and functional validation of drought inducible ascorbate peroxidase (Ec-apx1) from Eleusine coracana. Mol Biol Rep 40(2):1155–1165

Bino RJ, Hall RD, Fiehn O, Kopka J, Saito K, Draper J, Nikolau BJ, Mendes P, Roessner-Tunali U, Beale MH, Trethewey RN (2004) Potential of metabolomics as a functional genomics tool. Trends Plant Sci 9(9):418–425

Bray EA (2000) Response to abiotic stress. In: Biochemistry and molecular biology of plants. American Society of Plant Physiologists, Rockville, pp 1158–1203

Brhane H, Haileselassie T, Tesfaye K (2017) Genetic diversity and population structure of Ethiopian finger millet (Eleusine coracana (L.) Gaertn) genotypes using inter simple sequence repeat (ISSR) markers. Afr J Biotechnol 16(21):1203–1209

Brhane H, Haileselassie T, Tesfaye K, Hammenhag C, Ortiz R, Abreha KB, Geleta M (2021) Novel expressed sequence tag-derived and other genomic simple sequence repeat markers revealed genetic diversity in Ethiopian finger millet landrace populations and cultivars. Front Plant Sci 12:735610

Cakmak I (2000) Possible roles of zinc in protecting plant cells from damage by reactive oxygen species. New Phytol 146:185–205

Chandra AK, Pandey D, Tiwari A, Sharma D, Agarwal A, Sood S, Kumar A (2020) An omics study of iron and zinc homeostasis in finger millet: biofortified foods for micronutrient deficiency in an era of climate change? OMICS 24(12):688–705

Chandrasekara A, Shahidi F (2010) Content of insoluble bound phenolics in millets and their contribution to antioxidant capacity. J Agric Food Chem 58(11):6706–6714

Chandrashekar A (2010) Finger millet: Eleusine coracana. In: Advances in food and nutrition research, vol 59. Academic Press, New York, pp 215–262

Clarkson DT (1996) Marschner H. 1995. Mineral nutrition of higher plants. 889pp. London: Academic Press, £ 29.95 (paperback). Ann Bot 78(4):527–528

Cordeiro GM, Casu R, McIntyre CL, Manners JM, Henry RJ (2001) Microsatellite markers from sugarcane (Saccharum spp.) ESTs cross transferable to erianthus and sorghum. Plant Sci 160(6): 1115–1123

Crossa J, Pérez-Rodríguez P, Cuevas J, Montesinos-López O, Jarquín D, De Los CG, Burgueño J, González-Camacho JM, Pérez-Elizalde S, Beyene Y, Dreisigacker S (2017) Genomic selection in plant breeding: methods, models, and perspectives. Trends Plant Sci 22(11):961–975

Das S, Misra RC (2010) Assessment of genetic diversity among finger millet genotypes using RAPD markers. Indian J Agric Res 44(2):112–118

David RH, Ramakrishnan M, Maharajan T, BarathiKannan K, Babu GA, Daniel MA, Agastian P, Caesar SA, Ignacimuthu S (2021) Mining QTL and genes for root traits and biochemical parameters under vegetative drought in South Indian genotypes of finger millet (Eleusine coracana (L.) Gaertn) by association mapping and in silico comparative genomics. Biocatal Agric Biotechnol 32:101935

De Villiers SM, Michael VN, Manyasa EO, Saiyiorri AN, Deshpande S (2015) Compilation of an informative microsatellite set for genetic characterization of East African finger millet (Eleusine coracana). Electron J Biotechnol 18(2):77–82

Devi PB, Vijayabharathi R, Sathyabama S, Malleshi NG, Priyadarisini VB (2014) Health benefits of finger millet (Eleusine coracana L.) polyphenols and dietary fiber: a review. J Food Sci Technol 51(6):1021–1040

Dida MM, Ramakrishnan S, Bennetzen JL, Gale MD, Devos KM (2007) The genetic map of finger millet, Eleusine coracana. Theor Appl Genet 114(2):321–332

Dordas C (2009) Role of nutrients in controlling plant diseases in sustainable agriculture: a review. Agron Sustain Dev 28:33–46

Dwivedi S, Upadhyaya H, Senthilvel S, Hash C, Fukunaga K, Diao X et al (2012) Millets: genetic and genomic resources. In: Janick J (ed) Plant breeding reviews, vol 35. Wiley, Hoboken, pp 247–375

Fakrudin B, Shashidhar HE, Kulkarni RS, Hittalmani S (2004) Genetic diversity assessment of finger millet, Eleusine coracana (Gaertn), germplasm through RAPD analysis. PGR Newslett 138:50–54

FAO (2015) Food and Agriculture Organization. Agricultural production. http://www.faostat.fao.org/

FAO (2018) Food and Agriculture Organization. Agricultural production. http://www.faostat.fao.org/

FAO (2020) Food and Agriculture Organization. Agricultural production. http://www.faostat.fao.org

Fuchs D, Winkelmann I, Johnson IT, Mariman E, Wenzel U, Daniel H (2005) Proteomics in nutrition research: principles, technologies and applications. Br J Nutr 94(3):302–314

Gimode D, Odeny DA, de Villiers EP, Wanyonyi S, Dida MM, Mneney EE, Muchugi A, Machuka J, de Villiers SM (2016) Identification of SNP and SSR markers in finger millet using next generation sequencing technologies. PLoS One 11(7):e0159437

Godoy F, Olivos-Hernández K, Stange C, Handford M (2021) Abiotic stress in crop species: improving tolerance by applying plant metabolites. Plan Theory 10(2):186

Godwin ID, Aitken EA, Smith LW (1997) Application of inter simple sequence repeat (ISSR) markers to plant genetics. Electrophoresis 18(9):1524–1528

Gull A, Jan R, Nayik GA, Prasad K, Kumar P (2014) Significance of finger millet in nutrition, health and value-added products: a review. Magnesium (mg) 130(32):120

Guo S, Zheng Y, Joung JG, Liu S, Zhang Z, Crasta OR, Sobral BW, Xu Y, Huang S, Fei Z (2010) Transcriptome sequencing and comparative analysis of cucumber flowers with different sex types. BMC Genomics 11(1):1–3

Gupta S, Gupta SM, Gupta AK, Gaur VS, Kumar A (2014) Fluctuation of Dof1/Dof2 expression ratio under the influence of varying nitrogen and light conditions: involvement in differential regulation of nitrogen metabolism in two genotypes of finger millet (Eleusine coracana L.). Gene 546(2):327–335

Gupta SM, Arora S, Mirza N, Pande A, Lata C, Puranik S, Kumar J, Kumar A (2017) Finger millet: a "certain" crop for an "uncertain" future and a solution to food insecurity and hidden hunger under stressful environments. Front Plant Sci 8:643

Harland JI (2005) Nutrition and genetics-mapping individual health. ILSI Europe Concise Monograph Series, Brussels

Hatakeyama M, Aluri S, Balachadran MT, Sivarajan SR, Patrignani A, Grüter S, Poveda L, Shimizu-Inatsugi R, Baeten J, Francoijs KJ, Nataraja KN (2018) Multiple hybrid de novo genome assembly of finger millet, an orphan allotetraploid crop. DNA Res 25(1):39–47

Hema R, Vemanna RS, Sreeramulu S, Reddy CP, Senthil-Kumar M, Udayakumar M (2014) Stable expression of mtlD gene imparts multiple stress tolerance in finger millet. PLoS One 9(6): e99110

Hilu KW, De Wet JM (1976) Domestication of *Eleusine coracana*. Econ Bot 30(3):199–208

Hiremath C, Gowda J (2018) SSR based genetic diversity in blast resistant and susceptible accessions of finger millet (Eleusine coracana. L). Electron J Plant Breed 9(2):400–408

Hittalmani S, Mahesh HB, Shirke MD, Biradar H, Uday G, Aruna YR, Lohithaswa HC, Mohanrao A (2017) Genome and transcriptome sequence of finger millet (Eleusine coracana (L.) Gaertn.) provides insights into drought tolerance and nutraceutical properties. BMC Genomics 18(1):1–6

Imran QM, Falak N, Hussain A, Mun BG, Yun BW (2021) Abiotic stress in plants; stress perception to molecular response and role of biotechnological tools in stress resistance. Agronomy 11(8):1579

Jamra G, Agrawal A, Singh N, Sanyal SK, Kumar A, Pandey GK (2021) Ectopic expression of finger millet calmodulin confers drought and salinity tolerance in *Arabidopsis thaliana*. Plant Cell Rep 40(11):2205–2223

Jayasudha BG, Sushma AM, Hanjagi PS, Sashidhar VR (2014) An efficient in-vitro Agrobacterium-mediated transformation protocol for raising salinity tolerant transgenic plants in finger millet [Eleusine coracana (L.) Gaertn.]. Plant Arch 14(2):823–829

Joshi BK, Joshi D, Ghimire SK (2020) Genetic diversity in finger millet landraces revealed by RAPD and SSR markers. Nepal J Biotechnol 8(1):1–1

Kato H, Takahashi S, Saito K (2011) Omics and integrated omics for the promotion of food and nutrition science. J Tradit Complement Med 1(1):25–30

Komatsu S, Hossain Z (2013) Organ-specific proteome analysis for identification of abiotic stress response mechanism in crop. Front Plant Sci 4:71

Kulski JK (2016) Next-generation sequencing—an overview of the history, tools, and "Omic" applications. In: Next generation sequencing–advances, applications, and challenges, vol 3. IntechOpen, London, p 60

Kumar A, Metwal M, Kaur S, Gupta AK, Puranik S, Singh S, Singh M, Gupta S, Babu BK, Sood S, Yadav R (2016a) Nutraceutical value of finger millet [Eleusine coracana (L.) Gaertn.], and their improvement using omics approaches. Front Plant Sci 7:934

Kumar A, Sharma D, Tiwari A, Jaiswal JP, Singh NK, Sood S (2016b) Genotyping-by-sequencing analysis for determining population structure of finger millet germplasm of diverse origins. Plant Genome 9(2):2015-07. https://doi.org/10.3835/plantgenome2015.07.0058

Kumari K, Pande A (2010) Study of genetic diversity in finger millet (Eleusine coracana L. Gaertn) using RAPD markers. Afr J Biotechnol 9(29):4542–4549

Kussmann M, Raymond F, Affolter M (2006) OMICS-driven biomarker discovery in nutrition and health. J Biotechnol 124(4):758–787

Langridge P, Fleury D (2011) Making the most of 'omics' for crop breeding. Trends Biotechnol 29(1):33–40

Lee KJ, Yoon MS, Shin MJ, Lee JR, Cho YH, Lee HS, Ma KH, Lee GA (2017) Development of SSR markers and their use in studying genetic diversity and population of finger millet (*Eleusine coracana* L. Gaertn.). Plant Breed Biotechnol 5(3):183–191

Lesk C, Rowhani P, Ramankutty N (2016) Influence of extreme weather disasters on global crop production. Nature 529(7584):84–87

Li J, Wang Y, Wang L, Zhu J, Deng J, Tang R, Chen G (2021) Integration of transcriptomic and proteomic analyses for finger millet [Eleusine coracana (L.) Gaertn.] in response to drought stress. PLoS One 16(2):e0247181

Mahalakshmi S, Christopher GS, Reddy TP, Rao KV, Reddy VD (2006) Isolation of a cDNA clone (PcSrp) encoding serine-rich-protein from Porteresia coarctata T. and its expression in yeast and finger millet (Eleusine coracana L.) affording salt tolerance. Planta 224(2):347–359

Mahalle MD, Dey PC, Chetia SK, Baruah AR, Ahmed T, Sarma RN, Kaldate RC, Kumar A, Singh SK, Modi MK (2021) Association mapping for yield traits under drought stress in autumn rice germplasm collection of Assam. J Plant Biochem Biotechnol 30(1):26–36

Mariman EC (2006) Nutrigenomics and nutrigenetics: the 'omics' revolution in nutritional science. Biotechnol Appl Biochem 44(3):119–128

Morgante M, Hanafey M, Powell W (2002) Microsatellites are preferentially associated with nonrepetitive DNA in plant genomes. Nat Genet 30(2):194–200

Mundada PS, Ahire ML, Umdale SD, Barmukh RB, Nikam TD, Pable AA, Deshmukh RK, Barvkar VT (2021a) Characterization of influx and efflux silicon transporters and understanding their role in the osmotic stress tolerance in finger millet (Eleusine coracana (L.) Gaertn.). Plant Physiol Biochem 162:677–689

Mundada PS, Barvkar VT, Umdale SD, Kumar SA, Nikam TD, Ahire ML (2021b) An insight into the role of silicon on retaliation to osmotic stress in finger millet (Eleusine coracana (L.) Gaertn). J Hazard Mater 403:124078

Naga BL, Mangamoori LN, Subramanyam S (2012) Identification and characterization of EST-SSRs in finger millet (Eleusine coracana (L.) Gaertn.). J Crop Sci Biotechnol 15(1):9–16

Nageshbabu R, Jyothi MN, Sharadamma N, Rai DV, Devaraj VR (2013) Expression of miRNAs confers enhanced tolerance to drought and salt stress in finger millet (*Eleusine coracona*). J Stress Physiol Biochem 9(3):220–231

O'Kennedy MM, Grootboom A, Shewry PR (2006) Harnessing sorghum and millet biotechnology for food and health. J Cereal Sci 44(3):224–235

Pandian S, Ramesh M (2019) Decoding of finger millet genome: a milestone of millet genomics. Signal Transduct Insights 8:1178643418820541

Pandian S, Marichelvam K, Satish L, Ceasar SA, Pandian SK, Ramesh M (2018) SPAR markers-assisted assessment of genetic diversity and population structure in finger millet (*Eleusine coracana* (L.) Gaertn) mini-core collection. J Crop Sci Biotechnol 21(5):469–481

Panwar P, Saini RK, Sharma N, Yadav D, Kumar A (2010) Efficiency of RAPD, SSR and cytochrome P 450 gene based markers in accessing genetic variability amongst finger millet (Eleusine coracana) accessions. Mol Biol Rep 37(8):4075–4082

Parvathi MS, Nataraja KN (2017) Discovery of stress responsive TATA-box binding protein associated factor6 (TAF6) from finger millet (Eleusine coracana (L.) Gaertn). J Plant Biol 60(4):335–342

Parvathi MS, Nataraja KN, Yashoda BK, Ramegowda HV, Mamrutha HM, Rama N (2013) Expression analysis of stress responsive pathway genes linked to drought hardiness in an adapted crop, finger millet (Eleusine coracana). J Plant Biochem Biotechnol 22(2):193–201

Parvathi MS, Nataraja KN, Reddy YN, Naika MB, Gowda MC (2019) Transcriptome analysis of finger millet (Eleusine coracana (L.) Gaertn.) reveals unique drought responsive genes. J Genet 98(2):46

Puranik S, Kam J, Sahu PP, Yadav R, Srivastava RK, Ojulong H, Yadav R (2017) Harnessing finger millet to combat calcium deficiency in humans: challenges and prospects. Front Plant Sci 8:1311

Puranik S, Sahu PP, Beynon S, Srivastava RK, Sehgal D, Ojulong H, Yadav R (2020) Genome-wide association mapping and comparative genomics identifies genomic regions governing grain nutritional traits in finger millet (Eleusine coracana L. Gaertn.). Plants People Planet 2(6): 649–662

Pushpa BN, Kurnool K, Shankar AG (2020) Development of finger millet (Eleusine coracana (L.) Gaertn.) transgenic for salt tolerance by overexpressing antiporter gene OsSOS1 involved in sodium extrusion. Ind J Pure Appl Biosci 8(6):598–610

Rahman H, Jagadeeshselvam N, Valarmathi R, Sachin B, Sasikala R, Senthil N, Sudhakar D, Robin S, Muthurajan R (2014) Transcriptome analysis of salinity responsiveness in contrasting genotypes of finger millet (Eleusine coracana L.) through RNA-sequencing. Plant Mol Biol 85(4):485–503

Rahman H, Ramanathan V, Nallathambi J, Duraialagaraja S, Muthurajan R (2016) Over-expression of a NAC 67 transcription factor from finger millet (Eleusine coracana L.) confers tolerance against salinity and drought stress in rice. BMC Biotechnol 16(1):7–20

Ramakrishna C, Singh S, Raghavendrarao S, Padaria JC, Mohanty S, Sharma TR, Solanke AU (2018) The membrane tethered transcription factor EcbZIP17 from finger millet promotes plant growth and enhances tolerance to abiotic stresses. Sci Rep 8(1):1–4

Ramakrishnan M, Ceasar SA, Duraipandiyan V, Al-Dhabi NA, Ignacimuthu S (2016a) Assessment of genetic diversity, population structure and relationships in Indian and non-Indian genotypes of finger millet (Eleusine coracana (L.) Gaertn) using genomic SSR markers. Springerplus 5(1): 1–1

Ramakrishnan M, Ceasar SA, Duraipandiyan V, Al-Dhabi NA, Ignacimuthu S (2016b) Using molecular markers to assess the genetic diversity and population structure of finger millet (Eleusine coracana (L.) Gaertn.) from various geographical regions. Genet Resour Crop Evol 63(2):361–376

Ramakrishnan M, Ceasar SA, Vinod KK, Duraipandiyan V, Ajeesh Krishna TP, Upadhyaya HD, Al-Dhabi NA, Ignacimuthu S (2017) Identification of putative QTLs for seedling stage phosphorus starvation response in finger millet (*Eleusine coracana* L. Gaertn.) by association mapping and cross species synteny analysis. PLoS One 12(8):e0183261

Ramegowda V, Senthil-Kumar M, Nataraja KN, Reddy MK, Mysore KS, Udayakumar M (2012) Expression of a finger millet transcription factor, EcNAC1, in tobacco confers abiotic stress-tolerance. PLoS One 7(7):e40397

Ramegowda V, Gill US, Sivalingam PN, Gupta A, Gupta C, Govind G, Nataraja KN, Pereira A, Udayakumar M, Mysore KS, Senthil-Kumar M (2017) GBF3 transcription factor imparts drought tolerance in Arabidopsis thaliana. Sci Rep 7(1):1–3

Sakamma S, Umesh KB, Girish MR, Ravi SC, Satishkumar M, Bellundagi V (2018) Finger millet (Eleusine coracana L. Gaertn.) production system: status, potential, constraints, and implications for improving small farmer's welfare. J Agric Sci 10(1):162–179

Salimath SS, Oliveira AC, Bennetzen JL, Godwin ID (1995) Assessment of genome origins and genetic diversity in the genus Eleusine with DNA markers. Genome 38(4):757–763

Schubert R, Mueller-Starck G, Riegel R (2001) Development of EST-PCR markers and monitoring their intra-population genetic variation in Piceaabies (L.) Karst. Theor Appl Genet 103(8): 1223–1231

Scofield SR, Nelson RS (2009) Resources for virus-induced gene silencing in the grasses. Plant Physiol 149(1):152–157

Sen S, Kundu S, Dutta SK (2016) Proteomic analysis of JAZ interacting proteins under methyl jasmonate treatment in finger millet. Plant Physiol Biochem 108:79–89

Sharma D, Tiwari A, Sood S, Jamra G, Singh NK, Meher PK, Kumar A (2018) Genome wide association mapping of agro-morphological traits among a diverse collection of finger millet (Eleusine coracana L.) genotypes using SNP markers. PLoS One 13(8):e0199444

Shi R, Kumar C, Zougman A, Zhang Y, Podtelejnikov A, Cox J, Wiśniewski JR, Mann M (2007) Analysis of the mouse liver proteome using advanced mass spectrometry. J Proteome Res 6(8): 2963–2972

Singh Y, Kumar J (2010) Study of genomic fingerprints profile of Magnaporthe grisea from finger millet (Eleusine coracona) by random amplified polymorphic DNA-polymerase chain reaction (RAPD-PCR). Afr J Biotechnol 9(46):7798–7804

Singh S, Parihar P, Singh R, Singh VP, Prasad SM (2016) Heavy metal tolerance in plants: role of transcriptomics, proteomics, metabolomics, and ionomics. Front Plant Sci 6:1143

Singh RK, Phanindra MLV, Singh VK, Sonam, Raghavendrarao S, Solanke AU, Kumar PA (2014) Isolation and characterization of drought responsive EcDehydrin7 gene from finger millet (Eleusine coracana (L.) Gaertn). Indian J Genet. Plant Breed 74(4):456–462

Singh S, Chopperla R, Shingote P, Chhapekar SS, Deshmukh R, Khan S, Padaria JC, Sharma TR, Solanke AU (2021) Overexpression of EcDREB2A transcription factor from finger millet in tobacco enhances tolerance to heat stress through ROS scavenging. J Biotechnol 36:10–24
Sood S, Kumar A, Babu BK, Gaur VS, Pandey D, Kant L, Pattnayak A (2016) Gene discovery and advances in finger millet [Eleusine coracana (L.) Gaertn.] genomics—an important nutri-cereal of future. Front Plant Sci 7:1634
Sumner LW, Mendes P, Dixon RA (2003) Plant metabolomics: large-scale phytochemistry in the functional genomics era. Phytochemistry 62(6):817–836
Sun C, Li Y, Wu Q, Luo H, Sun Y, Song J, Lui EM, Chen S (2010) De novo sequencing and analysis of the American ginseng root transcriptome using a GS FLX titanium platform to discover putative genes involved in ginsenoside biosynthesis. BMC Genomics 11(1):1–2
Sun X, Pan B, Wang Y, Xu W, Zhang S (2020) Exogenous calcium improved resistance to Botryosphaeriadothidea by increasing autophagy activity and salicylic acid level in pear. Mol Plant Microbe Interact 33(9):1150–1160
Tiwari A, Sharma D, Sood S, Jaiswal JP, Pachauri SP, Ramteke PW, Kumar A (2020) Genome-wide association mapping for seed protein content in finger millet (Eleusine coracana) global collection through genotyping by sequencing. J Cereal Sci 91:102888
Tripathi B, Platel K (2010) Finger millet (Eleucine coracana) flour as a vehicle for fortification with zinc. J Trace Elem Med Biol 24(1):46–51
Varshney RK, Graner A, Sorrells ME (2005) Genic microsatellite markers in plants: features and applications. Trends Biotechnol 23(1):48–55
Varshney RK, Mohan SM, Gaur PM, Gangarao NV, Pandey MK, Bohra A, Sawargaonkar SL, Chitikineni A, Kimurto PK, Janila P, Saxena KB (2013) Achievements and prospects of genomics-assisted breeding in three legume crops of the semi-arid tropics. Biotechnol Adv 31(8):1120–1134
Venkatesan J, Ramu V, Sethuraman T, Sivagnanam C, Doss G (2021) Molecular marker for characterization of traditional and hybrid derivatives of Eleusine coracana (L.) using ISSR marker. J Genet Eng Biotechnol 19(1):1–2
Vetriventhan M, Upadhyaya HD, Dwivedi SL, Pattanashetti SK, Singh SK (2016) Finger and foxtail millets. In: Genetic and genomic resources for grain cereals improvement. Academic Press, pp 291–319
Wahid A, Gelani S, Ashraf M, Foolad MR (2007) Heat tolerance in plants: an overview. Environ Exp Bot 61(3):199–223
Wang J, Li D, Dangott LJ, Wu G (2006) Proteomics and its role in nutrition research. J Nutr 136: 1759–1762
Wang W, Vinocur B, Altman A (2007) Plant responses to drought, salinity and extreme temperatures towards genetic engineering for stress tolerance. Planta 218:1–14

Genome-Editing Approaches for Abiotic Stress Tolerance in Small Millets

13

Debasish Pati, Rishi Kesh, Vivekananda Mohanta, Ramesh Namdeo Pudake, Amitha Mithra Sevanthi, and Binod Bihari Sahu

Abstract

Abiotic factors are an important constraint in regulating plants growth, architecture, and survival in a particular geographical area. The abiotic components include temperature, water, and nutrients; any deviation from the above ideal condition concerning the plant leads to an adverse condition known as abiotic stress. Under abiotic stress plants' normal growth gets hindered, ultimately leading to lower crop yield or in extreme may lead to plant fatality. Throughout the process of evolution, plants have gained traits by novel mutation or acquired them from ancestors to counter stress. Small millets are one of the most adapted species belonging to the natural stress-resistant grass family. Along with the adaptability, small millets are a good source of nutrition, easy to cultivate and require lesser field input, along with the advantages of good market values. Thus, at present, it is the need to identify and characterize those novel traits in small millets to generate stress-resistant and high-yielding cultivars, which is not possible by conventional breeding approaches. With the advancement of long sequencing tools and multi-omics techniques, it is now easier to generate larger germplasm and genome database. The genome-editing techniques are now getting better with time, the novel techniques like homologous recombination, RNAi

Rishi Kesh and Vivekananda Mohanta contributed equally with all other contributors.

D. Pati · R. Kesh · V. Mohanta · B. B. Sahu (✉)
Laboratory of Molecular Genetics and Plant Immunity, Department of Life Science, National Institute of Technology Rourkela, Rourkela, Odisha, India
e-mail: sahub@nitrkl.ac.in

R. N. Pudake
Amity Institute of Nanotechnology, Amity University Uttar Pradesh, Noida, India

A. M. Sevanthi
ICAR - National Institute for Plant Biotechnology, New Delhi, India

R. N. Pudake et al. (eds.), *Omics of Climate Resilient Small Millets*,
https://doi.org/10.1007/978-981-19-3907-5_13

and CRISPR/Cas based genome editing have revolutionized the process. As the approaches are cost-effective, easy to use, and prone to negligible error, will help to counter the climate change challenges and to eradicate world food security along with may attract, public acceptance to genome engineered plants.

Keywords

Gene editing · RNAi · Abiotic stress tolerance · Small millets · Mutations · CRISPR/Cas

13.1 Introduction

Plants, like all organisms, flourish in their ideal natural habitat. Being sessile, the plants' growth, development, distribution, and survival depend upon many environmental constraints including biotic and abiotic factors. Throughout their lifespan, the abiotic factors are predominantly responsible for modulating plants' natural features and biomass; the abiotic stressors encountered by plants are temperature, cold, salt, metal, drought, and flooding stress (De Oliveira 2018). Any excess or limitation to the above constraint can cause adverse conditions to plants, and the consequences lead organisms to stress. The recurrences of this natural hazard are at present becoming common due to drastic changes in the environment, overpopulation, and human mismanagement ecological systems. The abiotic factors induce stress by regulating efficient nutrient mobilization by root, hampering plants' photosynthetic activity, altering cellular process, and inducing hyper-reactive oxygen species (ROS) generation, ultimately rendering plants' growth and reproductive efficiency, challenging global food security (Kheir 2020). That's why, throughout crop domestication, humans have retained and selected plants having suitable traits which are resulting from unique mutations or inherent variation from the ancestral species with adaptability to the ever-changing environment. The process of selection of plant varieties with desired characteristics for consumption and cultivation continues to present (Saleem et al. 2021).

Considering the prevailing environmental challenges for agriculture; and for future global food security, the millets with the ability to tolerate abiotic stresses can be the crops recommended for large-scale cultivation. Small millets represent the coarse cereals, and they are a diverse group of annual, small-seeded panicoid C4 grasses that belong to the Poaceae family; they include proso millet (*Panicum miliaceum*), fonio (*Digitaria exilis*), foxtail millet (*Setaria italica*), little millet (*Panicum sumatrense*), barnyard millet (*Echinochloa crus-Galli*), finger millet (*Eleusine coracana*), kodo millet (*Paspalum scrobiculate*), teff millet (*Eragrostis tef*), browntop millet (*Urochloa ramosa*), guinea millet (*Brachiaria deflexa*), and Job's tears (*Coix lacryma-jobi*) (Zhang et al. 2012a, b). Millets are wonder crops as they show minimal susceptibility to environmental stresses, adaptation to harsh ecological conditions, minimal nutritional requirements like non-reliance on synthetic fertilizers, as well as less water for better growth and productivity (Bandyopadhyay et al. 2017). To withstand water-limiting conditions, millets have evolved better nitrogen-use and water-use efficiencies. Many studies have shown that small millets can adapt to numerous abiotic factors; for example, in biochemical studies of foxtail millet (Lata et al. 2011), little millet (Ajithkumar and Panneerselvam 2014), and teff millet (Smirnoff and Colombe 1988), millets showed enhancement in the synthesis of osmolytes, ROS, levels of antioxidants, and their scavenging enzymes, as well as other stress-related proteins that have been reported to counter abiotic stresses. In *P. sumatrense*, an enlargement in root length was reported in one study (Ajithkumar and Panneerselvam 2014), and in *E. tef* (Balsamo et al. 2006) an enhanced leaf tensile strength during drought stress was observed. Small millets are a good source of micro/macronutrients, fiber, gluten-free, and high total protein, therefore easing the preparation of low glycemic content products (Muthamilarasan and Prasad 2021). Due to the above adaptations and inherent potential small millets can tackle wide range of abiotic stress (Fig. 13.1), as well it's a boon to Indian agricultural sector; which get challenged by population explosion. All of the small millets are underused at present in agricultural sectors, despite their useful attributes including climate resilience, enhanced market value, better taste and nutrition (Mayes et al. 2012). The novel and desired agronomic traits of small millets need to be integrated into the suitable cultivar to enhance benefits. For this "genome-editing tools" are required to carry out the task of integrating desired traits into crop genomes.

Plant breeding approaches in their preliminary phase, known as the first generation of plant breeding, completely relied upon the phenotype-based selection procedure for crop improvement. In the next phase of plant breeding, the molecular signature(s) was/were taken into consideration, and at present in the third generation of plant breeding major focus is on comprehensive genomic information for effective selection of phenotype and suitable traits. The shift to genome-mediated breeding is now only possible due to the availability of enormous nucleotide information. The plant breeding programs became modern and precise through the twenty-first century by the development of genome sequencing data, leading us to the post-genomic era. Identification of molecular markers like simple sequence repeats (SSR), random amplified polymorphic DNA (RAPD), restriction fragment length polymorphism (RFLP), sequence-based polymorphic markers (SBPs), and amplified fragment length polymorphism (AFLP) along with the advancement of transcriptome sequencing and whole-genome sequencing made it possible to generate comprehensive and high resolution genetic maps to determine potential quantitative traits loci (QTLs) of important traits in crop plants including small millets. Integrating the genomic approaches along with the conventional breeding tools has widened the path for genome-based breeding programs. The integration of QTL with microarrays and transcriptomic data have led to the development of multiple expression QTLs (Mohanta et al. 2017). With the availability of these genomic tools, the transfer of abiotic stress-responsive genes from millets to major crops can be a worthwhile approach to the outcome of global food security. Among all, at present few of the small millets like foxtail millet (Bennetzen et al. 2012), finger millet (Hittalmani et al. 2017), and teff millet (Cannarozzi et al. 2014) whole-genome sequences are available which render them as the experimental model for the genotyping (VanBuren et al. 2020). Genome editing is a reverse genetic technique and the availability of whole-genome sequence information and identification of causal genes in small millets can serve as a boon for editing and modifying them for their improvement. The current chapter is dedicated to citing the major advantages of genome-editing approaches in small millets to counter abiotic stress and the genomic potential of millets to produce stress-tolerant crops.

13.2 Adaptation of Small Millets for Tolerance to Abiotic Stress

The adaptation in small millets for tolerance to abiotic stress includes the general characteristics like shorter life span which is about 12–14 weeks, thickened cell wall, small leaf area, and the capability to form a dense fibrous root system; all the morphological characteristics directly or indirectly contribute toward stress tolerance. Dumbbell shaped guard cells allow faster stomatal responses leading to better water-use efficiency (Lawson and Vialet-Chabrand 2019). Other than the stomatal structure the stomatal density and distribution is also the adaptive parameter for better stress survival. The amphistomatic leaves arrangement in millets helps more efficient CO_2 diffusion into the leaf hence helping in maximizing the photosynthesis rate (Stebbins and Shah 1960). The other highly advantageous trait in millets is the C_4 photosynthesis system, in which CO_2 gets concentrated around ribulose-1,5-bisphosphate carboxylase/oxygenase (RuBisCo). Thus, the C_4 mechanism enhances water-use efficiency and nitrogen-use efficiency by regulating photosynthetic activity under higher CO_2 and warmer conditions (Aubry et al. 2011). Concerning the inherent traits, millets can also exhibit anatomical modification to adverse condition, for example in finger millet during waterlogging, adventitious root formation takes place as well as the plant adapt for anaerobic respiration (Matsuura et al. 2016; Schlader et al. 2019).

Along with the morphological adaptation plants' biomolecules are the important cues for stress survival. The biomolecules include hormones, secondary metabolites, proteins, and osmolytes. Among all secondary metabolites get synthesized and accumulated at greater concentration during abiotic stress. Studies showed that in millets the C_4 pathway metabolomics components like malate dehydrogenase, carbonic anhydrase, phosphoenolpyruvate, malic enzyme, and pyruvate orthophosphate dikinase get expressed at higher concentrations, thus helping in elevating adversaries (Lata et al. 2011).

The severity of stress depends upon another parameter of plants' efficiency in ROS neutralization. During an unfavorable environment, plants induce to produce ROS molecules, which damage the cellular process, to counter the toxicity plants have an arsenal of ROS scavenging enzymes (Ajithkumar and Panneerselvam 2014). For example, in foxtail millets and little millets during dehydration, it has been observed that CAT, APX, and GP activity increased many folds, whereas LP and EL

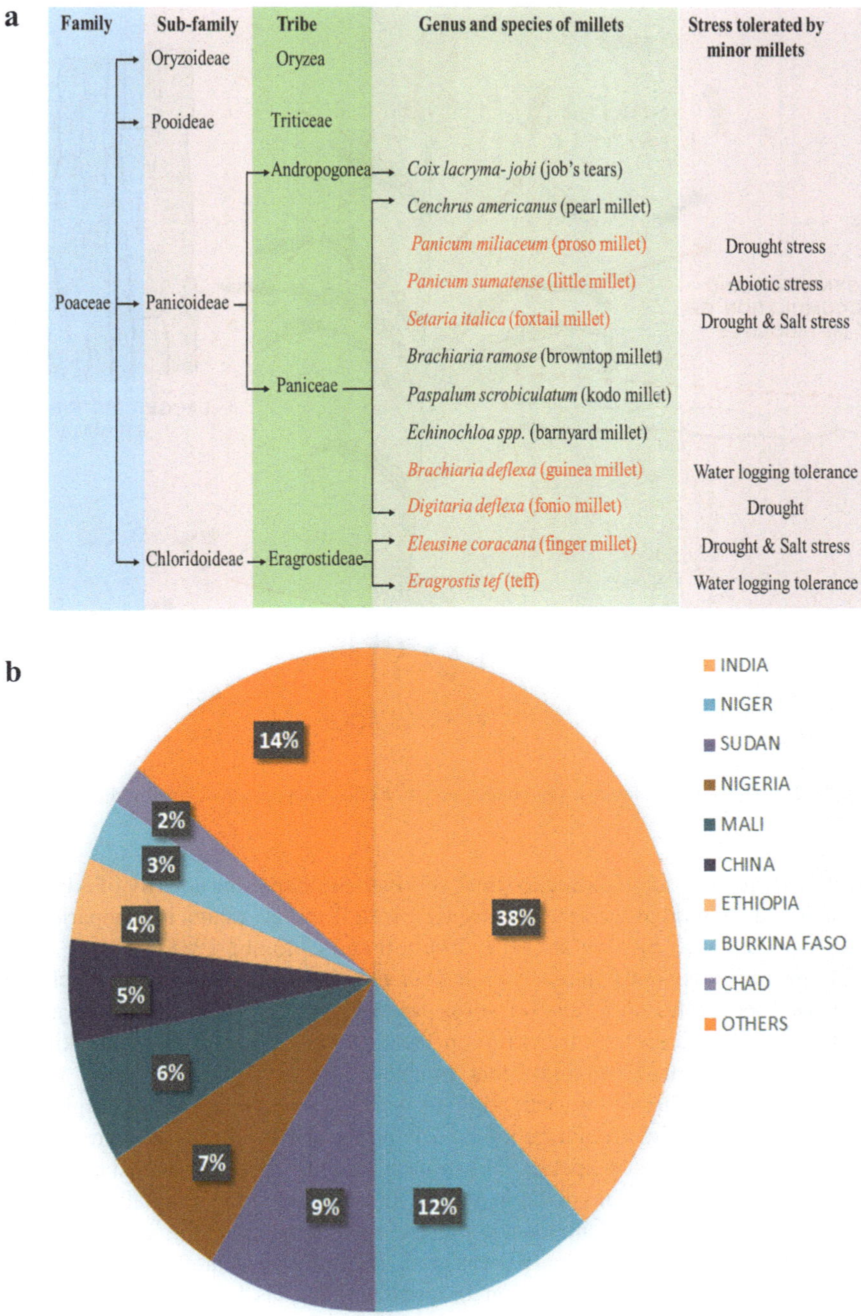

Fig. 13.1 (**a**) Diagram representing phylogeny of small millets. (**b**) Pie chart showing major countries' global status in millets cultivation. India holds the top position as the sub-tropical to temperate climate is most suitable for millet cultivation

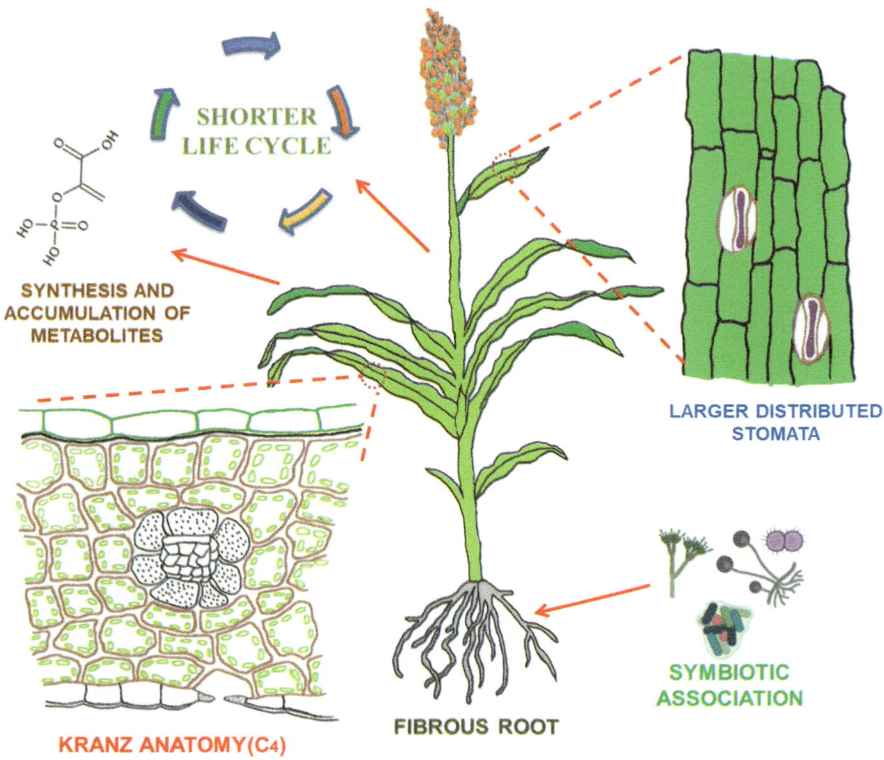

Fig. 13.2 General physiological, anatomical, and molecular adaptations in small millets to tackle abiotic stress

get downregulated, thus leading to stress survival. Similarly, the activity of APX and monodehydro-ascorbate reductase increases in "teff millet" plants in comparison to non-stressed plants (Sabir et al. 2011; Smirnoff and Colombe 1988).

At genetic level, structural genes are an important class of genes which can improve specific features of crop stress tolerance. Along with structural genes, regulatory proteins such as transcription factors (TFs), phosphatases, and kinases are another important candidate that modulate the several downstream genes' expression and achieve the activation of various stress signals. Cis-regulatory sequences of the downstream genes are of fundamental importance in regulating the expression of genes by specific TFs (transcription factors). Thus, the targeting of particular gene/sequence involved in abiotic stress regulation and mutating the gene or achieving the gain in function via precise genome-editing tools could improve abiotic stress tolerance in crops (Biswas et al. 2021; Zhang et al. 2000). The role of the gene in abiotic stress regulation in small millets has been well documented (Numan et al. 2021; Rana et al. 2021) and reviewed by other chapters in this book. Still there is a need of more studies to fully understand small millets' stress-responsive mechanisms (Fig. 13.2).

13.3 Progress Made in Genome Editing of Millets for Abiotic Stress Tolerance

Traditional breeding methods practiced since the early plant genomics era are time taking, tedious, often lead to loss of species' variation, often do not have control over genetic mutations, lead to the creation of unwanted traits, may cause loss of genetic diversity, and may pose some unwanted and unrecognized environmental risks. In recent times, genome editing through vector-based methods and genetic transformation via *Agrobacterium* have received a great attention, especially in agricultural plants. Genome-editing tools that are most used include the CRISPR/Cas9 system along with zinc finger nucleases (ZFNs), and transcription activator-like effector nucleases (TALENs). Among the methods available for gene editing, clustered regularly interspaced palindromic repeats (CRISPR) and CRISPR-associated protein 9 (Cas9) have gained a reputation as a genome-editing tool among a vast range of alternatives due to their precise editing mechanism. The CRISPR/Cas9 system can be efficiently utilized for genetic modification to accelerate crop improvement and provide accuracy in crop breeding programs to generate abiotic stress-tolerant crops (Zafar et al. 2020; Zhang et al. 2000). Though small millets have a great pool of nutritional value, it is found that very low importance is given for gene editing studies in these crops in the past and it is gaining importance in the recent days in unlocking the mechanism of climate resilience and nutrient fortification and improving millet research. However, the methods and protocols of transformation and selection for transgenic plants are already established in many small millets (as reviewed in Chap. 8) and can be readily utilized for speeding up the gene editing research with required modifications. We have discussed the few examples of recent major tools involved in genome editing and their application in millets.

13.3.1 CRISPR/Cas9 Targeting of Abiotic Stress Tolerance Genes

The CRISPR/Cas system was first identified in bacteria and archaea where it provides adaptive immunity against a wide range of foreign particles like viruses and plasmids. At present, the above approach is the most adaptive tool for genome-editing strategy. As the name suggests **"clustered regularly interspaced short palindromic repeat"** was discovered in *E. coli* having five direct repeats in the *iap* (isozyme of alkaline phosphatase) gene (Ishino et al. 2018). This identification paved the way for the identification of similar direct repeats in genome sequences of many distantly related prokaryotes. CRISPR/Cas9 consists of a locus called CRISPR which consists of short repetitive elements (repeats) separated by unique sequences (spacers). These spacers are usually derived from mobile genetic elements (MGEs) like transposons or plasmids and bacteriophages. The two components of the

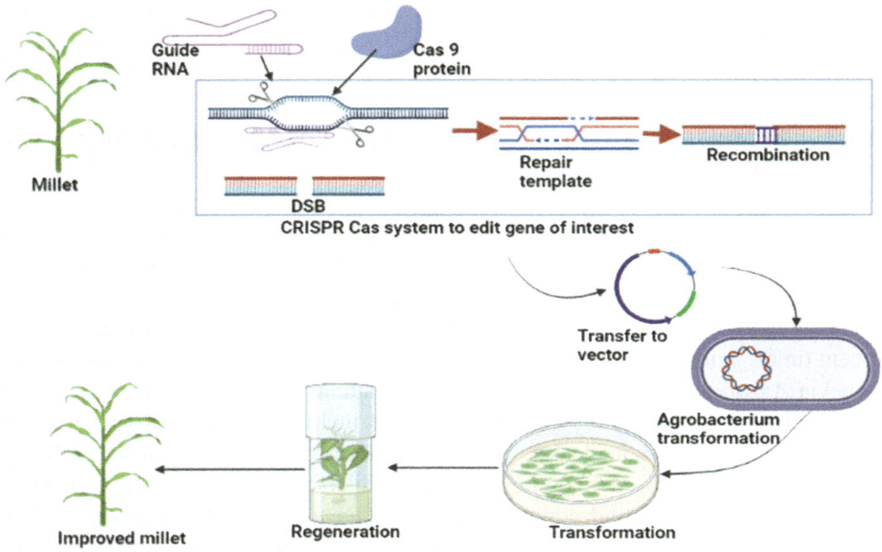

Fig. 13.3 Illustration of CRISPR/Cas9 mechanism. Here, sgRNA (single-guide RNA) binds to the target sequence by base-pairing upon recognition and Cas9 (DNA endonuclease) cleaves the DNA after binding with sgRNA. The cleaved site is repaired through the error-prone non-homologous end-joining pathway which leads to the formation of changed sites (by insertions or deletions at the sites of double-stranded breaks) in the gene. The CRISPR/Cas9 cassettes can be transferred to *Agrobacterium* through a vector. This transformed *Agrobacterium* acts as a medium to transform plants with genes of interest for improved tolerance against abiotic stresses

CRISPR/Cas9 system: the single-guide RNA (sgRNA), which binds to the target sequence through base-pairing upon recognition, and the DNA endonuclease Cas9, which cleaves the genomic DNA after binding with sgRNA. The fragmented genomic DNA resulting from the cleavage is mainly repaired by a non-homologous end-joining pathway in plants, as it is an error-prone strategy, which may lead to the formation of small insertions or deletions at the site of the double-strand break. Thus, CRISPR/Cas9 turns out to be an efficient technique to knock out a gene when the coding sequence can be targeted by single-guide RNA (sgRNA) in plants, and has been reviewed recently (Calvache et al. 2022; Maximiano and Franco 2022; Wada et al. 2022; Zhang et al. 2019). A diagrammatic illustration is given in Fig. 13.3, representing the detailed mechanism of the CRISPR/Cas9 approach toward improving crops.

This technology has been applied to crop improvement in many model plants and crops for acquired traits such as biotic and abiotic stress tolerance and improved productivity (Bhat et al. 2021; Ghosh and Dey 2022; Jaganathan et al. 2018; Rao and Wang 2021). The selection of a target gene is important for achieving the desired improvement in the trait. Trait improvement can be carried out by targeting mainly two categories of genes: regulatory and structural genes. Structural genes encoded proteins have a straight effect on traits, for example, while regulatory genes may act as an indirect regulator of other genes which may be involved in other cellular processes. Furthermore, abiotic stress tolerance is also controlled by cis-regulatory sequences. CRISPR/Cas9 technology has been recently applied toward the development of cereals crops like rice (Ahmad et al. 2022; Biswal et al. 2019), barley (Vlčko and Ohnoutkova 2020), and maize (Shi et al. 2017; Waqas et al. 2021) with an increase in the degree of tolerance to abiotic stresses. These studies have proved that CRISPR tools have the enormous ability in generating multi-stress-resistant cultivars through the multi-stack expression of many structural and regulatory genes in crop plants. Here we will discuss furthermore about the gene editing studies done in small millets.

Foxtail millet, due to its compact genome (~450 Mb), diploidy, and C_4 photosynthetic capability is considered a model crop among millets. Foxtail millet genome sequence was reported in 2012 (Zhang et al. 2012a) and the first genome-editing study in foxtail millet by CRISPR/CAS9 was carried out in 2018 in which the plasmid pCAMBIA1300-35s-Cas9-OsU3-SiPDS was used to target the phytoene desaturase gene of the foxtail millet. RFLP and sequencing were used to detect the targeted mutagenesis, sequencing results detected both deletions up to 43 bases and insertion of single bases (Lin et al. 2018). This was the first-ever successful application of the CRISPR/Cas9 system in foxtail millet.

Afterward few studies have been conducted in the foxtail millets, and in one of the studies CRISPR/Cas9 has been applied for the induction of double haploid (DH) lines by targeting the *S. italica* Matrilineal (SiMTL) gene. Two different gRNAs were used to target the exonic region of SiMTL and were expressed under the control of OsU3 promoter, and *Cas9* was kept under the influence of the ubiquitin promoter of maize (Cheng et al. 2021). In another study, genome editing was used to validate the gene that was responsible for reduced seed shattering. The locus—Less Shattering1 (SvLes1) identified through SNPs present in genomic resources of foxtail millets that was located on chromosome 5. The allele was made non-functional by insertion of one nucleotide and causing a frameshift mutation, thus generating non-shattering plants (Mamidi et al. 2020). Another study has successfully targeted *SiFMBP, SiDof4,* and *SiIPK1* genes and generated efficient heritable knockout mutants of foxtail millets by CRISPR/Cas9 (Liang et al. 2021). They used both Cytosine Base Editor (CBE) and Adenine Base Editor (ABE) tool that developed based on the CRISPR/Cas9 system. The system is composed of a nickase Cas9 (nCas9) and a cytosine or adenine deaminase, which results in precise replacement of nucleotides at specific sites (C to T or A to G). This doesn't result into double-strand break in DNA but result into change in amino acid that may change the function of the gene. A similar study was carried out in *S. viridis*, developed the protoplast-based assay for multiplex CRISPR/Cas9 gene-editing system, with the more efficient microhomology-mediated end joining as compared to non-homologous end-joining repair system (Weiss et al. 2020). They also demonstrated the co-expression with *Trex2* exonuclease which enhanced the mutagenic efficiency of the system by many folds.

Along with this, the role of *SiNPI* in microspore development in small foxtail millet was also confirmed by CRISPR/Cas9 gene-editing system. In the absence of

SiNPI the plant is rendered completely male sterility subsequent to pollen abortion (Zhang et al. 2021). Likewise the knockout line of *SiBoRI* gene in foxtail millet leads to reduction in boron content of panicle primary branches, causing boron deficiency syndromes like thickened cell wall and higher cell porosity as compared to wild type (Wang et al. 2022). These results and other studies have confirmed that the efficiency of CRISPR/Cas9 was reported to be higher and error-proof as compared to the other approaches like ribonucleoprotein complex and viral vector mediated approaches as well as superior in comparison to DNA-integration-based genome-editing method (Basso et al. 2021). CRISPR/Cas9 has the potential role in functional genomic analysis; and at present very few studies have been carried out in small millets for insight about unknown gene function. The base-editing technique of CRISPR/Cas9 can lead to producing desirable traits in a particular population by creating allelic variants in it, thus CRISPR/Cas9 genome-editing technology has immense potential, especially for highly neglected crops such as small millet which can be further lead to the development of environmental stress resilience and yield improvement (Ceasar 2021; Numan et al. 2021; Rana et al. 2021).

13.3.2 RNA Interference

In eukaryotic cells, the RNA interference acts as machinery to protect cells from foreign DNA. It is an evolutionary conserved, gene-regulating molecule. It also helps in producing epigenetic modifications, genomic stability, transcriptional and translational control over the cellular process. The gene expression can be regulated by short interfering RNA (siRNA) and microRNA in plants. Dicer or dicer-like enzymes, and multidomain endonuclease belonging to RNase III family, play role in cleaving long double-stranded RNA or pre-miRNA into siRNA or miRNA. Small non-coding RNA in association with RNA-induced silencing complex (RISC) and other effector proteins carry out the process of complex degradation of the target mRNA. Extensive studies have been carried out in the past couple of decades to understand the role of small non-coding RNA in gene silencing at the transcriptional and post-transcriptional levels. This exploration led to the discovery of wide classes of non-coding RNAs which include miRNA, siRNA, qiRNA, piRNA, etc. The miRNA can be derived from genomic DNA, whereas siRNA can have originated either by endogenous means, like cleavage of dsRNA or by exogenous means, by viruses, transgene, or transposons. The dsRNA or short hairpin precursor can induce gene silencing by RNAi; dsRNA perfectly binds to the target gene; dsRNA is cropped by the activity of dicer enzyme into short 21–24 nucleotide long fragments. After the recruitment of siRNA-induced silencing complex (siRISC) leads to degradation of the sense strand(which has the same sequence as that of the targeted mRNA) whereas, the antisense strand gets incorporated in a sequence-specific manner on mRNA which is targeted. AGO protein in association with siRISC and effector proteins cleaves the target mRNA or translation leading to post-transcriptional gene silencing (PTGS) (Saurabh et al. 2014) (Fig. 13.4). RNA interference technology can be used for the generation of transgenic cultivars that

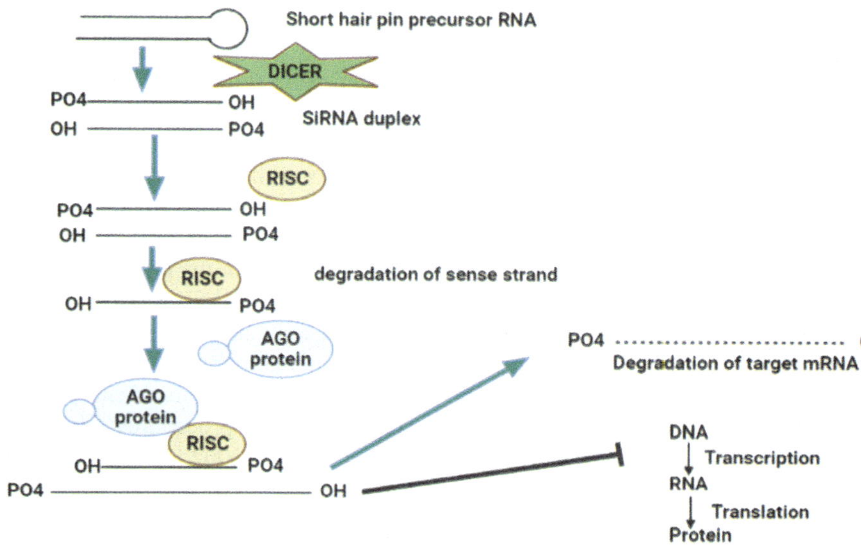

Fig. 13.4 Gene silencing by siRNA biogenesis. Precursor RNA leads tc the formation of siRNA duplex. Here siRNA duplex with overhangs at 3′-OH ends is formed because of the activity of dicer or dicer-like enzymes or dicer acts on the precursor RNA. Further, the siRNA antisense strand induces the silencing complex in association with RISC protein. Thereafter, incorporation of AGO and other effector proteins with siRISC facilitates the gene silencing through degradation of the target mRNA or translation inhibition

can deal with a broad range of abiotic stress. In *O. sativa*, a receptor for activated C-kinase (RAC-1) is a highly conserved scaffold protein that plays an important role in the regulation of growth and development in plants. RNAi-mediated downregulation of RAC-1 gene in transgenic *O. sativa* showed an increase in drought tolerance level than the non-transgenic lines (Li et al. 2009). Similarly, it was observed that rice plants perform well, in both vegetative and reproductive stages during drought, by the suppression of squalene synthase (SQS) gene (Manavalan et al. 2012).

But, as per our knowledge, there are no studies in small millets for achieving the abiotic stress tolerance by RNAi. However, there are studies that have established the methods to achieve the downregulation of genes that are important for the development of foxtail millets (Ceasar et al. 2017; Pan et al. 2016). These studies can be useful for future research in RNAi-mediated abiotic stress tolerance in foxtail millets.

13.4 Role of Biosafety and Regulation in Plant Genome Editing

Genome editing and genetic engineering play an important role in agriculture and are the promising approaches to meet the global need for food, enhance productivity, and counter numerous environmental challenges. Like all innovative technologies, genome-based editing approaches also have mixed outcomes, huge potential for crop improvement and functional genomics as well as safety and security issues. Off-target editing in the genome, which leads to unintended mutation is one of the major risks associated with genome editing. However, plants possess minimal risk as the impact of mutation is generally limited to the somatic changes and is less likely to affect critical tissue (Schmid-Siegert et al. 2017). As well breeding programs to develop new cultivars involved an intensive process of selection of individual plants, that will be helpful in eliminating undesirable mutation and phenotype. Therefore, an appropriate biosafety framework is necessary to be adopted for the proper development and sustainable use of genome editing.

India has a strong biosafety regulatory framework for transgenic under "Manufacture, use, import, export, and storage of hazardous microorganisms/genetically engineered organisms or cells, Rules 1989 (Rules 1989) under environment (protection) act (Epa).1986". The other applicable laws, acts, and procedures related to biosafety and biosecurity are: The Biological Diversity Act, 2002; Seed Act, 1966; Drugs and Cosmetic Act 1940; Food Safety and Standards Act, 2006; Protection of Plant Varieties and Farmer's Rights, 2001; The Unlawful Activities (Prevention) Act, 1967; Plant Quarantine Order 2003; Disaster Management Act, 20053; Weapons of Mass Destruction and Their Delivery System (Prohibition of Unlawful Activities) Act, 2005 (Gupta et al. 2021). It is always debatable between two groups of ideology; one claims to genome-edited plants to be considered as of low risk associated and thus negligible regulations should be applied, while other group of experts feel that the uncertainty and lack of complete knowledge, requires more strengthen regulations for GMOs (Agapito-Tenfen et al. 2018; Eckerstorfer et al. 2019).

At present, agriculture faces major challenges, where plant genome editing could play a key role in developing desired crops. But scientific research, policies, and regulation will not guarantee the success of this technology until social acceptance and political influence contribute positively toward the approaches (Gupta et al. 2021; Schiemann et al. 2020). Ultimately social consensus is of utmost importance for the sustainability of genome editing. The ethical principles, social need, and sustainable development should align research and innovation. Very recently, along with a few countries on the American continent, China and Indian government have exempted the certain types of genome-edited crops from the stringent regulations that are applicable on genetically modified or GM crops, thus giving a big boost to their further research and development (Turnbull et al. 2021) (https://www.isaaa.org/kc/cropbiotechupdate/ged/article/default.asp?ID=19392, Accessed on May 5, 2022). These kind of changes in regulation will encourage in discovering new breeding tools for crop improvement.

13.5 Conclusion

From the recent studies, it is now known that small millets can be a vital source of micronutrients, fibers, and complex starch composition that helps in better regulation of blood glucose for which they are gaining attention as equivalent to the major crops. Besides the advantages of high nutritional value, small millets are valuable owing to their innate ability to tolerate a wide array of pathogens and abiotic stress conditions. Higher inbreeding capacity, shorter life span, and smaller genome size make small millets suitable for pureline development. Targeted genomic cloning and desired traits engineering in suitable cultivars from small millets are now gaining attention to improve crop yield and stress tolerance crop varieties. It is now the need to investigate and identify the natural genetic variation among the small millet germplasms to exploit them for crop improvement for many agronomically and nutritionally important traits, not just for the improvement of small millets per se but also for other major cereal crops as they belong to the same Poaceae family. Among the techniques, CRISPR/Cas 9 is the most relevant method as it is cost-effective and has higher efficiency and yet, the extent of off-target mutation and cleavage efficiency of the targeted site need to be carefully investigated. Omics approaches have proved to be the most straightforward and potential technology for improving abiotic stress tolerance in crops giving rise to precision breeding.

References

Agapito-Tenfen SZ, Okoli AS, Bernstein MJ, Wikmark O-G, Myhr AI (2018) Revisiting risk governance of GM plants: the need to consider new and emerging gene-editing techniques. Front Plant Sci 9:1874

Ahmad S et al (2022) Harnessing CRISPR/Cas system for resistance against abiotic stresses in rice: recent advances, applications, challenges, and prospects. In: Plant abiotic stress physiology. Apple Academic Press, Oakville, pp 241–261

Ajithkumar IP, Panneerselvam R (2014) ROS scavenging system, osmotic maintenance, pigment and growth status of Panicum sumatrense roth. under drought stress. Cell Biochem Biophys 68: 587–595

Aubry S, Brown NJ, Hibberd JM (2011) The role of proteins in C3 plants prior to their recruitment into the C4 pathway. J Exp Bot 62:3049–3059

Balsamo R, Willigen CV, Bauer A, Farrant J (2006) Drought tolerance of selected Eragrostis species correlates with leaf tensile properties. Ann Bot 97:985–991

Bandyopadhyay T, Muthamilarasan M, Prasad M (2017) Millets for next generation climate-smart agriculture. Front Plant Sci 8:1266

Basso MF et al (2021) Efficient genome editing and gene knockout in Setaria viridis with CRISPR/ Cas9 directed gene editing by the non-homologous end-joining pathway. Plant Biotechnol 38: 227–238

Bennetzen JL et al (2012) Reference genome sequence of the model plant Setaria. Nat Biotechnol 30:555–561

Bhat MA, Mir RA, Kumar V, Shah AA, Zargar SM, Rahman S, Jan AT (2021) Mechanistic insights of CRISPR/Cas-mediated genome editing towards enhancing abiotic stress tolerance in plants. Physiol Plant 172:1255–1268

Biswal AK, Mangrauthia SK, Reddy MR, Yugandhar P (2019) CRISPR mediated genome engineering to develop climate smart rice: challenges and opportunities. In: Seminars in cell & developmental biology. Elsevier, New York, pp 100–106

Biswas D, Saha SC, Dey A (2021) CRISPR-Cas genome-editing tool in plant abiotic stress-tolerance plant. Gene 26:100286

Calvache C, Vazquez-Vilar M, Selma S, Uranga M, Fernández-del-Carmen A, Daròs J-A, Orzáez D (2022) Strong and tunable anti-CRISPR/Cas activities in plants. Plant Biotechnol J 20:399–408

Cannarozzi G et al (2014) Genome and transcriptome sequencing identifies breeding targets in the orphan crop tef (Eragrostis tef). BMC Genomics 15:1–21
Ceasar A (2021) Genome-editing in millets: current knowledge and future perspectives. Mol Biol Rep 49(1):773–781
Ceasar SA, Baker A, Ignacimuthu S (2017) Functional characterization of the PHT1 family transporters of foxtail millet with development of a novel Agrobacterium-mediated transformation procedure. Sci Rep 7:1–16
Cheng Z et al (2021) Establishing in planta haploid inducer line by edited SiMTL in foxtail millet (Setaria italica). Plant Biotechnol J 19:1089
De Oliveira A (2018) Abiotic and biotic stress in plants. IntechOpen. https://doi.org/10.5772/intechopen.77845
Eckerstorfer MF, Dolezel M, Heissenberger A, Miklau M, Reichenbecher W, Steinbrecher RA, Waßmann F (2019) An EU perspective on biosafety considerations for plants developed by genome editing and other new genetic modification techniques (nGMs). Front Bioeng Biotechnol 7:31
Ghosh S, Dey G (2022) Biotic and abiotic stress tolerance through CRISPR-Cas mediated genome editing. J Plant Biochem Biotechnol 31:227–238
Gupta S, Kumar A, Patel R, Kumar V (2021) Genetically modified crop regulations: scope and opportunity using the CRISPR-Cas9 genome editing approach. Mol Biol Rep 48:4851–4863
Hittalmani S et al (2017) Genome and transcriptome sequence of finger millet (Eleusine coracana (L.) Gaertn.) provides insights into drought tolerance and nutraceutical properties. BMC Genomics 18:1–16
Ishino Y, Krupovic M, Forterre P (2018) History of CRISPR-Cas from encounter with a mysterious repeated sequence to genome editing technology. J Bacteriol 200:e00580-00517
Jaganathan D, Ramasamy K, Sellamuthu G, Jayabalan S, Venkataraman G (2018) CRISPR for crop improvement: an update review. Front Plant Sci 9:985. https://doi.org/10.3389/fpls.2018.00985
Kheir AMS, Ding Z, Ali MGM, Feike T, Abdelaal AIN, Elnashar A (2020) Wheat crop modelling for higher production. Intechopen, London. https://doi.org/10.1007/978-981-15-4728-7_6
Lata C, Bhutty S, Bahadur RP, Majee M, Prasad M (2011) Association of an SNP in a novel DREB2-like gene SiDREB2 with stress tolerance in foxtail millet [Setaria italica (L.)]. J Exp Bot 62:3387–3401
Lawson T, Vialet-Chabrand S (2019) Speedy stomata, photosynthesis and plant water use efficiency. New Phytol 221:93–98
Li D-H, Hui L, Yang Y-L, Zhen P-P, Liang J-S (2009) Down-regulated expression of RACK1 gene by RNA interference enhances drought tolerance in rice. Rice Sci 16:14–20
Liang Z, Wu Y, Ma L, Guo Y, Ran Y (2021) Efficient genome editing in Setaria italica using CRISPR/Cas9 and base editors. Front Plant Sci 12:815946
Lin CS et al (2018) Application of protoplast technology to CRISPR/Cas9 mutagenesis: from single-cell mutation detection to mutant plant regeneration. Plant Biotechnol J 16:1295–1310
Mamidi S et al (2020) A genome resource for green millet Setaria viridis enables discovery of agronomically valuable loci. Nat Biotechnol 38:1203–1210
Manavalan LP, Chen X, Clarke J, Salmeron J, Nguyen HT (2012) RNAi-mediated disruption of squalene synthase improves drought tolerance and yield in rice. J Exp Bot 63:163–175
Matsuura A, An P, Murata K, Inanaga S (2016) Effect of pre-and post-heading waterlogging on growth and grain yield of four millets. Plant Prod Sci 19:348–359
Maximiano MR, Franco OL (2022) CRISPR/Cas: the new frontier in plant improvement. ACS Agricultural Science & Technology, Washington, DC
Mayes S, Massawe F, Alderson P, Roberts J, Azam-Ali S, Hermann M (2012) The potential for underutilized crops to improve security of food production. J Exp Bot 63:1075–1079
Mohanta TK, Bashir T, Hashem A, Abd Allah EF, Bae H (2017) Genome editing tools in plants. Genes 8:399
Muthamilarasan M, Prasad M (2021) Small millets for enduring food security amidst pandemics. Trends Plant Sci 26:33–40
Numan M, Serba DD, Ligaba-Osena A (2021) Alternative strategies for multi-stress tolerance and yield improvement in millets. Genes 12:739

Pan Y, Li J, Jiao L, Li C, Zhu D, Yu J (2016) A non-specific Setaria italica lipid transfer protein gene plays a critical role under abiotic stress. Front Plant Sci 7:1752

Rana S, Pramitha L, Muthamilarasan M (2021) Genomic designing for abiotic stress tolerance in foxtail millet (Setaria Italica L.). In: Genomic designing for abiotic stress resistant cereal crops. Springer, New York, pp 255–289

Rao MJ, Wang L (2021) CRISPR/Cas9 technology for improving agronomic traits and future prospective in agriculture. Planta 254:1–16

Sabir P, Ashraf M, Akram N (2011) Accession variation for salt tolerance in proso millet (Panicum miliaceum L.) using leaf proline content and activities of some key antioxidant enzymes. J Agron Crop Sci 197:340–347

Saleem S, Mushtaq NU, Shah WH, Rasool A, Hakeem KR, Rehman RU (2021) Morpho-physiological, biochemical and molecular adaptation of millets to abiotic stresses: a review. Phyton 90:1363

Saurabh S, Vidyarthi AS, Prasad D (2014) RNA interference: concept to reality in crop improvement. Planta 239:543–564

Schiemann J, Robienski J, Schleissing S, Spök A, Sprink T, Wilhelm RA (2020) Plant genome editing–policies and governance. Front Plant Sci 11:284

Schlader ZJ, Chapman CL, Benati JM, Gideon EA, Vargas NT, Lema PC, Johnson BD (2019) Renal hemodynamics during sympathetic activation following aerobic and anaerobic exercise. Front Physiol 9:1928

Schmid-Siegert E et al (2017) Low number of fixed somatic mutations in a long-lived oak tree. Nat Plants 3:926–929

Shi J et al (2017) ARGOS 8 variants generated by CRISPR-Cas9 improve maize grain yield under field drought stress conditions. Plant Biotechnol J 15:207–216

Smirnoff N, Colombe SV (1988) Drought influences the activity of enzymes of the chloroplast hydrogen peroxide scavenging system. J Exp Bot 39:1097–1108

Stebbins G, Shah S (1960) Developmental studies of cell differentiation in the epidermis of monocotyledons: II. Cytological features of stomatal development in the Gramineae. Dev Biol 2:477–500

Turnbull C, Lillemo M, Hvoslef-Eide TA (2021) Global regulation of genetically modified crops amid the gene edited crop boom—a review. Front Plant ence 12:258

VanBuren R et al (2020) Exceptional subgenome stability and functional divergence in the allotetraploid Ethiopian cereal teff. Nat Commun 11:1–11

Vlčko T, Ohnoutkova L (2020) Allelic variants of CRISPR/Cas9 induced mutation in an inositol trisphosphate 5/6 kinase gene manifest different phenotypes in barley. Plants 9:195

Wada N, Osakabe K, Osakabe Y (2022) Expanding the plant genome editing toolbox with recently developed CRISPR–Cas systems. Plant Physiol 188:1825–1837

Wang H et al (2022) The boron transporter SiBOR1 functions in cell wall integrity, cellular homeostasis, and panicle development in foxtail millet. Crop J 10(2):342–353

Waqas MA, Wang X, Zafar SA, Noor MA, Hussain HA, Azher Nawaz M, Farooq M (2021) Thermal stresses in maize: effects and management strategies. Plants 10:293

Weiss T et al (2020) Optimization of multiplexed CRISPR/Cas9 system for highly efficient genome editing in Setaria viridis. Plant J 104:828–838

Zafar SA et al (2020) Engineering abiotic stress tolerance via CRISPR/Cas-mediated genome editing. J Exp Bot 71:470–479

Zhang J, Klueva NY, Wang Z, Wu R, Ho T-HD, Nguyen HT (2000) Genetic engineering for abiotic stress resistance in crop plants. In Vitro Cell Dev Biol Plant 36:108–114

Zhang G et al (2012a) Genome sequence of foxtail millet (Setaria italica) provides insights into grass evolution and biofuel potential. Nat Biotechnol 30:549–554

Zhang J et al (2012b) Early mixed farming of millet and rice 7800 years ago in the middle Yellow River region, China. PLoS One 7:e52146

Zhang A et al (2019) Enhanced rice salinity tolerance via CRISPR/Cas9-targeted mutagenesis of the OsRR22 gene. Mol Breed 39:1–10

Zhang W et al (2021) Identification of no pollen 1 provides a candidate gene for heterosis utilization in foxtail millet (Setaria italica L.). Crop J 9:1309–1319

Integrating Genomics and Phenomics Tools to Dissect Climate Resilience Traits in Small Millets

14

Lydia Pramitha, Pooja Choudhary, Pronomita Das, Shriya Sharma, Vasundhara Karthi, Hindu Vemuri, and Mehanthar Muthamilarasan

Abstract

Deployment of advanced tools for trait improvement is imperative to sustain and improve the productivity of food crops. However, over-dependence on rice, wheat, and maize for catering to the nutritional requirements of more than 80% of the population turns out to be a bottleneck in addressing food and nutritional securities. Minor millets, otherwise called nutri-cereals, serve as better alternatives to major staples owing to their nutritional superiority to rice, wheat, and maize. Small millets possess better adaptability to the environmental conditions, higher market value due to their taste, texture, and nutritional properties, and require minimal agricultural inputs for producing these crops. Despite these merits, significantly less research has been pursued on these crops to study the beneficial traits. Dissecting the molecular determinants of climate-resilient traits will provide insights into the mechanisms underlying these traits. This would further expedite basic research on understanding the precise roles of genetic determinants and applied research on improving the traits using breeding, transgenic, and/or genome-editing approaches. In this context, the present chapter enumerates the different approaches used to dissect the climate-resilient traits in small millets, emphasizing the role of genomics and phenomics tools. Also, the chapter describes how to integrate genomics and phenomics for improving the climate resilience of these crops.

Lydia Pramitha and Pooja Choudhary contributed equally with all other contributors.

L. Pramitha (✉)
School of Agriculture and Biosciences, Karunya Institute of Technology and Sciences, Coimbatore, Tamil Nadu, India

P. Choudhary · P. Das · S. Sharma · V. Karthi · H. Vemuri · M. Muthamilarasan
Department of Plant Sciences, School of Life Sciences, University of Hyderabad, Hyderabad, Telangana, India

R. N. Pudake et al. (eds.), *Omics of Climate Resilient Small Millets*,
https://doi.org/10.1007/978-981-19-3907-5_14

Keywords

Small millets · Nutri-cereals · Stress biology · Climate resilience · Omics technologies · Crop improvement

14.1 Introduction

Small millets are known as wonder-cereals among the food crops as they possess the inherent potential to survive harsh conditions. Small millets being hardy crops are additionally rich in nutrients and serve as functional foods (Muthamilarasan and Prasad 2021). There are 11 species of small millets, and they have been considered the "lost crops" of the world. This includes finger millet (*Eleusine coracana*), foxtail millet (*Setaria italica*), proso millet (*Panicum miliaceum*), barnyard millet (*Echinochloa crus-galli*), kodo millet (*Paspalum scrobiculatum*), little millet (*Panicum sumatrense*), teff (*Eragrostis tef*), fonio (*Digitaria exilis*), job's tears (*Coix lacryma-jobi*), guinea millet (*Brachiaria deflexa*), and browntop millet (*Urochloa ramosa*) (Table 14.1; Muthamilarasan and Prasad 2021). Small millets have been utilized as one of the heritage crops before the green revolution and are free of gluten with higher crude protein. In addition to these factors, they are also known as less demanding crops and survive under poor fertile soils with low inputs providing higher income (Bhat et al. 2011). The current increase in hidden hunger on consuming the major staples realized the importance of these small millets, which are nutritionally superior to other cereals. Small millets are preferred as a major crop in rainfed zones by the marginal farmers, and they have multi-purpose usage. They are used for grain and forage purposes, enhancing their overall value in cultivation. These grains have also been demanded in markets as an input in the biscuit and confectionaries for value addition and processing (Gupta et al. 2012). Flaked, baked, popped, and instant food mixes have raised consumers' demand at a higher value. Recent ventures of a sustainable cropping system involving their intercropping with legumes are also a growing feature in sustainable agriculture to double the farmers' income. Altogether this sets a new boundary in the breeding and genomics of small millets to attain food and nutritional security (Goron and Raizada 2015).

Small millets are named because of their miniature grain size, and they are shorter in duration owing to the preference for cultivation. They have a higher storage value, and millets like kodo millet could be stored over the years without any loss of viability. Several new technologies are on board to minimize its farming difficulties and mechanization by using paddy harvesters in small millet fields to support the livelihood of farmers cultivating them. Being climate-resilient crops, they have been explored to understand the genetic resources that could be utilized in frontline crops (Gowda et al. 2008). Three of the 13 species have been sequenced, and comparative studies of these crops with major cereals have also retrieved several domestication events across species (Zhang et al. 2012; Hunt et al. 2014; Wallace et al. 2015).

Table 14.1 Characteristic features and importance of small millets

Common name	Botanical name	Progenitor	Races	Chromosome number	Application in food and feed	Reference
Finger millet	*Eleusine coracana*	*Eleusine indica × Eleusine floccifolia/E. tristachya*	1. Elongata	$2n = 4x = 36$	Grains in diets as well as in malting, bread biscuits, etc.	Gupta et al. (2017)
			(a) Laxa			
			(b) Reclusa			
			(c) Sparsa			
			2. Plana			
			(a) Seriata			
			(b) Confundere			
			(c) Grandigluma			
			3. Compacta			
Barnyard millet	*Echinochloa esculenta/E. frumentacea*	*Echinochloa crus-galli*	1. Japanese	$2n = 6x = 54$/ $2n = 4x = 36$	Grains, porridge to weight loss, fodder for cattle and instant mixes for babies	Sood et al. (2015)
		Echinochloa colona	(a) Utilis			
			(b) Intermedia			
			2. Indian			
			(a) Laxa			
			(b) Robusta			
			(c) Intermedia			
			(d) Stolonifera			
Foxtail millet	*Setaria italica*	*Setaria viridis*	1. Indica	$2n = 2x = 18$	Grain and fodder value also in fabric values and shoe manufacturing	Sharma et al. (2018)
			2. Maxima			
			3. Moharia			

(continued)

Table 14.1 (continued)

Common name	Botanical name	Progenitor	Races	Chromosome number	Application in food and feed	Reference
Little millet	*Panicum sumatrense*	*Panicum psilopodium*	1. Nana	$2n = 4x = 36$	As a whole grain and dry Stover as a cattle feed	Ganapathy et al. (2021)
			2. Robusta			
Proso millet	*Panicum miliaceaum*	*Panicum capillare*	1. Miliaceaum	$2n = 4x = 36$	As a whole grain and cattle fodder	Habiyaremye et al. (2017)
		Panicum repens	2. Patentissimum			
			3. Contactum			
			4. Compactum			
			5. Ovatum			
Kodo millet	*Paspalum scrobiculatum*	*Paspalum sanguinale*	1. Regularis	$2n = 4x = 40$	Whole grain and leaves as goat feed. Whole grain flour, pudding. Best as fodder for cattle, pigs, and sheep. And some species in lawns	Deshpande et al. (2015)
			2. Irregularis			
			3. Variabilis			
Fonio	*Digitaria exilis/iburua*	–	–	$2n = 4x = 36$	Whole grain and as cattle feed	Abrouk et al. (2020)
Teff	*Eragrostis tef*	–	–	$2n = 4x = 40$	As whole grain, bread, beer, porridge, cattle feed, fodder and as building material and in lawns	Burén et al. (2020)
Browntop millet	*Brachiaria ramosa/ Panicum ramosum*	–	–	$2n = 4x = 28$	As a palatable fodder, whole grain, straw production as grass cover	Maitra (2020)
Job's tears	*Coix lacryma-*jobi	–	–	$2n = 4x = 20$	Fodder and grain purpose	Shi et al. (2004)
Guinea millet	*Brachiaria deflexa*	–	–	$2n = 18, 36$	Fodder and milk extracted from it is used in milk and beef industries	Nithiyanantham et al. (2019)

– data not available

Hence, they could act as a major source for ideotype breeding and minimize the nutritional barriers among the growing population.

Small millets are typically annual grasses cultivated for their grain and fodder usage. Except for foxtail millet, the others are allotetraploid and possess a profuse tillering habit with jointed internodes (De Wet and Harlan 1975). The leaf sheaths are longer, broader than internodes, and the inflorescence is a panicle comprising spikelets. Belonging to the cereal family, the inflorescence consists of spikelets with two upper and lower glumes. The upper glumes are fertile, while the lower is sterile. The florets comprise three stamens and a bifid feathery stigma. Among the millets, foxtail millet contains a cylindrical drooping panicle with bristles, while finger millet has a digitate inflorescence with five fingers or spikes. Barnyard millet comprises a drooping elliptical dense raceme with no ligules in leaves. Little millet has the highest biomass among the small millets, and kodo millet comprises a higher fodder value due to its lecithin content. The leaves of kodo millet are narrower than the rest and resemble rice leaves (Vidya and Raundal 2018). All the small millets are highly self-pollinated with cleistogamous flowers and crossing them faces massive difficulty due to their smaller size. The fruit is a caryopsis, and it possesses around seven to eight layers of husk which establishes a difficulty in processing and threshing (Gupta et al. 2011; Sarita and Singh 2016).

Small millets have profound ethnobotanical importance due to their varied practice and predominant storage ability. These grains could be stored for a more extended period without losing any viability. Small millets consist of higher photosynthetic efficiency with a potential to ameliorate the CO_2 emissions owing to their C_4 mechanism. The seeds in small millets mature from top to bottom and can shatter at the time of harvest (Giussani et al. 2001). The origin of these crops comes from the ancient areas of civilization. In this aspect, foxtail millet, job's tears, and proso millet are from China, and their cultivation dates back to around 8000 years ago. Barnyard millet is found in Japan, while kodo millet, fonio millet, guinea millet, teff, and finger millet are from Africa. Their domestication and preference for cultivation in hilly tribes have led to its distribution, and its current presence is throughout the world with varying centers of diversity (Vavilov 1926).

Advancement in crop improvement approaches is essential to achieve the goal of sustainable food production for the continuously increasing human population. Currently, the changing food preferences and nutritional qualities of millets are making them a better replacement to the staple crops. Small millets are considered nutri-cereals due to their richness in minerals, vitamins, antioxidants, and climate-resilient features. However, the progress in breeding programs of these crops is lacking. Novel genomic tools have improved the new genomic-assisted breeding programs by detecting essential agronomic traits. Recently, with the development of high-throughput phenotyping, crop improvement programs are now entering the new era of green revolution. Integration of genomics and phenomics can enhance the efficiency of breeding programs targeting climate resilience. Phenomics has upgraded the development of climate-resilient crops by ensuring the selection of improved germplasms based on field performance, growth, and composition. Several advanced tools for phenotyping provide multidimensional data on phenotypes.

In this context, the present chapter enumerates the different approaches used to dissect the climate-resilient traits in small millets, emphasizing the role of genomics and phenomics tools. Also, the chapter describes how to integrate genomics and phenomics for improving the climate resilience of these crops.

14.2 Productivity and Major Producers of Small Millets

Considering the production of small millets globally, India serves as the world's largest producer of small millets, and Africa remains a significant consumer. Following India, Niger and China are following in small millets production. India, China, and Niger account for about 55% of the global production of small millets. Over the years, millet production has increased in India, but the per capita consumption has steadily decreased from 50 to 75%. In India, the use of millets for food dropped significantly during 2005; however, millets are being cultivated for feed and preparation of alcoholic beverages. Hence, small millets are a forgotten crop and are referred to as the orphan species. However, they have recently been rediscovered due to their varied nutritional attributes, certificated health advantages, adaptation to diverse environmental conditions, sustainability in low-input agriculture, and organic cultivation amenability. They are being cited as the primary source of energy and immunity in Asia and Africa's semi-arid tropics and drought-prone regions. Together, these continents account for 97% of the global millet production of about 29–30 million tonnes. After COVID-19, people have shifted their food habits towards boosting their immunity instead of junk foods. Hence, the forecast shows that the demand for small millets will increase in the coming years.

Millets, particularly small millets, are in a position of hardship in India. India has the third largest area under cultivation of small millets in the world. But for small millets the area under cultivation has, unfortunately, come down from 7.56 to 1.86 mha (Anbukkani et al. 2017; AICRPSM 2020). There are various factors and parameters that affect small millet productivity. As most of these minor millets are being cultivated in arid and semi-arid ecologies, the environmental and weather-related factors become the biggest limitation without any assured supply of irrigation water. India, a tropical country, suffers from unpredictable climate change, leading to inadequate and uneven distribution of rainfall, late-onset, and early cessation of monsoon rains. Also, prolonged dry spells during the crop growing season and excessive evaporation adversely limit millet productivity. Hence, overcoming these constraints by improving these crops could be a major solution for nutritional security in the future.

14.3 Nutritional Significance of Small Millets

Small millets are gaining the spotlight as the scientific fraternity is revealing their tremendous nutritional attributes to the world through recent research advances. Consumption of small millets has the potential to target global problems of food as well as that of nutritional insecurity (Muthamilarasan and Prasad 2021). Along with climate-resilient traits, small millets hold the crown for being nutritionally superior to the commonly consumed cereals like rice and wheat. Among the small millets, foxtail millet is nutritionally profound than others and holds double the amount of mineral content than wheat and rice combined. This is also rich in protein and phosphorous. Foxtail millet is also recognized for containing a higher resistant starch ranging from 13 to 15%, which is higher than all major cereals, and it renders a low glycemic index (Gopalan et al. 2007). This can be attributed to the fact that it has the lowest amount of RDS (rapidly digestible starch) among all millets, and it is extremely beneficial to type-II diabetes patients. Successively, barnyard millet also has a lower glycemic index and is well suited for diabetic patients. Its carbohydrates have a higher amylose retrogradation, leading to resistant starch formation. Hence, these could be an alternative dietary source for diabetic patients. All the minor millets possess a lower glycemic index ranging from 54 to 68 (Patil et al. 2015), and they can offer a possible dietary solution for the diabetic community.

Small millets are comparable to cereals with no significant difference in carbohydrate content. These, on the other side, gain the upper hand when other macro and micronutrients are considered. In terms of the major nutrients required by humans, small millets contain a higher amount of proteins, minerals, and fiber. Both foxtail and proso millet have double the content of proteins than rice. Foxtail millet has 12 mg/100 g compared to rice which has 6 mg/100 g (Kumari and Thayumanavan 1997). Among all, barnyard millet has tenfold the content of fiber when compared to rice or wheat. There is 13.60 mg of fiber in barnyard millet vs. 2 mg in wheat (Goron and Raizada 2015). Consuming a higher fiber is often associated with a lower risk of heart disease, stroke, type 2 diabetes, and bowel cancer. Thus, it is advised to include barnyard millet in regular diets.

Small millets are a reservoir of minerals, and among them, fonio millet has the highest iron content. This serves 8.5 mg of iron per serving and is also rich in calcium, magnesium, and zinc (Ballogou et al. 2013). Barnyard millet and finger millet also have the highest content of iron, which is tenfold greater than rice (Saleh et al. 2013). Finger millet has the highest amount of calcium (350 mg/100 g; Kumar et al. 2016a, b), and foxtail millet has the most content of thiamin (6 mg/100 g) among all the millets and staple cereals (Lata et al. 2013). In addition, foxtail millet and finger millet have twofold the content of phosphorous compared to rice. Little millet has nearly three times the range of zinc than rice (Patil 2018). Kodo millet and proso millet are the richest source of magnesium, a cofactor for a number of enzymatic reactions that regulate the secretion of glucose and insulin (Yadav et al. 2013; Kumar et al. 2016a, b).

Kodo millet is exceptionally high in Vitamin B complex, especially niacin, pyridoxine, and folic acid. Subsequently, it is also rich in minerals such as calcium, iron, potassium, magnesium, and zinc. Lysine is found higher in proso millet when compared with other cereals as it is the limiting amino acid (Upadhyaya et al. 2016). Successively, fonio millet also has an excellent source of essential amino acids like methionine, cysteine, and leucine. Fonio is regarded as the world's best-tasting cereal and is one of the highest nutritious grains among all the grains. This is considered as the grain of life in African regions. Other essential amino acids such as threonine are present in finger millet and foxtail millet. This is essential to create glycine and serine to produce collagen, elastin, and muscle tissue. Therefore, ethnobotanical records describe their role in treating various nervous system disorders, including spinal spasticity, multiple sclerosis, and familial spastic paraparesis (Gupta et al. 2012; Ugare et al. 2014).

Consuming small millets also requires some extent of processing due to the presence of anti-nutritional compounds such as phytates, tannins, and polyphenols. They inhibit the bioavailability of micronutrients; hence they must be eliminated by cooking and parboiling techniques. These anti-nutritional compounds are also termed nutraceuticals as they have a profound antioxidant activity rendering benefits in health, aging, and metabolic disorders (Hedge and Chandra 2005).

14.4 Abiotic Stress Tolerance in Small Millets

Environmental stresses are major challenges for plant growth and development. Small millets are nutrient-rich food sources traditionally grown and consumed by subsistence farmers in Asian and African zones. Since they are predominantly grown in rainfed and arid zones, environmental conditions have a crucial role in their growth and development. Understanding this regulatory mechanism would help in climate-resilient crop improvement programs.

14.4.1 Drought Stress

By nature, millets are drought-tolerant and capable of maintaining good yield under water-limited conditions. However, drought stress is a major limitation to their growth potential at their crucial developmental phase. Due to the short duration nature, these crops are best suited to withstand the prevailing rainfed situations. In addition to these, the core collections in foxtail millet were subjected to screening for drought tolerance by Krishnamurthy et al. (2016). The traits observed were total water use efficiency, transpiration efficiency, and harvest index. Among these traits, transpiration efficiency was assessed to contribute to yield under drought in foxtail millet. Surveying the races for conferring yield under water stress exhibited the race indica to have more water use efficiency than moharia and maxima. Further studies on enduring drought and yield also focus on minimizing small millets' lodging issues. The application of Paclobutrazol, a gibberellin (GA) inhibitor in teff and finger millet, developed non-lodging semi dwarf types with higher drought tolerance than GA mutants in rice (Plaza-Wüthrich et al. 2016).

Analyzing the genomic sequences of foxtail millet seedlings under drought revealed the key role of 252 upregulated proteins and 69 downregulated proteins in expression. These proteins had differential roles for drought in foxtail millet and were categorized into ROS scavenging enzymes, carbon metabolism, and photosynthesis. The key role of LEA proteins, aquaporins, osmoregulants like glycine betaine in foxtail millet also rendered more intensive research in this crop beneath its drought tolerance (Pan et al. 2018). Drought-stress responses also involve transcription factors of basic helix–loop–helix (bHLH). bHLH genes, such as AREB, MYB, or bHLH transcript factors, regulate ABA-dependent signalling pathways when the plant is under drought stress (Wang et al. 2018). Further genome-wide studies in foxtail millet for LIM genes dissected ten major LIM genes. These genes were observed to have MYB and MYC cis-acting elements linked to drought stress tolerance promoters. Among them, the gene SiWLIM2b was chosen for advanced study in rice. The overexpression of SiWLIM2b gene demonstrated a greater survival rate, higher relative water content, and less cell damage under drought stress than wild-type rice (Yang et al. 2019).

Drought in finger millet affects germination and early seedling growth. Screening of varieties for initial vigor tolerance rendered the varieties GBK042094 and GBK043137 to be tolerant. These varieties were found to have a higher relative water content, proline, chlorophyll, and lipid peroxidation than the rest (Ganapathy and Patil 2017). The genome sequence of finger millet and mRNA sequencing data is also subject to the presence of several drought-induced candidate genes. Thus, targeting the proteins and enzymes involved in these activities would identify key genes for coping water stress in other major cereals (Wambi et al. 2021). Barnyard millet has also been studied to possess a higher root length, relative water content, and lower leaf area index as a major concern for water stress tolerance among the minor millets. Targeting the key genes beneath these traits would benefit breeding programs (Senthil et al. 2018).

Higher antioxidants and antioxidant enzymes in kodo millet make it a prominent drought and salt-tolerant crop (Siddappa et al. 2016). Also, proso millet can adjust to drought stress quickly and give valuable knowledge and gene resources to improve drought tolerance. Differential regulation hormone signalling pathway in drought-tolerant cultivars compared to drought-sensitive cultivars may be one of the primary mechanisms underlying drought stress tolerance (Zhang et al. 2019). However, the detailed knowledge of these underlying functional mechanisms under drought stress is still in infancy. Therefore, more research is needed to explore them in small millets.

14.4.2 Heat Stress

Heat stress affects plants in various ways, including their biochemistry, physiology, and gene regulation mechanisms. High-temperature stress can disrupt the intermolecular connections essential for proper plant growth, resulting in improper plant development, leaf senescence, early or late fruit set, and inhibition of shoot and root growth (Janni et al. 2020). Three foxtail millet lines (448-Ames 21521, 463-P1391643, and 523-P1219619) were investigated under greenhouse circumstances at a high soil temperature, and their root zone physiology, metabolism, and root dynamics were evaluated. The resilience of millet roots to high soil temperatures was linked to specific changes in metabolites and physiological behavior. Compared to plants cultivated at the control soil temperature, all plant lines acclimatized to the high soil temperature, demonstrating significantly lower photosynthetic and stomatal conductance rates. Whereas in finger millet, heat-responsive gene, EcDREB2A has been investigated widely. This has a role in scavenging excess ROS in response to heat stress by regulating and modulating the enzyme activity of the CAT (catalase), APX (ascorbate peroxidase), SOD (superoxide dismutase), and POD (Class III peroxidase) proteins, resulting in improved heat stress tolerance. Overexpression of EcDREB2A enhanced the expression of several abiotic stress-responsive downstream genes for coping the heat stress. These findings showed that using a naturally truncated DREB2A gene from finger millet could improve plant resilience to heat stress, which could aid in the development of climate-smart crops. These two millets are the most investigated under heat stress, and the results of these study suggest that small millets have heat tolerance properties (Singh et al. 2021). Under heat stress conditions, other small millets such as proso millet, kodo millet, little millet, and barnyard millet need to be studied in more detail (Singh et al. 2016).

14.4.3 Salinity Stress

Soil salinization and alkalinization are widespread, and small millets bestow the ability to withstand these harsh conditions than major food crops. A comparative study was performed between foxtail millet and proso millet to assess the growth and physiological responses among foxtail millet and proso millet under saline and alkaline stress. The two neutral salts (NaCl and Na_2SO_4) and the two alkaline salts (NaCl and Na_2SO_4) were mixed ($NaHCO_3$ and Na_2CO_3). Here, the alkaline stress inhibited plant dry weight, relative growth rate (RGR), net assimilation rate (NAR), leaf area ratio (LAR), and relative water content (RWC) to a greater extent than salt stress, and the rates of reduction were more pronounced in foxtail millet, indicating that proso millet tolerance to alkaline stress was higher than foxtail millet (Islam et al. 2011). Therefore, it is necessary to look into the mechanical basis of salt stress tolerance in other small millets. Expression studies in foxtail millet for salinity indicated five- to ninefold expression of GT228170 in cultivars. After 24 h of stress, the abundance of GT228139 gene was 2.5-fold higher in cv. Prasad, while GT228255 gene expression was approximately 13-fold higher in cv. Lepakshi. Hence this differentially upregulating gene could improve saline tolerance in crops (Puranik et al. 2011).

The tolerance of finger millet to NaCl is moderate with a 50% reduction in shoot dry weight (DW), leaf surface area, and leaf number at 200 mM NaCl (Swati et al. 2011). Finger millet is sensitive to salinity at germination and early growth phases. And this phase under salinity is seen to have higher H_2O_2 concentrations and cell death. Thus, its sensitivity is suggested to be studied to reveal molecular processes that modulate salt stress tolerance (Satish et al. 2016). Another approach studied the effect of salinity on growth, protein content, and antioxidant enzymes in three kodo millet accessions viz., IC 426676, IC 382888, and IPS 145. The relative germination rate and relative water content of IC 382888 and IC 426676 were only slightly reduced by salinity. The overall protein content of all the plants under NaCl stress was reduced and the activity of reactive oxygen species (ROS) was elevated in all the accessions. IC 426676 has the highest enzyme activity of the three, followed by IC 382888 and IPS 145. The results showed that kodo millet seedlings responded by lowering their protein content when exposed to salinity. IC 426676 appeared to be the most stress-tolerant of the three accessions. Therefore, thorough screening in germplasm should be employed as a prospective to develop saline-tolerant varieties in small millets (Prasanthi Kumari and Vishnuvardhan 2015).

14.4.4 Flooding Stress

A comparative study with four small millets such as little millet, foxtail millet, pearl millet, and proso millet under waterlogging conditions suggests that little millet had a high tolerance to waterlogging. In contrast, pearl millet, proso millet, and foxtail millet were sensitive. The impact of pre- and post-heading waterlogging on grain yield and growth in four millets suggested that waterlogging affected the grain production of all millets at all stages of growth. Little millet displayed strong tolerance to waterlogging because of increased root growth and a higher proportion of lysigenous aerenchyma throughout the growing period. This root characteristic is thought to aid water and nutrient intake while sustaining oxygen delivery during waterlogging (Matsuura et al. 2016).

14.4.5 Chilling/Freezing Stress

The panicoid groups have a varying range of adaptability and have evolved in different conditions. The cold-tolerant species include foxtail millet, and cold-sensitive species comprise proso millet. Proso millet is usually considered a late seeded summer crop after wheat and barley in the US regions (Habiyaremye et al. 2017). These plants have evolved in various ways involving biochemical, physiological, genetic regulatory mechanisms to adapt to freezing conditions. The cold-responsive gene expressions have also been manifested by transposons, DNA transcriptional regulation, and epigenetic mechanisms. When foxtail millet and proso millet were grown under controlled freezing conditions, there was a comparatively slower electrolytic leakage than other crops. This study also confronted that the response to cold stress is not conserved in all species, and it completely varies from one accession to the other (Meng et al. 2021). Studies of cold stress in finger millet and foxtail millet have put forth the role of 1 aminocyclopropane-1-carboxylate (ACC) deaminase in alleviating the cold stress in small millets (Srinivasan et al. 2017). Despite this, the barnyard millet has its renowned center of diversity in the central Himalayan region. This diverse rich source revealed new varieties for different agroclimatic conditions. Superior lines in barnyard millet from these accessions with lipid peroxidation, malondialdehyde, and total ascorbate possessing higher yield and superoxide dismutase activity could further be empowered in incorporating cold tolerance in small millets (Trivedi et al. 2017).

14.4.6 Mineral-Deficient Soils

Small millets comprise all the weedy features, and they were predominantly considered major weeds among the mainstream crops. These are grown predominantly in poor soils that lack marginal farmers' basic nutrient requirements. Little millet is propounded to yield higher both in terms of forage and seed yield. This has the inherent ability to grow in soils with poor potassium, and reports on analyzing the fertilizer requirements of little millet state that under low fertilizer application, this crop sustains a higher yield compared to other food crops (Patil and Raundal 2018). Foxtail millet also has the inherent potential to withstand the poor nutritional status of the soil due to its weedy features. SiMYB3 is found to be specifically expressed in low K^+ conditions in foxtail. The overexpression of this gene rendered main root growth elongation with improved K^+ tolerance in Arabidopsis (Cao et al. 2019). In addition to these, finger millet is known to be grown in eastern Uganda's poor soils, and its genes are yet to be resolved (Ebanyat et al. 2021). Similar poor nutrient tolerance has been observed in barnyard millet and unraveling the genes beneath these traits could be used in improvising the adaptability of main food crops in un-arable lands (Maitra et al. 2020).

14.4.7 Metal Toxicity

Studies on the growth and tolerance of small millets in toxic soils have not yet been undertaken so far in all the crops. A study in cadmium-affected soils with proso millet exhibited the tolerance to a higher concentration of cadmium in soils. The core collections under hydroponic conditions established stress of cadmium ions in shoots, and it showed a positive association with zinc and magnesium. Thus, proso millet is a promising crop in cadmium bioremediation (Liu et al. 2021). The increase in nickel is an effect of industrialization, and finger millet was found to be adaptable in nickel-contaminated soils. Finger millet was comparably accounted to be tolerant towards nickel with elevated guaiacol peroxidase and superoxide dismutase levels. Among the major crops, finger millet had a higher capability to maintain homeostasis (Gupta et al. 2017).

14.5 Conventional and Molecular Breeding Approaches for Enhancing Stress Tolerance

Small millets are highly cleistogamous, and breeding approaches involve conventional techniques like pure line selection, introduction, and pedigree breeding. The genetic resources among the gene pool were analyzed by genetic diversity analysis, which also incorporates the utilization of molecular markers starting from Restriction Fragment Length Polymorphism (RFLP) to Single nucleotide polymorphisms (SNPs) (Vetriventhan et al. 2016). New techniques of crossing to favor the gene flow among the species have also been taking place, and methods such as approach method, contact method, and hot water method are implemented to bring about the desirable traits. Among all the species in small millets, foxtail millet is the most surveyed crop for both molecular and conventional techniques. Successively after foxtail millet, barnyard millet, finger millet, teff, fonio millet, and proso millet have been sequenced. Being a diploid crop, foxtail millet is considered a model crop in the panicoid group to dissect the genomic resources in the non-sequenced millets.

Molecular markers have been utilized in mapping populations to detect quantitative trait locus (QTLs) for important traits (Dida et al. 2007). QTLs for lysine, tryptophan, and protein content have been identified in finger millet (Babu et al. 2014). Their localization also exhibited the presence of RISBZ1 gene, which is from rice. Heterologous expression of genes such as NAC1, PIN1a, and PHT 1 confers drought and salinity tolerance in finger millet. The gene EcNAC67 from finger millet has been used in transgenic rice variety ASD 16 to confer drought and saline tolerance in rice (Rahman et al. 2016). Further, QTLs for disease resistance and plant defense mechanism like EcSAM, EcLOX, and EcPAL were also detected against pink stem borer in finger millet (Jadhao et al. 2020).

The utilization of markers such as Amplified Fragment Length Polymorphism (AFLP), Random Amplified Polymorphic DNA (RAPD), RFLP, SNPs, Insertion–deletion mutations (InDels), Simple Sequence Repeats (SSR), and expressed sequence tag (EST) is also renowned in barnyard millet. Construction of linkage maps for waxy traits with SNP markers identified three loci in bulk segregants (Tiwari et al. 2018). Little millet is yet to be explored for its key traits compared to other small millets. RAPD and SSR markers were implied in this crop for DNA fingerprinting (Wallace et al. 2015). Conversely, proso millet is well explored for its genetic and genomic resources. The gene GBSS I has shown a higher variation among this taxon. Fourteen linkage groups have been identified, and around 18 QTLs for major agronomic traits have been dissected in proso millet (Yang et al. 2019; Zou et al. 2019). Fonio millet and teff are also equally studied for their diversity through AFLP, SSR, and ISSR markers (Dansi et al. 2010; Barnaud et al. 2012).

The availability of complete genome sequence and a strong syntenic relationship with rice suggested the utilization of genetic evidences related to disease resistance in other grass species with sequenced genomes (Andersen and Nepal 2017). Almost all aspects of identifying syntenic regions for disease resistance genes, saline tolerance, drought tolerance, and quality aspects with key traits have been studied in foxtail millet. Comparative genomics and QTLs for branching have also identified the predominant domestication events in foxtail millet genome (Doust et al. 2005). The sequence of foxtail millet has been used to develop markers that are used in crops like kodo millet (Hariprasanna 2017). The interspecific crossability has been well utilized to tap the genetic potential of the non-sequenced crops. In the case of association studies, the mapping populations used in small millets, including F2 and

RILs, are the most utilized for identifying QTLs (Jia et al. 2013, Wang et al. 2010; Vetriventhan et al. 2016). Techniques for backcross and double haploid mapping populations are still yet to be implied in small millets owing to the constraints in standardizing hybridization and tissue culture techniques (Goron and Raizada 2015). Recent trends in genomic studies have involved mutation breeding in these crops to induce variability as they are highly self-pollinated. Several mutants from ethyl methanesulfonate (EMS) and gamma rays have been developed, and the mutant varieties have been released for cultivation among farmers. These mutants have also been sequenced, and targeted mutagenesis as a part of reverse genetics is also in progress to identify the key genes (Bhave et al. 2016; Mamidi et al. 2020).

14.6 Genomics-Assisted Breeding in Small Millets

Genomics-assisted breeding is an emerging technology that plays a crucial role in improving crop dynamics in unpredictable climatic scenarios. This comprises both phenotypic selection and genomic selection. This is also called third-generation plant breeding. Data mining and allele identification accompanied by advanced phenotyping select superior genotypes that are far reliable. As the genomic sequence of frontline crops are available and the third-generation sequencing technologies give numerous nucleotide information, these techniques could be well embodied in identifying desirable varietal lines (Yabe and Iwata 2020). Genomics-assisted breeding involves QTL identification, GWAS, and genomic selection. These techniques are far employed in staple crops to designing a model ideotype for crop improvement programs. Genomics-assisted breeding has been well empowered in rice and wheat (Varshney et al. 2021). Several superior lines like swarna-sub-1, Improved Sambha Masuri, and Improved Tapaswini resulted from GAB. Bi-parental mapping populations and QTL identifications are extensively studied in foxtail millet and high-density linkage maps for SSR and SNP markers are available (Jia et al. 2013; Wang et al. 2012). GWAS is also predominantly utilized in tapping the hidden potential of small millet diversity. Association mapping in foxtail millet core collections landraces with SSR and SNP markers have detected higher LDs and marker-trait associations (Goron and Raizada 2015). GWAS in proso millet identified ten marker-trait associations for seed morphology among the landraces (Boukail et al. 2021). Although the precise genomic selection techniques have not been implemented in small millets, a study on elucidating genomics-assisted breeding for fonio millet in West Africa was conducted by Ibrahim Bio Yerima and Achigan-Dako (2021). Here they analyzed the importance of genotyping by sequencing techniques to detect SNP and DArt markers in the diverse collections of fonio millet. Also, the study suggested the importance of GWAS to identify QTLs rather than bi-parental populations, which has to be framed from scratch. By analyzing these natural diversity sources, techniques to develop a training population to enumerate a design for genomic selection involving farmers in identifying superior varieties pave a varied dimension in crop improvement programs. This

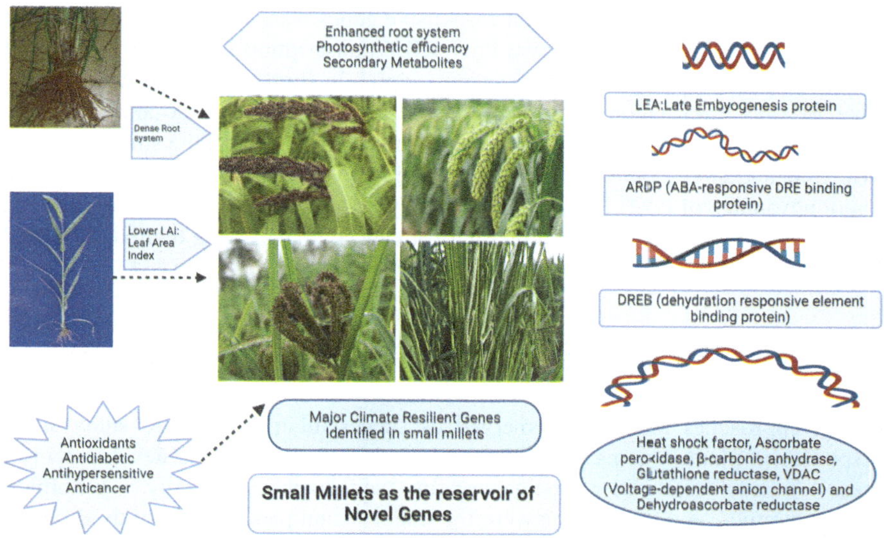

Fig. 14.1 Desirable features of small millets. A few known molecular determinants of climate-resilient traits in millets are shown. The data illustrates that millets could be a source of novel genes, alleles, and QTLs for climate resilience

action plan was framed to overrule seed shattering, seed size, and lodging in fonio millet.

The most crucial traits in small millets that have to be explored include earliness, plasticity, small leaf area, thickened cell wall, long dense root system, enhanced photosynthetic efficiency and secondary metabolites, and overexpressed genes in abiotic stresses. Thus, the identified genes beneath these traits could be utilized in other related cereals to reinstate the climate resilience in major staple crops. The major genes identified in small millets include ASR (abscisic acid ripening), AGO (argonaute protein encoding), ATG (autophagy), LEA (late embryogenesis abundant protein), ARDP (ABA-responsive DRE binding protein), DREB (dehydration responsive element binding protein), NAC transcription factor, aldose reductase, glutamine synthetase, pyyrroline-5-carboxylate reductase, OPR (12-oxophytodienoic acid reductase), WD-40, PHGPX (phospholipid hydroperoxide glutathione peroxidase), NAC transcription factors, bHLH transcription factor, dehydrin 7, heat shock factor, ascorbate peroxidase, β-carbonic anhydrase, glutathione reductase, VDAC (voltage-dependent anion channel), and dehydroascorbate reductase (Bandyopadhyay et al. 2017). In-depth studies involving these beneficial traits will develop a model population with all these novel genes in small millets. Further, this will be a source for incorporating favorable alleles in minor cereals and major staples (Fig. 14.1). Several bioinformatics tools and databases for these small millets are also available to identify key genes in them (Satyavathi et al. 2019). Yadav et al. (2016) and Muthamilarasan et al. (2014) also elaborated the role of micro RNAs and C2H2 zinc finger transcription factors in drought tolerance of foxtail millet. Around 2886 drought-responsive genes were detected in finger millet (Hittalmani et al. 2017). Thirty-two PmWRKY transcription factors in proso millet for abiotic stresses were identified in transcriptomic studies (Boukail et al. 2021). Thereby, this information could be endowed in initiating genomic and phenomic selection in small millets. Nowadays, the sequence of almost all the small millets is available, and GAB could be more amenable in exploring their essential genes. The major advantage of GAB is that it does not rely on a reference genome entirely. The genotyping by sequencing techniques with advanced phenotyping by imaging could be used to design climate-smart crops in the near future.

14.7 Phenomics and Its Implication in Crop Improvement

Omics approaches have been emerged as successful techniques to study stress responses in plant systems. The advancement has accelerated the development of these technologies in next-generation sequencing. Initially, genomics, transcriptomics, and proteomics were the major techniques used to study different aspects of crops. However, novel omics approaches such as metabolomics, ionomics, effectoromics, and phenomics have also been widely used in crop science. Phenomics has become an important genome selection tool, which is an important method of plant breeding. The high-throughput phenomics approach can highlight important phenotypic traits, which accelerates the process of pedigree selection and, therefore, improves and accelerates the breeding programs to develop climate-resilient and economically sustainable crops (Van Tassel et al. 2021).

The genome sequence of newly popularized crop candidates, such as millets, is unavailable, which impedes the rapid sequencing of their reference genome. However, the unavailability of funds and affordable genotyping pipelines for these crops restricts the assembly of their genomes. This makes millets orphans of any genotyping tool (Rincent et al. 2018). Interestingly, phenomic selection (PS) by using various high-throughput methods, such as near-infrared spectroscopy (NIRS), is a widely used method for phenotyping of millet crops to obtain variables. These variables can be utilized to evaluate the kinship matrix in the statistical models of genomic selection (GS). These statistical algorithms and workflows designed for GS and PS could improve the breeding programs of millets.

Recent development in the acquisition methods of important phenotypic traits further provides essential tools to unravel the potential of new crop candidates, including millets. These novel phenomics tools highlight important traits like yield to estimate breeding predictions (Van Tassel et al. 2021). One most important approach to understanding the phenotypic traits of crops is metabolomics, which provides a profile of metabolites from seed or seedlings. Also, the plant organs can be re-imagined by using these high-throughput tools to perform the morphometric analysis (Schlautman et al. 2020). Notably, the morphometric approaches of phenomics could enable the study of the response to environmental stresses in crops and determine the role of different accessions in agriculture (Schlautman et al. 2020). Recently several high-throughput phenotyping centers have been developed, for example, the Bellwether Phenotyping Platform at the Danforth Plant Science Center (https://www.danforthcenter.org/our-work/core-facilities/phenotyping/), which has improved the accuracy of phenotype measurement in plants.

Over the last few decades, plant phenomics has evolved successfully and provided multidimensional phenotypic data at several levels of plants, such as cell level, the organ level, plant level, and population level (Zhao et al. 2019). This trait-based breeding has been utilized in major cereal crops, wheat, and rice. However, this novel approach of breeding is not fully explored in millets. This opens the opportunity to accelerate the breeding programs of these nutri-cereals. In summary, phenomics could complement the breeding programs for new crop improvement and orphan crop breeding.

14.8 Integrating Genomics and Phenomics in Small Millets for Climate Resilience

Continuously changing climatic conditions and the ever-increasing human population has raised serious concern on food security. The cropping system for major cereal crops, such as rice and wheat, is drastically challenged by climatic changes. Recently, millets have gained wide importance due to their climate-resilient features and nutritional qualities. However, millet farming is still in infancy due to the lack of deployment of genetic resources for crops improvement and optimization of long-term storage conditions (Muthamilarasan and Prasad 2021). Recent advancements in omics approaches, such as phenomics and genomics, could provide a roadmap to identify the climate resilience traits in small millets, accelerating their cultivation under changing climatic conditions (Muthamilarasan et al. 2016). Combining genomics with phenomics could strengthen the crop improvement programs related to climate resilience in small millets. The genome sequence of most small millets is now available, namely, foxtail millet, proso millet, teff, finger millet, barnyard, and fonio millet. Therefore, it provides the opportunity to identify climate-resilient traits in small millets for their improvement, ensuring their sustainable cultivation. The advanced bioinformatic tools have made it possible to harness vast data from phenomics and genomics to develop climate-smart crops. Integration of information identified by these omics approaches identifies genomic markers underlying important climate-resilient traits (Fig. 14.2).

Advanced omics approaches enhanced understanding of important climate-resilient traits in millets. For example, several transcriptomics studies have unraveled the factors underlying stress tolerance in finger millet. However, the detailed genetics of agronomically important traits of finger millet needs further systematic work by employing advanced genomics and phenomics techniques. Also, more standardization in phenomics techniques is required in order to improve the understanding of trait variability. Advanced phenomics and its integration with genomics have not been successful in finger millet. Further, compared to major cereals, there are very few studies on finger millet using molecular markers. Nevertheless,

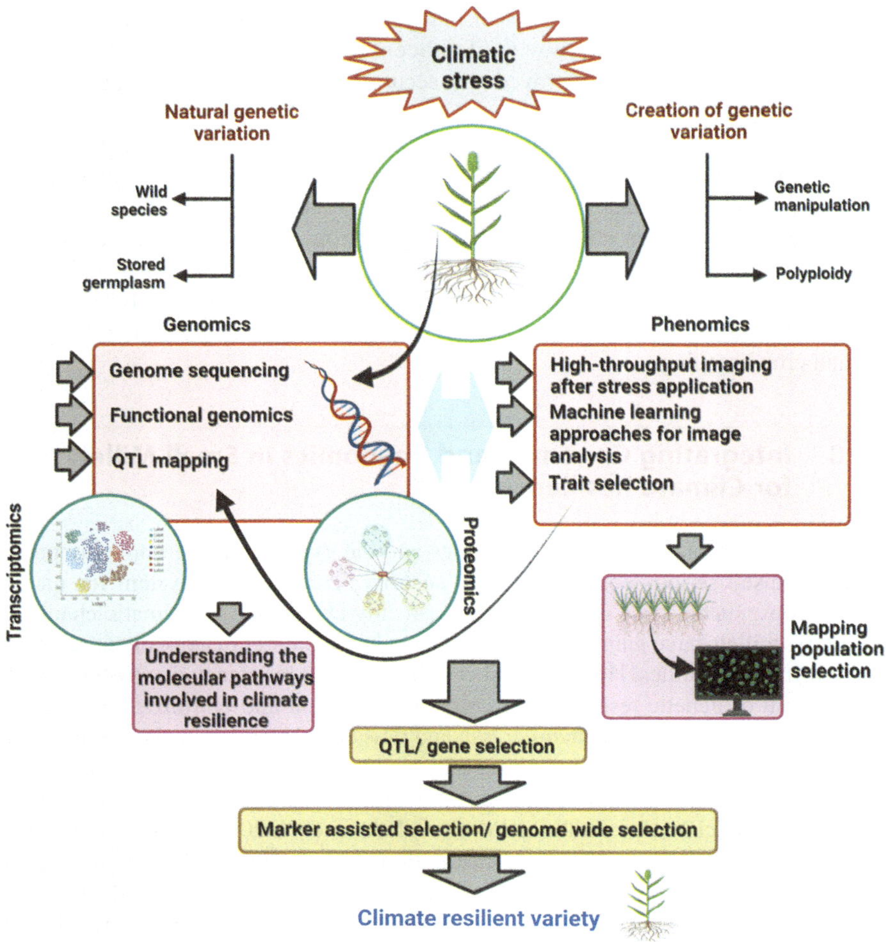

Fig. 14.2 Pipeline showing the procedure for integrating genomics and phenomics tools for developing climate-resilient varieties of small millets

association mapping is possible in finger millet due to great germplasm diversity. This allowed the identification of target multiple alleles at individual loci and their association with phenotypic traits (Sood et al. 2016; Babu et al. 2014). The complementation of genomics with phenotyping methods is also performed in proso millet to produce comprehensive data about various crucial traits (Boukail et al. 2021). This study in proso millet supported the fact that integration of genomics and phenomics rapidly identified marker-trait association to accelerate its breeding. Similarly, such integrated studies can be performed to identify climate-resilient traits in proso millet, which could strengthen the breeding programs for the development of climate-smart crop. To accelerate their breeding programs, such studies should be promoted in other small millets. The availability of advanced phenotyping methods facilitates the collection of phenotypic data in a non-destructive manner. Further, the detailed alignment of phenotypic data with genomics resources is essential for breeding and subsequently accelerates the development of climate-smart crops.

14.9 Conclusions and Future Perspectives

It is urgent and prudent to investigate the cultivation of sustainable crops which are climate-resilient and possess a well-known nutritional profile. Small millets hold immense potential as a solution for the impending crisis risking humankind in terms of hidden hunger, food security, water shortage, unstable agriculture, and climate change. They are adapted to grow under a wide range of harsh environmental conditions ranging from high temperature to drought and salinity. It does not demand proper irrigation, artificial soil nutrients, or insecticide. Through advancements in research, other economically important crops can be improved against environmental constraints through biotechnological approaches using small millet germplasms. In addition, the crop has excellent water- and nitrogen-use efficiency (WUE and NUE) and is phylogenetically adjacent to major cereals and biofuel grasses. These phenotypic and physiological attributes can be used for crop improvement through the genomic-assisted breeding or biotechnological approach. Small millets can serve as a model organism as they have small diploid genomes (490 Mb, about one fifth the size of the maize genome), short life cycles (50–90 days), self-pollination, small adult statures, and prolific seed production, for functional genomics of the Panicoideae, or for studying C_4 photosynthesis. With the availability of a high-quality reference genome sequence and other genomic data being continuously published, small millets can now indeed be considered a novel model plant system. They serve as an excellent resource material for abiotic stress-related gene function dissection and elite allele mining. In a recent experiment, the integration of the foxtail millet autophagy-associated gene, SiATG8a, into the rice genome improved tolerance to abiotic stress and nitrogen starvation (Li et al. 2016). The research in this arena has been slow, but it is gaining a steady impetus. More concerted efforts and support in terms of research and development are needed for sustaining the production of small millets. The year 2023 has been declared as the International Year of Millets by the United Nations, and by this time, concerted efforts should be invested in studying the small millets for climate-resilient traits.

Acknowledgements Authors' work in the area of millet genomics is funded by the DST INSPIRE Faculty Grant of Department of Science & Technology (DST), Ministry of Science & Technology, Government of India (File No. DST/INSPIRE/04/2016/002341), and Institute of Eminence grant (Project No.: UoH-IoE-RC2-21-014) awarded to the University of Hyderabad by Ministry of Education, Govt. of India (Ref. No.: F11/9/2019-U3(A)). Figures are created with BioRender.com.

References

Abrouk M, Ahmed HI, Cubry P, Šimoníková D, Cauet S et al (2020) Fonio millet genome unlocks African orphan crop diversity for agriculture in a changing climate. Nat Commun 11(1):4488. https://doi.org/10.1101/2020.04.11.037671

AICRPSM (2020) https://www.millets.res.in/aicsip20_sm.php

Anbukkani P, Balaji SJ, Nithyashree ML (2017) Production and consumption of minor millets in India–a structural break analysis. Ann Agric Res New Series 38(4):1–8

Andersen EJ, Nepal MP (2017) Genetic diversity of disease resistance genes in foxtail millet (*Setaria italica* L.). Plant Gene 10:8–16

Babu BK, Agrawal PK, Pandey D, Kumar A (2014) Comparative genomics and association mapping approaches for opaque2 modifier genes in finger millet accessions using genic, genomic and candidate gene-based simple sequence repeat markers. Mol Breed 34:1261–1279

Ballogou VY, Soumanou MM, Toukourou F, Hounhouigan JD (2013) Structure and nutritional composition of fonio (*Digitaria exilis*) grains: a review. Int Res J Biol Sci 2(1):73–79

Bandyopadhyay T, Muthamilarasan M, Prasad M (2017) Millets for next generation climate-smart agriculture. Front Plant Sci 8:1266. https://doi.org/10.3389/fpls.2017.01266

Barnaud A, Vignes H, Risterucci AM, Noyer JL, Pham JL et al (2012) Development of nuclear microsatellite markers for the fonio, *Digitaria exilis* (Poaceae), an understudied West African cereal. Am J Bot 99(3):e105–e107

Bhatt D, Negi M, Sharma P, Saxena SC, Dobriyal AK et al (2011) Responses to drought induced oxidative stress in five finger millet varieties differing in their geographical distribution. Physiol Mol Biol Plants 17(4):347–353

Bhave KG, Dalvi VV, Thaware BL, Mahadik SG, Kasture MC et al (2016) Mutagenesis in Proso millet (*Panicum millaceum* L.). Int J Sci Res 5(3):1635–1638

Boukail S, Macharia M, Miculan M, Masoni A, Calamai A et al (2021) Genome wide association study of agronomic and seed traits in a world collection of proso millet (*Panicum miliaceum* L.). BMC Plant Biol 21(1):330

Burén S, Jiménez-Vicente E, Echavarri-Erasun C, Rubio LM (2020) Biosynthesis of nitrogenase cofactors. Chem Rev 120(12):4921–4968

Cao X, Hu L, Chen X, Zhang R, Cheng D et al (2019) Genome-wide analysis and identification of the low potassium stress responsive gene SiMYB3 in foxtail millet (*Setaria italica* L.). BMC Genomics 20(1):1–13

Dansi A, Adoukonou-Sagbadja H, Vodouhe R (2010) Diversity, conservation and related wild species of Fonio millet (Digitaria spp.) in the northwest of Benin. Genet Resour Crop Evol 57(6):827–839

De Wet JM, Harlan JR (1975) Weeds and domesticates: evolution in the man-made habitat. Econ Bot 29(2):99–108

Deshpande SS, Mohapatra D, Tripathi MK, Sadvatha RH (2015) Kodo millet-nutritional value and utilization in Indian foods. J Grain Process Storage 2(2):16–23

Dida MM, Ramakrishnan S, Bennetzen JL, Gale MD, Devos KM (2007) The genetic map of finger millet, Eleusine coracana. Theor Appl Genet 114(2):321–332

Doust AN, Devos KM, Gadberry MD, Gale MD, Kellogg EA (2005) The genetic basis for inflorescence variation between foxtail and green millet (Poaceae). Genetics 169(3):1659–1672

Ebanyat P, de Ridder N, Bekunda M, Delve RJ, Giller K et al (2021) Efficacy of nutrient management options for finger millet production on degraded small holder farms in eastern Uganda. Front Sustain Food Syst 5:1–15

Ganapathy KN, Patil JV (2017) Improvement in finger millet: status and future prospects. In: Patil JV (ed) Millets and sorghum: biology and genetic improvement. Wiley, Chichester. https://doi.org/10.1002/9781119130765.ch6:87-111

Ganapathy KN, Hariprasanna K, Tonapi VA (2021) Breeding for enhanced productivity in millets. In: Millets and pseudo cereals. Woodhead Publishing, pp 39–63

Giussani LM, Cota-Sanchez JH, Zuloaga FO, Kellogg EA (2001) A molecular phylogeny of the grass sub-family Panicoideae shows multiple origins of C4 photosynthesis. Am J Bot 88(11): 1993–2012

Gopalan C, Ramasastri B, Balasubramanian S (2007) Nutritive value of Indian foods. National Institute of Nutrition (Indian council of Medical Research), Hyderabad

Goron TL, Raizada MN (2015) Genetic diversity and genomic resources available for the small millet crops to accelerate a new green revolution. Front Plant Sci 6:157

Gowda CL, Upadhyaya HD, Reddy VG, Singh S (2008) Diversity in small millets germplasm and enhancing its use in crop improvement. https://www.intlcss.org/files/icss/congress-proceedings/2008-papers/cs2-s1/cs2-s1-o2-laxmipathi-c-l-gowda.pdf. Last accessed 9 Mar 2022

Gupta A, Sood S, Agarwal PK, Bhatt JC (2011) Floral biology and pollination system in small millets. Eur J Plant Sci Biotechnol 174(4008):468–474

Gupta N, Srivastava AK, Pandey VN (2012) Biodiversity and nutraceutical quality of some Indian millets. Proc Natl Acad Sci India Sect B Biol Sci 82(2):262–273

Gupta V, Jatav PK, Verma R, Kothari SL, Kachwaha S (2017) Nickel accumulation and its effect on growth, physiological and biochemical parameters in millets and oats. Environ Sci Pollut Res 24(30):23915–23925

Habiyaremye C, Matanguihan JB, D'Alpoim Guedes J, Ganjyal GM, Whiteman MR et al (2017) Proso millet (*Panicum miliaceum* L.) and its potential for cultivation in the Pacific Northwest, US: a review. Front Plant Sci 7:1961

Hariprasanna K (2017) Kodo millet, Paspalum scrobiculatum L. In: Millets and sorghum: biology and genetic improvement, vol 1. Wiley, Chichester, pp 199–225

Hedge PS, Chandra TS (2005) ESR spectroscopic study reveals higher free radical quenching potential in kodo millet (*Paspalum scrobiculatum*) compared to other millets. Food Chem 92: 177–182

Hittalmani S, Mahesh HB, Shirke MD, Biradar H, Uday G et al (2017) Genome and transcriptome sequence of finger millet (*Eleusine coracana* (L.) Gaertn.) provides insights into drought tolerance and nutraceutical properties. BMC Genomics 18(1):1–16

Hunt HV, Badakshi F, Romanova O, Howe CJ, Jones MK, Heslop-Harrison JP (2014) Reticulate evolution in Panicum (Poaceae): the origin of tetraploid broomcorn millet, *P. miliaceaum*. J Exp Bot 65(12):3165–3175

Ibrahim Bio Yerima AR, Achigan-Dako EG (2021) A review of the orphan small grain cereals improvement with a comprehensive plan for genomics-assisted breeding of fonio millet in West Africa. Plant Breed 140(4):561–574

Islam MS, Akhter MM, El Sabagh A, Liu LY, Nguyen NT et al (2011) Comparative studies on growth and physiological responses to saline and alkaline stresses of Foxtail millet (Setaria italica L.) and Proso millet (*Panicum miliaceum* L.). Aust J Crop Sci 5(10):1269–1277

Jadhao KR, Bansal A, Rout GR (2020) Silicon amendment induces synergistic plant defense mechanism against pink stem borer (Sesamia inferens Walker.) in finger millet (*Eleusine coracana* Gaertn.). Sci Rep 10(1):1–15

Janni M, Gulli M, Maestri E, Marmiroli M, Valliyodan B et al (2020) Molecular and genetic bases of heat stress responses in crop plants and breeding for increased resilience and productivity. J Exp Bot 71(13):3780–3802

Jia G, Shi S, Wang C, Niu Z, Chai Y et al (2013) Molecular diversity and population structure of Chinese green foxtail [*Setaria viridis* (L.) Beauv.] revealed by microsatellite analysis. J Exp Bot 64(12):3645–3656

Krishnamurthy L, Upadhyaya H, Junichi K, Purushothaman R, Dwivedi S, Vadez V (2016) Variation in drought-tolerance components and their interrelationships in the core collection of foxtail millet (*Setaria italica*) germplasm. Crop Pasture Sci 67:834

Kumar A, Metwal M, Kaur S, Gupta AK, Puranik S et al (2016a) Genotyping by sequencing analysis for determining population structure of finger millet germplasm of diverse origins. Plant Genome 9(2):2015–2007

Kumar A, Metwal M, Kaur S, Gupta AK, Puranik S et al (2016b) Nutraceutical value of finger millet [*Eleusine coracana* (L.) Gaertn.], and their improvement using omics approaches. Front Plant Sci 7:934. https://doi.org/10.3389/fpls.2016.00934

Kumari SK, Thayumanavan B (1997) Comparative study of resistant starch from minor millets on intestinal responses, blood glucose, serum, cholesterol and triglycerides in rats. J Sci Food Agric 75(3):296–302

Lata C, Gupta S, Prasad M (2013) Foxtail millet: a model crop for genetic and genomic studies in bioenergy grasses. Crit Rev Biotechnol 33(3):328–343

Li W, Chen M, Wang E, Hu L, Malcom J et al (2016) Genome-wide analysis of autophagy-associated genes in foxtail millet (*Setaria italica* L.) and characterization of the function of *SiATG8a* in conferring tolerance to nitrogen starvation in rice. BMC Genomics 17:797. https://doi.org/10.1186/s12864-016-3113-4

Liu J, Zhang D, Yuan Y, Chen P, Zhang P et al (2021) A promising crop for cadmium-contamination remediation: broomcorn millet. Ecotoxicol Environ Saf 224:112669

Maitra S, Panda P, Panda SK, Behera D, Tanmoy S et al (2020) Relevance of barnyard millet (*Echinochloa frumentacea* L.) cultivation and agronomic management for production sustainability. Int J Bioinform Biol Sci 8:27–32

Mamidi S, Healey A, Huang P, Grimwood J, Jenkins J et al (2020) A genome resource for green millet Setaria viridis enables discovery of agronomically valuable loci. Nat Biotechnol 38(10): 1203–1210

Matsuura A, An P, Murata K, Inanaga S (2016) Effect of pre- and post-heading waterlogging on growth and grain yield of four millets. Plant Prod Sci 19(3):348–359

Meng X, Liang Z, Dai X, Zhang Y, Mahboub S et al (2021) Predicting transcriptional responses to cold stress across plant species. Proc Natl Acad Sci 118(10):e2026330118

Muthamilarasan M, Prasad M (2021) Small millets for enduring food security amidst pandemics. Trends Plant Sci 26(1):33–40

Muthamilarasan M, Bonthala VS, Mishra AK, Khandelwal R, Khan Y et al (2014) C2H2 type of zinc finger transcription factors in foxtail millet define response to abiotic stresses. Funct Integr Genomics 14(3):531–543

Muthamilarasan M, Dhaka A, Yadav R, Prasad M (2016) Exploration of millet models for developing nutrient rich graminaceous crops. Plant Sci 242:89–97

Nithiyanantham S, Kalaiselvi P, Mahomoodally MF et al (2019) Nutritional and functional roles of millets—a review. J Food Biochem 43(7):e12859

Pan J, Li Z, Wang Q, Garrell AK, Liu M, Guan Y, Zhou W, Liu W (2018) Comparative proteomic investigation of drought responses in foxtail millet. BMC Plant Biol 18:315

Patil HE (2018) Breeding for quality improvement in small millets: a review. Int J Genet 10(9): 507–510

Patil VU, Raundal PU (2018) Nutrient uptake studies in little millet as influenced by varieties and levels of fertilizers. Int J Recent Sci Res 9(3):25474–25476

Patil KB, Chimmad BV, Itagi S (2015) Glycemic index and quality evaluation of little millet (*Panicum miliare*) flakes with enhanced shelf life. J Food Sci Technol 52(9):6078–6082. https://doi.org/10.1007/s13197-014-1663-5

Plaza-Wüthrich S, Blösch R, Rindisbacher A, Cannarozzi G, Tadele Z (2016) Gibberellin deficiency confers both lodging and drought tolerance in small cereals. Front Plant Sci 7:643

Prasanthi Kumari R, Vishnuvardhan Z (2015) Effect of salinity on growth, protein and antioxidant enzymes in three Kodo millet (*Paspalum scrobiculatum*) germplasm. Int J Curr Microbiol Appl Sci 4(6):475–483

Puranik S, Bahadur RP, Srivastava PS, Prasad M (2011) Molecular cloning and characterization of a membrane associated NAC family gene, SiNAC from foxtail millet [Setaria italica (L.) P. Beauv]. Mol Biotechnol 49(2):138–150

Rahman H, Ramanathan V, Nallathambi J, Duraialagaraja S, Muthurajan R (2016) Over-expression of a NAC 67 transcription factor from finger millet (*Eleusine coracana* L.) confers tolerance

against salinity and drought stress in rice. BMC Biotechnol 16:35. https://doi.org/10.1186/s12896-016-0261-1
Rincent R, Charpentier JP, Faivre-Rampant P et al (2018) Phenomic selection is a low-cost and high-throughput method based on indirect predictions: proof of concept on wheat and poplar. G3 (Bethesda) 8(12):3961–3972
Saleh AS, Zhang Q, Chen J, Shen Q (2013) Millet grains: nutritional quality, processing, and potential health benefits. Compr Rev Food Sci Food Saf 12:281–295. https://doi.org/10.1111/1541-4337.12012
Sarita ES, Singh E (2016) Potential of millets: nutrients composition and health benefits. J Sci Innov Res 5(2):46–50
Satish L, Rathinapriya P, Rency AS, Ceasar SA, Prathibha M, Pandiana S, Rameshkumar R, Ramesh M (2016) Effect of salinity stress on finger millet (Eleusine coracana (L.) Gaertn): histochemical and morphological analysis of coleoptile and coleorhizae. Flora: Morphol Distrib Funct Ecol Plants 222:111–120
Satyavathi CT, Solanki RK, Kakani RK, Bharadwaj C, Singhal T et al (2019) Genomics assisted breeding for abiotic stress tolerance in millets. In: Genomics assisted breeding of crops for abiotic stress tolerance II. Springer, Cham, pp 241–255
Schlautman B, Diaz-Garcia L, Barriball S (2020) Reprint of: Morphometric approaches to promote the use of exotic germplasm for improved food security and resilience to climate change: a kura clover example. Plant Sci 295:110415
Senthil A, Ashok S, Sritharan N, Punitha S, Divya K et al (2018) Physiological efficiency of small millets under drought condition. Madras Agric J 105(7–9(2)):363–367. https://doi.org/10.29321/MAJ.2018.000161
Sharma N, Goyal SK, Alam T, Fatma S, Niranjan K (2018) Effect of germination on the functional and moisture sorption properties of high–pressure-processed foxtail millet grain flour. Food Bioprocess Technol 11(1):209–222
Shi CK, Chiang W, Kuo ML (2004) Effects of adlay on azoxymethane-induced colon carcinogenesis in rats. Food Chem Toxicol 42(8):1339–1347
Siddappa N, Raghu GK, Varadahally DR (2016) Identification of drought-responsive transcripts in Kodo millet *(Paspalum scrobiculatum L.)*. Int J Innov Res Dev 5(11):140–143
Singh RK, Jaishankar J, Muthamilarasan M, Shweta S, Dangi A et al (2016) Genome-wide analysis of heat shock proteins in C4 model, foxtail millet identifies potential candidates for crop improvement under abiotic stress. Sci Rep 6:32641
Singh S, Chopperla R, Shingote P, Chhapekar SS, Deshmukh R et al (2021) Overexpression of EcDREB2A transcription factor from finger millet in tobacco enhances tolerance to heat stress through ROS scavenging. J Biotechnol 336:10–24. https://doi.org/10.1016/j.jbiotec.2021.06.013
Sood S, Khulbe RK, Gupta AK, Agarwal PK, Upadhyaya H (2015) Barnyard millet—a potential food and feed crop of future. Plant Breed 134(2):135–147
Sood S, Kumar A, Babu BK, Gaur VS, Pandey D et al (2016) Gene discovery and advances in finger millet [*Eleusine coracana* (L.) Gaertn.] genomics—an important nutri-cereal of future. Front Plant Sci 7:1634
Srinivasan R, Mageswari A, Subramanian P, Maurya VK, Suganthi C et al (2017) Exogenous expression of ACC deaminase gene in psychrotolerant bacteria alleviates chilling stress and promotes plant growth in millets under chilling conditions. Indian J Exp Biol 55(7):463–468
Swati P, Sarita J, Prem S, Nese S, Manoj P (2011) Comparative transcriptome analysis of contrasting foxtail millet cultivars in response to short-term salinity stress. J Plant Physiol 168:280–287
Tiwari S, Yadav SK, Sahu VK, Tripathi MK (2018) Current status and future prospects of marker assisted breeding for genetic improvement of minor millets. Int J Curr Microbiol Appl Sci 7(12): 2587–2590

Trivedi AK, Arya L, Verma SK, Tyagi RK, Hemantranjan A (2017) Evaluation of barnyard millet diversity in central Himalayan region for environmental stress tolerance. J Agric Sci 155(10): 1497–1507

Ugare R, Chimmad B, Naik R, Bharati P, Itagi S (2014) Glycemic index and significance of barnyard millet (*Echinochloa frumentacae*) in type II diabetics. J Food Sci Technol 51(2): 392–395

Upadhyaya HD, Bajaj D, Narnoliya L, Das S, Kumar V, Gowda CL, Sharma S, Tyagi AK, Parida SK (2016) Genome-wide scans for delineation of candidate genes regulating seed-protein content in chickpea. Front Plant Sci 7:302

Van Tassel DL, DeHaan LR, Diaz-Garcia L et al (2021) Re-imagining crop domestication in the era of high throughput phenomics [published online ahead of print, 2021 Dec 6]. Curr Opin Plant Biol 65:102150

Varshney RK, Bohra A, Yu J, Graner A, Zhang Q, Sorrells ME (2021) Designing future crops: genomics-assisted breeding comes of age. Trends Plant Sci 26(6):631–649

Vavilov NI (1926) Centers of origin of cultivated plants. Bull Appl Bot 16:1–248

Vetriventhan M, Upadhyaya HD, Dwivedi SL, Pattanashetti SK, Singh SK (2016) Finger and foxtail millets. In: Genetic and genomic resources for grain cereals improvement. Academic, New York, pp 291–319

Vidya PU, Raundal PU (2018) Nutrient uptake studies in little millet as influenced by varieties and levels of fertilizers. Int J Rec Sci Res 9(3):25474–25476

Wallace JG, Upadhyaya HD, Vetriventhan M, Buckler ES, Tom Hash C et al (2015) The genetic makeup of a global barnyard millet germplasm collection. Plant Genome 8(1):1–7

Wambi W, Otienno G, Tumwesigye W, Mulumba J (2021) Genetic and genomic resources for finger millet improvement: opportunities for advancing climate-smart agriculture. J Crop Improv 35(2):204–233

Wang T, Picard JC, Tian X, Darmency H (2010) A herbicide-resistant ACCase 1781 Setaria mutant shows higher fitness than wild type. Heredity 105(4):394–400

Wang C, Jia G, Zhi H, Niu Z, Chai Y et al (2012) Genetic diversity and population structure of Chinese foxtail millet [*Setaria italica* (L.) Beauv.] landraces. G3 (Bethesda) 2(7):769–777

Wang P, Wang H, Wang Y, Ren F, Liu W (2018) Analysis of bHLH genes from foxtail millet (*Setaria italica*) and their potential relevance to drought stress. PLoS One 13(11):e0207344

Yabe S, Iwata H (2020) Genomics-assisted breeding in minor and pseudo-cereals. Breed Sci 70(1): 19–31

Yadav N, Chaudhary K, Singh A, Gupta A (2013) Evaluation of hypoglycaemic properties of kodo millet based food products in healthy subjects. IOSR J Pharm 3:14–20

Yadav A, Khan Y, Prasad M (2016) Dehydration-responsive miRNAs in foxtail millet: genome-wide identification, characterization and expression profiling. Planta 243:749–766

Yang R, Chen M, Sun JC, Yu Y, Min DH et al (2019) Genome-wide analysis of LIM family genes in foxtail millet (*Setaria italica* L.) and characterization of the role of *SiWLIM2b* in drought tolerance. Int J Mol Sci 20(6):1303

Zhang G, Liu X, Quan Z, Cheng S, Xu X et al (2012) Genome sequence of foxtail millet (*Setaria italica*) provides insights into grass evolution and biofuel potential. Nat Biotechnol 30(6): 549–554

Zhang Y, Gao X, Li J, Gong X, Yang P et al (2019) Comparative analysis of proso millet (*Panicum miliaceum* L.) leaf transcriptomes for insight into drought tolerance mechanisms. BMC Plant Biol 19(1):397

Zhao C, Zhang Y, Du J, Guo X, Wen W et al (2019) Crop phenomics: current status and perspectives. Front Plant Sci 10:714

Zou C, Li L, Miki D, Li D, Tang Q et al (2019) The genome of broomcorn millet. Nat Commun 10(1):1–11

Abiotic Stress Tolerant Small Millet Plant Growth Regulation by Long Non-coding RNAs: An Omics Perspective

15

Navonil Mal and Chittabrata Mal

Abstract

Millets are a group of annual, panicoid grasses with C_4 mode of photosynthesis. With the ease of their *Arabidopsis*-like lifecycle, tremendous stress-tolerance capacity, and huge nutritional value, millets can be the genetic and nutritional epitome for worldwide food security as well as for the development of C_4 model plant system. Recent sophistications in "omics" approaches may enlighten the path for improvement of the existing genetic resources to sustain over the ongoing environmental dynamics and to incorporate these climate-resilient genetic elements within the established food crops. With the advent of different omics-perspective, long non-coding RNAs (lncRNAs) get huge attention particularly in the prospect for plants' stress response. The aim of this chapter is to discuss the present state of knowledge about how small millets can be established as the climate-sustainable model system, and how lncRNAs involve in abiotic stress management. It also discusses the utilization of several bioinformatics pipelines to help in unravelling the molecular mystery behind it.

Keywords

LncRNA · Millet · Abiotic stress · Bioinformatics · ChIP sequencing · Agricultural sustainability

N. Mal
Department of Botany, University of Calcutta, Kolkata, West Bengal, India

C. Mal (✉)
Department of Bioinformatics, Maulana Abul Kalam Azad University of Technology (Formerly Known as West Bengal University of Technology), Nadia, West Bengal, India

R. N. Pudake et al. (eds.), *Omics of Climate Resilient Small Millets*,
https://doi.org/10.1007/978-981-19-3907-5_15

15.1 Introduction

An ideal plant model not only gears up the development of agronomic blueprint for sustainable crop improvement practices, but also provides an ease of study for gene functionalities from different omics-perspective. Since the last two decades, *Arabidopsis thaliana* (the dicot candidate) and *Oryza sativa* (monocot candidate) serve as the genetic models for unravelling the genetic background of different existing characteristics and to design the genetic basis for novel agricultural traits (Provart et al. 2016). But the C_3 photosynthetic nature of both the model candidates imposes thrive for their alternate counterparts to mitigate with the recent, rapid environmental dynamics. In addition, C_4 candidates are much more efficient CO_2 fixers with their involvement in Hatch and Slack pathway beside their elite nitrogen absorption efficacies. Despite of being only 3% among the flowering plants, they account for approximately 30% agricultural grain production and about 25% of terrestrial primary productivity at global scale (Sage 2004). On the flip side, elaborate life cycle and a very large stature retract the attractions of plant biotechnologists for developing a C_4 model. However, a C_4 model plant can easily fine-tune with the momentum of the rapid environmental fluctuations and by understanding the molecular background via omics approaches, these molecular intricacies can be incorporated within C_3 cells of agricultural interest (Peng and Zhang 2020).

15.2 Millet as a Climate-Smart Model Candidate for Future Agricultural Sustainability: An Overview

Small millets collectively refer to a number of annual Panicoid grasses with C_4 mode of photosynthesis, including foxtail millet (*Setaria italica*), finger millet (*Eleusine coracana*), pearl millet (*Pennisetum glaucum*), proso millet (*Panicum miliaceum*), and many more, mostly used as fodder and food for poor marginal people in dry regions of temperate, subtropical, and tropical regions of the world (Dwivedi et al. 2012; Lata et al. 2013). Most interestingly, small millets are not only nutritionally equivalent to many other cereal grains with huge proportions of carbohydrates, proteins, and fats, but also very much climate-resilient and resistant to a number of biotic as well as abiotic stress factors, so it can be epitomized as a poor man's crop with immense nutritional security. According to FAOSTAT, 2014 (http://faostat.fao.org/site/567/DesktopDefault.aspx?PageID=567#ancor), India dominates the global millet production with an annual range of about 11 million tonnes in 2013. Hence, plant breeding approaches in association with advance cutting-edge genomics and other omics-based methods for intensifying their stress tolerance efficacies for environmentally sustainable agricultural practices in dry, arid regions must be appreciated (Lata 2015).

Among all the other candidates, Foxtail millet (*Setaria italica*), one of the most ancient domesticated crops (around >10,000 years ago), is of great interest for scientific investigations due to their extreme genetic simplicity with only 9 pairs of chromosomes involving about 490 Mb. Ethyl methane sulfonate (EMS)-based mutations have been incorporated in one good cultivar, namely Jingu21, and the mutational status has been identified and the mutated line was named as *xiaomi*, with reduced life cycle and higher seed production than the wild type, which makes the mutant population more suitable for studying gene functionalities (Yang et al. 2020). Noteworthy, the authors have successfully established the whole genome information from multiomics perspective and a temporal transcriptomic database, which is one of the mandatory criteria to be a model species. This *xiaomi* appears to be a point mutation in the phytochrome C gene, involved in photoperiodic floral induction. Interestingly, they have also successfully standardized the protocol of *Agrobacterium*-mediated transformation for *xiaomi* mutant population, with transformation rate of about 23.28%, very close to other model candidates (Wang et al. 2020).

With the advent of recent refine technologies like CRISPR/Cas9 (clustered regularly interspaced short palindromic repeat (CRISPR)-associated protein-9 (Cas9)), the precision making for genetic manipulation for individual genes becomes more sophisticated. In a nutshell, a reduced lifespan, shorter plant size along with the whole genome information make the *xiaomi* population a premier C_4 candidate for studying several molecular, physiological, biochemical backgrounds for the traits of agronomic interest (Peng and Zhang 2020). Small millets involve a group of 10–12 small-grained, annual, C4 grass, each with their unique trait value. Among this, finger millet is mostly sound from research perspective, in addition other five varieties like foxtail millet [*Setaria italica* (L.) *Beauv.*], kodo millet [*Paspalum scrobiculatum* L.], little millet [*Panicum sumatrense* Roth. ex. Roem. & Schult.], proso millet [*Panicum miliaceum* L.], and barnyard millet [*Echinochloa colona* (L.) Link] and [*Echinochloa crus-galli* (L.) P. Beauv.] are important from agricultural prospect. As climate-resilient food crops, small millets become staple food grain for millions of poor smallholders and households in the developing world's harshest. Climate resilience provides a great source for genetic traits which can strengthen the agricultural security to face with environmental dynamics. As temperature rises and climates get hotter and drier, small millets will become increasingly well suited for dryland, arid areas. Small millets are very high in iron, zinc, and calcium content, and have other dietary qualities which can help in the diagnosis of celiac disease, anemia, and diabetes. This high nutritional value—coupled with the impressive hardiness of small millets—makes them desirable food security crops, as well as a good sources of fodder and feed in mixed crop/livestock systems. Here we elaborate the role of lncRNAs in stress tolerant millet production and the advancements of omics technologies to identify the regulatory roles of the lncRNAs in these mechanisms (Fig. 15.1).

15.3 Long Non-coding RNAs in Abiotic Stress Mitigation

Long non-coding RNA (lncRNA) is a cumulative term, involving the different kinds of non-coding RNA, one of the major key-player of transcriptional regulation at both transcriptional and post-transcriptional level, as well as for different epigenetic

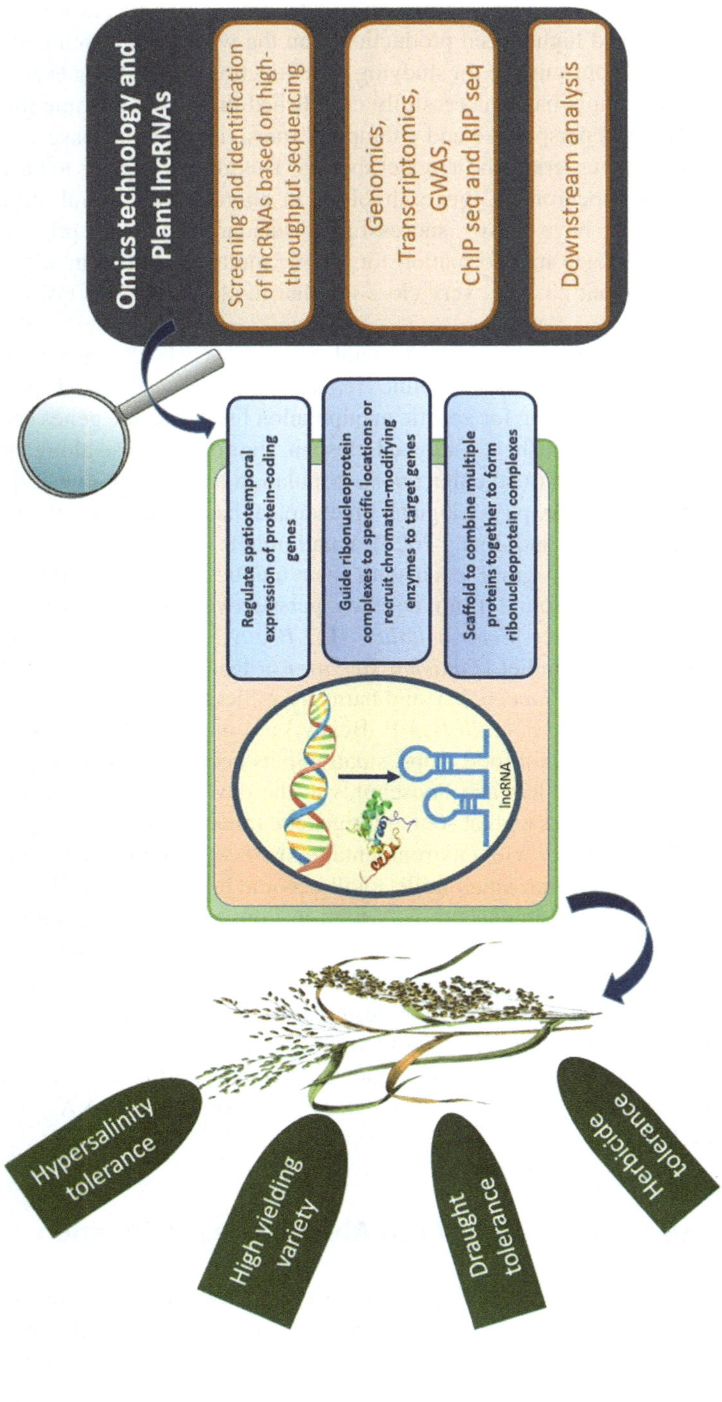

Fig. 15.1 Overview of omics technologies to reveal the regulatory role of small millet's lncRNAs under stress conditions

modifications like chromatin remodelling, RNA silencing, and many more (Jha et al. 2020; Waseem et al. 2021). Mostly the lncRNAs are of a length greater than 200 nucleotides, temporally and spatially segregated within nucleus as well as in cytoplasm and with intricate structural complexity at the secondary level. According to their structural organization, lncRNAs are categorized into different groups like bidirectional, overlapping, cis-antisense, and intronic, whereas their functional role subdivided them as backbone molecules, guide molecules, decoy molecules, and signal molecules. Usually, RNA polymerase II do the synthesis of lncRNAs from their corresponding genomic sources. Like typical mRNAs, most of the lncRNAs possess 5′ cap as well as 3′ poly adenosine tail. LncRNAs show high degree of sequence variation at inter-specific level, whereas they are highly conserved intra-specifically (Deng et al. 2018).

Plants being sessile organism involve an array of regulatory strategies to maintain their cellular cell homeostasis during different adverse conditions. In addition to stress-responsive mechanisms, plants' reproductive developmental stages are also interfered by different lncRNAs. In plant system, the lncRNA-mediated regulatory mechanism has been reported in the developmental stages of *Medicago* sp. (Crespi et al. 1994). They have identified one lncRNA enod40 (early nodulin 40), acting as riboregulator. Liu et al. (2012) have unravelled various precision-making stages, interfered by lncRNAs, upon exposure to various abiotic stresses like drought, hypersalinity, mineral deficiency, etc. and reported 1832 lncRNAs in *Arabidopsis thaliana*, involved in stress responses. Fukuda et al. (2019) have reported the involvement of 60 differentially expressed lncRNAs in *Arabidopsis* during nutrient deficiency (Fukuda et al. 2019). Zhang et al. (2018) identified 16,551 novel lncRNAs, including 4554 annotated lncRNAs (targeting 3574 genes), and 11,997 unknown lncRNAs in switchgrass under dehydration stress. Ochogavía et al. (2018) found that *PN_LNC_N13,* a lncRNA like molecule is expressed only in apomictic plants and displays quantitative representation variation across reproductive developmental stages. Interplay of miRNA-lncRNA-coding target transcript modules has been reported to maintain boron homeostasis in barley (Unver and Tombuloglu 2020). Negative impact of one antisense heat-inducible lncRNA, as HSFB2a over the gametophytic development of *Arabidopsis* has been reported by Wunderlich et al. (2014). Differential expression of 1614 lncRNAs has been reported in *Brassica juncea* in response to heat and temperature stress (Bhatia et al. 2020). Similarly, differential expression of several cold-responsive lncRNAs has been reported in *Arabidopsis* (Calixto et al. 2019) and in grape (Wang et al. 2019). In this prospect, the two premiere candidates are COOLAIR and COLDAIR, identified through silencing of FLOWERING LOCUS C (FLC), involved in vernalization. The expression product of the first intron of FLC gene is the COLDAIR lncRNA, which interacts Polycomb Repressive Complex 2 (PRC2) and ultimately promotes histone methylation at the FLC locus (Kim et al. 2017); whereas being an antisense transcript of FLC, COOLAIR promotes histone methylation at the FLC locus through plant homeo-domain (PHD)-PRC2 (Swiezewski et al. 2009).

With the advent of high-throughput technological advancements, the precise roles of these "transcriptional noise" have ultimately come under spotlight. These so-called junk sequences have gained great biological interests (Urquiaga et al. 2021). As the expression of lncRNAs are very less in quantitative scale compared to mRNAs, highly sensitive techniques like real-time quantitative polymerase chain reaction (qRT-PCR) and RNA fluorescence in situ hybridization (RNA FISH) are required for their expression profiling (Wu et al. 2020). Recent genome-editing tools like clustered regularly interspaced short palindromic repeats associated nucleases (CRISPR/Cas) and further manipulation of lncRNAs become possible due to their large size, it is much easier to find out corresponding PAM sequence (Zhang and Zhang 2020). In addition, optimization of lncRNA expression can also be done by virtue of RNAi technologies to promote GMO independent editing events in lncRNA sequences (Werner et al. 2020).

15.4 LncRNA-Associated Omics Technologies

The lncRNAs are found in different types of plants. They are regulating different biological processes like growth and developmental processes, including root organogenesis, photomorphogenesis, flowering time and flower development, sexual reproduction, fruit development and ripening, and stress response. However, in most of the plants, their regulatory roles remain largely unknown (Zhao et al. 2020). Advanced technologies like tiling arrays and deep sequencing have been used to discover many classes of ncRNAs. Re-annotation of microarrays by bioinformatics approaches has also led to the identification of novel lncRNAs in human and mouse (Liao et al. 2011). Bioinformatic analysis of RNA-seq and ChIP-seq data are also routinely applied to search for unexpected transcripts without protein-coding potential (Young et al. 2012). However, till date, a few research works using omics technologies have been performed to know the regulatory role of lncRNAs in small millets. Omics studies, including genomics, transcriptomics, and GWAS (genome-wide association study), have enormous potential to study the role of lncRNAs in different types of millets. Further such studies will help to identify the functional elements embedded in lncRNAs such as miRNA binding sites, transposable elements, and highly structured motifs (Bai et al. 2015).

15.4.1 Genomics

As explained earlier, foxtail millet is one of the most suitable plant for whole genome sequencing and analysis. Draft genome of the foxtail millet strain "Zhang gu" was also assembled by whole genome shotgun combined with next-generation sequencing. With the advancement of genome sequencing techniques, now it is easy to know about specific genomic regions or specific genes which are associated with agronomically important traits including grain and fodder yield. Large-scale resequencing projects are also helpful in this purpose (Zhang et al. 2012). Resequencing of the pearl millet inbred germplasm association panel (PMiGAP) set revealed that small structural rearrangements such as insertions and deletions in the genome have occurred throughout the evolution of pearl millet (Zhang et al. 2012). However, one of the major problems encountered in the genome assembly of finger millet was its polyploidy, which hampers genome assembly compared with a diploid genome. To overcome this problem, Hatakeyama et al. (2018) sequenced the genome of finger millet using diverse technologies with sufficient coverage. They assembled it via a novel multiple hybrid assembly workflow that combines next-generation with single-molecule sequencing, followed by whole genome optical mapping using the Bionano Irys® system.

15.4.2 Transcriptomics

With the rapid development of high-throughput sequencing technologies, numerous lncRNAs have been identified under stress conditions in many species by transcriptome re-assembly (Qi et al. 2013). After exposure to simulated drought stress, a genome-wide transcriptome of foxtail millet has been generated by deep sequencing approach. A large number of differentially expressed genes (DEGs) were identified and the role of long non-coding RNAs (lncRNAs) to regulate the DEGs under drought condition was also investigated. Recent studies provide new insights into drought-induced lncRNAs in the foxtail millet gene expression. Qi et al. (2013) have characterized a total of 584 lncRNAs (494 lincRNAs and 90 NATs) within foxtail millet transcriptome. Among which, 17 lincRNAs and 2 NATs (natural antisense transcripts) appear to be responsible for PED-induced drought tolerance in foxtail millet, supported by the data from differential expression analysis, with one lnc transcript having synteny with its counterpart *Sorghum bicolor*. But statistical result (P value $= 0.9834$) suggests there is no general agreement about the functional link between the lincRNAs and the drought-responsive genes, in spite of their physical proximity. Lengths of the lncRNAs vary from 210 to 4168 nt with a mean of 891 nt, which was shorter than the foxtail millet protein-coding genes (1400 nt). The mean expression levels of the lncRNAs were about twofold lower than the mean expression levels of the protein-coding genes. Two identified NATs appear to be an antisense transcript having overlaps of Si003758m and Si038715m (Fig. 15.2). Salleh et al. (2012) have pointed out the homology of Si038715m with *Arabidopsis* SAG21, functions in oxidative stress responses and therefore denoted as drought-responsive genes. Si003758m had two predicted antisense transcripts that contained overlapping regions at both terminals of the sense partner, while the antisense Si038715m transcript was a 30 tail-to-tail transcript that ended within the second exon of Si038715m and overlapped with the last 74 nucleotides of the sense transcript.

In another study, strand-specific RNA-seq has been used to detect the expression changes of lncRNAs in both resistant and sensitive cultivars of foxtail millet exposed to imazapic herbicide (Wang et al. 2020). Strand-specific deep sequencing technology is used with Illumina Hiseq 4000 platform. Strand-specific deep sequencing has advantages over the non-strand-specific sequencing, which can be used to distinguish the non-sense strand from a sense strand that gives more confidence on

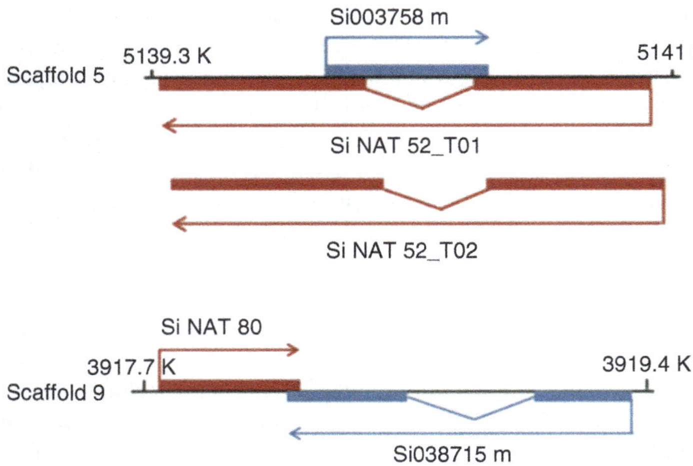

Fig. 15.2 Schematic representation of two NATs and their sense overlapping transcripts. (Adapted from Qi et al. 2013)

identifying lncRNAs. Besides the high-confidence known lncRNAs, novel lncRNAs can be identified based on some strict criteria. These lncRNAs were unevenly distributed across all nine chromosomes. Majority of lncRNAs are located in the intergenic region and contain 1–3 exons. Such lncRNAs have shorter length and lower expression levels compared to the protein-coding genes. LncRNAs may function through interacting with mRNAs and miRNAs. lncRNA-mRNA pairs can be identified in trans between the DELs (Differentially Expressed LncRNAs) and DEGs (Differentially Expressed Genes) which indicates that the target genes in trans-regulatory relationships may be more related to herbicide stress response. However, all trans lncRNA-mRNA pairs have positive expressional correlations.

LncRNAs can act as precursors of miRNAs and thus repress target mRNAs. Some lncRNAs can be targeted by miRNAs. Such microRNA–lncRNA regulatory interactions and their role in gene expression patterns have been reviewed by Yoon et al. (2014).

15.4.3 GWAS and Other Next Generation Populations

Genome-wide association study (GWAS) is widely known to be an important approach for genetic dissection on complex traits. GWAS has been successfully conducted in cereal crops like wheat, rice, sorghum, maize, etc. for a number of traits (Bandyopadhyay et al. 2017; Wang et al. 2015; Ma et al. 2021). For untangling genetic determinants for nutritional elements, QTL mapping has been conducted in a number of crops like wheat, rice, barley, sorghum, beans, etc. (Bandyopadhyay et al. 2017). However, in the case of millets, few such studies have been conducted only in pearl millet (Kumar et al. 2018; Anuradha et al. 2017). Recombinant inbred lines

(RILs) are used to identify QTLs and phenotypic variations. In foxtail millet, the genetic determinant of micronutrients is unknown yet. 10K SNP markers are utilized to reveal the genetic architecture of nutritionally important elements. Jaiswal et al. identified genomic regions associated with ten nutritionally important elements in foxtail millet through GWAS which may prove promising to understand the genetic architecture of traits (Jaiswal et al. 2019). Candidate genes thus identified may be considered for further characterization in the context of elemental pyramiding in grains. For complex traits, the background genome is always playing a critical role for a particular phenotype. In such cases, single locus analysis may give a biased result since the background genome is totally ignored. Thus, use of the multilocus GWAS approach (Liu et al. 2016) that takes the background genome into consideration while conducting association tests will be more advantageous. Haplotypes analysis using significant SNPs present within LD range would also be required. Significant pyramiding effect and identification of desirable genotypes may prove useful for the development of biofortified foxtail millet variety through molecular breeding.

The International Crops Research Institute for the Semi-Arid Tropics (ICRISAT) has already developed high-throughput and precise phenotyping platforms along with NGS technologies. Such tools are very efficient to identify the genetic basis of agriculturally important traits and for predicting the breeding value of individuals in a plant breeding population (Varshney et al. 2014). Two most potent next-generation populations to dissect genomics of complex traits are nested-association mapping (NAM) populations (Yu et al. 2008; Buckler et al. 2009) and backcross-derived NAM (BCNAM) populations (Jordan et al. 2012). These new tools are able to find out novel variability in comparatively elite backgrounds. Such new-generation populations are required for linking phenotypic variations with sequence variations at high resolution. However, lack of data management skills and database resources are the major limitations of such cutting-edge techniques. Currently NGS-based genotyping platforms such as genotyping by-sequencing (GbS) could be applied for background selection in traditional introgression breeding and genomic selection (GS). The key advantages for its application are availability of low-cost high-throughput genotyping platforms and access to high-throughput phenotyping facilities.

15.4.4 ChIP Seq, RIP Seq, and CHART

Identification of RNA–DNA and RNA–protein interactions are important to know about the detailed mechanism of action of lncRNAs (Bai et al. 2015). Advanced high-throughput techniques like ChIP-seq (Chromatin Immunoprecipitation Sequencing), RIP-seq (Ribonucleoprotein (RNP) immunoprecipitation followed by high-throughput sequencing), and CHART (Capture Hybridization Analysis of RNA Targets) have significantly increased the progress in this field. Co-immunoprecipitation of lncRNAs involves immunoprecipitation of a protein from cross-linked cell lysate followed by reverse-cross-linking, isolation, and deep sequencing of RNAs, leading to the identification of all lncRNAs that are associated with a specific protein complex (Moran et al. 2012a). Recently Ribonucleoprotein (RNP) immunoprecipitation followed by high-throughput sequencing (RIP-seq) has been applied to discover RNA transcripts that interact with a specific protein or protein complex. A modified RIP-seq method without cross-linking has been identified that a large number of lncRNAs are associated with several chromatin-modifying complexes including PRC2, CoREST, and SMCX (Khalil et al. 2009; Zhao et al. 2010).

CHART (capture hybridization analysis of RNA targets) is a hybridization-based purification strategy that can be used to map the genomic binding sites for endogenous RNAs, which is analogous to chromatin immunoprecipitation (ChIP) for proteins (Simon et al. 2011). CHART is a new technique to purify lncRNAs together with their targets (proteins and DNA fragments) in order to determine the genome-wide localization of a specific lncRNA in chromatin as well as the protein content by Western blot analysis. CHART was successfully applied to lncRNAs of different lengths from human and fruit fly (Simon et al. 2011). At the same time, another technique was developed for chromatin isolation by RNA purification to allow high-throughput discovery of DNA–RNA–protein interactions. In this method, specific lncRNAs bound with protein(s) and DNA sequences are retrieved by tiling oligonucleotides, and followed by high-throughput sequencing (Chu et al. 2011). However, in case of small millet research, application of such novel techniques have not been found till date though these techniques are very much promising to identify the regulatory roles of lncRNAs and associated proteins in stress response in case of small millets.

15.5 Role of lncRNAs in Abiotic Stress Tolerance of Small Millets

To combat an array of environmental challenges, plants metamorphose transcriptomic blueprints as per their need, which sometimes becomes successful, leading to adaptability against the stress response. Therefore, transcriptional profiling may be an important aid to understanding different genetic strategies of plants during pathological problems, secondary metabolism induction as well as during several environmental nuisances.

15.5.1 Drought Stress

Intergovernmental Panel on Climate Change (IPCC) advised about an alarming issue regarding the global temperature increase of about 0.2 °C per decade, which is expected to be 1.8–4.0 °C greater than the present level by 2100 (Zhao et al. 2017). Therefore, the rate of average rainfall has been expected to decrease in subtropics and mid-latitudes, assuming to constrain the crop yield. So, improvisation of drought tolerance is necessary among crops of agricultural interest to overcome this impediment. To develop drought-tolerant model plant system, foxtail millet (*Setaria italica*) is mostly sound C_4 panicoid grass. In this prospect lncRNAs are mostly involved in epigenetic modifications and chromatin remodelling and they are mainly assorted into three categories based on their position over the genome-natural antisense transcript (NAT), intervening lncRNA (lincRNA), and intronic lncRNA (Moran et al. 2012b). Qi et al. (2013) have explored the transcriptome profile of foxtail millet behind their drought tolerance potential. Significant findings of their work have been already illustrated in Sect. 15.4.3. Sun et al. (2020) have studied the transcriptomic basis of heat stress in pearl millet (*Pennisetum glaucum* (L.) *R. Br.*). They have used CNCI (Coding-Non-Coding-Index), PLEK SVM classifier, Pfam-scan tools, and CPC (Coding Potential Calculator) software to predict the responded lncRNAs. According to the data from CNCI, 4266 detected transcripts appear to be lncRNAs, whereas Pfam-scan (default parameters of -E 0.001--domE 0.001) result showed 20,233 transcripts appear to be long non-coding proteins. Jayakodi et al. (2019) have worked with Indian barnyard millet (*Echinochloa frumentacea*). They have functionally annotated 31,789 lncRNAs in barnyard millet transcriptomics. They have performed differential expression analysis and found that 2258 long non-coding transcripts appear to drought response in Indian barnyard millet, probably interfering with their photosynthetic machinery.

15.5.2 Salinity Stress

Salt stress is one of the major abiotic impediments to worldwide agricultural productivity. Identification, analysis, and expression profiling of salinity-responsive novel genes from salt-tolerant wild cultivars may be the stepping-stones for the development of salt resistance among the crops of our agronomic interest. This kind of study has been carried out by a number of research groups with the advent of several omics technologies. For instance, Mishra et al. (2007) have worked with *Pennisetum glaucum*, Mehta et al. (2005) with *Avicennia marina*, Garg et al. (2014) with *Porteresia coarctata*. Transcriptome profiling through RNA sequencing or microarray analysis is mostly sound approach in this prospect to understand the intricacy of spatial and temporal expression of stress-responsive genes under hyper-saline growth environment (Kogenaru et al. 2012). During salt stress plants activate a plethora of genetic networks to maintain their intracellular salt-homeostasis.

15.5.3 Herbicide Tolerance

Recently Wang et al. (2020) have deciphered the lncRNA expression profile of foxtail millet (*Setaria italica*) upon herbicide treatment using refined bioinformatics tools. They have elucidated the involvement of 1760 novel lncRNAs and 787 previously known lncRNAs, distributed among 9 chromosomes. Mostly the trans-acting lncRNAs regulate the expression of herbicide-tolerant genes, as they have confirmed the positive expression correlation between 64 lncRNA-mRNA pairs in trans between 24 DELs and 46 DEGs. They have also applied the GO term and KEGG pathway annotations over these trans-regulatory genes and the result showed lncRNA accumulation in the presence of DNA damage stimulus, thereby activating several DNA-repair mechanisms like nucleotide excision repair, homologous recombination, etc. They have illustrated the interaction between one lncRNAs (MSTRG.10187.1) and one stress-responsive miRNA targets of sit-miR399. As lncRNA MSTRG.10187.1 were differentially expressed during the herbicide treatment, it can be inferred that miR399 and its corresponding lncRNA MSTRG.10187.1 may play a role during foxtail millet response to herbicide treatment (Wang et al. 2020).

15.6 Sample Preparation and Bioinformatic Pipeline for lncRNA Identification

Target plant samples should be collected and processed with precautions. The samples need to be immediately frozen in liquid nitrogen and then stored at −80 °C for total RNA extraction. Ideally, each treatment should be sampled at least three times as three biological replications at each time point for each cultivar. Total RNAs are extracted using the reagent kits following the manufacturer's protocol. The purity, concentration, and integrity of RNAs must be assessed. Then the sequencing library for each sample is prepared and finally, the PCR amplified products are purified and library quality is assessed. After that, sequence clusters are generated, and using suitable platforms (e.g., Illumina Hiseq 4000) paired-end reads (150 bp) are generated. Following bioinformatics pipeline (Fig. 15.3) can be used to obtain potential lncRNAs.

Quality of the paired-end RNA-Seq reads is determined initially using FastQC software with a quality score Q20 (Phred score). Sequenced reads from the libraries

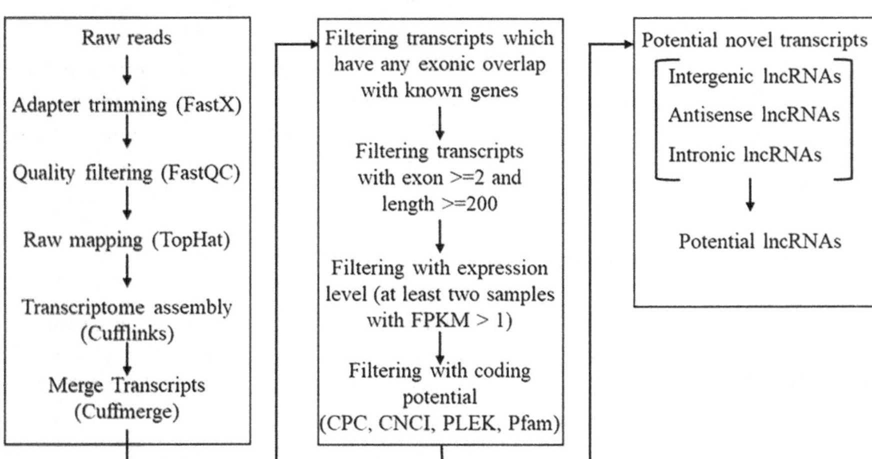

Fig. 15.3 Bioinformatics pipeline for identification of lncRNAs

are mapped to the millet genome using Tophat v2.0.4. Using the SAM files generated by Tophat as input, the Cufflinks software package is employed to assemble the transcripts and calculate their expression levels (FPKM, fragments per kilobase of transcript per million fragments) with the millet genome annotation file. The assembled transcripts produced by Cufflinks are merged together according to millet genome using Cuffmerge software in the Cufflinks package. The class codes in the Cuffmerge output provided information that is useful for identifying transcript types, namely, known transcript, novel isoform, and intergenic transcript. Finally, Cuffdiff is used to identify differentially expressed genes and novel transcripts (Qi et al. 2013).

15.6.1 Downstream Analysis of Differentially Expressed lncRNAs

Differentially expressed lncRNAs can be obtained by using edgeR software with a log2(fold change) ≥ 2 and false discovery rate (FDR) ≤ 0.01 (Wang et al. 2020). Cluster analysis and heatmap can be generated by *R* and bioconductor packages. The cis-target genes of lncRNA can be identified by searching the protein-coding genes 10 Kb upstream and downstream of lncRNA by BEDTools intersect (Quinlan and Hall 2010). On the other hand, Pearson's correlation coefficients (*R*) between lncRNAs and mRNAs can be calculated using *R* tool, and the genes with $|R| \geq 0.95$ are identified as trans-target genes. The lncRNAs can function as precursors of miRNAs. The precursor sequences of known miRNAs in miRBase (http://www.mirbase.org) are aligned with lncRNAs and precursor lncRNAs can be identified using BLAST. A lncRNA harboring a miRNA precursor sequence with 100% query identity and *E*-value less than 0.01 is defined as a precursor of that miRNA. Further, the lncRNA-miRNA–mRNA interaction networks can be visualized with Cytoscape software (Wang et al. 2020). Quantitative real-time PCR (qRT-PCR) is often used to validate the results obtained by bioinformatics prediction.

15.7 Smart Small Millet Production Using lncRNAs

Besides being potentially resistant to various stress conditions, millets, particularly foxtail millet, involve a plethora of bioactive compounds like amino acids, minerals, and many more with immense health benefits. Very small genome size (490 Mb only), genetic parsimony to several other staple bioenergy sources like napier-grass (*Pennisetum purpureum*), switchgrass (*Panicum virgatum*), and pearl millet (*Pennisetum glaucum*) make the ideal to be a climate-resilient model plant system with C_4 photosynthesis. Although lncRNAs are responsible for expression of several traits with agronomic interests, there are very little evidence about the multifactorial importance of lncRNAs in crop productivity. As millets are cultivated mostly in developing countries as a minor food grain, research on millet transcriptome is still in its juvenility. Twelve thousand three hundred seventy-eight novel lncRNAs have been isolated from the young inflorescence of foxtail millet and deciphered their role in the regulation of crop yield. In foxtail millet, the lncRNAs appear to be more AU rich than that of the protein-coding genes, and also their level of expression is much more inferior compared to other coding sequences. They have studied the transcriptomic status of the immature spikelet as this is the determinant stage for grain yield, and grain quality. They have characterized the expression profile of four different lncRNAs—270, 330, 437, and 537 lncRNAs—with the advent of differential expression analysis in high-yield varieties (5695 and 56,229), showing higher expression levels in the former strains as compared to other wild varieties like JG31 and JG32, indicating the role of lncRNAs over grain production, by interfering the target mRNAs at their post-transcriptional state. In addition, understanding the genetic background behind the stress-responsive characters can trigger the introduction of novel genetic elements within the established food crops either through genetic engineering or conventional breeding approaches. Taken together, all these results may provide new insights about the role of these so-called junk transcriptomic game-changers over the molecular basis for different climate-resilient traits for future agricultural sustainability (Zhao et al. 2020).

15.8 Conclusion

RNA-based technologies specifically lncRNA-based agricultural toolkits have a great potential to improve the quality and quantity of small millets though the high-throughput techniques associated with lncRNAs are so far not applied in most of the small millets. To build extreme abiotic stress-resistant millet plants, lncRNAs may play a great role. To combat an ever-changing environment and limited resources, small millets are good candidate plants for future research. We hope that a more detailed analysis of genomics, transcriptomics, and/or ChIP-based omics technologies will definitely help to accelerate the progress of smart small millet production.

References

Anuradha N, Satyavathi CT, Bharadwaj C, Nepolean T, Sankar SM, Singh SP, Meena MC, Singhal T, Srivastava RK (2017) Deciphering genomic regions for high grain iron and zinc content using association mapping in pearl millet. Front Plant Sci 8:412

Bai Y, Dai X, Harrison AP, Chen M (2015) RNA regulatory networks in animals and plants: a long noncoding RNA perspective. Brief Funct Genomics 14(2):91–101

Bandyopadhyay T, Jaiswal V, Prasad M (2017) Nutrition potential of foxtail millet in comparison to other millets and major cereals. In: The foxtail millet genome. Springer, Cham, pp 123–135

Bhatia G, Singh A, Verma D, Sharma S, Singh K (2020) Genome-wide investigation of regulatory roles of lncRNAs in response to heat and drought stress in Brassica juncea (Indian mustard). Environ Exp Bot 171:103922

Buckler ES, Holland JB, Bradbury PJ, Acharya CB, Brown PJ, Browne C, Ersoz E, Flint-Garcia S, Garcia A, Glaubitz JC, Goodman MM (2009) The genetic architecture of maize flowering time. Science 325(5941):714–718

Calixto CP, Tzioutziou NA, James AB, Hornyik C, Guo W, Zhang R, Nimmo HG, Brown JW (2019) Cold-dependent expression and alternative splicing of Arabidopsis long non-coding RNAs. Front Plant Sci 10:235

Chu C, Qu K, Zhong FL, Artandi SE, Chang HY (2011) Genomic maps of long noncoding RNA occupancy reveal principles of RNA-chromatin interactions. Mol Cell 44(4):667–678

Crespi MD, Jurkevitch E, Poiret M, d'Aubenton-Carafa Y, Petrovics G, Kondorosi E, Kondorosi A (1994) enod40, a gene expressed during nodule organogenesis, codes for a non-translatable RNA involved in plant growth. EMBO J 13(21):5099–5112

Deng P, Liu S, Nie X, Weining S, Wu L (2018) Conservation analysis of long non-coding RNAs in plants. Sci China Life Sci 61(2):190–198

Dwivedi SL, Upadhyaya HD, Senthilvel S, Hash CT, Fukunaga K, Diao X, Santra D, Baltensperge D, Prasad M (2012) Millets: genetic and genomic resources. In: Janick J (ed) Plant breeding reviews. Springer, New York. https://doi.org/10.1002/9781118100509.ch5

Fukuda M, Nishida S, Kakei Y, Shimada Y, Fujiwara T (2019) Genome-wide analysis of long intergenic noncoding RNAs responding to low-nutrient conditions in Arabidopsis thaliana: possible involvement of trans-acting siRNA3 in response to low nitrogen. Plant Cell Physiol 60(9):1961–1973

Garg R, Verma M, Agrawal S, Shankar R, Majee M, Jain M (2014) Deep transcriptome sequencing of wild halophyte rice, Porteresia coarctata, provides novel insights into the salinity and submergence tolerance factors. DNA Res 21(1):69–84

Hatakeyama M, Aluri S, Balachadran MT, Sivarajan SR, Patrignani A, Grüter S, Poveda L, Shimizu-Inatsugi R, Baeten J, Francoijs KJ, Nataraja KN (2018) Multiple hybrid de novo genome assembly of finger millet, an orphan allotetraploid crop. DNA Res 25(1):39–47

Jaiswal V, Bandyopadhyay T, Gahlaut V, Gupta S, Dhaka A, Ramchiary N, Prasad M (2019) Genome-wide association study (GWAS) delineates genomic loci for ten nutritional elements in foxtail millet (Setaria italica L.). J Cereal Sci 85:48–55

Jayakodi M, Madheswaran M, Adhimoolam K, Perumal S, Manickam D, Kandasamy T, Yang TJ, Natesan S (2019) Transcriptomes of Indian barnyard millet and barnyard grass reveal putative genes involved in drought adaptation and micronutrient accumulation. Acta Physiol Plant 41(5): 1–11

Jha UC, Nayyar H, Jha R, Khurshid M, Zhou M, Mantri N, Siddique KH (2020) Long non-coding RNAs: emerging players regulating plant abiotic stress response and adaptation. BMC Plant Biol 20(1):1–20

Jordan DR, Hunt CH, Cruickshank AW, Borrell AK, Henzell RG (2012) The relationship between the stay-green trait and grain yield in elite sorghum hybrids grown in a range of environments. Crop Sci 52(3):1153–1161

Khalil AM, Guttman M, Huarte M, Garber M, Raj A, Morales DR, Thomas K, Presser A, Bernstein BE, van Oudenaarden A, Regev A (2009) Many human large intergenic noncoding RNAs associate with chromatin-modifying complexes and affect gene expression. Proc Natl Acad Sci 106(28):11667–11672

Kim DH, Xi Y, Sung S (2017) Modular function of long noncoding RNA, COLDAIR, in the vernalization response. PLoS Genet 13(7):e1006939

Kogenaru S, Yan Q, Guo Y, Wang N (2012) RNA-seq and microarray complement each other in transcriptome profiling. BMC Genomics 13(1):1–16

Kumar S, Hash CT, Nepolean T, Mahendrakar MD, Satyavathi CT, Singh G, Rathore A, Yadav RS, Gupta R, Srivastava RK (2018) Mapping grain iron and zinc content quantitative trait loci in an Iniadi-derived immortal population of pearl millet. Genes 9(5):248

Lata CH (2015) Advances in omics for enhancing abiotic stress tolerance in millets. Proc Indian Natl Sci Acad 81:397–417

Lata C, Gupta S, Prasad M (2013) Foxtail millet: a model crop for genetic and genomic studies in bioenergy grasses. Crit Rev Biotechnol 33(3):328–343

Liao Q, Liu C, Yuan X, Kang S, Miao R, Xiao H et al (2011) Large-scale prediction of long non-coding RNA functions in a coding–non-coding gene co-expression network. Nucleic Acids Res 39(9):3864–3878

Liu J, Jung C, Xu J, Wang H, Deng S, Bernad L et al (2012) Genome-wide analysis uncovers regulation of long intergenic noncoding RNAs in Arabidopsis. Plant Cell 24(11):4333–4345

Liu X, Huang M, Fan B, Buckler ES, Zhang Z (2016) Iterative usage of fixed and random effect models for powerful and efficient genome-wide association studies. PLoS Genet 12(2): e1005767

Ma P, Zhang X, Luo B, Chen Z, He X, Zhang H, Li B, Liu D, Wu L, Gao S, Gao D (2021) Transcriptomic and genome-wide association study reveal long noncoding RNAs responding to nitrogen deficiency in maize. BMC Plant Biol 21(1):1–19

Mehta PA, Sivaprakash K, Parani M, Venkataraman G, Parida AK (2005) Generation and analysis of expressed sequence tags from the salt-tolerant mangrove species Avicennia marina (Forsk) Vierh. Theor Appl Genet 110(3):416–424

Mishra RN, Reddy PS, Nair S, Markandeya G, Reddy AR, Sopory SK, Reddy MK (2007) Isolation and characterization of expressed sequence tags (ESTs) from subtracted cDNA libraries of Pennisetum glaucum seedlings. Plant Mol Biol 64(6):713–732

Moran VA, Niland CN, Khalil AM (2012a) Co-immunoprecipitation of long noncoding RNAs. In: Genomic imprinting. Humana Press, Totowa, pp 219–228

Moran VA, Perera RJ, Khalil AM (2012b) Emerging functional and mechanistic paradigms of mammalian long non-coding RNAs. Nucleic Acids Res 40(14):6391–6400

Ochogavía A, Galla G, Seijo JG, González AM, Bellucci M, Pupilli F, Barcaccia G, Albertini E, Pessino S (2018) Structure, target-specificity and expression of PN_LNC_N13, a long non-coding RNA differentially expressed in apomictic and sexual *Paspalum notatum*. Plant Mol Biol 96(1):53–67

Peng R, Zhang B (2020) Foxtail millet: a new model for C4 plants. Trends Plant Sci 26(3):199–201

Provart NJ, Alonso J, Assmann SM, Bergmann D, Brady SM, Brkljacic J et al (2016) 50 years of Arabidopsis research: highlights and future directions. New Phytol 209(3):921–944

Qi X, Xie S, Liu Y, Yi F, Yu J (2013) Genome-wide annotation of genes and noncoding RNAs of foxtail millet in response to simulated drought stress by deep sequencing. Plant Mol Biol 83(4–5):459–473

Quinlan AR, Hall IM (2010) BEDTools: a flexible suite of utilities for comparing genomic features. Bioinformatics 26(6):841–842

Sage RF (2004) The evolution of C4 photosynthesis. New Phytol 161(2):341–370

Salleh FM, Evans K, Goodall B, Machin H, Mowla SB, Mur LA, Runions J, Theodoulou FL, Foyer CH, Rogers HJ (2012) A novel function for a redox-related LEA protein (SAG21/AtLEA5) in root development and biotic stress responses. Plant Cell Environ 35(2):418–429

Simon MD, Wang CI, Kharchenko PV, West JA, Chapman BA, Alekseyenko AA, Borowsky ML, Kuroda MI, Kingston RE (2011) The genomic binding sites of a noncoding RNA. Proc Natl Acad Sci 108(51):20497–20502

Sun M, Huang D, Zhang A, Khan I, Yan H, Wang X, Zhang X, Zhang J, Huang L (2020) Transcriptome analysis of heat stress and drought stress in pearl millet based on Pacbio full-length transcriptome sequencing. BMC Plant Biol 20(1):1–15

Swiezewski S, Liu F, Magusin A, Dean C (2009) Cold-induced silencing by long antisense transcripts of an Arabidopsis Polycomb target. Nature 462(7274):799–802

Unver T, Tombuloglu H (2020) Barley long non-coding RNAs (lncRNA) responsive to excess boron. Genomics 112(2):1947–1955

Urquiaga MC, Thiebaut F, Hemerly AS, Ferreira PC (2021) From trash to luxury: the potential role of plant LncRNA in DNA methylation during abiotic stress. Front Plant Sci 11:2130

Varshney RK, Terauchi R, McCouch SR (2014) Harvesting the promising fruits of genomics: applying genome sequencing technologies to crop breeding. PLoS Biol 12(6):e1001883

Wang H, Niu QW, Wu HW, Liu J, Ye J, Yu N, Chua NH (2015) Analysis of non-coding transcriptome in rice and maize uncovers roles of conserved lnc RNA s associated with agriculture traits. Plant J 84(2):404–416
Wang P, Dai L, Ai J, Wang Y, Ren F (2019) Identification and functional prediction of cold-related long non-coding RNA (lncRNA) in grapevine. Sci Rep 9(1):1–15
Wang T, Song H, Wei Y, Li P, Hu N, Liu J et al (2020) High throughput deep sequencing elucidates the important role of lncRNAs in foxtail millet response to herbicides. Genomics 112(6): 4463–4473
Waseem M, Liu Y, Xia R (2021) Long non-coding RNAs, the dark matter: an emerging regulatory component in plants. Int J Mol Sci 22(1):86
Werner BT, Gaffar FY, Schuemann J, Biedenkopf D, Koch AM (2020) RNA-spray-mediated silencing of Fusarium graminearum AGO and DCL genes improve barley disease resistance. Front Plant Sci 11:476
Wu L, Liu S, Qi H, Cai H, Xu M (2020) Research progress on plant long non-coding RNA. Plan Theory 9(4):408
Wunderlich M, Groß-Hardt R, Schöffl F (2014) Heat shock factor HSFB2a involved in gametophyte development of Arabidopsis thaliana and its expression is controlled by a heat-inducible long non-coding antisense RNA. Plant Mol Biol 85(6):541–550
Yang Z, Zhang H, Li X, Shen H, Gao J, Hou S, Zhang B, Mayes S, Bennett M, Ma J, Wu C (2020) A mini foxtail millet with an Arabidopsis-like life cycle as a C4 model system. Nat Plants 6(9): 1167–1178
Yoon JH, Abdelmohsen K, Gorospe M (2014) Functional interactions among microRNAs and long noncoding RNAs. In: Seminars in cell & developmental biology, vol 34. Academic Press, Cambridge, pp 9–14
Young RS, Marques AC, Tibbit C, Haerty W, Bassett AR, Liu JL, Ponting CP (2012) Identification and properties of 1,119 candidate lincRNA loci in the Drosophila melanogaster genome. Genome Biol Evol 4(4):427–442
Yu J, Holland JB, McMullen MD, Buckler ES (2008) Genetic design and statistical power of nested association mapping in maize. Genetics 178(1):539–551
Zhang D, Zhang B (2020) SpRY: engineered CRISPR/Cas9 harnesses new genome-editing power. Trends Genet 36(8):546–548
Zhang G, Liu X, Quan Z, Cheng S, Xu X, Pan S, Xie M, Zeng P, Yue Z, Wang W, Tao Y (2012) Genome sequence of foxtail millet (Setaria italica) provides insights into grass evolution and biofuel potential. Nat Biotechnol 30(6):549–554
Zhang C, Tang G, Peng X, Sun F, Liu S, Xi Y (2018) Long non-coding RNAs of switchgrass (*Panicum virgatum* L.) in multiple dehydration stresses. BMC Plant Biol 18(1):1–15
Zhao J, Ohsumi TK, Kung JT, Ogawa Y, Grau DJ, Sarma K, Song JJ, Kingston RE, Borowsky M, Lee JT (2010) Genome-wide identification of polycomb-associated RNAs by RIP-seq. Mol Cell 40(6):939–953
Zhao C, Liu B, Piao S, Wang X, Lobell DB, Huang Y, Huang M, Yao Y, Bassu S, Ciais P, Durand JL (2017) Temperature increase reduces global yields of major crops in four independent estimates. Proc Natl Acad Sci 114(35):9326–9331
Zhao Z, Liu D, Cui Y, Li S, Liang D, Sun D, Wang J, Liu Z (2020) Genome-wide identification and characterization of long non-coding RNAs related to grain yield in foxtail millet [*Setaria italica* (L.) P. Beauv.]. BMC Genomics 21(1):1–13

Omics of Climate Change on Nutritional Quality of Small Millets

16

P. Rajendrakumar

Abstract

Small millets are small seeded nutri-rich crops that are widely grown in diverse environments under low fertility and minimum inputs. They are considered as inherently stress tolerant and hence it will be interesting to study the impact of climate change on their nutritional quality. The genetic resources of these crops possess wide variations for agronomical, nutritional and stress tolerant traits, which need to be exploited in the crop improvement programme. The generation of genomic resources in small millets is attaining rapid pace with the availability of whole genome sequences of finger millet, foxtail millet, proso millet and barnyard grass along with the repertoire of DNA markers for further exploitation in gene/QTL mapping approaches. Currently, the transcriptome studies are picking-up pace towards the identification of differentially expressed genes/transcripts associated with abiotic stress tolerance and nutritional traits, which adds to the knowledge accumulated through genomic studies. However, proteomic and metabolomic studies are lacking in small millets, which have to be given priority to have a comprehensive understanding of the mechanisms and interaction of various factors in the manifestation of climate resilience. Therefore, it will be worthwhile to focus on an integrated approach involving genetic resources, genomics, phenomics, transcriptomics, proteomics, metabolomics, and crop modelling to understand the impacts of climate change on nutritional quality of small millets. This book chapters focusses on the above-mentioned aspects and brings about a comprehensive review of work done across the globe aimed towards the development of nutri-rich and climate resilient cultivars in small millets.

P. Rajendrakumar (✉)
ICAR-Indian Institute of Millets Research (ICAR-IIMR), Hyderabad, India
e-mail: rajendra@millets.res.in

R. N. Pudake et al. (eds.), *Omics of Climate Resilient Small Millets*,
https://doi.org/10.1007/978-981-19-3907-5_16

Keywords

Small millets · Abiotic stress · Nutritional quality · Genomics · Climate change · Phenomics · Transcriptomics · Proteomics · Metabolomics

16.1 Introduction

Small millets are considered as gluten-free nutri-rich crops for healthy living since they are good sources of slow digestible carbohydrate, proteins, dietary fibre, minerals and vitamins. They are widely cultivated in semi-arid tropics of Asia and Africa under poor and marginal soils with limited rainfall exhibiting inherent resilience to climate change. Presently, agriculture is experiencing a serious challenge from climate change, which is a global phenomenon impacting the mankind today due to continuous rise in temperatures, triggering a cascade of extreme weather events like heat waves, drought, and flooding (Feulner 2017). Recent reports and simulation studies globally have predicted an adverse effect on the yields of all major crops due to changing climate (Varshney et al. 2011; Zhao et al. 2017). Apart from this, such studies also predicted reduction in the nutritional quality in many crops due to increased levels of CO_2 in the earth's atmosphere, while in some crops the impact will lead to toxic alterations in the chemical composition of tissues during their development (Dwivedi et al. 2013). Such a challenging situation warrants the development of nutri-rich climate-smart crops by enriching with superior alleles for climate resilience and nutritional quality through the application of genomics and molecular breeding approaches. Regulation of physiological, cellular and molecular activities of crops plays a major role in their adaptation to survive adverse stress conditions (Chinnusamy et al. 2004; Ahuja et al. 2010). It is evident that the strong regulation of environmental acclimatization plays a crucial role in the survival of the crop (Lopez et al. 2008; Muthamilarasan and Prasad 2015; Kissoudis et al. 2014). Hence, deciphering the mechanisms of stress tolerance and crop adaptation with the advancements in omics technologies and functional characterization of genes becomes important.

Advances in genomics approaches and sequencing technologies offer a platform to formulate effective and prospective strategies for the genetic improvement of target traits associated with climate change. These approaches help in understanding the molecular mechanism of abiotic stress tolerance through integration of gene, protein or metabolite profile and their target phenotype. Genomics offers excellent opportunities to enhance crop yield and quality as well as climate resilience through molecular breeding approaches. However, such studies are still very limited in other small millets. Understanding the molecular mechanism behind the regulation of genes associated with nutritional and climate resilience traits demands the use of multi-omics strategy involving genomics, transcriptomics, phenomics, proteomics and metabolomics. Integration of multi-omics, molecular breeding and gene editing strategies will bring about the genetic improvement in small millets leading to the development of cultivars with superior nutrient content, climate resilience and adaptation to local agro-ecological conditions. This chapter discusses the impacts of climate change on nutritional quality, exploitation of genetic resources, multi-omics strategies for climate resilience and integration of multi-omics strategies on the impact of climate change in nutritional quality of small millets.

16.2 Impacts of Climate Change on the Nutritional Quality

In the current scenario of climate change, there is a debate among the researchers and nutritionists globally on whether the nutritional quality of the food crops be affected due to changing climatic conditions. There is adequate evidence to show that crop yield and quality will be affected by climate change (Shindell et al. 2012). Even though the impact of climate change on nutritional quality of vegetable/legume and fruit crops remains unclear due to limited reports (Scheelbeek et al. 2018), there is overwhelming evidence to show the clear impact of climate change in the case of cereals (Beleggia et al. 2018; Chaturvedi et al. 2017; Giri et al. 2016; Manderscheid et al. 1995; Scheelbeek et al. 2018; Smith and Myers 2018). In support of this, growing C_3 crops such as wheat, rice and soybeans under higher CO_2 concentration has shown to lower the concentrations of zinc, iron and protein and increase starch and sugar contents (Vermeulen et al. 2012).

A meta-analysis in C_3 (wheat, rice, field peas, soybean) and C_4 (maize, sorghum) crops by Myers et al. (2014) revealed that zinc, iron and protein concentrations were significantly lower in C_3 crops when they were grown at elevated CO_2 levels while no significant reductions were observed, except for iron concentration in maize (5.8%). It can be hypothesized that elevated CO_2 due to the effects of climate change leads to micronutrient deficiencies in crop plants (Müller et al. 2014; Myers et al. 2014, 2015; Uttam et al. 2017). Although there are no substantial studies on the impact of climate change on nutritional quality in small millets, these crops may be resilient to climatic changes since they are cultivated by the farmers over generations under marginal soils and harsh environments, thereby developing an inherent capacity to tide over unpredictable environment conditions. However, there is an urgent need for undertaking comparative studies for understanding the effect of climate change on the nutritional quality of small millets to confirm that nutritional quality may be resilient to climate shocks.

Meeting the food and nutritional demand of the ever increasing population globally under the changing climate, it is important to mainstream millets into the existing food systems rather than relying heavily on fine cereals. In addition, growing of millets could lead to reduction of greenhouse gas emissions in the atmosphere, thereby reducing the levels of environmental contaminants due to agricultural production. Small millets possess important attributes such as short stature, short life cycle, small leaf area, thickened cell walls, C_4 photosynthesis, minimum vulnerability to stresses, ability to grow in water deficit conditions due to dense root system and low input requirement, which make them the perfect "climate-smart crop" fitting into the sustainable agriculture in future involving the components such as crop diversification, climate resilience and environment-friendly food production for food and nutritional security of the global population.

16.3 Genetic Resources: A Reservoir of Valuable Alleles for Climate Resilience and Nutritional Quality

Crop germplasm accessions are the storehouses of valuable alleles for important target traits. To achieve genetic improvement for climate resilience and nutritional quality in small millets, new donors possessing superior alleles for crop adaptation under changing climate need to be identified from the genetic resources that comprise of germplasm accessions, landraces and wild relatives. However, the utility of germplasm collections in crop improvement over the years is very limited. Constitution of a smaller set of germplasm accessions known as "core set" was proposed by Frankel and Brown (1984) to promote efficient and increased use of germplasm in the crop improvement programmes. The core set comprises of only 10% of accessions of the total germplasm collection representing the genetic diversity of the total collection. In the case where the core set is large and hard to manage, a mini-core set was proposed with reduced size (~1% of the total collection), representing the total diversity of the crop species (Upadhyaya et al. 2010). Core, mini-core and reference collections have been developed for all the millets (Table 16.1). Evaluation of these genetic resources will help in the identification of novel sources of variation for morpho-agronomic, nutritional and climate resilience traits. Core and mini-core sets can be used as a diversity panel for the assessment of population structure, genetic diversity, allelic richness and for association mapping purposes (Upadhyaya 2015).

Mapping populations are also considered as valuable genetic resources and bi-parental populations have been extensively used in sorghum and pearl millet over the last two decades for the quantitative trait loci (QTL) identification and mapping for important target traits. Due to the small seeded nature of the small millets, development of bi-parental population was considered as challenging due to the difficulty in performing emasculation and pollination. However, with the

Table 16.1 Core and mini-core collections of small millets

Type of collection	No. of accessions	Crop	Reference
Core	622	Finger millet	Upadhyaya et al. (2006)
	155, 107, 152, 128	Foxtail millet	Upadhyaya et al. (2008), Lata et al. (2011a, b), Liu et al. (2011)
	106	Proso millet	Upadhyaya et al. (2011b)
	89	Barnyard millet	Upadhyaya et al. (2014)
	75	Kodo millet	Upadhyaya et al. (2014)
	56	Little millet	Upadhyaya et al. (2014)
Mini-core	80	Finger millet	Upadhyaya et al. (2010)
	35	Foxtail millet	Upadhyaya et al. (2011a)

standardization and refinements of emasculation and pollination techniques in recent years, it becomes important to focus on the development of bi-parental mapping population using contrasting genotypes for the traits of interest so that they could be employed for the identification and mapping of QTL. Advancements in next generation sequencing (NGS) technology over the last decade leading to rapid development of large-scale DNA markers followed by a paradigm shift in the mapping approaches resulted in the prominent application of association mapping (AM) for the detection of stable and effective marker-trait associations for the traits of interest with a greater resolution since it accounts for the recombination events accumulated during the evolutionary time. Moreover, specialized breeding populations such as nested association mapping (NAM) populations (Yu et al. 2008) and multi-parent advanced generation inter-cross (MAGIC) populations (Cavanagh et al. 2008) are used nowadays to improve the power and efficiency of association studies since they take into account both the evolutionary and recent recombination events. Such specialized populations have been successfully used in sorghum (Bouchet et al. 2017; Ongom and Ejeta 2018; Boatwright et al. 2021) for the identification and mapping of QTL for the traits of utmost importance. Attempts should be made towards the development of such specialized mapping populations in small millets, which will go a long way in the effective identification of marker-trait associations for important target traits.

16.4 Genomics for Climate Resilience and Nutritional Quality

Precise and high-throughput genotyping and phenotyping of individuals of the mapping/breeding population either for trait mapping or cultivar development is vital for the identification of effective genomic regions or QTL for the target traits and also for the development of superior lines with climate resilience. Crop improvement through molecular breeding has achieved remarkable success over the past decade due to the availability and application of diverse DNA marker technologies (Gupta et al. 2014; Anuradha et al. 2017; Uttam et al. 2017). With the advances in next generation sequencing (NGS) technologies, the whole genome sequences of small millets such as finger millet (Hittalmani et al. 2017; Hatakeyama et al. 2018), foxtail millet (Zhang et al. 2012; Bennetzen et al. 2012), proso millet (Zou et al. 2019) and barnyard grass (Guo et al. 2017) are available, which led to the identification of genome-wide DNA markers such as SSRs, In-Dels, ILPs and SNPs. Advances in the NGS technologies have revolutionized the crop genomics research by the development of sequence-based and chip-based genotyping platforms leading to rapid and high-throughput genotyping with reduced time and cost of genotyping. The details of DNA markers developed in small millets are given in Table 16.2. These DNA markers are powerful tools for precision breeding through marker-assisted selection since they are useful for the construction of linkage maps, identification of QTL and their mapping leading to DNA markers tightly linked to the QTL for the traits of interest.

Table 16.2 DNA-based markers developed in small millets

Type of DNA markers	Crop	No. of markers	Reference
Genomic SSRs	Finger millet	10,501	Dida et al. (2007), Gimode et al. (2016), Musia (2013)
	Foxtail millet	395	Jia et al. (2009), Heng et al. (2011), Gupta et al. (2012, 2013)
	Proso millet	525	Cho et al. (2010), Liu et al. (2016)
Genic SSRs	Finger millet	862	Arya et al. (2009), Naga et al. (2012), Jagadeesh Selvam et al. (2015), Obidiegwu et al. (2014), Nirgude et al. (2014), Nnaemeka (2009), Kalyana Babu et al. (2014a, b)
	Foxtail millet	17,105	Jia et al. (2007), Kumari et al. (2013), Xu et al. (2013)
	Proso millet	43,139	Yue et al. (2016), Jiang et al. (2018)
	Little millet	4443	Desai et al. (2021)
	Barnyard millet	10,881	Murukarthick et al. (2019)
Genome-wide SSRs	Finger millet	1,14,083	Hittalmani et al. (2017)
	Foxtail millet	29,075	Zhang et al. (2014)
Genome-wide SNPs	Finger millet	46,285	Gimode et al. (2016), Kumar et al. (2016)
	Foxtail millet	42,61,599	Bai et al. (2013), Jia et al. (2013)
	Proso millet	69,981	Rajput et al. (2016)
	Little millet	2245	Johnson et al. (2018)
	Kodo millet	3641	Johnson et al. (2018)
	Barnyard millet	10,816	Wallace et al. (2015)

SSR simple sequence repeats, *SNPs* single nucleotide polymorphisms

Drought and salt stress, together with high and low temperature stress, are major challenges in agriculture, which prevents the plant from realizing its complete genetic potential. Salt stress affects crop production, particularly in irrigated lands (Epstein et al. 1980). But, the drought stress is more persistent and causes economical damage (Boyer 1982; Wang et al. 2001). Abiotic stresses pose a major threat to agricultural production due to changes in climate and degradation of environment caused by human interference. Plants respond to abiotic stress with a number of molecular, cellular and physiological changes and adapt to avoid yield reduction. A better understanding of the plant response to abiotic stress will help in formulating strategies for improving abiotic stress tolerance through conventional as well as molecular breeding approaches. Drought tolerance is a complex and quantitatively inherited trait that is manifested by the involvement of morphological traits as well as physiological mechanisms. In finger millet, the gene CBL interacting protein kinase31 (EcCIPK31-like) responsible for drought tolerance was identified and characterized (Nagarjuna et al. 2016). Under drought stress, a gene associated with TATA box binding protein-associated factors (*EcTAF6*) was identified in the finger millet cultivar, GPU 28.

Salinity affects crop growth, causing nutritional constraints by lowering the uptake of phosphorus, potassium, nitrate and calcium along with cytotoxicity and osmotic stress. A clear understanding of the mechanism involved in plants' response to salinity stress is essential for the genetic improvement of salinity tolerance in crop plants. Salinity tolerance is also a complex trait and the QTL associated with this trait plays a significant role in understanding the plant response to stress (Gorantla et al. 2005). Identifying genes associated with plant response to salinity stress and their functional characterization is important for the development of cultivars possessing salinity stress tolerance through genetic engineering or molecular breeding (Nongpiur et al. 2016). In finger millet, the *NAC* gene (*EcNAC1*) was highly upregulated under salinity stress, suggesting its involvement in tolerance to salinity and other abiotic stresses (Joshi et al. 2017). Two transcription factors belonging to *bZIP* family (*EcbZIP60*) and basic helix-loop-helix (*bHLH*) family (*EcbHLH57*) that are responsive to drought, osmotic and salinity stresses were identified in the finger millet cultivar, GPU 28 (Babitha et al. 2015a, b). Under high salinity stress, the transcripts of hydrogen peroxide scavenging enzymes, particularly phospholipid hydroperoxide glutathione peroxidase (PHGPX), ascorbate peroxidase and catalase 1 were upregulated in a salinity tolerant foxtail millet genotype and a 25 kDa PHGPX protein was induced significantly at high salinity level (250 mmol/L), suggesting an important role of this protein in salinity tolerance (Yang et al. 2020). In a study involving the foxtail millet cultivars, Prasad (salinity tolerant) and Lepakshi (salinity susceptible), the accumulation of proline was positively correlated with the increased activity of glutamine synthetase and pyrroline-5-carboxylate (Preston and Kellogg 2006). A significant increase in the expression of *SiDREB2* gene under salinity stress was noticed and an allele-specific marker was developed by targeting the synonymous SNP (A/T) in this gene at the 558th position that was identified in a core set of 45 germplasm accessions in foxtail millet (Lata et al. 2011a). Overexpression of a novel atypical late embryogenesis abundant (*LEA*) gene, *SiLEA14*, through transgenic *Arabidopsis* and foxtail millet revealed a higher tolerance to salt and osmotic stress as compared to their wild type (Wang et al. 2014). In the future scenario of unpredictable changes in the climatic conditions encompassing precipitation, drought, temperature fluctuations, etc., it is essential to develop cultivars with multiple resistance to abiotic stresses with the introgression of effective and desirable QTL through molecular breeding, thereby imparting the climate resilience in these millets.

16.5 Phenomics for Climate Resilience and Nutritional Quality

Phenotyping is a critical component for the genetic dissection of certain target traits, especially those associated with yield, nutritional quality and climate resilience. Crop phenomics is described as the multi-disciplinary study involving high-throughput phenotyping (HTP), accurate data acquisition and analysis of multi-dimensional crop phenotypes performed on an organism-wide level (Yang et al. 2020). Generally, yield and its component traits (primary trait) are targeted for direct selection while the traits such as root architecture, stay green, leaf rolling, etc. (secondary traits) are crucial for drought tolerance and also indirectly contributing to final yield (Araus et al. 2011; Cabrera et al. 2020; Khadka et al. 2020). Traditionally, phenotyping of secondary traits has been done through easily observable and scorable phenological traits such as plant height, days to flowering, leaf number, etc. (Zenda et al. 2021). Phenotyping of several target traits such as crop establishment, phenology and flowering, abiotic and biotic stress tolerance, lodging and yield components are performed manually in majority of the breeding programmes (Masuka et al. 2017a, b). Since these manual measurements lack repeatability and are prone to error, the data is collected on more replications or multiple environments to improve trait heritability. This situation warrants robust, accurate and high-throughput phenotyping platforms for measuring the whole range of component traits that contribute to quantitative phenotypic variation of the target traits.

The pace of advances in phenotyping methods could not cope-up with the rapid advancements in high-throughput genotyping techniques (Araus and Cairns 2014; Ghanem et al. 2015). Without accurate phenotyping data, the massive genotyping data generated cannot be used for an effective dissection of important target traits. The main goal of high-throughput phenotyping platforms is the ability to measure "breeder-preferred" traits through an automated and non-destructive manner, which allows for observation of the change in phenotypic traits for the same plant throughout its growth and development. In addition to the common traits such as grain yield, biomass and tolerance to pests and diseases, the breeders are largely interested in phenotypic (leaf, stem, root, inflorescence, etc.) and physiological traits (early vigour, photosynthetic rate, transpiration rate and water use efficiency, resistance to abiotic stress). Measurement of these key traits manually is time consuming and involves more manpower. The crop phenotyping has been revolutionized with the development of HTP platforms due to the recent advances in robotics, informatics, and data extraction and analysis, in combination with systems integration (Yang et al. 2020), which will help in enhancing the crop breeding efficiency.

The HTP platforms will enable to record trait data via sophisticated camera-assisted non-invasive imaging (red-green-blue imaging, magnetic resonance imaging, hyperspectral imaging, laser imaging, 3D imaging) remote-sensing, near-infrared (NIR) spectroscopy, digital image analysis, robotics, high-performance computational facilities, phenomics tools and databases. High-throughput phenotyping platforms have been developed by combining controlled-environment crop growth chambers with VIS imaging systems possessing blue, green and red sensors to measure crop growth, development, and plant responses to environmental conditions (Tisné et al. 2013; Honsdorf et al. 2014; Chen et al. 2014; Yang et al. 2014). Abiotic stresses such as drought and chilling stress can be effectively detected by chlorophyll fluorescence imaging (Baker 2008; Jansen et al. 2009; Chen et al. 2014). Autonomous ground-based phenotyping platforms and aerial phenotyping platforms equipped with multiple sensors can capture images of the experimental area in real time several times in a day, or over the entire crop season, resulting in large amounts of phenotypic data for analysis and storage (Wang et al. 2014).

Several HTP platforms have been developed in recent years based on the plant architecture, phenology and production systems. Till date, there are no HTP platforms available for accurate phenotyping of small millets. However, HTP platforms suitable for phenotyping of sorghum crop have been developed in recent years. Characterization of plant architecture parameters in sorghum was performed using a phenotyping system possessing Microsoft Kinect cameras and the ability for 3D reconstruction of single potted plants to identify QTLs related to shoot height and leaf area that colocalized with earlier reported ones (McCormick et al. 2016). Phenobot 1.0, an auto-steered and self-propelled field-based high-throughput phenotyping platform suitable for tall dense canopy crops, such as sorghum was developed recently, equipped with stereo RCB cameras and tested algorithms for the 3D reconstruction and estimation of plant height and stem diameter (Fernandez et al. 2017). A low cost, high-throughput root phenotyping platform comprising of 500 soil filled root chambers (50 × 45 × 0.3 cm in size) equipped with an imaging box containing 2 remote-controlled digital cameras for in situ imaging was developed for sorghum, which is suited to dissect the genetic control of nodal root angle (Joshi et al. 2017). Phenofield® is a platform developed with large plot capacity for characterization and management of abiotic stresses such as drought, thereby helping in tolerance evaluation and other important physiological analysis (Beauchêne et al. 2019). Some of the popular HTP platforms are LemnaTec (http://www.lemnatec.com/plant-phenotyping/), Digital Phenotyping-KeyGene (http://www.keygene.com/products-tech/digital-phenotyping/), The international plant Phenotyping Network (https://www.plant-phenotyping.org/), LEPSE-Montpellier Plant Phenotyping Platform (https://www6.montpellier.inra.fr/lepse/Presentation-generale/Montpellier-Plant-Phenotyping-Platform-M3P), PhenoFab, Wageningen (http://bioweb.supagro.inra.fr/phenopsis/InfoBDD.php), the Biotron, Canada (KeyGene + LemnaTec) (https://www.keygene.com/technology/2-the-digital-phenotype/) and The Australian Plant Phenomics Facility (https://www.uwo.ca/sci/research/biotron/). Use of such platforms for trait phenotyping on bi-parental and/or association mapping populations, evaluated in harsh conditions, in simulated climate change conditions will help in the identification of genes and superior lines with climate resilience.

16.6 Transcriptomics for Climate Resilience and Nutritional Quality

Transcriptomics has been employed popularly by the researchers for studying the alterations in the gene expression in relation to stress situations in the climate change scenario. The techniques such as microarrays or the more recent RNA-Seq have been used regularly to analyse the shifts in mRNA abundance, which are triggered in response to the changes in vital environmental factors such as drought, salinity and high or low temperature stress. In foxtail millet, a set of 29 differentially expressed proteins having a crucial role in stress-related metabolism, signal transduction and photosynthesis were upregulated and downregulated in the seedlings, which seems to contribute towards salinity stress tolerance (Veeranagamallaiah et al. 2008). A study of transcriptome changes under PEG-induced drought stress in a drought tolerant foxtail millet genotype resulted in the identification of 327 differentially expressed transcripts, of which, 86 were upregulated transcripts mostly involved in metabolism, regulation of transcription, signalling and protein degradation, indicating their likely role in dehydration adaption. Through qRT-PCR, the study also detected a 5–11-fold induction of the Dehydration Responsive Element Binding type protein (DREB2) (Lata et al. 2010). In tolerant and foxtail millet cultivars, about 115 out of 159 differentially expressed transcripts exhibiting >twofold change in expression were upregulated while 44 were downregulated in response to salinity stress. Moreover, transcription factors and signalling genes were expressed at a higher level in the tolerant genotype as compared to the susceptible one, thereby contributing to signal perception mechanisms under salinity stress (Puranik et al. 2011). A total of 3066 differentially expressed genes (DEGs) were identified using a drought tolerant cultivar of finger millet, of which, 1404 were upregulated and 1662 were downregulated, which contributed to the regulatory networks involved in photosynthesis, signal transduction, osmo-regulation, redox regulation, hormonal signalling and improved drought tolerance (Shi et al. 2018). In a recent study, about 8887 and 12,249 DEGs were detected in the salinity tolerant and susceptible varieties of foxtail millet, respectively. Among these DEGs, in the tolerant variety, about 4830 and 4057 genes were upregulated and downregulated, respectively; however, in the susceptible variety, about 6339 and 5910 genes were upregulated and downregulated, respectively (Zenda et al. 2021). Transcriptome analysis in little millet by Das et al. (2020) revealed that the WRKY transcription factors viz., WRKY1 and WRKY2 exhibited 13-fold and 7-fold higher expressions in leaves under drought stress, respectively while WRKY3 exhibited 50-fold upregulation exclusively in root tissue under both salinity and drought. Similarly, the ABC transporters viz., ABC1, ABC2 and ABC3 showed sixfold higher expression under drought stress in leaf tissues and >threefold change in root tissues as compared to salinity stress while ABC4 and ABC5 showed 14-fold upregulation in leaves under drought stress. Significant expression change of several DEGs involved in regulation of abscisic acid triggered signalling pathway, ethylene triggered signalling, protein ubiquitination, redox and ion homeostasis, cell wall organization, and developmental processes with a potential role in mediating the dehydration stress response has been identified in kodo millet by Suresh et al. (2022). Interestingly, DEGs involved in the regulation of phytohormone signalling like ABA pathway genes (TRINITY_DN744_c0_g1, TRI-NITY_DN9015_c1_g1, TRINITY_DN4365_c1_g1, TRINITY_DN138_c0_g2), ethylene-responsive genes (TRINITY_DN1078_c1_g1, TRINI-TY_DN5570_c0_g1, TRINITY_DN1799_c0_g1) were expressed differentially under dehydration stress. More importantly, an ABA pathway gene such as probable lysophospholipase BODYGUARD 1 (BDG1) (TRINITY_DN15179_c0_g2) showed upregulation at an early stage while a gene involved in the oxidative degradation of ABA [Abscisic acid 8′-hydroxylase 3 (TRINITY_DN744_c0_g1)] showed upregulation, indicating their potential role in the regulation of ABA level in response to dehydration stress. Very limited studies are being reported towards the understanding of the network of genes and their role in conferring tolerance to abiotic stress though small millets are considered to have inbuilt tolerance to abiotic stresses due to their cultivation over a long period of time under low input and harsh environments. Hence, with the advent of long read sequencing supported by a robust validation through qRT-PCR in combination with automated simulation platforms representing the climate change parameters, there is an immense need to conduct many such studies to bring about a clear understanding of the genes, gene networks and the pathways involved in the manifestation of abiotic stress tolerance in small millets.

16.7 Integrated Multi-omics Approaches

Plants will respond with the abiotic stress conditions by various self-protection mechanisms brought about by genetic, physiological and biochemical changes. Therefore, an effective improvement in climate resilience needs a clear understanding of regulation of gene expression and mechanism of interaction among proteins in response to stress. Rapid advancements of high-throughput sequencing technology during the past decade has considerably increased the understanding about the genotype to phenotype relationship. However, similar understanding is not there for the gene to protein relationships. Even though transcriptome analysis and high-throughput liquid chromatography-tandem mass spectrometry (LC-MS/MS) have led to a comprehensive knowledge on the role of DEGs in various biological processes and protein level changes happening in cell metabolism and other processes, there is an immense need for an integrated approach for a better understanding of the transcriptome and proteome level changes to identify key regulatory genes and proteins operating during stress. In line with this, a recent study in finger millet by Li et al. (2021) resulted in the identification of 188 matching DEGs with differentially expressed proteins (DEPs) referred as DEGs-DEPs that were associated with drought stress. Co-expression network analysis helped in the identification of two hub DEGs-DEPs viz., *BAMY2* (barley K-amylase 2) (CL10421.Contig11_All_Gene.108706) and *ISA1* (isoamylase-type starch debranching enzyme) (Unigene12111_All_Gene.147034) upstream of drought-regulated genes with possible regulatory roles. The *BAMY2* gene impedes dextrinase in germinating barley seeds (Kristensen et al. 1999) while the *ISA1* gene in addition involves in the biosynthesis and crystallization of starch (Lin et al. 2013). A significant upregulation of these two genes under drought stress helps in the restriction of starch and dextrin production and delaying the vegetative growth thereby increasing stress tolerance. Similarly, it is also important to integrate proteomics and metabolomics approaches to decipher the mechanisms behind the ability of the plant to tolerate or overcome biotic stress through a better understanding of the changes in proteins and metabolites occurring at the cellular level as well as their interactions. The integrated

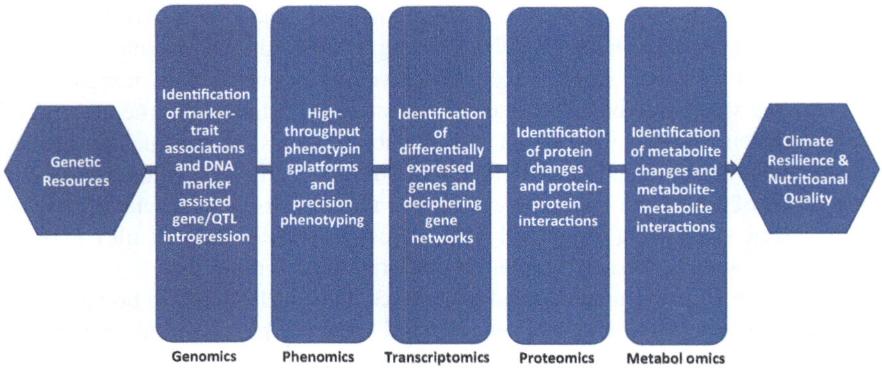

Fig. 16.1 Integrated multi-omics approach to study the effect of climate change on the nutritional quality of small millets

multi-omics approach for understanding the effect of climate change on the nutritional quality of small millets is represented in Fig. 16.1. Vast genetic resources in small millets can be effectively exploited for trait improvement by adopting an integrated approach involving high-resolution phenotyping, high-throughput multi-omics approaches and crop simulation modelling. Such futuristic research framework will help in understanding the impacts of climate change on nutritional quality.

16.8 Future Prospects

The millets are considered as the lifeline of poor farmers from semi-arid tropics and tribal regions due to low input agriculture and their resilience to extreme climatic conditions, especially drought. Even though a lot of progress has been made on the genomics and transcriptomics aspects in millets such as sorghum, pearl millet, finger millet and foxtail millet, there is an immense need for such research in other small millets. In addition, development of HTP platforms for rapid and accurate phenotyping, changes in protein folding in response to stress and the production and regulation of metabolites in enabling the crop survival during stress need more emphasis to understand the complete network of genes/proteins/metabolites involved in abiotic stress response. Transcriptome analysis through RNA-seq has revolutionized the gene expression studies helping in the identification of DEGs in various cellular and biological processes. High-throughput LC-MS/MS on various aspects of protein such as expression levels of protein, post-translational modifications and protein–protein interactions have helped in a thorough understanding of cell metabolism and other biological processes at the protein level. In short, an integrated approach involving genomics, phenomics, transcriptomics, proteomics and metabolomics approaches will be effective in understanding the molecular mechanisms underlying climate resilience and nutritional quality traits in addition to yield. Effective dissection of the mechanisms can be achieved through the crop simulation modelling studies that link the genotype to the responsive phenotype under climate change situations (Van Eeuwijk et al. 2019). Hence, a combination of multi-omics approaches and crop simulation modelling will help in a clear understanding of the mechanisms operating behind climate resilience for designing efficient strategies towards the development of nutri-rich climate smart small millets in the future.

References

Ahuja KG, Nath P, Swamy KRM (2010) Foods and nutrition, 1st edn. Studium Press, New Delhi

Anuradha N, Satyavathi CT, Bharadwaj C, Nepolean T, Sankar SM, Singh SP, Meena MC, Singhal T, Srivastava RK (2017) Deciphering genomic regions for high grain iron and zinc content using association mapping in pearl millet. Front Plant Sci 8:412

Araus JL, Cairns JE (2014) Field high-throughput phenotyping: the new crop breeding frontier. Trends Plant Sci 19:52–61

Araus JL, Sa'nchez C, Edmeades GO (2011) Phenotyping maize foradaptation to drought. In: Monneveux P, Ribaut JM (eds) Drought phenotyping in crops: from theory to practice. Generation Challenge Programme, pp 259–282

Arya L, Verma M, Gupta VK, Karihaloo JL (2009) Development of EST-SSRs in finger millet (*Eleusine coracana* ssp. *coracana*) and their transferability to pearl millet (*Pennisetum glaucum*). J Plant Biochem Biotechnol 18(1):97–100

Babitha KC, Ramu SV, Nataraja KN, Sheshshayee MS, Udayakumar M (2015a) *EcbZIP 60*, a basic leucine zipper transcription factor from *Eleusine coracana* L. improves abiotic stress tolerance in tobacco by activating unfolded protein response pathway. Mol Breed 35:181–197

Babitha KC, Vemanna RS, Nataraja KN, Udayakumar M (2015b) Overexpression of *EcbHLH57* transcription factor from *Eleusine coracana* L. in tobacco confers tolerance to salt, oxidative and drought stress. PLoS One 10:e0137098

Bai H, Cao Y, Quan J, Dong L, Li Z, Zhu Y, Zhu L, Dong Z, Li D (2013) Identifying the genome-wide sequence variations and developing new molecular markers for genetics research by re-sequencing a landrace cultivar of foxtail millet. PLoS One 8(9):e73514

Baker NR (2008) Chlorophyll fluorescence: a probe of photosynthesis in vivo. Annu Rev Plant Biol 59:89–113

Beauchêne K, Leroy F, Fournier A, Huet C, Bonnefoy M, Lorgeou J, de Solan B, Piquemal B, Thomas S, Cohan JP (2019) Management and characterization of abiotic stress via PhénoField®, a high-throughput feld phenotyping platform. Front Plant Sci 10:904

Beleggia R, Fragasso M, Miglietta F, Cattivelli L, Menga V, Nigro F, Pecchioni N, Fares C (2018) Mineral composition of durum wheat grain and pasta under increasing atmospheric CO2 concentrations. Food Chem 242:53–61

Bennetzen JL, Schmutz J, Wang H, Percifield R, Hawkins J, Pontaroli AC, Estep M, Feng L, Vaughn JN, Grimwood J, Jenkins J, Barry K, Lindquist E, Hellsten U, Deshpande S, Wang X, Wu X, Mitros T, Triplett J, Yang X, Ye CY, Mauro-Herrera M, Wang L, Li P, Li P, Sharma R, Ronald PC, Panaud O, Elizabeth A, Kellogg BTP, Doust AN, Tuskan GA, Rokhsar D, Devos KM (2012) Reference genome sequence of the model plant *Setaria*. Nat Biotechnol 30:555–561

Boatwright JL, Brenton ZW, Boyles RE, Sapkota S, Myers MT, Jordan KE, Dale SM, Shakoor N, Cooper EA, Morris GP, Kresovich S (2021) Genetic characterization of a *Sorghum bicolor* multiparent mapping population emphasizing carbon-partitioning dynamics. G3 (Bethesda) 11(4):jkab060

Bouchet S, Olatoye MO, Marla SR, Perumal R, Tesso T, Yu J, Tuinstra M, Morris GP (2017) Increased power to dissect adaptive traits in global sorghum diversity using a nested association mapping population. Genetics 6(2):573–585

Boyer JS (1982) Plant productivity and environment. Science 218:443–448
Cabrera CJ, Boter M, Onate-Sánchez L, Pernas M (2020) Root growth adaptation to climate change in crops. Front Plant Sci 11:544
Cavanagh C, Morell M, Mackay IJ, Powell W (2008) From mutations to MAGIC; resources for gene discovery, validation and delivery in crop plants. Curr Opin Plant Biol 11:215–221
Chaturvedi AK, Bahuguna RN, Pal M, Shah D, Maurya S, Jagadish KSV (2017) Elevated CO2 and heat stress interactions affect grain yield, quality and mineral nutrient composition in rice under feld conditions. Field Crop Res 206:149–157
Chen D, Neumann K, Friedel S, Kilian B, Chen M, Altmann T, Klukas C (2014) Dissecting the phenotypic components of crop plant growth and drought responses based on high-throughput image analysis. Plant Cell 26(12):4636–4655
Chinnusamy V, Schumaker K, Zhu J-K (2004) Molecular genetic perspectives on cross-talk and specificity in abiotic stress signalling in plants. J Exp Bot 55:225–236
Cho YI, Chung JW, Lee GA, Ma KH, Dixit A, Gwag JG, Park YJ (2010) Development and characterization of twenty-five new polymorphic microsatellite markers in proso millet (*Panicum miliaceum* L.). Genes Genomics 32(3):267–273
Das RR, Pradhan S, Parida A (2020) De-novo transcriptome analysis unveils diferentially expressed genes regulating drought and salt stress response in *Panicum sumatrense*. Sci Rep 10:21251
Desai H, Hamid R, Ghorbanzadeh Z, Bhut N, Padhiyar SM, Kheni J, Tomar RS (2021) Genic microsatellite marker characterization and development in little millet (*Panicum sumatrense*) using transcriptome sequencing. Sci Rep 11:20620
Dida MM, Srinivasachary, Ramakrishnan S, Bennetzen JL, Gale MD, Devos KM (2007) The genetic map of finger millet, *Eleusine coracana*. Theor Appl Genet 114:321–332
Dwivedi SL, Sahrawat KL, Upadhyaya HD, Ortiz R (2013) Food, nutrition and agrobiodiversity under global climate change. Adv Agron 120:1–128
Epstein E, Norlyn JD, Rush DW, Kingsbury RW, Kelly DB (1980) Saline culture of crops: a genetic approach. Science 210:399–404
Fernandez MGS, Bao Y, Tang L, Schnable PS (2017) A high-throughput, field-based phenotyping technology for tall biomass crops. Plant Physiol 174:2008–2022
Feulner G (2017) Global challenges: climate change. Global Chall 1:5–6
Frankel OH, Brown AHD (1984) Plant genetic resources today: a critical appraisal. In: Holden JHW, Williams JT (eds) Crop genetic resources: conservation & evaluation. George Allen & Unwin, London, pp 249–257
Ghanem ME, Marrou H, Sinclair TR (2015) Physiological phenotyping of plants for crop improvement. Trends Plant Sci 20:139–144
Gimode D, Odeny DA, de Villiers EP, Wanyonyi S, Dida MM, Mneney EE, Muchugi A, Machuka J, de Villiers SM (2016) Identification of SNP and SSR markers in finger millet using next generation sequencing technologies. PLoS One 11(7):e0159437
Giri A, Armstrong B, Rajashekar CB (2016) Elevated carbon dioxide level suppresses nutritional quality of lettuce and spinach. Am J Plant Sci 07:246–258
Gorantla M, Babu PR, Reddy VBL, Feltus FA, Paterson AH, Reddy AR (2005) Functional genomics of drought stress response in rice: transcript mapping of annotated unigenes of an indica rice (Oryza sativa L. cv. Nagina 22). Curr Sci 89:496–514
Guo L, Qiu J, Ye CY, Jin G, Lingfeng M, Zhang H et al (2017) *Echinochloa crus-galli* genome analysis provides insight into its adaptation and invasiveness as a weed. Nat Commun 8:1031
Gupta S, Kumari K, Sahu PP, Vidapu S, Prasad M (2012) Sequence based novel genomic microsatellite markers for robust genotyping purposes in foxtail millet [*Setaria italica* (L.) P. Beauv.]. Plant Cell Rep 31:323–337
Gupta S, Kumari K, Muthamilarasan M, Subramanian A, Prasad M (2013) Development and utilization of novel SSRs in foxtail millet [*Setaria italica* (L.) P. Beauv.]. Plant Breed 132: 367–374

Gupta S, Kumari K, Muthamilarasan M, Parida SK, Prasad M (2014) Population structure and association mapping of yield contributing agronomic traits in foxtail millet. Plant Cell Rep 33: 881–893

Hatakeyama M, Aluri S, Balachadran MT, Sivarajan SR, Patrignani A, Grüter S, Poveda L, Shimizu-Inatsugi R, Baeten J, Francoijs KJ, Nataraja KN, Reddy YAN, Phadnis S, Ravikumar RL, Schlapbach R, Sreeman SM, Shimizu KK (2018) Multiple hybrid de novo genome assembly of finger millet, an orphan allotetraploid crop. DNA Res 25(1):39–47

Heng L, Chih C, Song C, Chang K (2011) Development of simple sequence repeats (SSR) markers in *Setaria italica* (Poaceae) and cross-amplification in related species. Int J Mol Sci 12:7835–7845

Hittalmani S, Mahesh HB, Shirke MD, Biradar H, Uday G, Aruna YR, Lohithaswa HC, Mohanrao A (2017) Genome and transcriptome sequence of finger millet (*Eleusine coracana* (L.) Gaertn.) provides insights into drought tolerance and nutraceutical properties. BMC Genomics 18:465

Honsdorf N, March TJ, Berger B, Tester M, Pillen K (2014) High-throughput phenotyping to detect drought tolerance QTL in wild barley introgression lines. PLoS One 9:e97047

Jagadeesh Selvam N, Muthukumar M, Rahman H, Senthil N, Raveendran M (2015) Development and validation of SSR markers in finger millet [*Eleusine coracana* (L.)]. Indian J Trop Agric 33(3):2055–2066

Jansen M, Gilmer F, Biskup B, Nagel KA, Rascher U, Fischbach A, Briem S, Dreissen G, Tittmann S, Braun S, Jaeger ID, Metzlaff M, Schurr U, Scharr H, Walter A (2009) Simultaneous phenotyping of leaf growth and chlorophyll fluorescence via GROWSCREEN FLUORO allows detection of stress tolerance in *Arabidopsis thaliana* and other rosette plants. Funct Plant Biol 36:902–914

Jia XP, Shi YS, Song YC, Wang GY, Wang TY, Li Y (2007) Development of EST-SSR in foxtail millet (*Setaria italica*). Genet Resour Crop Evol 54:233–236

Jia XP, Tan XJ, Li YX, Wang TY, Li Y (2009) A study on the genetic diversity of foxtail millet cultivars by SSR markers. Acta Agriculturae Universitatis Jiangxiensis 31(4):633–638

Jia G, Shi S, Wang C, Niu Z, Chai Y, Zhi H, Diao X (2013) Molecular diversity and population structure of Chinese green foxtail [*Setaria viridis* (L.) Beauv.] revealed by microsatellite analysis. J Exp Bot 64:3645–3656

Jiang Y, Li H, Zhang J, Xiang J, Cheng R, Liu G (2018) Whole genomic EST-SSR development based on high-throughput transcript sequencing in proso millet (Panicum miliaceum). Int J Agric Biol 20:617–620

Johnson M, Deshpande S, Vetriventhan M, Upadhyaya HD, Wallace JG (2018) Genome-wide population structure analyses of three minor millets: kodo millet, little millet, and proso millet. bioRxiv. https://doi.org/10.1101/499087

Joshi DC, Singh V, Hunt C, Mace E, van Oosterom E, Sulman R, Jordan D, Hammer G (2017) Development of a phenotyping platform for high throughput screening of nodal root angle in sorghum. Plant Methods 13:1–12

Kalyana Babu B, Pandey D, Agrawal PK, Sood S, Kumar A (2014a) *In-silico* mining, type and frequency analysis of genic microsatellites of finger millet (*Eleusine coracana* (L.) Gaertn.): a comparative genomic analysis of NBS-LRR regions of finger millet with rice. Mol Biol Rep 41: 3081–3090

Kalyana Babu B, Pandey D, Agrawal PK, Sood S, Chandrashekara C, Bhatt JC, Kumar A (2014b) Comparative genomics and association mapping approaches for blast resistant genes in finger millet using SSRs. PLoS One 9(6):e99182

Khadka K, Earl HJ, Raizada MN, Navabi A (2020) A physio-morphological trait-based approach for breeding drought tolerant wheat. Front Plant Sci 11:715

Kissoudis C, van de Wiel C, Visser RGF, van der Linden G (2014) Enhancing crop resilience to combined abiotic and biotic stress through the dissection of physiological and molecular crosstalk. Front Plant Sci 5:Article 207

Kristensen M, Lok F, Véronique P, Svendsen I, Leah R, Svensson B (1999) Isolation and characterization of the gene encoding the starch debranching enzyme limit dextrinase from germinating barley. Biochim Biophys Acta 1431(2):538–546

Kumar A, Sharma D, Tiwari A, Jaiswal JP, Singh NK, Sood S (2016) Genotyping-by-sequencing analysis for determining population structure of finger millet germplasm of diverse origins. Plant Genome 9:1–15

Kumari K, Muthamilarasan M, Misra G, Gupta S, Subramanian A, Parida SK, Chattopadhyay D, Prasad M (2013) Development of eSSR-markers in *Setaria italica* and their applicability in studying genetic diversity, cross-transferability and comparative mapping in millet and non-millet species. PLoS One 8:e67742

Lata C, Sahu PP, Prasad M (2010) Comparative transcriptome analysis of differentially expressed genes in foxtail millet (*Setaria italica* L.) during dehydration stress. Biochem Biophys Res Commun 393(4):720–727

Lata C, Bhutty S, Bahadur RP, Majee M, Prasad M (2011a) Association of an SNP in a novel DREB2-like gene *SiDREB2* with stress tolerance in foxtail millet [*Setaria italica* (L.)]. J Exp Bot 62:3387–3401

Lata C, Jha S, Dixit V, Sreenivasulu N, Prasad M (2011b) Differential antioxidative responses to dehydration-induced oxidative stress in core set of foxtail millet cultivars [*Setaria italica* (L.)]. Protoplasma 248:817–828

Li J, Wang Y, Wang L, Zhu J, Deng J, Tang R et al (2021) Integration of transcriptomic and proteomic analyses for finger millet [*Eleusine coracana* (L.) Gaertn.] in response to drought stress. PLoS One 16(2):e0247181

Lin Q, Facon M, Putaux JL, Dinges JR, Wattebled F, D'Hulst C, Hennen-Bierwagen TA, Myers AM (2013) Function of isoamylase-type starch debranching enzymes ISA1 and ISA2 in the Zea mays leaf. New Phytol 200:1009–1021

Liu ZL, Bai GH, Zhang DD, Zhu CS, Xia XY, Cheng RH, Shi ZG (2011) Genetic diversity and population structure of elite foxtail millet [*Setaria italica* (L.) Beauv.] germplasm in China. Crop Sci 51(4):1655–1663

Liu M, Xu Y, Zhang S, Wang Y, Lu P (2016) Genetic diversity and population structure of broomcorn millet (*Panicum miliaceum* L.) cultivars and landraces in China based on microsatellite markers. Int J Mol Sci 17:370–388

López MA, Bannenberg G, Castresana C (2008) Controlling hormone signaling is a plant and pathogen challenge for growth and survival. Curr Opin Plant Biol 11:420–427

Manderscheid R, Bender J, Jäger H-J, Weigel HJ (1995) Efects of season long CO2 enrichment on cereals. II. Nutrient concentrations and grain quality. Agric Ecosyst Environ 54:175–185

Masuka B, Atlin GN, Olsen M, Magorokosho C, Labuschagne M, Crossa J, Bänziger M, Pixley KV, Vivek BS, von Biljon A, Macrobert J, Alvarado G, Prasanna BM, Makumbi D, Tarekegne A, Das B, Zaman-Allah M, Cairns JE (2017a) Gains in maize genetic improvement in eastern and southern Africa: I. CIMMYT hybrid breeding pipeline. Crop Sci 57:168–179

Masuka B, Magorokosho C, Olsen M, Atlin GN, Bänziger M, Pixley KV, Vivek BS, Labuschagne M, Matemba-Mutasa R, Burgueño J, Macrobert J, Prasanna BM, Das B, Makumbi D, Tarekegne A, Crossa J, Zaman-Allah M, van Biljon A, Cairns JE (2017b) Gains in maize genetic improvement in eastern and southern Africa: II. CIMMYT open-pollinated variety breeding pipeline. Crop Sci 57:180–191

McCormick RF, Truong SK, Mullet JE (2016) 3D sorghum reconstructions from depth images identify QTL regulating shoot architecture. Plant Physiol 172:823–834

Müller C, Elliott J, Levermann A (2014) Food security: fertilizing hidden hunger. Nat Clim Change 4:540–541

Murukarthick J, Manimekalai M, Karthikeyan A, Sampath P, Dhasarathan M, Thangaraj K, Yang TJ, Senthil N (2019) Transcriptomes of Indian barnyard millet and barnyardgrass reveal putative genes involved in drought adaptation and micronutrient accumulation. Acta Physiol Plantarum 41:66

Musia GD (2013) Identification of microsatellite markers for finger millet (*Eleusine coracana*) by analysis of Roche 454 GS-FLX Titanium sequence data. MSc thesis, School of Pure and Applied Sciences, Kenyatta University, Nairobi

Muthamilarasan M, Prasad M (2015) Advances in Setaria genomics for genetic improvement of cereals and bioenergy grasses. Theor Appl Genet 128(1):1–14

Myers SS et al (2014) Increasing CO_2 threatens human nutrition. Nature 510:139–142

Myers SS, Wessells KR, Kloog I, Zanobetti A, Schwartz J (2015) Effect of increased concentrations of atmospheric carbon dioxide on the global threat of zinc deficiency: a modelling study. Lancet Global Health 3:e639–e645

Naga BLRI, Lakshmi Narasu M, Sivaramakrishnan S (2012) Identification and characterization of EST-SSRs in finger millet (*Eleusine coracana* (L.) Gaertn.). J Crop Sci Biotechnol 15(1):9–16

Nagarjuna KN, Parvathi MS, Sajeevan RS, Pruthvi V, Mamrutha HM, Nataraja KN (2016) Full-length cloning and characterization of abiotic stress responsive *CIPK31*-like gene from finger millet, a drought-tolerant crop. Curr Sci 111:890–894

Nirgude M, Babu BK, Shambhavi Y, Singh UM, Upadhyaya HD, Kumar A (2014) Development and molecular characterization of genic molecular markers for grain protein and calcium content in finger millet (*Eleusine coracana* (L.) Gaertn.). Mol Biol Rep 41:1189–1200

Nnaemeka OO (2009) Development of microsatellite markers for genetic characterization of a core subset of foxtail millet (Setaria italica) and finger millet (Eleusine coracana) germplasm, MS thesis. University of Hohenheim, Stuttgart

Nongpiur RC, Singla-Pareek SL, Pareek A (2016) Genomics approaches for improving salinity stress tolerance in crop plants. Curr Genomics 17(4):343–357

Obidiegwu ON, Parzies H, Obidiegwu JE (2014) Development and genotyping potentials of EST-SSRs in finger millet (*E. coracana* (L.) Gaertn.). Int J Genet Genomics 2:42–46

Ongom PO, Ejeta G (2018) Mating design and genetic structure of a multi-parent advanced generation intercross (MAGIC) population of sorghum [*Sorghum bicolor* (L.) Moench]. G3 (Bethesda) 8(1):331–341

Preston JC, Kellogg EA (2006) Reconstructing the evolutionary history of paralogous APETALA1/ FRUITFULL-like genes in grasses (Poaceae). Genetics 174:421–437

Puranik S, Bahadur RP, Srivastava PS, Prasad M (2011) Molecular cloning and characterization of a membrane associated NAC family gene, *SiNAC* from foxtail millet [*Setaria italica* (L.) P. Beauv.]. Mol Biotechnol 49(2):138–150

Rajput SG, Santra DK, Schnable J (2016) Mapping QTLs for morpho-agronomic traits in proso millet (*Panicum miliaceum* L.). Mol Breed 36:1–18

Scheelbeek PFD, Bird FA, Tuomisto HL, Green R, Harris FB, Joy EJM, Chalabi Z, Allen E, Haines A, Dangour AD (2018) Effect of environmental changes on vegetable and legume yields and nutritional quality. Proc Natl Acad Sci 115:6804–6809

Shi WP, Cheng JY, Wen JX, Wang JX, Shi GY et al (2018) Transcriptomic studies reveal a key metabolic pathway contributing to a well-maintained photosynthetic system under drought stress in foxtail millet (*Setaria italica* L.). Peer J 6:e4752

Shindell D, Kuylenstierna JCI, Vignati E, van Dingenen R, Amann M, Klimont Z, Anenberg SC, Muller N, Janssens-Maenhout G, Raes F, Schwartz J, Faluvegi G, Pozzoli L, Kupiainen K, HöglundIsaksson L, Emberson L, Streets D, Ramanathan V, Hicks K, Oanh NTK, Milly G, Williams M, Demkine V, Fowler D (2012) Simultaneously mitigating near-term climate change and improving human health and food security. Science 335:183–189

Smith MR, Myers SS (2018) Impact of anthropogenic CO2 emissions on global human nutrition. Nat Clim Chang 8:834–839

Suresh BV, Choudhary P, Aggarwal PR, Rana S, Singh RK, Ravikesavan R, Prasad M, Muthamilarasan M (2022) De novo transcriptome analysis identifies key genes involved in dehydration stress response in kodo millet (*Paspalum scrobiculatum* L.). Genomics 114:110347

Tisné S, Serrand Y, Bach L, Gilbault E, Ben Ameur R, Balasse H, Voisin R, Bouchez D, Durand-Tardif M, Guerche P, Chareyron G, Da Rugna J, Camilleri C, Loudet O (2013) Phenoscope: an automated large-scale phenotyping platform offering high spatial homogeneity. Plant J 74:534–544

Upadhyaya HD (2015) Establishing core collections for enhanced use of germplasm in crop improvement. Ekin J Crop Breed Genet 1(1):1–12

Upadhyaya HD, Gowda CLL, Pundir RPS, Reddy VG, Singh S (2006) Development of core subset of finger millet germplasm using geographical origin and data on 14 quantitative traits. Genet Resour Crop Evol 53:679–685

Upadhyaya HD, Pundir RPS, Gowda CLL, Reddy VG, Singh S (2008) Establishing a core collection of foxtail millet to enhance the utilization of the germplasm of an underutilized crop. Plant Genet Resour 7:177–184

Upadhyaya HD, Sarma NDRK, Ravishankar CR, Albrecht T, Narsimhudu Y, Singh SK, Varshney SK, Reddy VG, Singh S, Dwivedi SL, Wanyera N, Oduori COA, Mgonja MA, Kisandu DB, Parzies HK, Gowda CLL (2010) Developing a mini core collection in finger millet using multilocation data. Crop Sci 50:1924–1931

Upadhyaya HD, Ravishankar CR, Narasimhudu Y, Sarma NDRK, Singh SK, Varshney SK, Reddy VG, Singh S, Parzies SK, Dwivedi SL, Nadaf HL, Sahrawat KL, Gowda CLL (2011a) Identification of trait-specific germplasm and developing a mini core collection for efficient use of foxtail millet genetic resources in crop improvement. Field Crops Res 124:459–467

Upadhyaya HD, Sharma S, Gowda CLL, Reddy VG, Singh S (2011b) Developing proso millet (*Panicum miliaceum* L.) core collection using geographic and morpho-agronomic data. Crop Pasture Sci 62:383–389

Upadhyaya HD, Dwivedi SL, Singh SK, Singh S, Vetriventhan M, Sharma S (2014) Forming core collections in barnyard, kodo, and little millets using morpho-agronomic descriptors. Crop Sci 54:2673–2682

Uttam GA, Praveen M, Rao YV, Tonapi VA, Madhusudhana R (2017) Molecular mapping and candidate gene analysis of a new epicuticular wax locus in sorghum (Sorghum bicolor L. Moench). Theor Appl Genet 130:2109–2125

Van Eeuwijk FA, Bustos-Korts D, Millet EJ et al (2019) Modelling strategies for assessing and increasing the effectiveness of new phenotyping techniques in plant breeding. Plant Sci 282:23–39

Varshney RK, Bansal KC, Aggarwal PK, Datta SK, Craufurd PQ (2011) Agricultural biotechnology for crop improvement in a variable climate: hope or hype? Trends Plant Sci 16:363–371

Veeranagamallaiah G, Jyothsnakumari G, Thippeswamy M, Reddy PCO, Surabhi GK et al (2008) Proteomic analysis of salt stress responses in foxtail millet (*Setaria italica* L. cv. Prasad) seedlings. Plant Sci 175(5):631–641

Vermeulen SJ, Campbell B, Ingram JS (2012) Climate change and food systems. Annu Rev Environ Resour 37:195–222. SSRN. https://doi.org/10.1146/annurev-environ-020411-1306084

Wallace JG, Upadhyaya HD, Vetriventhan M, Buckler ES, Hash CT, Ramu P (2015) The genetic makeup of a global barnyard millet germplasm collection. Plant Genome 8(01):1–7

Wang WX, Vinocur B, Shoseyov O, Altman A (2001) Biotechnology of plant osmotic stress tolerance: physiological and molecular considerations. Acta Horticult 560:285–292

Wang M, Li P, Li C, Pan Y, Jiang X, Zhu D, Zhao Q, Yu J (2014) SiLEA14, a novel atypical LEA protein, confers abiotic stress resistance in foxtail millet. BMC Plant Biol 14:290

Xu J, Li Y, Ma X, Ding J, Wang K, Wang S, Tian Y, Zhang H, Zhu XG (2013) Whole transcriptome analysis using next-generation sequencing of model species *Setaria viridis* to support C_4 photosynthesis research. Plant Mol Biol 83:77–87

Yang W, Guo Z, Huang C, Duan L, Chen G, Jiang N, Fang W, Feng H, Xie W, Lian X, Wang G, Luo Q, Zhang Q, Liu Q, Xiong L (2014) Combining high-throughput phenotyping and genome-wide association studies to reveal natural genetic variation in rice. Nat Commun 5:5087

Yang W, Feng H, Zhang X, Zhang J, Doonan JH, Batchelor WD, Xiong L, Yan J (2020) Crop phenomics and high-throughput phenotyping: past decades, current challenges, and future perspectives. Mol Plant 13:187–214

Yu J, Holland JB, McMullen MD, Buckler ES (2008) Genetic design and statistical power of nested association mapping in maize. Genetics 178:539–551

Yue H, Wang L, Liu H, Yue W, Du X, Song W, Nie X (2016) De novo assembly and characterization of the transcriptome of broomcorn millet (*Panicum miliaceum* L.) for gene discovery andmarker development. Front Plant Sci 7:1083

Zenda T, Liu S, Dong A, Duan H (2021) Advances in cereal crop genomics for resilience under climate change. Life 11:502

Zhang G, Liu X, Quan Z, Cheng S, Xu X, Pan S, Xie M, Zeng P, Yue Z, Wang W, Tao Y, Bian C, Han C, Xia Q, Peng X, Cao R, Yang X, Zhan D, Hu J, Zhang Y, Li H, Li H, Li N, Wang J, Wang C, Wang R, Guo T, Cai Y, Liu C, Xiang H, Shi Q, Huang P, Chen Q, Li Y, Wang J, Zhao Z, Wang J (2012) Genome sequence of foxtail millet (*Setaria italica*) provides insights into grass evolution and biofuel potential. Nat Biotechnol 30:549–554

Zhang S, Tang CJ, Zhao Q, Li J, Yang L, Qie L, Fan X, Li L, Zhang N, Zhao M, Liu X, Chai Y, Zhang X, Wang H, Li Y, Li W, Zhi H, Jia G, Diao X (2014) Development of highly polymorphic simple sequence repeat markers using genome-wide microsatellite variant analysis in foxtail millet [*Setaria italica* (L.) P. Beauv.]. BMC Genomics 15:78

Zhao C, Liu B, Piao S, Wang X, Lobell DB, Huang Y, Huang M, Yao Y, Bassu S, Ciais P, Durand JL, Elliott J, Ewert F, Janssens IA, Li T, Lin E, Liu Q, Martre P, Müller C, Peng S, Peñuelas J, Ruane AC, Wallach D, Wang T, Wu D, Liu Z, Zhu Y, Zhu Z, Asseng S (2017) Temperature increase reduces global yields of major crops in four independent estimates. Proc Natl Acad Sci U S A 114:9326–9329

Zou C, Li L, Miki D, Li D, Tang Q, Xiao L et al (2019) The genome of broomcorn millet. Nat Commun 10:436

Exploring Genome-Wide Analysis of Heat Shock Proteins (HSPs) in Small Millets as Potential Candidates for Development of Multistress Tolerant Crop Plants

17

Sharmistha Barthakur and Nandita Bharadwaj

Abstract

With climate change a reality today, towards maintaining a sustainable and enhanced production we require crop plants well suited to stressful environments. Small millets, hitherto relegated away from the mainstream agriculture are slowly and steadily making a comeback in recent times. This diverse group of crop plants are a veritable powerhouse of genomic resource for mining stress tolerant genes and alleles with their natural inherent resilience. One such group of molecules are the multifaceted molecular chaperones called heat shock proteins (HSPs). From their initial discovery in Drosophila in 1962, HSPs have come a long way as a multi-functional group of proteins with demonstrated critical roles in plant life cycle, hormonal biology as well as regulatory role in plant abiotic and biotic defence. In this chapter we have described the structural and functional attributes of HSP in model plants and small millets, thereby highlighting their potential roles. The research can be utilized towards improvement of small millets as well as other crop plants for multiple stress tolerance.

Keywords

Heat shock proteins · Multistress tolerance · Small millets · Molecular chaperones

S. Barthakur (✉)
ICAR-National Institute for Plant Biotechnology, New Delhi, India

N. Bharadwaj
Department of Environmental Science, Tezpur University, Tezpur, Assam, India

R. N. Pudake et al. (eds.), *Omics of Climate Resilient Small Millets*,
https://doi.org/10.1007/978-981-19-3907-5_17

17.1 Introduction

All organisms including sessile plants encounter physical stress from their environment which presents an existential threat. Plant physiology is perturbed not only by various inimical climatic variables such as irradiation, temperature, and drought but also other plausible causes like herbivores, pathogens, soil salinity and biotic factors. All these variables exert the plant with multiple stresses (Levitt 1980; Nguyen et al. 2016). Additionally, fluctuations in toxins, presence of heavy metals, temperature extremes and the oxidants released consequential to unpropitious human exploits exert furthermore stress on plants (Vierling 1991). Basic stresses such as water stress (Farooq et al. 2017a; Bharadwaj et al. 2018), salt stress (Farooq et al. 2017a), high temperature (Farooq et al. 2017b), chemical pollutants (Bora et al. 2021), etc. have heretofore been behemothic bottlenecks impinging the plants leading to cellular injury (Wang et al. 2003). To counter this, organisms respond with a coordinated manner at the tissue and cellular level with various effector molecules incorporating multiple adaptations at the morphological, biochemical, physiological and molecular level (Bharadwaj et al. 2019).

Various stresses alter the levels of abundant soluble or structural proteins and maintaining the functional conformations and preventing the aggregation of non-native proteins are of paramount importance for adaptation (Hasan and Barthakur 2014). A wide range of proteins called as molecular chaperones and many of which were originally identified as heat shock proteins play important roles to regulate this catastrophe (Lindquist and Craig 1988). The general role of HSPs is to act as molecular chaperones regulating the folding and accumulation of proteins, maintaining their functional structure as well as localization, transport and degradation in all plants and animal species. This chaperone activity is responsible for minimizing the irreversible aggregation of proteins and maintaining cellular homeostasis under both optimal and adverse developmental conditions (Reddy et al. 2010).

The year 2022 marks the 60 years of the pioneering discovery of the heat shock response by the Italian scientist Ferruccio Ritossa. While studying nucleic acid synthesis in puffs of Drosophila salivary glands at the Genetics Institute in Pavia and at the International Institute of Genetics and Biophysics in Naples, Italy he published the first observation that cells could mount very strong transcriptional activity when exposed to elevated temperatures, which was termed as the heat shock response (HSR) (Ritossa 1962; De Maio et al. 2012). This discovery led to the identification of heat shock proteins 12 years later in the laboratory of Herschel K. Mitchell (Tissiéres et al. 1974). The initial response to this phenomenon was not encouraging from the scientific community and later Susan Lindquist, one of the leading investigators in the field also faced questions on her choice of work in this area. Ritossa's observations reported in 1962 were later described as "the first known environmental stress acting directly on gene activity". Later, heat shock responses were reported in other organisms and came to be recognized as a universal phenomenon. Since then, rapid progressive research on HSR and HSPs has shown their impact in numerous areas of current biology and medicine and has created new avenues for more exciting discoveries which is continuing as an ongoing process till date.

Heat shock proteins (HSPs) are ubiquitously found proteins in all living organisms from bacteria to human beings (Parsell and Lindquist 1993; Vierling 1991; Gupta et al. 2010). Initially the induction and synthesis of HSPs were believed to be generated only in response to heat shock; however, they were later known to be produced under exposure to cold, UV light, wound healing, tissue remodelling or biotic stresses (Boston et al. 1996; Lindquist and Craig 1988; Vierling 1991). This established the fact that exposure to an array of stress leads to transcriptional regulation and gene expression culminating to production of heat shock proteins in cells (Feige et al. 1996; De Maio 1999). Initial reports of HSPs from several organisms include *Saccharomyces cerevisiae* (McAlister and Finkelstein 1980) *Escherichia coli* (Yamamori and Yura 1982), and in plants the first report came from soybean (Lin et al. 1984). A comparison of the response in different organisms has shown that HSPs are conserved highly across organisms (Bharti and Nover 2004). The evolutionary conservation of the heat shock response shows that the production of HSPs is a fundamental and essential process in all organisms (Kotak et al. 2007). Thus even though the name heat shock protein is a misnomer from its functional context today, the name and nomenclature continue since its initial discovery.

17.2 Heat Shock Response (HSR) Transcriptional Network in Plants and HSPs

Classically, the induction of molecular chaperones following a sudden increase in temperature is defined as the heat shock response (HSR). Today HSR is implicated in almost all the cellular stress cues encountered by an organism. Defects in protein folding, ribosome biogenesis, aggregation of misfolded proteins and organellar targeting of nascent proteins lead to HSR induction which may lead to protein degradation by the ubiquitin proteasome system and autophagy. Upon heat shock or any other stress as evident in recent times, the above aberrant processes may serve as ligands that activate the HSR leading to the transcriptional induction of the full arsenal of cellular chaperones to restore protein homeostasis (Pincus 2020). The general HSR pathway is described in the literature as initial heat perception by the plasma membrane through calcium channels such as cyclic nucleotide gated channels (CNGCs) leading to activation of transcription factor called heat shock factors (HSFs). CNGCs are plasma membrane thermosensors. HSFs bind to heat shock elements in HSP as well as other stress inducible gene promoters, mediate dissociation of bound histones, leading to gene transcription. The HSR is very critical to cellular homeostasis as it provides cells with an enhanced ability to endure proteotoxic insults and plays a crucial role in determining subsequent cell death or survival (Lang et al. 2021).

The 2021 Nobel Prize in Physiology and Medicine was awarded for the discovery of the animal thermosensory channel TRPV1. Guihur et al. (2022) have described the notable shared features of TRPV1 with the higher plant thermosensory channel, cyclic nucleotide gated channel (CNGC). In a similar way both these channels sense temperature-induced changes in plasma membrane fluidity and respond by hyperphosphorylating the heat shock factor (HSF1) transcription factor vide a specific heat-signalling cascade.

17.3 Classification of HSFs

Heat shock factors (HSFs) are an integral part of the transcription controlling machinery regulating the activity of stress protective genes. HSFs are transcriptional regulators of genes which encode heat shock proteins (HSPs) as well as other stress-responsive genes to maintain protein homeostasis (Akerfelt et al. 2010). In plants large gene families with 18–52 members encode HSFs, which belong to A, B and C classes, whereas yeast and Drosophila have a single HSF and mammalian cells express four HSFs (Scharf et al. 2012; Andrási et al. 2021).

HSFs have a modular structure with several conserved domains: (1) the N-terminal containing DNA-binding domain (DBD) which recognizes heat shock elements (HSEs, 5′-nGAAnnTTCn-3′) in the target promoters (Schultheiss et al. 1996; Scharf et al. 2012); (2) the oligomerization domain (OD or HR-A/B motif) is responsible for protein–protein interactions and trimerization during transcriptional activation. HSFs exhibit a nuclear localization signal (NLS) and a nuclear export signal (NES) at the C-terminus which aid in shuttling of HSFs in between the cytoplasm and the nucleus. Class A HSFs possess activator motifs or aromatic and hydrophobic amino acid residues (AHAs) close to the NES at the C-terminal region, which is needed for transcriptional activation, interaction with the basal transcription complex, TATA-binding protein (TBP) or transcription factor II B (TFIIB) (Kotak et al. 2004). Class B HSFs have a highly conserved tetrapeptide LFGV repressor motif (Czarnecka-Verner et al. 2004). The precise function of Class C HSFs is not yet known even though in several monocots, multigene members point towards specialized tasks in these plants (Scharf et al. 2012; Guo et al. 2014). The Heatster database (http://www.cibiv.at/services/hsf/) currently contains 848 HSF sequences from 33 plant species.

17.4 HSPs as Ideal Candidates for Developing Multistress Tolerance in Plants

Waddington (1961) described the process termed as canalization which leads to phenotypic stability in spite of genetic and environmental variations. Considerable progress has been made in development of single stress resistance in crop plants through genetic transformation. Under continuously changing climate patterns, the need of hour today is multistress resistant plants as in nature several stresses appear together. Reports have documented that response of plants to multiple and co-occurring stresses are strikingly different even when individual responses are added together (Jacob et al. 2017).

Different approaches of engineering multistress tolerance in plants can be internal cellular manipulation as well as targeting the outermost shield such as cuticles which are the primary defence interface between plant and the environment. Various reports have stated targeting cuticular structures which are made up of various lipid components such as primary alcohol producer fatty acyl–coA reductase (FARs) (Wang et al. 2015) and 3-keto acyl Co-A synthase (KCS) as suitable candidates (Hooker et al. 2002). Secondly targeting membrane fluidity, which is susceptible to almost all the stresses, specifically temperature extremes can also offer multistress resistance. ROS scavengers and compatible solutes which accumulate under stress as osmoprotectants have also been studied to improve multistress tolerance. Molecular chaperones and HSPs function in maintaining cellular protein homeostasis as a universal salvation system in all living organisms under all kind of stressors and have important function in plant defence. Besides due to their wide presence in plants relative to other organisms, HSPs make the most attractive group of proteins for functional modifications towards multistress resistance development.

17.5 Types of HSPs and Their Structural Details

As mentioned earlier, HSPs are ubiquitously found in all living organisms. They are generally divided into five groups on the basis of their molecular weight: Hsp100, Hsp90, Hsp70, Hsp60, and small Hsps (Marrs et al. 1993; Schirmer et al. 1994; Krishna et al. 1995; Mayer and Bukau 2005; Waters 2013). Majority of the HSPs are biotic and abiotic stress responsive and are found in the cell cytoplasm (Boston et al. 1996; Vierling 1991). Other than cytoplasm, HSPs are also located in organelles viz. the endoplasmic reticulum, mitochondria, chloroplasts and nucleus, reemphasizing the fact that they have profound role to play in protein homeostasis (Boston et al. 1996; Vierling 1991).

17.5.1 Small Heat Shock Proteins (sHSPs)

Amongst the five different families of HSPs, small heat shock proteins (sHSPs) are the most common in plants. Reports suggested sHSPs to be the very "first responders" in response to cellular stress. They have a common C-terminal alpha-crystalline domain (ACD) of 90-amino acids (Vierling 1991; Waters et al. 1996; Mu et al. 2013) and are distributed into several subclasses viz. (class CI to class CVI). Members of CI, CII and CIII are found in the nucleus or cytosol and the remaining three subclasses CIV, CV and CVI are found in the plastids, ER and mitochondria, respectively (Low et al. 2000; Ma et al. 2006; Siddique et al. 2008; Jiang et al. 2009).

One of the most structurally conserved sHsp is HSP20. A conserved C-terminal region, variable N-terminal region and an extension region at the C-terminal constitute the HSP20s (Kriehuber et al. 2010; Poulain et al. 2010; Waters 2013). The variable N-terminal region sequences vary greatly in different subfamilies, but they show high similarity sequences within the same subfamily (Bondino et al. 2012), whereas the conserved C-terminal region has a conserved alpha-crystallin domain (ACD), made up of around 90 amino acids, subdivided into CRI and CRII regions which have a 6-loop junction in the middle region. Hsp20s function as ATP-independent chaperones. They bind to altered or denatured proteins of similar molecular weight to make stable compounds, preventing their aggregation without changing the activity of the protein (Hartl et al. 2011).

Around 150 million years ago, even before the differentiation of the dicots and monocots, small heat shock protein gene families came into existence (Doyle and Donoghue 1993). There is a diverse evolutionary lineage in the small and large HSPs. One of those evolutionary mechanisms could be gene replacement where eubacterial genes replaced pre-existing nuclear counterparts (Huang et al. 2008). The HSPs in plants did not fall in the category of HSPs from prokaryotes or any homologues of prokaryotes. Hence, it becomes convenient enough to stick to the fact that plant small heat shock proteins didn't originate from the prokaryotes with the aid of gene transfer during the early evolution (Huang et al. 2008). Sequence closeness and evolutionary analysis patterns revealed that the gene families of small heat shock proteins (sHSPs) came into existence by doubling of genes followed by sequence preceding divergence to the radiation of angiosperms.

17.5.2 Heat Shock Protein 70 (HSP70)

The most abundant and highly conserved family of proteins are the ATP-dependent 70-kDa heat shock proteins HSP70, showing the highest sequence similarity within a large evolutionary distance (Borchiellini et al. 1998). A major HSP multigene family made up of the cytosolic Hsp70s and the mitochondrial Hsp70 are known to encode these HSP70 genes. DnaK protein or prokaryotic HSP70 shares ~50% amino acid identity with eukaryotic HSP70s. Plant HSP70s share at least two of the four structural features of the archetype bacterial HSP70 DnaKs: an N-terminal, 45 kDa nucleotide-binding domain (NBD), followed by a 15 kDa substrate-binding domain (SBDβ), a 10 kDa helical lid domain (SBDα), and a disordered C-terminal tail of variable length which frequently ends with a characteristic charged motif (EEVD) that interacts with specific co-chaperones containing TPR repeats such as HSP-organizing protein (HOP) involved in transfer of unfolded HSP70 substrates to foldases, HSP-interacting protein (HIP) which stabilizes the high-affinity ADP-bound state of HSP70, and C-terminus of HSP70-interacting protein (CHIP) involved in ubiquitin mediated proteasome degradation of unfolded HSP70 substrates (Berka et al. 2022).

Hsp70s are capable of interacting with proteins or substrates proteins under both unstressed and stressed conditions in their hydrophobic peptide segments in an ATP-dependent manner which significantly demarcates their specific roles (Mayer and Bukau 2005). Both NBD and SBD domains are bridged with the help of a conserved linker, with vital activity for inter-domain communication (Jiang et al.

2005). The crystal structures of the SBD of HSP70 had been elucidated by complexing to a model peptide (NRLLLTG). A single peptide binding site connected to a hydrophobic channel surrounded by an arch facilitating access of peptides (Zhu et al. 1996).

The HSP70 binds to protein substrates in the ADP-bound state with a significantly increased affinity, and ATP hydrolysis is essential for its chaperone function. However, the intrinsic ATPase rate is very low (Kityk et al. 2012, 2018) and requires co-chaperones such as J-domain proteins (JDP) which accelerate ATP hydrolysis and nucleotide exchange factors. HSP70s are also essential for driving the movement of proteins across the membrane and protein translocation into the endoplasmic reticulum, mitochondria and chloroplasts with organellar HSP70s being more critical in the translocation processes (Berka et al. 2022).

17.5.3 Heat Shock Protein 90 (HSP90)

Under normal conditions HSP90 protein levels represent 1–2% of the total cellular proteins, expressing constitutively, while during stress conditions, their levels rise as their transcription and translation are enhanced (Lindquist 1986). The function of HSP90s is considered indispensable for the maintenance of cellular homeostasis under natural growth conditions, but their activity becomes even more crucial when stress conditions compromise protein stability. There are three structural domains in HSP90 which comprises an N-terminal nucleotide-binding domain (NBD) which can also bind HSP90 inhibitors (Prodromou et al. 1997); a middle segment which interacts with client proteins; and the C-terminus, which plays a role in homodimerization (Harris et al. 2004). HSP90s physically interact and cooperate with co-chaperones which have diverse structures, with some of the interactions modulating the ATPase activity, while others affect substrate specificity (Catlett and Kaplan 2006). Binding to all protein subdomains of HSP90s is facilitated or abolished by other co-chaperones mediating the maturation of client proteins that comprise about 10–20% of the cellular proteome (Taipale et al. 2012; Tichá et al. 2020).

17.5.4 Heat Shock Protein 40 (HSP40)

In similar grounds, HSP40/DnaJ, which is also named as J-domain-containing protein (J-protein) functions to surge HSP70 affinity for clients as a co-chaperone component in the HSP70 system (Kampinga and Craig 2010). HSP40 is observed to associate with the nucleotide-binding domain (NBD) of HSP70 and is known to encompass a conserved 70-amino acid J-domain.

17.6 Functional Component of Plant HSPs

17.6.1 Thermotolerance and Other Abiotic Stresses

Abiotic stresses including heat stress (high temperature) affect the metabolism and structure of plants, especially cell membranes and many basic physiological processes including photosynthesis, respiration and water relations (Wahid et al. 2007). As ramifications of these stresses, certain essential metabolic process such as DNA replication-repair, transcription, mRNA export and translation are affected dramatically until the cells recover (Biamonti and Caceres 2009). Plants combat these stressors with several mechanisms of defence (Farooq et al. 2017b) for survival and adaptation. A number of vital processes like osmotic regulation, scavenging of reactive oxygen species (ROS) secreted during oxidative stress, cell membrane stability maintenance, production of antioxidants, activation of Ca-dependent kinase proteins and other stress inducible kinases become active and enhance the transcriptional, translational machinery. Various signal transduction machineries along with the molecular chaperones work at optimum level, thereby restoring and levelling normal functioning of the plant system (Wahid et al. 2007).

The multiple stresses to which plants are exposed result in the disruption of protein homeostasis. Maintenance of proteins in their functional native conformations while preventing aggregation of non-native proteins is very crucial for survival. This marks the role of HSPs as molecular chaperones. Under stress conditions, HSPs have a vital role to play in protein folding, assembly, translocation, and degradation. The HSP chaperones and their co-chaperones bind to partially denatured membrane proteins, prevent irreversible aggregation, and hence function in stabilizing proteins (Wang et al. 2004). HSPs are abundantly and heterogeneously found because of which they have a significant role to play in stress tolerance (Sun et al. 2012). Reports suggested several types of Hsp70s being induced under heat stress tolerance in different plants like *Oryza sativa* (Sarkar et al. 2013), *Nicotiana tabacum* (Cho and Hong 2004), *Capsicum annuum* (Guo et al. 2014) *Dactylis glomerata* (Cha et al. 2012), *Triticum aestivum* (Duan et al. 2011), etc. Researchers have also observed enhanced heat tolerance on overexpression of the *Arabidopsis* cytosolic Hsc70-1. Not only heat stress, but there are also HSPs such as chrysanthemum CgHsp70 in the presence of which alleviated tolerance to various other abiotic stresses was observed in *A. thaliana*, which in turn shielded the plant from severe oxidative damage (Song et al. 2014).

17.6.2 HSPs in Plant Immunity

Plants innate immune response to pathogen invasion occurs by two separate mechanisms mediated by pattern recognition receptors (PRRs) and resistance (R) proteins. In both the cases HSPs role as molecular chaperones in the quality control of plasma membrane resident PRRs and intracellular R proteins against potential invaders has been shown to be indispensable (Park and Seo 2015). Even though the role of HSPs in regulation of biotic interactions is much less characterized as compared to reports from their animal counterparts, the role of HSP70 in plant biotic defence against pathogens is slowly accumulating.

Below we discuss some of the reports on involvement of HSP70 in various post-transcriptional and post-translational biotic stress regulation. *Arabidopsis* expresses four *HSP70* genes and three of these are strongly upregulated in response to different pathogens. Noel and his team (2007) with limited evidence showed that these HSP70 genes are at least partially redundant, and experiments with a loss-of-function mutation of individual HSP70 isoforms 1, 2, and 3 did not reveal any difference in the resistance against pathogens. The critical role of cytosolic HSP70s in plant defence is supported by the fact that these proteins are targeted by pathogen effector proteins. One example cites HopI1, a virulence effector of pathogenic *Pseudomonas syringae*, which binds directly to the host HSP70 and recruits it to chloroplasts, the site of HopI1 localization (Jelenska et al. 2010). The interaction was confirmed by co-immunoprecipitation for all four *Arabidopsis* isoforms (1–4) and for chloroplastic HSP70-6. Further, plants with depleted cytosolic HSP70–1 were more susceptible to *P. syringae* (Jelenska et al. 2010), and silencing of cytosolic HSP70 in pepper increased the plant's susceptibility to *Xanthomonas campestris* (Kim and Hwang 2015). The effector Pi23226 of was co-immunoprecipitated with HSP70s, and transient overexpression of these HSP70s in *Nicotiana benthamiana* inhibited *Phytophthora* growth (Lee et al. 2018).

17.6.3 Endoplasmic Reticulum Stress Response

Whenever inaccurately folded proteins accumulate in the ER, a major site of secretory protein synthesis, ER stress response (ERSR) pathway or unfolded protein response pathway (UPR) gets activated to rectify the process. Plant pathogens contain the ability to bind to host ERSR pathway and manipulate it to their advantage which can lead to PCD (Jing and Wang 2020). ER resident binding proteins (BiPs), a group of HSP70s get induced which participate as central regulator in plant immunity and stabilize misfolded intermediates, arrest aggregation, and aid in the regular subsequent protein folding and assembly. In contrast to mammals and yeast, flowering plants contain several BiP proteins (HSP11–13 in *Arabidopsis*; Noh et al. 2003). Mutants of bip2 in *Arabidopsis* showed enhanced fungal colonization rates, and BiP protein levels were also reduced in colonized roots. Upregulation of BiP was found in many viruses' infective tissues including potato virus X, turnip crinkle virus, garlic virus X, and sugarcane mosaic virus (Berka et al. 2022). Overexpressed lines of BiPs displayed an accelerated hypersensitive response to *Pseudomonas syringae* pv. tomato in soybean and tobacco (Carvalho et al. 2014). Various other reports also showed that HSP70 has a positive correlation to *Phytophthora parasitica* in tomato (Naveed and Ali 2018), and faster induction of HSP70 genes was observed with enhanced resistance to barley infected by *Blumeria graminis* (Molitor et al. 2011).

17.6.4 HSPs in Hormone Biology

Multiple roles of HSP90 and its co-chaperones vide its unique ATP binding and hydrolysis mechanism in plant hormone biology have been reported in recent times. Plant hormones and the signalling crosstalk play a key role in integrating external and internal cues, enabling the plants to acclimate accordingly. HSP90 and several of its co-chaperones known as pleiotropic factors are involved in the hormonal signalling pathways by modulating substrates or client proteins which comprise a diverse and unrelated clientele (di Donato and Geisler 2019). A few examples of HSP90 involvement in hormone transduction pathways are mentioned below.

Auxin is involved in temperature-induced adaptation in combination with phytochrome-interacting factor 4. HSP90-mediated auxin receptor stabilization has been reported in literature thereby indicating that various auxin related phenotypes under temperature fluctuations are also regulated by HSP90 (Wang et al. 2016). HSP90 function has also been implicated in polar auxin transport (Vanneste and Friml 2009). Brassinosteroid (BR) and auxin signalling outputs are known to have an extensive transcriptional overlap and HSP90 has been reported to be involved in maintenance of various Brassinosteroid signalling pathway effector molecules (Shigeta et al. 2015). There are also reports of HSP90s involvement in broad spectrum physiological processes of other hormones such as cytokinins and jasmonic acid (di Donato and Geisler 2019).

17.6.5 HSPs in General Growth and Development

The general constitutive function and expression of HSPs in plant growth and development is well established now. Various developmental stages of the plant like germination, embryogenesis, root development, hypocotyls elongation, shoot apical meristem development, pollen development and fruit maturation are regulated by the induction of HSPs (Sun et al. 2001, 2002; Chauhan et al. 2012). Experimentally also function of HSPs in general growth and development has been reiterated using *hsp90* mutants. The phenotypic abnormalities showed various morphological defects in cotyledon development, root hair development, impaired inflorescence and flower development, higher number of rosette leaves together with the differential expression of cytoplasmic HSP90s throughout ontogenesis (Samakovli et al. 2007; Xu et al. 2012).

Additionally, the role of HSP70/HSP90 in the regulation of stomatal aperture under changing environmental conditions or in the response to abscisic acid was also demonstrated by genetic and pharmacological evidence (Clément et al. 2011). HSP90 has also been shown to be involved in transition from vegetative to reproductive stage. Well-characterized client proteins of HSP90 such as kinases, E3 ligases and wide-ranging transcription factors exhibit epistatic interaction with HSP90 which can inhibit or activate the expression of numerous genes. This specific form of regulation can provide the plants with a genetic buffering system crucial for fighting internal and external cues during plant growth and development (Tichá et al.

2020). Another distinguishing feature of HSP90 chaperone machinery has been attributed as genetic buffering leading to phenotypic vigour is protecting the organisms from hidden mutations by efficient and correct folding of these mutated proteins (Rutherford and Lindquist 1998; Queitsch et al. 2002).

17.7 HSPs in Major and Minor Millets

17.7.1 Sorghum and Pearl Millet: Major Millets

Molecular characterization of the HSP70 gene from Sorghum *cv* MN1618 was reported and designated as SbHsp70-1 (Mulaudzi-Masuku et al. 2015). SbHsp70 showed high homology with other plant Hsp70s, especially grain crops. It is a cytoplasmic protein, and its expression had been found to be significantly upregulated upon heat shock at 42 °C. Results also indicated that *E. coli* cells overexpressed with SbHsp70-1 were able to survive heat stress even up to higher temperatures like 65 °C, thereby shedding light on the role of SbHsp70-1 in conferring heat stress tolerance and moulding it as a promising target to study stress tolerance in crops (Mulaudzi-Masuku et al. 2015).

In another study, genome-wide identification and transcriptional profiling of sHSP protein gene family was carried out under diverse abiotic stress conditions in a sorghum cultivar which revealed the presence of total 47 sHSPs distributed across 10 subfamilies. The major subfamily was identified as chloroplastic group with 17 genes. Promoter analysis indicated that most of the HSPs are associated with both biotic and abiotic stresses, as well as plant development (Nagaraju et al. 2020).

Thermotolerance behaviour analysis of HSP20s was performed by cloning a PgHSP16.97 from a thermotolerant cultivar of pearl millet cv. WGI-126. The study also identified 28 different HSP20 genes of pearl millet involved stress tolerance. The authors also showed and suggested that transcript expression profiling of HSPs can be used as an effective tool for high-throughput screening for identification of thermo-resistant cultivars (Mukesh Sankar et al. 2021). In another study in pearl millet reported by Divya and co-workers (2019), genome-wide screening identified 18 PgHsp70 genes in the pearl millet genome which have been categorized into 4 subfamilies depending on their cellular localization as cytosolic or organellar. Phylogenetic analysis of Hsp70s revealed that they are closely related to Sorghum Hsp70s. Promoter analysis showed the presence of *cis* acting elements associated with abiotic stress conditions emphasizing the involvement of these genes in the abiotic stress.

17.7.2 Small Millet and HSPs

The term "small millets" is a generic term and covers almost all the coarse cereals including finger millet (*Eleusine coracana*), foxtail millet (*Setaria italica*), kodo millet (*Paspalum scrobiculatum*), little millet (*Panicum sumatrense*), barnyard millet (*Echinochloa crus-galli*), proso millet (*Panicum miliaceum*), teff (*Eragrostis tef*), fonio (*Digitaria exilis*), Job's tears (*Coix lacryma-jobi*) and guinea millet (Muthamilarasan et al. 2019; Goron and Raizada 2015).

Till date genome-wide analysis of HSPs has been reported from only one small millet, i.e. foxtail millet (Singh et al. 2016). The authors reported 20, 9, 27, 20 and 37 genes belonging to *SiHSP100, SiHSP90, SiHSP70, SiHSP60* and *SisHSP* families, respectively. In silico analysis and gene expression profiling under diverse abiotic stresses showed alleviated levels of expression of several HSPs genes in a tolerant foxtail millet cultivar. Over-expression of SisHSP-27 in yeast system also demonstrated augmented tolerance to several abiotic stresses (Singh et al. 2016). DNA methylation is an epigenetic mechanism utilized by organisms to control gene expression in specific conditions. Several abiotic stress responsive genes from *SiHSP* group showed lesser methylation in the tolerant cultivar relative to the susceptible cultivar which can reveal potential role of methylation in stress response in this small millet after further functional investigations.

In another report, sHSP gene HSP17.8 had been isolated and characterized from a small millet viz. finger millet. With a complete Open Reading Frame (ORF) of 489 bp and a peptide length of 162 amino acids, the gene *EcHSP17.8* was induced by heat, oxidative stress, and mannitol (200 mM), with the highest upregulation of 40 folds observed under heat stress (Chopperla et al. 2018). Results dictated the role of the gene as an early responsive gene under heat stress claiming that the alleles of HSP17.8 from finger millet would serve as suitable candidate for producing thermo-resistant plants.

17.8 Potential Mining and Utilization of HSPs from Small Millet: Future Perspectives

In the above discussion we have briefly touched upon the structure and functional aspects of plant HSP multigene family members underlining their crucial role in various aspects of plant growth, development and physiology during regular as well as under stressful environment. Most of the basic studies have been done in *Arabidopsis* with a few notable exceptions of tomato and rice. Plant HSP research is lagging behind compared to animal and mammalian counterparts, and our knowledge about post-transcriptional and post-translational control is also rudimentary. A comprehensive understanding of their structure, organization, evolution and expression profiling in a naturally stress tolerant crop is necessary to delineate their precise roles in stress-responsive molecular machinery.

There are instances in cereals as well as other crop plants genetically overexpressing a single gene imparting stress resistance. However, very few studies have dissected the exact mode of action delineating interacting partners and mechanism of imparting stress tolerance. The ideal way to move forward in mining small millet HSPs should be based on already available information and designing experiments with precision techniques. The mechanism of gene interaction and response under a certain environment also differs from species to species and amongst cultivars within a species. Today, with precision genome engineering and editing techniques available at our disposal along with new generation sequencing technology, high resolution omics tools, the climate resilience of each member of small millet can be deciphered based on their inherent characteristics to find out the key responsible for these special traits. Recent advances in phenotyping and phenomics platforms, together with utilization of available germplasm diversity, could be harnessed in small millets improvement. Instead of randomly selecting a prospective HSP gene, it would be more useful in mechanistically analysing the functional aspects of imparting stress tolerance in a selected small millet species based on both the gene and crop specific traits and thereafter deploying them to trap their functional features.

There are various aspects of plant HSPs still to be unravelled. For example, animal HSP70s play a significant signalling role in the extracellular space, particularly in the inflammatory and immune responses (Mambula et al. 2007), whereas upon analysis it was revealed that none of the *Arabidopsis* HSP70s contains a targeting signal for secretion. The role of extracellular HSP70 could be evolutionarily conserved and could provide an additional explanation for the observed accumulation in response to pathogens. Plants exhibit a secretory pathway and can secrete proteins without the consensus N-terminal peptide sequence via alternative, unconventional protein secretory pathways into the extracellular space relative to animal systems despite the presence of cell walls. The exact extent of these alternative pathways in plants has not yet been fully elucidated, and the role of plant HSP70s in the extracellular space is not known, even though at least nine *Arabidopsis* HSP70s have been detected in the extracellular space, and it has been confirmed that the extracellular vesicles in the plant secretome contribute to the plant defence system (Regente et al. 2017).

Most of the HSPs are also engaged within many functionally relevant protein interaction networks relative to any other group of proteins. To cite an example, HSP90s interact with approximately 2000 proteins and just a few interactions with their clients have been functionally characterized. Thus, analysing the HSP client protein interactome will reveal their functional aspects.

Comprehensive reference genome sequence assemblies are of great importance and provide the foundation to promote molecular breeding and functional genomics research. They help in understanding genetic relationships, isolation of excellent allelic variants, comprehensive genetic exploration, detect quantitative trait locus (QTL) with specific traits such as stress resistance, yield and quality (Chen et al. 2019; Hillary and Ceasar 2019).

Till date, five small millet genomes, namely foxtail millet, finger millet, proso millet, teff and barnyard millet, have been sequenced. Amongst these the genome of foxtail millet is the smallest (423–510 Mb) and finger millet (1.5 Gb) is the largest. One of the low downs in small millet research is availability of limited genomic resources as well as efficient transformation systems for genetic manipulation and genome editing.

As compared to major cereals like rice, wheat and maize small millets are short duration crop and also show more plasticity under adverse climate conditions

(Ceasar 2021) demanding less maintenance in the field. They are also low maintenance and require minimum water and other inputs. Most of these traits are yet to be functionally characterized at the genetic level. Yang et al. (2020) developed a mini foxtail millet variety with a shorter life cycle with a point mutation in the Phytochrome C gene naming it Xiaomi. The variety showed a heading date of 39 days versus 82 days for wild-type plants. The authors have also further optimized a transformation system and assembled the genome and transcriptome of the Xiaomi variety. This variety could serve as a model for foxtail millet also for comparative analysis in other small millets to facilitate genetic studies.

17.9 Conclusions

Complex characteristics and diverse subcellular location make analyses of plant HSPs a demanding task. Along with functional redundancy due to the multigenic nature of most HSPs and developmental stage specific expression, it is also difficult to get phenotypic distinctions with single gene mutations. However, considering their indispensable role in plant general growth and development, involvement in environmental signalling networks, research on HSPS and their co-chaperones and client protein modulation offers tremendous opportunities for closing the gaps in understanding of various basic plant physiological processes. Till date small millet HSP study is very, very limited and hitherto unexplored. Thus, taking advantage of all the information on HSPs available from other plant systems as well as animal and mammalian systems, well defined, structured programmes can be initiated in small millets. To solve recurrent stress issues and develop a climate proof agriculture dissecting the contribution of specific HSP members from these ideal small millet plants and their functional specification is a task for future plant research.

References

Akerfelt M, Morimoto RI, Sistonen L (2010) Heat shock factors: integrators of cell stress, development and lifespan. Nat Rev Mol Cell Biol 11:545–555

Andrási N, Rigó G, Zsigmond L, Pérez-Salamó I, Papdi C, Klement E, Pettkó-Szandtner A, Baba AI, Ayaydin F, Dasari R, Cséplő Á (2021) Corrigendum to: The mitogen-activated protein kinase 4-phosphorylated heat shock factor A4A regulates responses to combined salt and heat stresses. J Exp Bot 72(15):5781

Berka M, Kopecká R, Berková V, Brzobohatý B, Černý M (2022) Regulation of heat shock proteins 70 and their role in plant immunity. J Exp Bot 73(7):1894–1909

Bharadwaj N, Gogoi N, Barthakur S, Basumatary N (2018) Morpho-physiological responses in different mungbean genotypes under drought stress. Res J Recent Sci 7(7):10–18

Bharadwaj N, Barthakur S, Biswas AD, Kumar Das M, Kour M, Ramteke A, Gogoi N (2019) Transcript expression profiling in two contrasting cultivars and molecular cloning of a SKP-1 like gene, a component of SCF-ubiquitin proteasome system from mungbean Vigna radiate L. Sci Rep 9(1):1–7

Bharti K, Nover L (2004) Heat stress response in plants: a complex game with chaperones and more than twenty heat stress transcription factors. J Biosci 29:471–487

Biamonti G, Caceres JF (2009) Cellular stress and RNA splicing. Trends Biochem Sci 34(3): 146–153
Bondino HG, Valle EM, Ten Have A (2012) Evolution and functional diversification of the small heat shock protein/alpha-crystallin family in higher plants. Planta 235:1299–1313. https://doi.org/10.1007/s00425-011-1575-9
Bora MS, Devi U, Bharadwaj N, Sharma P, Kalita S, Baruah S, Das A, Sarma KP (2021) Pollution and health risk assessment of toxic metals in solid waste dumping site soil and its impact on groundwater: a case study. Int J Environ Anal Chem 10:1–21
Borchiellini C, Boury-Esnault N, Vacelet J, Le Parco Y (1998) Phylogenetic analysis of the Hsp70 sequences reveals the monophyly of Metazoa and specific phylogenetic relationships between animals and fungi. Mol Biol Evol 15:647–655
Boston RS, Viitanen PV, Vierling E (1996) Molecular chaperones and protein folding in plants. Plant Mol Biol 32:191–222
Carvalho HH, Silva PA, Mendes GC, Brustolini OJB, Pimenta MR, Gouveia BC, Valente MAS, Ramos HJO, Soares-Ramos JRL, Fontes EPB (2014) The endoplasmic reticulum binding protein BiP displays dual function in modulating cell death events. Plant Physiol 164:654–670
Catlett MG, Kaplan KB (2006) Sgt1p is a unique co-chaperone that acts as a client adaptor to link Hsp90 to Skp1p. J Biol Chem 281:33739–33748
Ceasar A (2021) Genome-editing in millets: current knowledge and future perspectives. Mol Biol Rep 26:1–9
Cha JY, Su'udi M, Kim WY et al (2012) Functional characterization of orchard grass cytosolic Hsp70 (DgHsp70) and the negative regulation by Ca2+/AtCaM2 binding. Plant Physiol Biochem 58:29–36
Chauhan H, Khurana N, Nijhavan A, Khurana JP, Khurana P (2012) The wheat chloroplastic small heat shock protein (sHSP26) is involved in seed maturation and germination and imparts tolerance to heat stress. Theor Appl Genet 35:1912–1931
Chen K, Wang Y, Zhang R et al (2019) CRISPR/Cas genome editing and precision plant breeding in agriculture. Annu Rev Plant Biol 70:667–697
Cho EK, Hong CB (2004) Molecular cloning and expression pattern analyses of heat shock protein 70 genes from *Nicotiana tabacum*. J Plant Biol 47(2):149–159
Chopperla R, Singh S, Tomar R, Mohanty S, Khan S, Reddy N, Padaria JC, Solanke AU (2018) Isolation and allelic characterization of finger millet (*Eleusine coracana* L.) small heat shock protein echsp17.8 for stress tolerance. Indian J Genet Plant Breed 78:95–103
Clément M, Leonhardt N, Droillard MJ, Reiter I, Montillet JL, Genty B, Lauriere C, Nussaume L, Noël LD (2011) The cytosolic/nuclear HSC70 and HSP90 molecular chaperones are important for stomatal closure and modulate abscisic acid-dependent physiological responses in Arabidopsis. Plant Physiol 156(3):1481–1492
Czarnecka-Verner E, Pan S, Salem T, Gurley WB (2004) Plant class B HSFs inhibit transcription and exhibit affinity for TFIIB and TBP. Plant Mol Biol 56:57–75
De Maio A (1999) Heat shock proteins: facts, thoughts, and dreams. Shock 11:1–12
De Maio A, Santoro MG, Tanguay RM, Hightower LE (2012) Ferruccio Ritossa's scientific legacy 50 years after his discovery of the heat shock response: a new view of biology, a new society, and a new journal. Cell Stress Chaperones 17:139–143
di Donato M, Geisler M (2019) HSP 90 and co-chaperones: a multitaskers' view on plant hormone biology. FEBS Lett 593(13):1415–1430
Divya K, Kavi Kishor PB, Maraka N, Bhatnagar-Mathur P, Singam P, Vadez V, Reddy PS (2019) Genome-wide identification and characterization of Hsp70 gene family in Pearl millet (*Pennisetum glaucum*). Curr Trends Biotechnol Pharm 13(2):102–111
Doyle JA, Donoghue MJ (1993) Phytogenies and angiosperm diversification. Paleobiology 19: 141–167
Duan YH, Guo J, Ding K et al (2011) Characterization of a wheat Hsp70 gene and its expression in response to stripe rust infection and abiotic stresses. Mol Biol Rep 38(1):301–307

Farooq M, Gogoi N, Barthakur S, Baroowa B, Bharadwaj N, Alghamdi SS, Siddique KH (2017a) Drought stress in grain legumes during reproduction and grain filling. J Agron Crop Sci 203(2): 81–102

Farooq M, Gogoi N, Hussain M, Barthakur S, Paul S, Bharadwaj N, Migdadi HM, Alghamdi SS, Siddique KH (2017b) Effects, tolerance mechanisms and management of salt stress in grain legumes. Plant Physiol Biochem 118:199–217

Feige U, Morimoto RI, Yahara I, Polla BS (eds) (1996) Stress-inducible cellular responses. Birkhauser, Basel

Goron TL, Raizada MN (2015) Genetic diversity and genomic resources available for the small millet crops to accelerate a new green revolution. Front Plant Sci 6:157

Guihur A, Rebeaud ME, Bourgine B, Goloubinoff P (2022) How do humans and plants feel the heat?. Trends Plant Sci S1360-1385(22)00059-0. https://doi.org/10.1016/j.tplants.2022.03.006

Guo M, Zhai YF, Lu JP et al (2014) Characterization of CaHsp70-1, a pepper heat-shock protein gene in response to heat stress and some regulation exogenous substances in Capsicum annuum L. Int J Mol Sci 15(11):19741–19759

Gupta SC, Sharma A, Mishra M, Mishra R, Chowdhuri DK (2010) Heat shock proteins in toxicology: how close and how far? Life Sci 86:377–384

Harris SF, Shiau AK, Agard DA (2004) The crystal structure of the carboxy-terminal dimerization domain of htpG, the Escherichia coli Hsp90, reveals a potential substrate binding site. Structure 12(6):1087–1097

Hartl FU, Bracher A, Hayer-Hartl M (2011) Molecular chaperones in protein folding and proteostasis. Nature 475:324332. https://doi.org/10.1038/nature10317

Hasan M, Barthakur S (2014) Hsp70 based gene expression biomarker shows growth stage specific genotypic diversity in Indian wheat (*Triticum aestivum* L.) cultivars. Ann Agric Res 35(3): 233–243

Hillary VE, Ceasar SA (2019) Application of CRISPR/Cas9 genome editing system in cereal crops. Open Biotechnol J 13(1):173–179

Hooker TS, Millar AA, Kunst L (2002) Significance of the expression of the CER6 condensing enzyme for cuticular wax production in Arabidopsis. Plant Physiol 129(4):1568–1580

Huang LH, Wang HS, Kang L (2008) Different evolutionary lineages of large and small heat shock proteins in eukaryotes. Cell Res 18(10):1074–1076

Jacob P, Heribert H, Abdelhafid B (2017) The heat-shock protein/chaperone network and multiple stress resistance. Plant Biotechnol J 15:405–414

Jelenska J, van Hal JA, Greenberg JT (2010) *Pseudomonas syringae* hijacks plant stress chaperone machinery for virulence. Proc Natl Acad Sci U S A 107:13177–13182

Jiang J, Prasad K, Lafer EM, Sousa R (2005) Structural basis of interdomain communication in the Hsc70 chaperone. Mol Cell 20(4):513–524

Jing M, Wang Y (2020) Plant pathogens utilize effectors to hijack the host endoplasmic reticulum as part of their infection strategy. Engineering 6:500–504

Jiang C, Xu J, Zhang H, Zhang X, Shi J, Li M, Ming F (2009) A cytosolic class I small heat shock protein, RcHSP17.8, of Rosa chinensis confers resistance to a variety of stresses to Escherichia coli, yeast and Arabidopsis thaliana. Plant Cell Environ 32:1046–1059

Kampinga HH, Craig EA (2010) The HSP70 chaperone machinery: J proteins as drivers of functional specificity. Nat Rev 11:579–592

Kim NH, Hwang BK (2015) Pepper heat shock protein 70a interacts with the type III effector AvrBsT and triggers plant cell death and immunity. Plant Physiol 167:307–322

Kityk R, Kopp J, Sinning I, Mayer MP (2012) Structure and dynamics of the ATP-bound open conformation of Hsp70 chaperones. Mol Cell 48:863–874

Kityk R, Kopp J, Mayer MP (2018) Molecular mechanism of J-domain-triggered ATP hydrolysis by Hsp70 chaperones. Mol Cell 69:227–237.e4

Kotak S, Port M, Ganguli A, Bicker F, von Koskull-Döring P (2004) Characterization of C-terminal domains of Arabidopsis heat stress transcription factors (Hsfs) and identification of a new

signature combination of plant class A Hsfs with AHA and NES motifs essential for activator function and intracellular localization. Plant J 39:98–112
Kotak S, Larkindale J, Lee U, von Koskull-Döring P, Vierling E, Scharf KD (2007) Complexity of the heat stress response in plants. Curr Opin Plant Biol 10(3):310–316
Kriehuber T, Rattei T, Weinmaier T, Bepperling A, Haslbeck M, Buchner J (2010) Independent evolution of the core domain and its flanking sequences in small heat shock proteins. FASEB J 24:3633–3642. https://doi.org/10.1096/fj.10-156992
Krishna P, Sacco M, Cherutti JF, Hill S (1995) Cold-induced accumulation of hsp90 transcripts in Brassica napus. Plant Physiol 107:915–923. https://doi.org/10.1104/pp.107.3.915
Lang BJ, Guerrero ME, Prince TL, Okusha Y, Bonorino C, Calderwood SK (2021) The functions and regulation of heat shock proteins; key orchestrators of proteostasis and the heat shock response. Arch Toxicol 95(6):1943–1970
Lee JH, Lee SE, Oh S, Seo E, Choi D (2018) HSP70s enhance a *Phytophthora infestans* effector-induced cell death via an MAPK cascade in *Nicotiana benthamiana*. Mol Plant-Microbe Interact 31:356–362
Levitt J (1980) Responses of plants to environmental stresses. In: Water, radiation, salt and other stresses, vol II, 2nd edn. Academic Press Inc., New York
Lin CY, Roberts JK, Key JL (1984) Acquisition of thermotolerance in soybean seedlings: synthesis and accumulation of heat shock proteins and their cellular localization. Plant Physiol 74(1): 152–160
Lindquist S (1986) The heat-shock response. Annu Rev Biochem 55:1151–1191
Lindquist S, Craig EA (1988) The heat-shock proteins. Annu Rev Genet 22:631–677
Low D, Brandle K, Nover L, Forreiter C (2000) Cytosolic heat-stress proteins Hsp 17.7 class I and Hsp17.3 class II of tomato act as molecular chaperones in vivo. Planta 211:575–582
Ma C, Haslbeck M, Babujee L, Jahn O, Reumann S (2006) Identification and characterization of a stress inducible and a constitutive small heat-shock protein targeted to the matrix of plant peroxisomes. Plant Physiol 141:47–60
Mambula SS, Stevenson MA, Ogawa K, Calderwood SK (2007) Mechanisms for Hsp70 secretion: crossing membranes without a leader. Methods 43:168–175
Marrs KA, Casey ES, Capitant SA, Bouchard RA, Dietrich PS, Mettler IJ, Sinibaldi RM (1993) Characterization of two maize HSP90 heat shock protein genes: expression during heat shock, embryogenesis, and pollen development. Dev Genet 14:2741. https://doi.org/10.1002/dvg.1020140105
Mayer MP, Bukau B (2005) Hsp70 chaperones: cellular functions and molecular mechanism. Cell Mol Life Sci 62:670–684
McAlister L, Finkelstein DB (1980) Heat shock proteins and thermal resistance in yeast. Biochem Biophys Res Commun 93:819–824
Molitor A, Zajic D, Voll LM, Pons-Kühnemann J, Samans B, Kogel KH, Waller F (2011) Barley leaf transcriptome and metabolite analysis reveals new aspects of compatibility and *Piriformospora indica*-mediated systemic induced resistance to powdery mildew. Mol Plant-Microbe Interact 24:1427–1439
Mu C, Zhang S, Yu G, Chen N, Li X (2013) Overexpression of small heat shock protein LimHSP16.45 in Arabidopsis enhances tolerance to abiotic stresses. PLoS One 8:e82264
Mukesh Sankar S, Tara Satyavathi C, Barthakur S, Singh SP, Bharadwaj C, Soumya SL (2021) Differential modulation of heat-inducible genes across diverse genotypes and molecular cloning of a sHSP from Pearl millet [Pennisetum glaucum (L.) R. Br.]. Front Plant Sci 2021:1333
Mulaudzi-Masuku T, Mutepe RD, Mukhoro OC, Faro A, Ndimba B (2015) Identification and characterization of a heat-inducible Hsp70 gene from Sorghum bicolor which confers tolerance to thermal stress. Cell Stress Chaperones 20(5):793–804
Muthamilarasan M et al (2019) Multi-omics approaches for strategic improvement of stress tolerance in underutilized crop species: a climate change perspective. Adv Genet 103:1–38

Nagaraju M, Reddy PS, Kumar SA, Kumar A, Rajasheker G, Rao DM, Kishor PK (2020) Genome-wide identification and transcriptional profiling of small heat shock protein gene family under diverse abiotic stress conditions in Sorghum bicolor (L.). Int J Biol Macromol 142:822–834

Naveed ZA, Ali GS (2018) Comparative transcriptome analysis between a resistant and a susceptible wild tomato accession in response to *Phytophthora parasitica*. Int J Mol Sci 19:3735

Nguyen D, Rieu I, Mariani C, van Dam NM (2016) How plants handle multiple stresses: hormonal interactions underlying responses to abiotic stress and insect herbivory. Plant Mol Biol 91(6): 727–740

Noh S-J, Kwon C, Oh D-H, Moon, jae sun & Chung, Won-Il. (2003) Expression of an evolutionarily distinct novel BiP gene during the unfolded protein response in Arabidopsis thaliana. Gene 311:81–91

Noël LD, Cagna G, Stuttmann J, Wirthmüller L, Betsuyaku S, Witte CP, Bhat R, Pochon N, Colby T, Parker JE (2007) Interaction between SGT1 and cytosolic/nuclear HSC70 chaperones regulates Arabidopsis immune responses. Plant Cell 19(12):4061–4076

Park CJ, Seo YS (2015) Heat shock proteins: a review of the molecular chaperones for plant immunity. Plant Pathol J 31(4):323–333

Parsell PA, Lindquist S (1993) The function of heat-shock proteins in stress tolerance. Degradation and reactivation of damaged proteins. Annu Rev Genet 27:437–496

Pincus D (2020) Regulation of Hsf1 and the heat shock response. Adv Exp Med Biol 1243:41–50

Poulain P, Gelly JC, Flatters D (2010) Detection and architecture of small heat shock protein monomers. PLoS One 5:e9990. https://doi.org/10.1371/journal.pone.0009990

Prodromou C, Roe SM, O'Brien R, Ladbury JE, Piper PW, Pearl LH (1997) Identification and structural characterization of the ATP/ADP-binding site in the Hsp90 molecular chaperone. Cell 90(1):65–75

Queitsch C, Sangster TA, Lindquist S (2002) Hsp90 as a capacitor of phenotypic variation. Nature 417:618–624

Reddy PS, Mallikarjuna G, Kaul T, Chakradhar T, Mishra RN, Sopory SK, Reddy MK (2010) Molecular cloning and characterization of gene encoding for cytoplasmic Hsc70 from Pennisetum glaucum may play a protective role against abiotic stresses. Mol Gen Genet 283:243–254

Regente M, Pinedo M, San Clemente H, Balliau T, Jamet E, de la Canal L (2017) Plant extracellular vesicles are incorporated by a fungal pathogen and inhibit its growth. J Exp Bot 68:5485–5495

Ritossa F (1962) A new puffing pattern induced by heat shock and DNP in Drosophila. Experientia 18:571–573

Rutherford SL, Lindquist S (1998) Hsp90 as a capacitor for morphological evolution. Nature 396: 336–342

Samakovli D, Thanou A, Valmas C, Hatzopoulos P (2007) Hsp 90 canalizes developmental perturbation. J Exp Bot 58:3513–3524

Sarkar NK, Kundnani P, Grover A (2013) Functional analysis of Hsp70 superfamily proteins of rice (Oryza sativa). Cell Stress Chaperones 18:427–437

Scharf KD, Berberich T, Ebersberger I, Nover L (2012) The plant heat stress transcription factor (Hsf) family: structure, function and evolution. Biochim Biophys Acta 1819:104–119

Schirmer EC, Lindquist S, Vierling E (1994) An Arabidopsis heat shock protein complements a thermotolerance defect in yeast. Plant Cell 6:1899–1909

Schultheiss J, Kunert O, Gase U, Scharf KD, Nover L, Rüterjans H (1996) Solution structure of the DNA-binding domain of the tomato heat stress transcription factor HSF24. Eur J Biochem 236: 911–921

Shigeta T, Zaizen Y, Sugimoto Y, Nakamura Y, Matsuo T, Okamoto S (2015) Heat shock protein 90 acts in brassinosteroid signaling through interaction with BES1/BZR1 transcription factor. J Plant Physiol 178:69–73

Siddique M, Gernhard S, von Koskull-Döring P, Vierling E, Scharf KD (2008) The plant sHSP superfamily: five new members in Arabidopsis thaliana with unexpected properties. Cell Stress Chaperones 13:83–197

Singh RK, Jaishankar J, Muthamilarasan M, Shweta S, Dang A, Prasad M (2016) Genome-wide analysis of heat shock proteins in C4 model, foxtail millet identifies potential candidates for crop improvement under abiotic stress. Sci Rep 6(1):1–14
Song A, Zhu X, Chen F et al (2014) A chrysanthemum heat shock protein confers tolerance to abiotic stress. Int J Mol Sci 15(3):5063–5078
Sun W, Bernard C, van de Cotte B, Van Montagu M, Verbruggen N (2001) At-HSP17.6A, encoding a small heat-shock protein in Arabidopsis, can enhance osmotolerance upon overexpression. Plant J 27:407–415
Sun W, Van Montagu M, Verbruggen N (2002) Small heat shock proteins and stress tolerance in plants. Biochem Biophys Acta 1577:1–9
Sun L, Liu Y, Kong X, Zhang D, Pan J, Zhou Y, Wang L, Li D, Yang X (2012) ZmHSP16. 9, a cytosolic class I small heat shock protein in maize (Zea mays), confers heat tolerance in transgenic tobacco. Plant Cell Rep 31:1473–1484
Taipale M, Krykbaeva I, Koeva M, Kayatekin C, Westover KD, Karras GI, Lindquist S (2012) Quantitative analysis of HSP90-client interactions reveals principles of substrate recognition. Cell 150:987–1001
Tichá T, Samakovli D, Kuchařová A, Vavrdová T, Šamaj J (2020) Multifaceted roles of HEAT SHOCK PROTEIN 90 molecular chaperones in plant development. J Exp Bot 71(14): 3966–3985
Tissiéres A, Mitchell HK, Tracy UM (1974) Protein synthesis in salivary glands of Drosophila melanogaster: relation to chromosome puffs. J Mol Biol 84(3):389–398
Vanneste S, Friml J (2009) Auxin: a trigger for change in plant development. Cell 136:1005–1016
Vierling E (1991) The role of heat shock proteins in plants. Annu Rev Plant Biol 42:579–620
Waddington CH (1961) Genetic assimilation. Adv Genet 10:257–293
Wahid A, Gelani S, Ashraf M, Foolad MR (2007) Heat tolerance in plants: an overview. Environ Exp Bot 61(3):199–223
Wang W, Vinocur B, Altman A (2003) Plant responses to drought, salinity and extreme temperatures: towards genetic engineering for stress tolerance. Planta 218(1):1–4
Wang W, Vinocur B, Shoseyov O, Altman A (2004) Role of plant heat-shock proteins and molecular chaperones in the abiotic stress response. Trends Plant Sci 9:244–252
Wang Y, Wang M, Sun Y, Wang Y, Li T, Chai G, Jiang W, Shan L, Li C, Xiao E, Wang Z (2015) FAR5, a fatty acyl-coenzyme A reductase, is involved in primary alcohol biosynthesis of the leaf blade cuticular wax in wheat (Triticum aestivum L.). J Exp Bot 66(5):1165–1178
Wang R, Zhang Y, Kieffer M, Yu H, Kepinski S, Estelle M (2016) HSP90 regulates temperature dependent seedling growth in Arabidopsis by stabilizing the auxin co-receptor F-box protein TIR1. Nat Commun 7:10269
Waters ER (2013) The evolution, function, structure, and expression of the plant sHSPs. J Exp Bot 64:391–403. https://doi.org/10.1093/jxb/ers355
Waters ER, Lee GJ, Vierling E (1996) Evolution, structure, and function of the small heat shock proteins in plants. J Exp Bot 47:325–338
Yamamori T, Yura T (1982) Genetic control of heat-shock protein synthesis and its bearing on growth and thermal resistance in Escherichia coli K-12. Proc Natl Acad Sci U S A 79:860–864
Yang Z, Zhang H, Li X et al (2020) A mini foxtail millet with an Arabidopsis-like life cycle as a C4 model system. Nat Plants 6:1167–1178
Xu ZS, Li ZY, Chen Y, Chen M, Li LC, Ma YZ (2012) Heat shock protein 90 in plants: molecular mechanisms and roles in stress responses. Int J Mol Sci 13:15706–15723
Zhu X, Zhao X, Burkholder WF et al (1996) Structural analysis of substrate binding by the molecular chaperone DnaK. Science 272(5268):1606–1614

The manufacturer's authorised representative in the EU is Springer Nature Customer Service Centre GmbH, Europaplatz 3, 69115 Heidelberg, Germany. If you have any concerns regarding our products, please contact ProductSafety@springernature.com

Printed and bound by CPI Group (UK) Ltd, Croydon, CR0 4YY
30/09/2026
02225250-0011